MATHEMATICAL TOOLS FOR APPLIED MULTIVARIATE ANALYSIS

STUDENT EDITION

MATHEMATICAL TOOLS FOR APPLIED MULTIVARIATE ANALYSIS

STUDENT EDITION

Paul E. Green

Wharton School
University of Pennsylvania
Philadelphia, Pennsylvania

with contributions by

J. Douglas Carroll

Bell Laboratories
Murray Hill, New Jersey

ACADEMIC PRESS New York San Francisco London 1976

A Subsidiary of Harcourt Brace Jovanovich, Publishers

ACADEMIC PRESS, INC.
111 Fifth Avenue, New York, New York 10003

United Kingdom Edition published by
ACADEMIC PRESS, INC. (LONDON) LTD.
24/28 Oval Road, London NW1 7DX

Library of Congress Cataloging in Publication Data

Green, Paul E.
Mathematical tools for applied multivariate analysis.

Bibliography: p.
Includes index.
1. Multivariate analysis. I. Carroll, J. Douglas.
II. Title.
QA278.G73 1978 519.5'3 78-547
ISBN 0-12-297552-9

PRINTED IN THE UNITED STATES OF AMERICA

80 81 82 9 8 7 6 5 4 3 2

To Betty and Sylvia

Contents

Preface

The student willing to learn something about multivariate analysis will find no dearth of textbooks and monographs on the subject. From introductory to advanced, theoretical to applied, general to specific, the field has been well covered.

However, most of these books assume certain mathematical prerequisites—typically matrix algebra and introductory calculus. Single-chapter reviews of the topics are usually provided but, in turn, presuppose a fair amount of advance preparation. What appears to be needed for the student who has received less exposure is a somewhat more elementary and leisurely approach to developing the necessary mathematical foundations of applied multivariate analysis.

The present book has been prepared to help students with those aspects of transformational geometry, matrix algebra, and the calculus that seem most relevant for the study of multivariate analysis. Since the author's interest is in applications, both the material selected for inclusion and the point of view from which it is presented reflect that orientation.

The book has been prepared for students who have either taken no matrix algebra at all or, if they have, need a refresher program that is between a full-fledged matrix algebra course and the highly condensed review chapter that is often found in multivariate textbooks. The book can serve as a textbook for courses long enough to permit coverage of precursory mathematical material or as a supplement to general textbooks on multivariate analysis.

The title was chosen rather carefully and helps demarcate what the book is not as much as what it is. First, those aspects of linear algebra, geometry, and the calculus that are covered here are treated from a pragmatic viewpoint—as tools for helping the applications researcher in the behavioral and business disciplines. In particular, there are virtually no formal proofs. In some cases outlines of proofs have been sketched, but usually small numerical examples of the various concepts are presented. This decision has been deliberate and it is the author's hope that the instructor will complement the material with more formal presentations that reflect his interests and perceptions of the technical backgrounds of his students.

The book consists of six chapters and two appendixes. Chapter 1 introduces the topic of multivariate analysis and presents three small problems in multiple regression, principal components analysis, and multiple discriminant analysis to motivate the mathematics that subsequent chapters are designed to supply. Chapter 2 presents a fairly standard treatment of the mechanics of matrix algebra including definitions and operations on vectors, matrices, and determinants. Chapter 3 goes through much of this same material but from a geometrically oriented viewpoint. Each of the main ideas in matrix algebra is illustrated geometrically and numerically (as well as algebraically).

Chapters 4 and 5 deal with the central topics of linear transformations and eigenstructures that are essential to the understanding of multivariate techniques. In Chapter 4, the theme of Chapter 3 receives additional attention as various matrix transformations are illustrated geometrically. This same (geometric) orientation is continued in Chapter 5 as eigenstructures and quadratic forms are described conceptually and illustrated numerically.

Chapter 6 completes the cycle by returning to the three applied problems presented in Chapter 1. These problems are solved by means of the techniques developed in Chapters 2–5, and the book concludes with a further discussion of the geometric aspects of linear transformations.

Appendix A presents supporting material from the calculus for deriving various matrix equations used in the book. Appendix B provides a basic discussion on solving sets of linear equations and includes an introduction to generalized inverses. Numerical exercises appear at the end of each chapter and represent an integral part of the text. With the student's interest in mind, solutions to all numerical problems are provided. (After all, it was those *even*-numbered exercises that used to give us all the trouble!) The student is urged to work through these exercises for purposes of conceptual as well as numerical reinforcement.

Completion of the book should provide both a technical base for tackling most applications-oriented multivariate texts and, more importantly, a geometric perspective for aiding one's intuitive grasp of multivariate methods. In short, this book has been written for the student in the behavioral and administrative sciences—not the statistician or mathematician. If it can help illuminate some of the material in current multivariate textbooks that are designed for this type of reader, the author's objective will have been well satisfied.

Acknowledgments

Many people helped bring this book to fruition. Literally dozens of masters and doctoral students provided critical reviews of one or more chapters from the most relevant perspective of all—their's. Those deserving special thanks are Ishmael Akaah, Alain Blancbrude, Frank Deleo, J. A. English, Pascal Lang, and Gunter Seidel. Professor David K. Hildebrand, University of Pennsylvania, prepared a thorough and useful critique of the full manuscript. Helpful comments were also received from Professor Joel Huber, Purdue University.

Production of the book was aided immeasurably by the competent and cheerful help of Mrs. Joan Leary, who not only typed drafts and redrafts of a difficult manuscript, but managed to do the extensive art work as well. The editorial staff of Academic Press also deserves thanks for their production efforts and general cooperative spirit.

The author's biggest debt of gratitude is to J. Douglas Carroll of Bell Laboratories. His imprint on the book's organization and exposition goes well beyond the role of reviewer. While the specific words in the book are mine, its general intent and orientation are fully shared with Dr. Carroll. (However, he is to be held blameless for any of the words that did not come out quite right.)

P. E. G.

CHAPTER 1

The Nature of Multivariate Data Analysis

1.1 INTRODUCTION

Stripped to their mathematical essentials, multivariate methods represent a blending of concepts from matrix algebra, geometry, the calculus, and statistics. In function, as well as in structure, multivariate techniques form a unified set of procedures that can be organized around a relatively few prototypical problems. However, in scope and variety of application, multivariate tools span all of the sciences.

This book is concerned with the mathematical foundations of the subject, particularly those aspects of matrix algebra and geometry that can help illuminate the structure of multivariate methods. While behavioral and administrative applications are stressed, this emphasis reflects the background of the author more than any belief about special advantages that might accrue from applications in these particular fields.

Multivariate techniques are useful for:

1. discovering regularities in the behavior of two or more variables;
2. testing alternative models of association between two or more variables, including the determination of whether and how two or more groups (or other entities) differ in their "multivariate profiles."

The former pursuit can be regarded as exploratory research and the latter as confirmatory research. While this view may seem a bit too pat, multivariate analysis *is* concerned with both the discovery and testing of patterns in associative data.

The principal aim of this chapter is to present motivational material for subsequent development of the requisite mathematical tools. We start the chapter off on a somewhat philosophical note about the value of multivariate analysis in scientific research generally. Some of the major characteristics of multivariate methods are introduced at this point, and specific techniques are briefly described in terms of these characteristics.

Application of multivariate techniques is by no means confined to a single discipline. In order to show the diversity of fields in which the methods have been applied, a number of examples drawn from the behavioral and administrative sciences are briefly described. Comments are also made on the trends that are taking place in multivariate analysis itself and the implications of these developments for future application of the methodology.

We next turn to a description of three small, interrelated problems that call for multivariate analysis. Each problem is described in terms of a common, miniature data bank with integer-valued numbers. As simple as the problems are, it turns out that developing the apparatus necessary to solve them covers most of the mathematical concepts in multivariate analysis that constitute the rest of the book.

1.2 MULTIVARIATE METHODS IN RESEARCH

It is difficult to imagine any type of scientific inquiry that does not involve the recording of observations on one or more types of objects. The objects may be things, people, natural or man-made events. The selected objects–white rats, model airplanes, biopsy slides, x-ray pictures, patterns of response to complex stimulus situations, ability tests, brand selection behavior, corporate financial activities–vary with the investigator's discipline. The process by which he codifies the observations does not.

Whatever their nature, the objects themselves are never measured in total. Rather, what is recorded are observations dealing with *characteristics* of the objects, such as weight, wind velocity, cell diameter, location of a shadow on the lung, speed or latency of response, number of correctly answered questions, specific brand chosen, previous year's sales, and so on. It is often the case that two or more characteristics (e.g., weight, length, and heartbeat) will be measured at the same time on each object being studied. Furthermore, it would not be unusual to find that the measured characteristics were associated in some way; that is, values taken on by one variable are frequently related to values taken on by another variable.

As a set of statistical techniques, multivariate data analysis is strategically neutral. Techniques can be used for many purposes in the behavioral and administrative sciences–ranging from the analysis of data obtained from rigidly controlled experiments to teasing out relationships assumed to be present in a large mass of survey-type data. What can be said is that multivariate analysis is concerned with *association among multiple variates* (i.e., many variables).[1]

Raymond Cattell (1966) has put the matter well. Historically, empirical work in the behavioral sciences–more specifically, experimental psychology–has reflected two principal traditions: (a) the manipulative, typically bivariate approach of the researcher viewed as controller and (b) the nonmanipulative, typically multivariate approach of the researcher viewed as observer.

Cattell points out three characteristics that serve to distinguish these forms of strategic inquiry:

1. bivariate versus multivariate in the type of data collected,
2. manipulative versus noninterfering in the degree of control exercised by the researcher,
3. simultaneous versus temporally successive in the time sequence in which observations are recorded.

[1] Analysis of bivariate data can, of course, be viewed as a special case of multivariate analysis. However, in this book our discussion will emphasize association among more than two variables. One additional point–some multivariate statisticians restrict the term *multivariate* to cases involving more than a single criterion variable. Here, we take a broader view that includes multiple regression and its various extensions as part of the subject matter of multivariate analysis.

In recent years, bivariate analysis and more rigid forms of controlled inquiry have given way to experiments and observational studies dealing with a comparatively large number of variables, not all of which may be under the researcher's control. However, if one takes a broad enough view of multivariate data analysis, one that includes bivariate analysis as a special case, then the concepts and techniques of this methodology can be useful for either stereotype. Indeed, Cattell's definition of an experiment as:

> . . .A recording of observations, quantitative or qualitative, made by defined operations under defined conditions, and designed to permit non-subjective evaluation of the existence or magnitude of relations in the data. It aims to fit these relations to parsimonious models, in a process of hypothesis creation or hypothesis checking, at least two alternatives being logically possible in checking this fit. . . . (p. 9)

says quite a bit about the purview of multivariate analysis. That is, the process of scientific inquiry should embrace the search for naturalistic regularities in phenomena as well as their incorporation into models for subsequent testing under changed conditions. And in this book we shall be as much, if not more so, interested in using multivariate analysis to aid the process of discovery (hypothesis creation) as to aid the process of confirmation (hypothesis testing).

The heart of any multivariate analysis consists of the data matrix, or in some cases, matrices.[2] The data matrix is a rectangular array of numerical entries whose informational content is to be summarized and portrayed in some way. For example, in univariate statistics the computation of the mean and standard deviation of a single column of numbers is often done simply because we are unable to comprehend the meaning of the entire column of values. In so doing we often (willingly) forego the full information provided by the data in order to understand some of its basic characteristics, such as central tendency and dispersion. Similarly, in multivariate analysis we often use various summary measures—means, variances, covariances—of the raw data. Much of multivariate analysis is concerned with placing in relief certain aspects of the association among variables at the expense of suppressing less important details.

In virtually all applied studies we are concerned with variation in some characteristic, be it travel time of a white rat in a maze or the daily sales fluctuations of a retail store. Obviously, if there is no variation in the characteristic(s) under study, there is little need for statistical methods.

In multivariate analysis we are often interested in accounting for the variation in one variable or group of variables in terms of *covariation* with other variables. When we analyze associative data, we hope to "explain" variation according to one or more of the following points of view:

1. determination of the nature and degree of association between a set of *criterion* variables and a set of *predictor* variables, often called "dependent" and "independent" variables, respectively;

2. finding a function or formula by which we can estimate values of the criterion variable(s) from values of the predictor variable(s)—this is usually called the *regression* problem;

[2] Much of this section is drawn from Green and Tull (1975).

3. assaying the statistical "confidence" in the results of either or both of the above activities, via tests of statistical significance, placing confidence intervals on parameter estimates, or other ways.

In some cases of interest, however, we have no prior basis for distinguishing between criterion and predictor variables. We may still be interested in their interdependence as a whole and the possibility of summarizing information provided by this interdependence in terms of other variables, often taken to be linear composites of the original ones.

1.3 A CLASSIFICATION OF TECHNIQUES FOR ANALYZING ASSOCIATIVE DATA

The field of associative data analysis is vast; hence it seems useful to enumerate various descriptors by which the field can be classified. The key notion underlying the classification of multivariate methods is the *data matrix*. A conceptual illustration is shown in Table 1.1. We note that the table consists of a set of objects (the m rows) and a set of measurements on those objects (the n columns). Cell entries represent the value X_{ij} of object i on variable j. The objects are any kind of entity with characteristics capable of being measured. The variables are characteristics of the objects and serve to define the objects in any specific study. The cell values represent the state of object i with respect to variable j. Cell values may consist of nominal, ordinal, interval, or ratio-scaled measurements, or various combinations of these, as we go across columns.

By a nominal scale we mean categorical data where the only thing we know about the object is that it falls into one of a set of mutually exclusive and collectively exhaustive categories that have no necessary order vis à vis one another. Ordinal data are ranked data where all we know is that one object i has more, less, or the same amount of some variable j than some other object i'. Interval scale data enable us to say how much more one object has than another of some variable j (i.e., intervals between scale values are meaningful). Ratio scale data enable us to define a natural origin (e.g., a case in which

TABLE 1.1

Illustrative Data Matrix

	Variables							
Objects	1	2	3		j		n
1	X_{11}	X_{12}	X_{13}	...	X_{1j}	...	X_{1n}
2	X_{21}	X_{22}	X_{23}	...	X_{2j}	...	X_{2n}
3	X_{31}	X_{32}	X_{33}	...	X_{3j}	...	X_{3n}
⋮	⋮	⋮	⋮		⋮		⋮
i	X_{i1}	X_{i2}	X_{i3}		X_{ij}		X_{in}
⋮	⋮	⋮	⋮		⋮		⋮
m	X_{m1}	X_{m2}	X_{m3}	...	X_{mj}	...	X_{mn}

object i has zero amount of variable j), and ratios of scale values are meaningful. Each higher scale type subsumes the properties of those below it. For example, ratio scales possess all the properties of nominal, ordinal, and interval scales, in addition to a natural origin.

There are many descriptors by which we can characterize methods for analyzing associative data.[3] The following represent the more common bases by which the activity can be classified:

1. purpose of the study and the types of assertions desired by the researcher—what kinds of statements does he wish to make about the data or about the universe from which the data were drawn?
2. focus of research emphasis—statements regarding the objects (i.e., the whole profile or "bundle" of variables), specific variables, or both;
3. nature of his prior judgments as to how the data matrix should be partitioned in terms of the type and number of subsets of variables;
4. number of variables in each of the partitioned subsets;
5. type of association under study—linear in the parameters, transformable to linear, or "inherently" nonlinear in the parameters;
6. scales by which variables are measured—nominal, ordinal, interval, ratio, mixed.

All of these descriptors relate to certain decisions required of the researcher. Suppose he is interested in studying certain descriptive relationships among variables. If so, he must make decisions about how he wants to partition the set of columns (see Table 1.1) into subsets. Often he will call one subset "criterion" variables and the other subset "predictor" variables.[4] He must also decide, however, on the number of variables to include in each subset and on what type of functional relationship is to hold among the parameters in his statistical model.

Most decisions about associative data analysis are based on the researcher's "private model" of how the variables are related and what features are useful for study.[5] His choice of various "public models" for analysis—multiple regression, discriminant analysis, etc.—is predicated on his prior knowledge of the characteristics of the statistical universe from which the data were obtained and his knowledge of the assumption structure of each candidate technique.

1.3.1 Researcher's Objectives and Predictive Statements

We have already commented that the researcher may be interested in (a) measuring the nature and degree of association between two or more variables; (b) predicting the values of one or more criterion variables from values of one or more predictor variables; or (c)

[3] An excellent classification, based on a subset of the descriptors shown here, has been provided by M. M. Tatsuoka and D. V. Tiedeman (1963).

[4] As Horst (1961) has shown, relationships need not be restricted to two sets.

[5] To some extent this is true even of the scales along which the data are measured. The researcher may wish to "downgrade" data originally expressed on interval scales to ordered categories, if he feels that the quality of the data does not warrant the "strength" of scale in which it is originally expressed. In other cases he may "upgrade" data in order to use some statistical technique that assumes a type of measurement that is absent originally.

assessing the statistical reliability of an association between two or more variables. In a specific study all three objectives may be pursued. In using other techniques (i.e., those dealing mainly with factor and cluster analysis), the researcher may merely wish to portray association in a more parsimonious way without attempting to make specific predictions or inferential statements.

1.3.2 Focus of Research Interest

Some multivariate techniques (e.g., multiple regression) focus on association among variables; objects are treated only as replications. Other techniques (e.g., cluster analysis) focus on association among objects; information about specific variables is usually, although not necessarily, suppressed. In still other instances one may wish to examine interrelationships among variables, objects, and object–variable combinations, as well.

1.3.3 Nature of Assumed Prior Judgments or Presuppositions

In many cases the investigator is able to partition the data matrix into subsets of columns (or rows) on the basis of prior judgment. For example, suppose the first column of Table 1.1 is average weekly consumption of coffee by households, and the other columns consist of various demographic measurements of the m households. The analyst may wish to predict average weekly consumption of coffee from some linear composite of the $n - 1$ remaining variables. If so, he has used his presuppositions regarding how the dependence is to be described and, in this instance, might employ multiple regression.

In most cases the number of subsets developed from the data matrix partitioning will be two, usually labeled as criterion and predictor variable subsets. However, techniques have been designed to summarize association in cases involving more than two subsets of data.

Finally, we may have no reasonable basis for partitioning the data matrix into criterion or predictor variables. Our purpose here may be merely to group objects into "similar" subsets, based on their correspondence over the whole profile of variables. Alternatively, we may wish to portray the columns of the data matrix in terms of a smaller number of variables, such as linear combinations of the original set, that retain most of the information in the original data matrix. Cluster analysis and factor analysis, respectively, are useful techniques for these purposes.

1.3.4 Number of Variables in Partitioned Subsets

Clearly, the term "association" implies at least two characteristics—for example, a single criterion and a single predictor variable, usually referred to as bivariate data. In other cases involving two subsets of variables, we may wish to study association between a single criterion and more than one predictor. Or we may wish to study association between composites of several criterion variables and composites of several predictor variables. Finally we may want to study the relationship between several criterion variables and a single predictor variable.

Of course, we may elect not to divide the variables at all into two or more subsets, as would be the case in factor analysis. Furthermore, if we do elect to partition the matrix and end up with two or more variables in a particular subset, what we are usually concerned with are various *linear composites* of the variables in that subset and each composite's association with other variables.

1.3.5 Type of Association

Most of the models of multivariate analysis emphasize linear relationships among the variables. The assumption of linearity, in the parameters, is not nearly so restrictive as it may seem.[6] First, various preliminary transformations (e.g., square root, logarithmic) of the data are possible in order to achieve linearity in the parameters.[7] Second, the use of "dummy" variables, coded, for example, as elementary polynomial functions of the "real" variables, or indicating category membership by patterns of zeroes and ones, will enable us to handle certain types of nonlinear relationships within the framework of a linear model. Third, a linear model is often a good approximation to a nonlinear one, at least over restricted ranges of the variables in question.

1.3.6 Types of Scales

Returning to the data matrix of Table 1.1, we now are concerned with the scales by which the characteristics are represented. Since all of the multivariate statistical techniques to be discussed in this book require no stronger form of measurement than an interval scale, we shall usually be interested in the following types: (a) nominal, (b) ordinal, and (c) interval. In terms of nominal scaling we shall find it useful to distinguish between dichotomies and (unordered) polytomies, the latter categorization involving more than two classes.

This distinction is important for three reasons. First, many of the statistical techniques for analyzing associative data are amenable to binary-coded (zero–one) variables but *not* to polytomies. Second, any polytomy can be recoded as a set of dichotomous "dummy" variables; we shall describe how this recoding is done in the next section. Third, when we discuss geometrical representations of variables and/or objects, dichotomous variables can be handled within the same general framework as interval-scaled variables.

Finally, mention should be made of cases in which the analyst must contend with *mixed* scales in the criterion subset, predictor subset, or both. Many multivariate techniques—if not modified for this type of application—lead to rather dubious results under such circumstances.

[6] By linear in the parameters is meant that the b_j's in the expression $y = b_1x_1 + b_2x_2 + \cdots + b_nx_n$ are each of the first degree. Similarly, $z = b_1x_1^2 + \cdots + b_nx_n^{n+1}$ is still linear in the parameters since each b_j continues to be of the first degree even though x_j is not.

[7] For example, the complicated expression $y = ax^be^{cx}$ (with both a, $x > 0$) can be "linearized" as $\ln y = \ln a + b \ln x + cx$ and, as shown by Hoerl (1954), is quite flexible in approximating many diverse types of curves. On the other hand, the function $y = 1/(a + b^{-cx})$ is inherently nonlinear in the parameters and cannot be "linearized" by transformation.

1.4 ORGANIZING THE TECHNIQUES

In most textbooks on multivariate analysis, three of the preceding characteristics are often used as primary bases for technique organization:

1. whether one's principal focus is on the objects or on the variables of the data matrix;
2. whether the data matrix is partitioned into criterion and predictor subsets, and the number of variables in each;
3. whether the cell values represent nominal, ordinal, or interval scale measurements.

This schema results in four major subdivisions of interest:

1. *single criterion, multiple predictor association,* including multiple regression, analysis of variance and covariance, and two-group discriminant analysis;
2. *multiple criterion, multiple predictor association,* including canonical correlation, multivariate analysis of variance and covariance, multiple discriminant analysis;
3. *analysis of variable interdependence,* including factor analysis, multidimensional scaling, and other types of dimension-reducing methods;
4. *analysis of interobject similarity,* including cluster analysis and other types of object-grouping procedures.

The first two categories involve dependence structures where the data matrix *is* partitioned into criterion and predictor subsets; in both cases interest is focused on the variables. The last two categories are concerned with interdependence—either focusing on variables or on objects. Within each of the four categories, various techniques are differentiated in terms of the type of scale assumed.

1.4.1 Scale Types

Traditionally, multivariate methods have emphasized two types of variables:

1. more or less continuous variables, that is, interval-scaled (or ratio-scaled) measurements;
2. binary-valued variables, coded zero or one.

The reader is no doubt already familiar with variables like length, weight, and height that can vary more or less continuously over some range of interest.

Natural dichotomies such as sex, male or female, or marital status, single or married, are also familiar. What is perhaps not as well known is that any (unordered) polytomy, consisting of three or more mutually exclusive and collectively exhaustive categories, can be recoded into dummy variables that are typically coded as one or zero. To illustrate, a person's occupation, classified into five categories, could be coded as:

	Dummy variable			
Category	1	2	3	4
Professional	1	0	0	0
Clerical	0	1	0	0
Skilled laborer	0	0	1	0
Unskilled laborer	0	0	0	1
Other	0	0	0	0

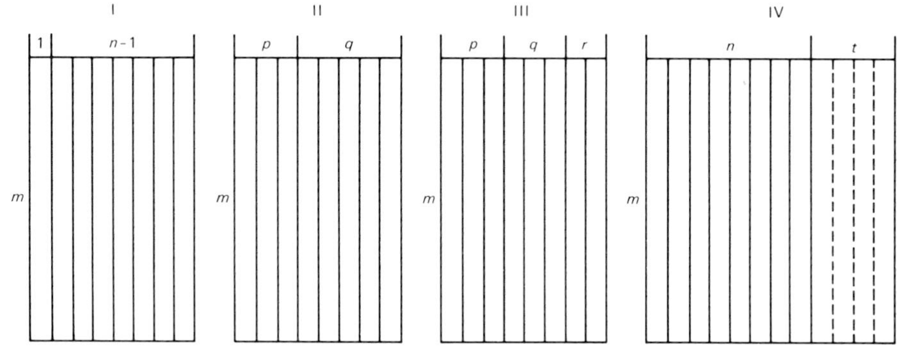

Fig. 1.1 Illustrative partitionings of data matrix.

For example, if a person falls into the professional category, he is coded 1 on dummy variable 1 and 0 on dummies 2 through 4. In general, a k-category polytomy can be represented by $k - 1$ dummy variables, with one category—such as the last category—receiving a value of zero on all $k - 1$ dummies.[8]

Multivariate techniques that are capable of dealing with some or all variables at the ordinally scaled level are of more recent vintage. With few exceptions our attention in this book will be focused on either continuous or binary-valued variables.[9]

Figure 1.1 shows some of the major ways in which the data matrix can be viewed from the standpoint of technique selection.

1.4.2 Single Criterion, Multiple Predictor Association

In Panel I of the figure we note that the first column of the matrix has been singled out as a criterion variable and the remaining $n - 1$ variables are considered as predictors. For example, the criterion variable could be average weekly consumption of beer by the ith individual. The $n - 1$ predictors could represent various demographic variables, such as the individual's age, years of education, income, and so on. This is a prototypical problem for the application of multiple regression in which one tries to predict values of the criterion variable from a linear compositie of the predictors. The predictors, incidentally, can be either continuous variables or dummies, such as marital status or sex. Alternatively, the single criterion variable could represent the prior categorization of each individual as heavy beer drinker (coded one, arbitrarily) or light beer drinker (coded zero), based on some designated amount of average weekly beer consumption. If our purpose is to develop a linear composite of the predictors that enables us to classify each individual into either heavy or light beer drinker status, then we would employ two-group discriminant analysis. The critical distinction here is that the criterion variable is expressed as a single dummy variable rather than as a continuous one.

[8] Not only is greater parsimony achieved by using only $k - 1$ (rather than k) categories, but, as will be shown in later chapters, this type of coding device permits the matrix to be inverted by regular computational methods.

[9] While a classification *could* be coded as 1, 2, 3, . . . , k in terms of a single variable, the resulting analysis would assume that *all classes are ordered and equally spaced*—a rather dubious assumption in most kinds of classificatory data.

Another possibility can arise in which the criterion variable continues to be average weekly beer consumption, but the predictor set consists of a classification of each individual into some occupational group, coded as a set of dummy or design variables. This represents an instance in which the technique of analysis of variance could be used. On the other hand, if the predictor set consists of a classification of occupations as well as annual income in dollars, then the latter variable could be treated as a covariate. In this case we would be interested in whether average weekly beer consumption differs across occupations once the effect of income is controlled for statistically.

1.4.3 Multiple Criterion, Multiple Predictor Association

In Panel II of Fig. 1.1 the $p = 3$ criterion variables could denote individual consumption of beer, wine, and liquor, and the remaining variables could denote demographic characteristics. If we were interested in the linear association between the two *batteries* of variables, we could employ the technique of canonical correlation.

Suppose, alternatively, that all individuals had been previously classified into one of four groups: (a) malt beverage drinker only, (b) drinker of spirits (liquor or wine) other than malt beverages, (c) drinker of both spirits and malt beverages, and (d) drinker of neither. We could then develop linear functions of the demographics that would enable us to assign each individual to one of the four groups in some "best" way (to be defined later). This is an illustration of multiple discriminant analysis; note that four mutually exclusive groups are classifiable in terms of $p = 3$ criterion dummies.

Alternatively, we could continue to let the criterion variables denote individual consumption levels of beer, wine, and liquor, but now assume that the predictors represent dummies based on an occupational classification. If so, multivariate analysis of variance is the appropriate procedure. If income is again included as a covariate, we have an instance of multivariate analysis of covariance.

Panel III of Fig. 1.1 shows a data structure involving association among three batteries of variables. Generalized canonical correlation can be employed in this type of situation. In this case we would be interested in what all three batteries exhibit in common and also in the strength of association between all distinct pairs of batteries as well.

1.4.4 Dimension-Reducing Methods

Panel IV of Fig. 1.1 shows a set of t appended columns, each of which is expressed as a linear composite of the original n variables. Suppose we want to portray the association across the m individuals in terms of fewer variables than the original n variables. If so, we might employ factor analysis, multidimensional scaling, or some other dimension-reduction method to represent the original set of n correlated variables as linear (or nonlinear) composites of a set of t ($t < n$) underlying or "latent" variables in such a way as to retain as much of the original information as possible. The composites themselves might be chosen to obey still other conditions, such as being mutually uncorrelated.

Thus, if the original n variables are various demographics characterizing a set of beer drinkers, we might be able to find a set of more basic dimensions—social class, stage in life cycle, etc.—so that linear composites of these basic dimensions account for the observable demographic variables.

1.4.5 Interobject Similarity

So far we have confined our attention to the columns of the matrices in Fig. 1.1. Suppose now that the n columns represent consumption of various kinds of alcoholic beverages—beers, ales, red wines, white wines, liquors, after-dinner cordials—over some stated time period. Each individual's consumption profile could be compared with every other individual's, and we could develop a measure of interindividual similarity with respect to patterns of alcoholic beverage drinking.

Having done so, we could then proceed to cluster individuals into similar groups on the basis of the overall similarity of their consumption profiles. Note here that information on specific variables is lost in the computation of interindividual similarity measures. Since our focus of interest is on the objects rather than on the variables, we may be willing to discard information on separate variables in order to grasp the notion of *overall* interobject similarity (and the "clusteriness" of objects) more clearly.

All of these techniques—and others as well—have been employed in the behavioral and administrative sciences. As suggested above, the tools of multivariate analysis form a unified set, based on a relatively few descriptors for distinguishing specific techniques.

1.5 ILLUSTRATIVE APPLICATIONS

Any empirically grounded discipline has need on occasion to use various types of multivariate techniques. Indeed, in some fields like psychometrics and survey research, multivariate analysis represents the methodological cornerstone.

Although multivariate analysis can be, and has been, used in the physical and life sciences, increasing applications are being made in the behavioral and administrative sciences. Even at that, the view is a broad one, as the following list of behavioral and administrative examples illustrate.

Example 1 Two economists, Quandt and Baumol (1966), were interested in predicting the demand for alternative modes of travel between various pairs of cities. They developed a characterization of each mode of travel (e.g., airplane, train, bus, private car) as a service profile varying in levels of cost, departure frequency, convenience to the traveler, speed, and so on.

A travel-demand forecasting model for each mode was then prepared which utilized a linear function of the logarithms of service profile levels. Traffic volumes, involving sixteen different city pairs, were available for each of the abovementioned modes to serve as criterion variables. The parameters of their demand forecasting model were estimated by multiple regression.

Example 2 A group of psychologists, Rorer *et al.* (1967), were concerned with the modeling of clinical judgment and, in particular, how subjects combined various information cues into an overall judgment. The subjects of their experiment were small groups of physicians, nurses, psychologists, and social workers. The judgment to be made concerned the subject's probability of granting a weekend pass to each of 128 (presumed real) patients. Each "patient" was described according to a six-component profile, involving such characteristics as (a) whether he had a drinking problem; (b) whether he had abused privileges in the past; (c) whether his personal appearance was neat, and so on. The patient was described simply as to whether he displayed the characteristic or not.

The six characteristics used by the researchers were formulated in a 2^6 design of all possible combinations, and two replications were made up of each combination, leading to a total of 128 that were presented (in random order) to each subject. An analysis was made of each subject's response data separately, using an analysis of variance model applicable to a full factorial design. The researchers found evidence that subjects used cues interactively in arriving at an overall judgment. The relative importance of these interactions was measured as well as the separate main-effect contributions to the overall judgment regarding the subjective probability of granting a pass.

Example 3 A political scientist, R. J. Rummel (1970), was interested in a cross-national comparison of some 82 different countries, measured according to 230 characteristics. In particular, he wished to see what underlying factors or dimensions might account for various observed relationships across such characteristics as the nations' trade levels, memberships in international organizations, production of various commodities, and so on.

A variety of factor analyses and cluster analyses were performed on the data, leading to a set of underlying dimensions, identified principally as the nation's political orientation, economic development, and degree of foreign conflict.

Two mathematical psychologists, Wish and Carroll (1973), were also interested in national similarities and differences but, in this case, as subjectively perceived by U.S. and foreign students. Their methodology emphasized multidimensional scaling and, in particular, individual differences models of perception. They found that different subjects gave different importances to perceptual dimensions, depending upon the subject's attitude toward U.S. involvement in the Vietnam conflict.

Example 4 Two educational psychologists, Cooley and Lohnes (1971), were engaged in a massive sampling survey, called Project TALENT, involving measurement on a large number of personality and ability variables of a representative sample of American high school students. The purpose of the study was to examine interrelationships among these variables and various environmental variables in order to predict the students' motivations involving subsequent career and higher education activities.

A variety of multivariate techniques were employed in the analysis of these data. For example, one analysis used canonical correlation to examine the association between a set of eleven ability-type factors (e.g., verbal knowledge, mathematics, visual reasoning) and a set of eleven factors dealing with career motives (e.g., interest in science, interest in business). In this case the canonical correlation followed a preliminary factor analysis of each separate battery of variables.

Example 5 A group of survey researchers, Morgan, Sirageldin, and Baerwaldt (1965), were engaged in a large-scale survey in which the criterion variable of interest was hours spent on do-it-yourself activities by heads of families and their spouses. A sample size of 2214 households provided the data, and the predictor variables included a large number of demographic characteristics.

Not surprisingly, the authors found that marital status was the most important predictor variable. Single men and women spent relatively little time on do-it-yourself activities. On the other hand, married couples with large families who had higher-than-average education, lived in single-family structures in rural areas, with youngest child between two and eight years, devoted a large amount of time to do-it-yourself activities. The researchers used a multivariate technique, known as Automatic Interaction Detection, to develop a sequential branching of groups. At each stage the program selects a predictor variable that accounts for the most variation in the criterion variable and splits the sample into two subgroups, according to their values on that predictor variable. The result is a sequential branching "tree" of groups that are most homogeneous with regard to the criterion variable of interest.

Example 6 A group of management scientists, Haynes, Komar, and Byrd (1973), were interested in the comparative performance of three heuristic rules that had been proposed for sequencing production jobs that incur setup changes. The objective of each of the rules was to minimize machine downtime over the whole production sequence. For example, Rule 1 involved selecting as the next job that one which has the least setup time relative to the job last completed, of all jobs yet unassigned. The researchers were interested in how the competing heuristics would perform under variations in setup time distributions and total number of jobs to be sequenced.

The researchers set up an experimental design in which application of the three rules was simulated in a computer under different sets of distribution times and numbers of jobs. The factorial design employed by the authors to test the behavior of the rules was then analyzed by analysis of variance procedures. The experiment indicated that a composite of the three heuristics might perform better than any of the three rules taken singly.

Example 7 Two marketing researchers, Perry and Hamm (1969), believed that consumers might ascribe higher importance to personal sources (in the selection of products involving high socioeconomic risk) than to impersonal sources of product information. They set up an experiment in which consumers rated a set of 25 products on degree of perceived social risk and degree of economic risk. Each respondent was also asked to rate the significance of various sources of influence (e.g., advertisements, *Consumer's Reports,* a friend's recommendations) on one's choice of brand within each product class.

The authors used canonical correlation to relate the two sets of measures. They found that the higher the risk, particularly the social risk, the greater the perceived importance of personal influence on brand choice. The authors concluded that in the advertising of high-risk products (e.g., color TV, automobiles, sports jackets), advertisers should try to reach prospective buyers through personal channels, such as opinion leaders, rather than through general media. Moreover, advertisers should emphasize the social, rather than the economic, benefits of the purchase.

As the preceding examples suggest, virtually any discipline in the behavioral and administrative sciences can find applications for multivariate tools. It is not surprising why this is so, given that multivariate techniques can be used in both controlled experiments and observational studies. And, in the latter case at least, data refuse to come in neat and tidy packages. Rather, the predictor variables are usually correlated themselves, and one needs statistical tools to assist one in finding out what is going on.

Even in controlled experiments it is usually not possible to control for *all* variables. Various experimental devices, such as blocking and covariance adjustment, are often used to increase precision as well as to reduce some of the sources of statistical bias.

Moreover, it seems to be the nature of things in both the behavioral and administrative sciences that the possible explanatory variables of some phenomenon of interest are myriad and difficult to measure (as well as interrelated). It should come as no surprise that methods are needed to reveal whatever patterns exist in the data as well as to help the analyst test hypotheses regarding his content area of interest. Thus, multivariate techniques are becoming as familiar to the marketing researcher, production engineer, and corporate finance officer as they are to the empirically oriented psychologist, sociologist, political scientist, and economist.

In addition to the diffusion among disciplines, multivariate techniques themselves are increasing in variety and sophistication. Researchers' past emphasis on multiple regression and factor analysis has given way to application of whole new classes of techniques–canonical correlation, multiple discriminant analysis, cluster analysis, and multidimensional scaling, to name a few. Methods are being extended to deal with multiway matrices and time-dependent observations. Computing routines have incorporated still-recent developments in nonlinear optimization and other forms of numerical analysis. Methods typically used for measured variables have been modified and extended to cope with data that are expressed only as ranks or in some cases only in terms of category membership.

In short, multivariate data analysis has become a vigorous field methodologically and a catholic field substantively. Indeed, it is difficult to think of any behavioral or administrative discipline in which multivariate methods have no applicability.

1.6 SOME NUMERICAL EXAMPLES

To a large extent, the study of multivariate techniques is the study of linear transformations. In some techniques the whole data matrix–or some matrix derived from it–may undergo a linear transformation. Other methods involve various transformations of submatrices obtained from partitioning the original matrix according to certain presuppositions about the substantive data of interest.

In the chapters that follow we shall be discussing those aspects of linear algebra and transformational geometry that underlie all methods of multivariate analysis. The arithmetic operations associated with vectors and matrices, determinants, eigenstructures, quadratic forms, and singular value decomposition are some of the concepts that will be presented.

As motivation for the study of these tools, let us consider three problems that could arise in an applied research area. While almost any field could supply appropriate examples, suppose we are working in the field of personnel research. In particular, imagine that we are interested in certain aspects of employee absenteeism. We shall assume that all employees are male clerks working in an insurance company.

Absenteeism records have been maintained for each employee over the past year. Personnel records also indicate how long each employee has worked for the company. In addition, each employee recently completed a clinical interview with the company psychologist and was scored by the psychologist on a 1-to-13 point rating scale, with "1"

TABLE 1.2

Personnel Data Used to Illustrate Multivariate Methods

	Number of days absent			Attitude rating			Years with company		
Employee	Y	Y_d	Y_s	X_1	X_{d1}	X_{s1}	X_2	X_{d2}	X_{s2}
a	1	−5.25	−0.97	1	−5.25	−1.39	1	−3.92	−1.31
b	0	−6.25	−1.15	2	−4.25	−1.13	1	−3.92	−1.31
c	1	−5.25	−0.97	2	−4.25	−1.13	2	−2.92	−0.98
d	4	−2.25	−0.41	3	−3.25	−0.86	2	−2.92	−0.98
e	3	−3.25	−0.60	5	−1.25	−0.33	4	−0.92	−0.31
f	2	−4.25	−0.78	5	−1.25	−0.33	6	1.08	0.36
g	5	−1.25	−0.23	6	−0.25	−0.07	5	0.08	0.03
h	6	−0.25	−0.05	7	0.75	0.20	4	−0.92	−0.31
i	9	2.75	0.51	10	3.75	0.99	8	3.08	1.03
j	13	6.75	1.24	11	4.75	1.26	7	2.08	0.70
k	15	8.75	1.61	11	4.75	1.26	9	4.08	1.37
l	16	9.75	1.80	12	5.75	1.53	10	5.08	1.71
Mean	6.25			6.25			4.92		
Standard deviation	5.43			3.77			2.98		

indicating an extremely favorable attitude and "13" indicating an extremely unfavorable attitude toward the company. (The 13 scale points were chosen arbitrarily.)

For purposes of illustration, a sample of 12 employees was selected for further study.[10] The "raw" data on each of the three variables are shown in Table 1.2. Figure 1.2 shows each pair of variables in scatter plot form.

From Fig. 1.2 we note the tendency for all three variables to be positively associated. That is, absenteeism increases with unfavorableness of attitude toward the firm and number of years with the company. Moreover, unfavorableness of attitude is positively associated with number of years of employment with the firm (although one might question the reasonableness of this assumed relationship).

Table 1.2 also shows the means and sample standard deviations of each of the three variables. By subtracting out the mean of each variable from that variable's original observation we obtain three columns of deviation (or mean-corrected) scores, denoted by Y_d, X_{d1}, and X_{d2} in Table 1.2. To illustrate:

$$\boxed{Y_{di} = Y_i - \bar{Y}}$$

where $\bar{Y}$, denoting the mean of Y, is written as

$$\boxed{\bar{Y} = \sum_{i=1}^{m} Y_i/m}$$

[10] Obviously, the small sample size of only 12 employees is for illustrative purposes only; moreover, all data are artificial.

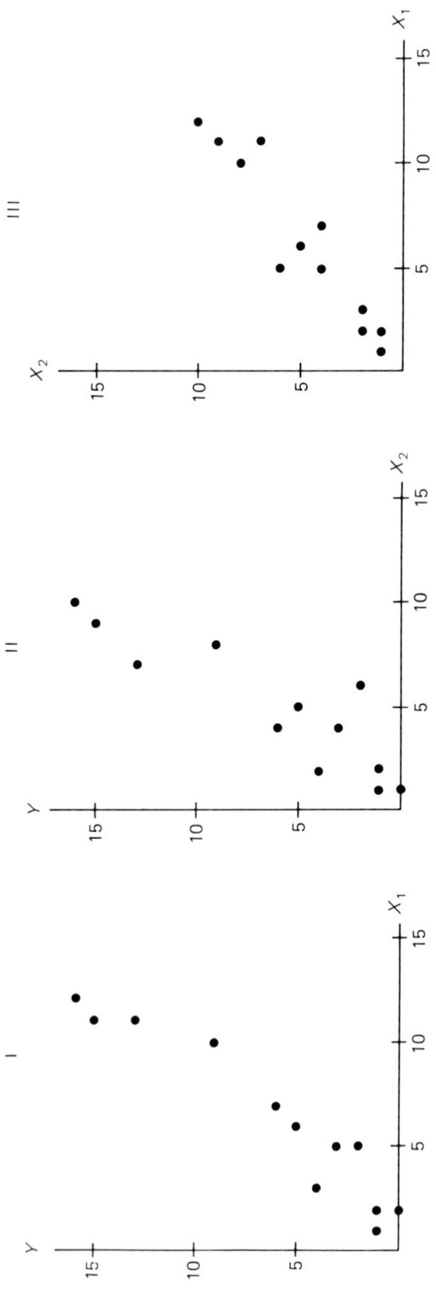

Fig. 1.2 Two-variable scatter plots of data from sample problem. Key: I, Y vs X_1; II, Y vs X_2; III, X_2 vs X_1.

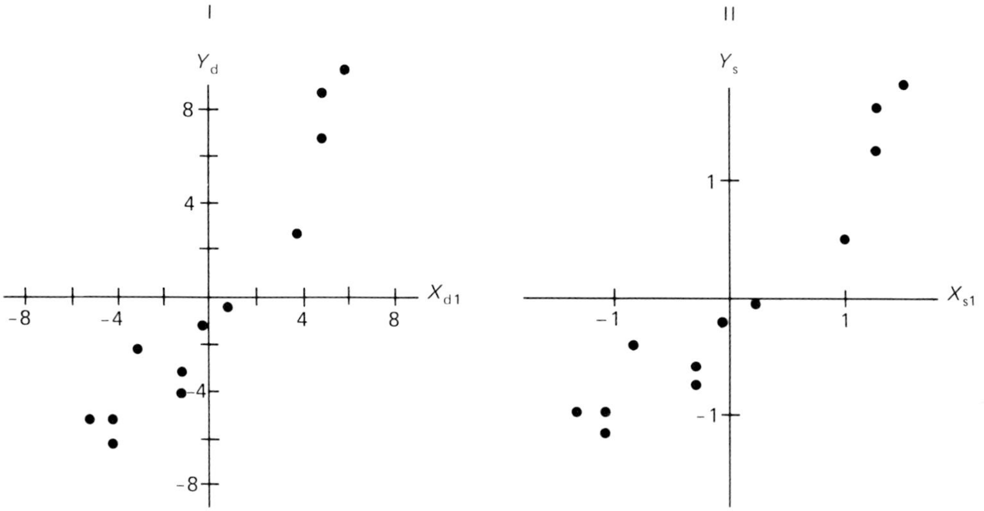

Fig. 1.3 Illustrative scatter plots of (I) mean-corrected and (II) standardized data.

Standardized scores, denoted by Y_s, X_{s1}, and X_{s2}, are obtained by dividing each mean-corrected score by the sample standard deviation of that variable.[11] To illustrate:

$$\boxed{Y_{si} = Y_{di}/s_y}$$

where s_y, in turn, is defined as

$$\boxed{s_y = \left[\sum_{i=1}^{m} (Y_i - \bar{Y})^2/m\right]^{1/2}}$$

Figure 1.3 shows, illustratively, the scatter plot of mean-corrected and standardized scores involving the criterion variable Y versus the first predictor X_1. We note that the computation of deviation scores merely changes the origin of the plot to an average of zero on each dimension. Interpoint distances do not change, and the configuration of points looks just like the configuration shown in the leftmost panel of Fig. 1.2.

Standardization of the data, however, does change the shape of the configuration, as well as shifting the origin. If the right-hand panel of Fig. 1.3 is compared to the left-hand panel, we see that the vertical axis, or ordinate, is compressed, relative to the horizontal axis, or abcissa. This is because the Y_d values are being divided by a larger constant (5.43) than the X_{d1} values, the latter being divided by 3.77, the sample standard deviation of X_1.

[11] The reader will note that s_y denotes the *sample* standard deviation rather than an estimate of the universe standard deviation. In this latter case the divisor would be $m - 1$, rather than m, as used here.

1.6.1 Research Questions

After this preliminary examination of the data, suppose the researcher raises the following questions:

1. How does Y relate to changes in X_1 and X_2?
   a. Can an equation be developed that will enable us to predict values of Y as a linear function of X_1 and X_2?
   b. How strong is the overall relationship of Y with X_1 and X_2?
   c. Is the overall relationship statistically significant?
   d. What is the relative influence of X_1 and X_2 on variation in Y and are these separate influences statistically significant?
2. Next, considering the relationship between the predictor variables X_1 and X_2, some further questions can be asked:
   a. Can the 12 scores on X_1 and X_2 be replaced by scores on a single variable that represents a linear composite of the two separate scores? That is, do X_1 and X_2 really reflect just a single underlying factor, or are there two separate factors operating?
   b. What is the association of X_1 and X_2, respectively, with this linear composite?
   c. How much of the total variation in X_1 and X_2 is accounted for by the single linear composite?
3. One additional thing that we might do is to split the sample of employees into three groups: Group 1–employees a, b, c, d; Group 2–employees e, f, g, h; Group 3–employees i, j, k, l. We could call these three groups low-, intermediate-, and high-absenteeism groups, respectively.[12]

Having classified the 12 respondents in this way, we would then raise the questions:
   a. How do we go about defining a linear composite of X_1 and X_2 that maximally separates the three groups?
   b. Is this linear composite statistically significant?
   c. How well does the linear composite assign individuals to their correct groups?
   d. What is the relative influence of X_1 and X_2 on group assignment and are their separate contributions statistically significant?
   e. How could we find a second linear composite, uncorrelated with the first, that does the next best job of separating the groups (and so on)?

Each set of questions describes a particular multivariate technique which we now consider.

1.6.2 Multiple Regression

The first set of questions pertain to a problem in multiple regression. This, in turn, involves the subproblems of developing an estimating equation, computing its strength of

[12] This assumes, of course, that specific numerical data on number of days absent are being supplanted by interest only in the three groups of the classification: low, intermediate, and high absenteeism. This disregard for numerical information on number of days absent is strictly for motivating the discussion of multiple discriminant analysis, although it could be rationalized on other grounds, such as possible nonlinear association.

relationship and statistical significance, and examining the contribution of each predictor to changes in the criterion variable.

Insofar as the first problem is concerned, we shall want to find parameter values for the linear equation:

$$\boxed{\hat{Y} = b_0 + b_1 X_1 + b_2 X_2}$$

where $\hat{Y}$ denotes predicted values of Y; b_0 denotes the intercept term when X_1 and X_2 are both 0; and b_1 and b_2 denote the partial regression coefficients of X_1 and X_2, respectively. The partial regression coefficient measures the change in $\hat{Y}$ per unit change in some specific predictor, with other predictors held constant.

In terms of the data of Table 1.2 we shall want to find the parameter values b_0, b_1, b_2 and the 12 predicted values:

$$\begin{aligned}
\hat{Y}_1 &= b_0 + b_1(1) + b_2(1) \\
\hat{Y}_2 &= b_0 + b_1(2) + b_2(1) \\
\hat{Y}_3 &= b_0 + b_1(2) + b_2(2) \\
&\vdots \\
\hat{Y}_{12} &= b_0 + b_1(12) + b_2(10)
\end{aligned}$$

(where the numbers in parentheses are actual numerical values of X_1 and X_2, from Table 1.2). As will be shown in subsequent chapters, we shall find the specific values of b_0, b_1, and b_2, according to the *least-squares* principle. This entails minimizing the sum of the squared errors e_i :

$$\sum_{i=1}^{12} e_i^2 = \sum_{i=1}^{12} (Y_i - \hat{Y}_i)^2 = \sum_{i=1}^{12} (Y_i - b_0 - b_1 X_{i1} - b_2 X_{i2})^2$$

The least-squares principle leads to a set of linear equations which, when solved, provide the desired parameter values.

The second question, concerning strength of the (linear) relationship, is answered by computing R^2, the squared multiple correlation. R^2 measures how much of the variation in Y, as measured about its mean $\bar{Y}$, is accounted for by variation in X_1 and X_2. R^2 can be expressed quite simply as

$$\boxed{R^2 = 1 - \frac{\sum_{i=1}^{12} e_i^2}{\sum_{i=1}^{12} (Y_i - \bar{Y})^2}}$$

where the denominator represents the sum of squares in Y as measured about its own mean. As can be observed, if the sum of squared errors is zero, then the $\hat{Y}_i$'s predict their respective Y_i's perfectly and $R^2 = 1$. However, if the inclusion of X_1 and X_2 in the estimating equation does no better than use of the $\bar{Y}$ alone, then the numerator of the fraction equals the denominator and $R^2 = 0$, denoting no variance accounted for, beyond using the criterion variable's mean.

The third question entails a test of the null hypothesis of no linear association between Y and X_1 and X_2, as considered together. This can be expressed either as

$$\boxed{R_{\mathrm{p}} = 0}$$

where R_{p} denotes the population multiple correlation, or as

$$\boxed{\beta_1 = \beta_2 = 0}$$

where β_1 and β_2 denote population partial regression coefficients. In Chapter 6 these tests will actually be carried out in terms of the sample problem of Table 1.2.

The fourth question concerns what are called partial correlation coefficients. One interpretation of a partial correlation coefficient considers it as a measure of the linear association between the criterion variable and some predictor when both have been adjusted for their linear association with the remaining predictors. Although the question of determining the relative influence of predictors is an ambiguous one, we shall comment on partial correlations, and their associated tests of significance, in Chapter 6.

Multiple regression, aside from being the most popular multivariate technique in applied research, provides a vehicle for subsequent discussion of all basic matrix operations and, in particular, the topics of determinants, matrix inversion, and matrix rank. These aspects of matrix algebra are essential in understanding the procedures for solving simultaneous equations, as appearing in multiple regression and other multivariate procedures.

1.6.3 Factor Analysis

The second set of questions refers to a topic in multivariate analysis that is generically called factor analysis. In factor analysis we are interested in the interdependence among sets of variables and the possibility of representing the objects of the investigation in terms of fewer dimensions than originally expressed.

To illustrate, let us plot the mean-corrected scores of the predictors, X_2 versus X_1, as shown in Fig. 1.4. Also shown in the same figure is a new axis, labeled z_1, which makes an angle of 38° with the horizontal axis. Now suppose we drop perpendiculars, represented by dotted lines, from each point to the axis z_1. Assuming the same scale for z_1 as used for $X_{\mathrm{d}1}$ and $X_{\mathrm{d}2}$, we could compute the variance of the 12 projected scores on z_1 as

$$\boxed{\operatorname{Var}[z_{i(1)}] = \sum_{i=1}^{12} (z_{i1} - \bar{z}_1)^2/12}$$

where $\bar{z}$ is the mean (which, because the X's are in deviation form, is zero) of the 12 scores on z_1.

The idea behind this procedure—called principal components and representing one type of factor analysis—is to find the axis z_1 so that the variance of the 12 projections

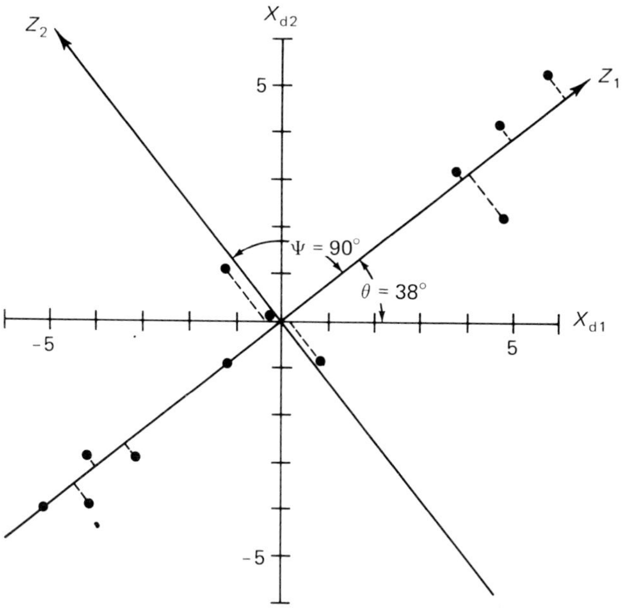

Fig. 1.4 Scatter plot of mean-corrected predictor variables.

onto it is maximal. As such the 12 mean-corrected scores, X_{d1} and X_{d2}, can be represented by a linear composite:

$$z_{i(1)} = t_1 X_{di1} + t_2 X_{di2}$$

and, hence, each pair of scores, X_{di1} and X_{di2} for each observation i, is replaced by a single score $z_{i(1)}$

In the present example, not much parsimony would be gained by merely replacing two scores with one score. In larger-scale problems, consisting of a large number of variables, considerable data reduction might be obtained. Moreover, principal components analysis allows the researcher to find additional axes, each at right angles to previously found axes and all with the property of maximum variance (subject to being at right angles to previously found axes).

This idea is also illustrated in Fig. 1.4 via the second axis z_2. Note that this axis has been drawn, as it should be, at right angles to z_1. One could, of course, project the 12 points onto this second axis and obtain a second set of scores. In this case, however, no parsimony would be gained, although the two axes z_1 and z_2 would be at right angles to each other and z_1 would contribute, by far, the greater variation in the pair of derived composites.

The second question, concerning the association of X_{d1} and X_{d2} with z_1, can be answered by computing product-moment correlations, X_{d1} with z_1 and X_{d2} with z_1. These are called component loadings and are measures of the association between each contributing (original) variable and the linear composite variable z_1 that is derived from them. Component loadings could also be computed for z_2.

The third question regarding how much of the total variation in X_{d1} and X_{d2} is accounted for by z_1 is also found from the principal components technique. In this example it happens to be 98 percent (leaving only 2 percent for z_2). That is, almost all of the original variation in X_{d1} and X_{d2} is retained in terms of the single composite variable z_1. This is evident by noting from Fig. 1.4 that the original 12 points lie close to the new axis z_1, and little information would be lost if their projections onto z_1 were substituted for their original values on X_{d1} and X_{d2}.

Subsequent chapters will discuss a number of important concepts from transformational geometry and matrix algebra–rotations, quadratic forms, eigenstructures of symmetric matrices–that pertain to solution procedures for principal components. Finally, in Chapter 6 the solution for the present problem will be described in detail.

1.6.4 Multiple Discriminant Analysis

Multiple discriminant analysis also entails a maximization objective. To illustrate, Fig. 1.5 shows a plot of X_{d2} versus X_{d1}. This time, however, each of the three groups–low, intermediate, and high absenteeism–is represented by different symbols. The first axis w_1 is the one, in this case, that maximizes among-group variation relative to average within-group variation.

That is, we wish to find a linear composite

$$\boxed{w_{i(1)} = v_1 X_{di1} + v_2 X_{di2}}$$

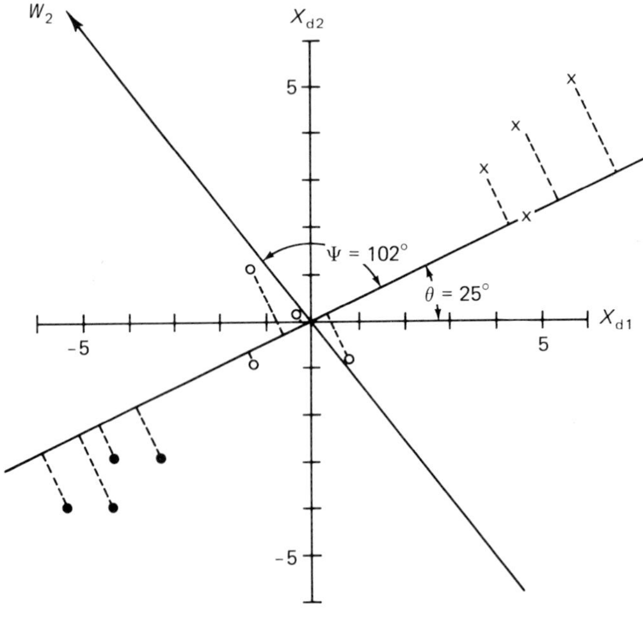

Fig. 1.5 A discriminant transformation of the mean-corrected predictor variables. Key: • Group 1; ○ Group 2; x Group 3.

with the property of maximizing the variation across group means relative to their pooled within-group variation.

Note that the first of these axes, w_1 in Fig. 1.5, makes an angle of 25° with the horizontal and, hence, is a different axis than the first principal component of Fig. 1.4. Note, also, that the second axis, w_2 in Fig. 1.5, is *not* at right angles to the first. In multiple discriminant analysis we can often find additional axes by which the groups can be discriminated, but these axes need not make right angles with each other.

In the case of multiple discriminant analysis, the points are projected onto w_1, and this axis exhibits the property of maximally separating the three group means, relative to their pooled within-group variation. Each point would be assigned to the closest group mean on the w_1 discriminant axis (i.e., the average of the four w_{i1}'s making up that group). As it turns out in this particular illustration, if this rule were followed, no point would be misclassified.[13]

The remaining questions dealing with statistical significance of the discriminant function(s), classification accuracy, and the relative importance of the predictor variables (X_{d1} and X_{d2}) are answered in ways analogous to the multiple regression case. We consider these, and related, questions in Chapter 6.

However, from the standpoint of transformational geometry and matrix algebra, multiple discriminant analysis involves such concepts as general linear transformations, simultaneous diagonalization of two different quadratic forms, and the eigenstructure of nonsymmetric matrices. Moreover, multiple discriminant analysis can also be related to principal component analysis in a space in which the points have previously been transformed (i.e., "spherized") to a pooled within-groups variance of unity.

In short, the three preceding problems provide sufficient motivational material for all of the remaining chapters, including the appendixes. Moreover, a full understanding of how these three problems can be solved will serve the applied researcher well insofar as understanding almost any other multivariate technique he may encounter.

1.7 FORMAT OF SUCCEEDING CHAPTERS

Succeeding chapters of the book are designed to develop the necessary concepts from transformational geometry and matrix algebra to deal with most problems in multivariate analysis, including the sample problems outlined in the preceding section. Obviously, no definitive treatment of the subject has been attempted. What we have tried to do is to select those aspects of matrix algebra that are most relevant for subsequent discussion of multivariate procedures.

Chapter 2 discusses definitions and operations on vectors and matrices. Here our emphasis is on the *mechanics* of working with vectors and matrices, rather than their geometric conceptualization. Elementary material on determinants is also presented as well as a demonstration of how various computations in multivariate analysis—for example, sums of squares and cross products—can be compactly expressed in matrix notation.

[13] Such would not be the case for the second discriminant w_2. As will be shown in Chapter 6, this second function is not statistically significant and would not be used for classification purposes anyway.

Chapter 3 is concerned with the conceptual aspects of vectors and matrices. Geometric representations are employed extensively as we discuss various operations on vectors and matrices, including scalar products and related topics in Euclidean distance geometry. Common statistical measures—standard deviation, covariance and correlation—are also portrayed from a geometric viewpoint.

Much of multivariate analysis is concerned with linear transformations, and this is the focus of Chapter 4. Each type of matrix transformation is described geometrically as we discuss such topics as rotations, stretches, and other transformations that have simple, geometric representations. Matrix inverses and the notion of matrix rank are also introduced here. We conclude the chapter with a description of the geometric effect of composite transformations that represent the matrix product of simpler ones.

Chapter 5 takes the opposite (and complementary) point of view. Here we are concerned with decomposing general matrix transformations into the product of simpler ones. The topic of matrix eigenstructure is introduced and related to the idea of changing basis vectors in order to bring out simpler geometric interpretations of the space. Matrix rank—first introduced in Chapter 4—is discussed more thoroughly in the context of the basic structure of a matrix. Quadratic forms are also introduced and related to matrix eigenstructures. Again, geometric analogy is used wherever it can help illuminate the algebraic operations.

Chapter 6 completes the cycle by taking the reader back to multivariate methods per se and, in particular, to the three sample problems introduced in the present chapter. Each of these problems—centering around multiple regression, principal components analysis, and multiple discriminant analysis—is described from the standpoint of concepts developed in Chapters 2–5. Numerical solutions are obtained for each case, and several geometric aspects of the methods are illustrated. We conclude the chapter by presenting a complementary framework for multivariate technique classification in terms of the nature of the transformations characterizing the matching of one set of numbers with some other set or sets. As such, this classification descriptor serves as both a way to unify earlier material and as a prologue for various textbooks on the general topic of multivariate analysis.

The appendixes cover more advanced material relevant to the general topic of multivariate analysis. Included here are such concepts as symbolic differentiation, constrained optimization, generalized inverses, and other special topics of relevance to multivariate analysis.

1.8 SUMMARY

The purpose of this chapter has been to set the stage for later material dealing with aspects of transformational geometry and matrix algebra of interest to multivariate analysis. The topic of multivariate analysis was introduced as a set of procedures for dealing with association among multiple variables. A classification system based on aspects of the data matrix and the researcher's objectives and presuppositions was described as a way of matching problem with technique.

We then described briefly a number of substantive applications of multivariate methods so as to give the reader some flavor of their breadth and diversity of use.

Following this, three prototypical problems, calling for various types of multivariate analysis, were described in terms of a miniature and common data bank. These problems will serve to motivate subsequent discussion of algebraic and geometric tools, leading to the solution of the sample problems in the concluding chapter of the book.

REVIEW QUESTIONS

1. What other systems for classifying multivariate techniques can you find in the literature of your field?

    a. How would you compare these classifications with the one presented here?

    b. Criticize the present classification and indicate how you would modify it for purposes of research in your own discipline.

2. Examine the literature of your field and select a number of examples using multivariate analysis.

    a. In each example, what was the substantive problem of interest?

    b. How did the author's use of the technique(s) relate to the content side of the problem?

    c. Do other techniques suggest themselves for the specific problem examined by the author?

3. In terms of your own research try to formulate a problem that appears suitable for multivariate analysis.

    a. How would you classify the problem in terms of the system described in this chapter?

    b. What multivariate procedures are suggested by your classification of the problem?

CHAPTER 2

Vector and Matrix Operations for Multivariate Analysis

2.1 INTRODUCTION

Facility in the arithmetic of vectors and matrices, just like skill in applying ordinary arithmetic to daily affairs, is essential in multivariate analysis. In this chapter our purpose is to review the fundamentals of vector and matrix operations and the concept of the determinant of a matrix. The emphasis here is on defining vector and matrix operations and illustrating the mechanics of their application.

We begin the chapter with a description of vectors as ordered n-tuples of numbers that are subject to certain manipulative rules. Selected arithmetic operations on vectors are defined and illustrated numerically. A number of special-purpose vectors, such as the null vector, unit vector, and sign vector, are also described.

Matrices are then introduced and discussed from the same kind of viewpoint. Also, we describe various kinds of special matrices, such as symmetric, diagonal, scalar, and identity matrices and illustrate their application via small numerical examples.

The determinant of a matrix plays an important role in more advanced topics, such as matrix inversion, rank, and quadratic forms, that are introduced in later chapters. For this reason it seems appropriate to discuss determinants and some of their numerical properties at an early stage, and we do so in this section of the chapter.

We conclude the chapter with a discussion of certain matrices of particular interest to multivariate analysis, namely mean-corrected sums of squares and cross product (SSCP) matrices, covariance matrices, and correlation matrices. Computation of these major types of statistical matrices is carried out as a demonstration of how concise the matrix formulation of various arithmetic operations can be.

A word on notation: boldface lowercase letters, **a**, **b**, **c**, etc., will be used to denote vectors and boldface capitals, **A**, **B**, **C**, etc., will be used for matrices. The determinant of a matrix **A** will be expressed as $|\mathbf{A}|$. A prime, for example, $\mathbf{a}'$ or $\mathbf{A}'$, will denote the transpose of a vector or matrix, respectively. (The concept of transpose is taken up later in the chapter.)

The material of this chapter is presented rather crisply since our purpose here is to provide only the mechanics of vector and matrix operations before introducing the more conceptually oriented material of later chapters. However, sufficient numerical examples are presented to illustrate the computational aspects in some detail.

2.2 VECTOR REPRESENTATION

Multivariate analysis makes liberal use of vector concepts from linear algebra. Vectors can be defined in four major ways: (a) as strictly abstract entities on which certain relations and operations are specified, (b) as directed line segments in a geometric space, (c) as coordinate representations of points in a geometric space, or (d) as ordered n-tuples of numbers. We adopt the last viewpoint in this chapter in order to demonstrate the kinds of operations that can be performed on vectors. The geometric representations of (b) and (c), which can be illustrated graphically if two or three dimensions are involved, are discussed in Chapters 3–5.

2.3 BASIC DEFINITIONS AND OPERATIONS ON VECTORS

A vector a *of order $n \times 1$ is an ordered set of n real*[1] *numbers (called scalars), which we can write as*

$$\mathbf{a} = \begin{bmatrix} a_1 \\ a_2 \\ \vdots \\ a_n \end{bmatrix}$$

The a's denote real numbers and are called components, elements, or entries of a. The form above is called a *column* vector and consists of n rows and 1 column of elements (from which the designation $n \times 1$ derives). Alternatively we can write a vector a′ of order $1 \times n$ as

$$\mathbf{a}' = (a_1, a_2, \ldots, a_n)$$

and call this a *row* vector, consisting of 1 row and n columns of elements.

We shall use the notation a to denote a column vector and the notation a′, which is called the *transpose* of a, to denote a row vector. By vector transpose, generally, is meant that a column vector of order n by 1 becomes a row vector, involving the same ordered set of entries, but now of order 1 by n. Similarly, the transpose of a row vector of 1 by n is a column vector, involving the same ordered set of entries, but now of order n by 1.

Examples of column vectors are

$$\begin{bmatrix} 3 \\ 1 \end{bmatrix}; \quad \begin{bmatrix} 4 \\ 2.6 \\ 5 \\ 0 \end{bmatrix}; \quad \begin{bmatrix} 2.718 \\ 5 \\ 1 \end{bmatrix}; \quad \begin{bmatrix} 18 \\ 21 \\ 0 \\ 0 \end{bmatrix}; \quad \mathbf{x} = \begin{bmatrix} x_1 \\ x_2 \\ x_3 \end{bmatrix}$$

Examples of row vectors are

$$(3, 1); \quad (18, 42, 6); \quad (\sqrt{\pi}, 13, 0, 5.2); \quad (0, 0, 2, 7); \quad \mathbf{t}' = (t_1, t_2)$$

[1] Throughout the book we shall always assume that the scalars are drawn from the set of real (as opposed to complex) numbers.

We can transpose the 3 x 1 column vector

$$\mathbf{a} = \begin{bmatrix} 1 \\ 2 \\ 3 \end{bmatrix}$$

to get the 1 x 3 row vector

$$\mathbf{a}' = (1, 2, 3)$$

Similarly, we can then find the transpose of the 1 x 3 row vector $\mathbf{a}' = (1, 2, 3)$ as follows:

$$(\mathbf{a}')' = \mathbf{a} = \begin{bmatrix} 1 \\ 2 \\ 3 \end{bmatrix}$$

and note that we are back where we started, that is, where $\mathbf{a}$ is a 3 x 1 column vector.

2.3.1 Null, Unit, Sign, and Zero–One Vectors

If all components of a vector are zero, we shall call this a *null* or zero vector, denoted as $\mathbf{0}$. This should not be confused with the scalar 0. If all components of a vector are 1, this type of vector is called a *unit* vector, denoted as $\mathbf{1}$. If the components consist of either 1's or −1's (with at least one of each type present), this is called a *sign* vector. If the components consist of either 1's or 0's (with at least one of each type present), this is called a zero–one vector. To illustrate:

	Column vectors	*Row vectors*
Null vectors	$\begin{bmatrix} 0 \\ 0 \\ 0 \end{bmatrix}$	$(0, 0, 0, 0)$
Unit vectors	$\begin{bmatrix} 1 \\ 1 \end{bmatrix}$	$(1, 1, 1)$
Sign vectors	$\begin{bmatrix} -1 \\ 1 \\ -1 \end{bmatrix}$	$(1, -1)$
Zero–one vectors	$\begin{bmatrix} 0 \\ 1 \\ 1 \end{bmatrix}$	$(0, 0, 1, 1, 0)$

As will be shown in subsequent chapters, the zero vector frequently plays a role that is analogous to the scalar 0 in ordinary arithmetic. The unit vector is useful in certain kinds of summations, as is illustrated in Section 2.8. Sign and zero–one vectors are also useful in various kinds of operations involving either algebraic sums or the isolation of rows, columns, or elements of an array of numbers.

2.3.2 Vector Equality

Two vectors of the same order (either both n × 1 or both 1 × n) are equal if they are equal component by component. Let

$$\mathbf{a} = \begin{bmatrix} a_1 \\ a_2 \\ \vdots \\ a_n \end{bmatrix} \quad \text{and} \quad \mathbf{b} = \begin{bmatrix} b_1 \\ b_2 \\ \vdots \\ b_n \end{bmatrix}$$

Then

$$\boxed{\begin{gathered} \mathbf{a} = \mathbf{b} \\ \text{if and only if} \\ a_i = b_i \qquad (i = 1, 2, \ldots, n) \end{gathered}}$$

For example,

$$\mathbf{a} = \begin{bmatrix} 3 \\ 0 \\ 4 \end{bmatrix} = \mathbf{b} = \begin{bmatrix} 3 \\ 0 \\ 4 \end{bmatrix}$$

But

$$\mathbf{a} \neq \mathbf{c} = \begin{bmatrix} 3 \\ 0 \\ 2 \end{bmatrix}; \qquad \mathbf{a} \neq \mathbf{d} = \begin{bmatrix} 3 \\ 0 \\ 4 \\ 9 \end{bmatrix}$$

$$\mathbf{a} \neq \mathbf{e} = \begin{bmatrix} 3 \\ 4 \\ 0 \end{bmatrix}; \qquad \mathbf{a} \neq \mathbf{a}' = (3, 0, 4)$$

In the last case $\mathbf{a}$ and $\mathbf{a}'$ are not of the same order, since the first is a column vector and the second is a row vector.

Throughout most of this chapter, we shall present definitions in terms of column vectors, although our remarks will also hold true for row vectors.[2] Moreover, in discussing various operations on vectors, it will be assumed, unless otherwise specified, that the vectors are of common order—either all are $n \times 1$ or all are $1 \times n$.

[2] When row vectors are employed as numerical examples, emphasis is primarily on conserving space. The reader should remember that we could just as appropriately describe the operations in terms of column vectors.

2.3.3 Vector Addition and Subtraction

Two or more vectors of the same order can be added by adding correspondent components. That is

$$\mathbf{a} + \mathbf{b} = \begin{bmatrix} a_1 + b_1 \\ a_2 + b_2 \\ \vdots \\ a_n + b_n \end{bmatrix}$$

Examples are

$$\mathbf{a} = \begin{bmatrix} 1 \\ 2 \\ 14 \end{bmatrix}; \qquad \mathbf{b} = \begin{bmatrix} 3 \\ 0 \\ 9 \end{bmatrix}; \qquad \mathbf{a} + \mathbf{b} = \begin{bmatrix} 4 \\ 2 \\ 23 \end{bmatrix}$$

$$\mathbf{c}' = (1, 3); \qquad \mathbf{d}' = (4, 13); \qquad \mathbf{c}' + \mathbf{d}' = (5, 16)$$

But we cannot add

$$\mathbf{e} = \begin{bmatrix} 4 \\ 2 \\ 13 \end{bmatrix} \text{ to } \mathbf{f} = \begin{bmatrix} 2 \\ 1 \end{bmatrix}; \qquad \text{or} \qquad \mathbf{g} = \begin{bmatrix} 5 \\ 1 \\ 2 \\ 7 \end{bmatrix} \text{ to } \mathbf{h}' = (6, 3, 0, 2)$$

since in each case the order differs.

The difference between two vectors **a** *and* **b**, *of the same order, is defined to be that vector,* **a** – **b**, *which, when added to* **b**, *yields the vector* **a**. Again, subtraction is performed componentwise.[3] That is,

$$\mathbf{a} - \mathbf{b} = \begin{bmatrix} a_1 - b_1 \\ a_2 - b_2 \\ \vdots \\ a_n - b_n \end{bmatrix}$$

Examples are

$$\mathbf{a} = \begin{bmatrix} 1 \\ 2 \\ 14 \end{bmatrix}; \qquad \mathbf{b} = \begin{bmatrix} 3 \\ 0 \\ 9 \end{bmatrix}; \qquad \mathbf{a} - \mathbf{b} = \begin{bmatrix} -2 \\ 2 \\ 5 \end{bmatrix}$$

$$\mathbf{c}' = (1, 3); \qquad \mathbf{d}' = (4, 13); \qquad \mathbf{c}' - \mathbf{d}' = (-3, -10)$$

[3] A more rigorous presentation would first define multiplication of a vector by a scalar (specifically multiplication by –1), followed by vector addition. Here, however, we follow the more natural presentation order of traditional arithmetic in which subtraction follows addition.

But we cannot subtract

$$\mathbf{e} = \begin{bmatrix} 4 \\ 2 \\ 13 \end{bmatrix} \quad \text{from} \quad \mathbf{f} = \begin{bmatrix} 2 \\ 1 \end{bmatrix}; \qquad \text{or} \qquad \mathbf{g} = \begin{bmatrix} 5 \\ 1 \\ 2 \\ 7 \end{bmatrix} \quad \text{from} \quad \mathbf{h}' = (6, 3, 0, 2)$$

since in each case the order differs.

The operation of vector *addition*—of either column or row vectors—possesses the following properties:

1. The sum of two vectors **a** and **b** is a unique third vector **c**.

$$\text{Let} \quad \mathbf{a} = \begin{bmatrix} 1 \\ 3 \end{bmatrix} \quad \text{and} \quad \mathbf{b} = \begin{bmatrix} 2 \\ 4 \end{bmatrix}$$

$$\text{Then} \quad \mathbf{a} + \mathbf{b} = \mathbf{c} = \begin{bmatrix} 3 \\ 7 \end{bmatrix} \quad \text{is unique.}$$

2. Vector addition is commutative.

$$\mathbf{a} + \mathbf{b} = \mathbf{b} + \mathbf{a}$$

$$\begin{bmatrix} 1 \\ 3 \end{bmatrix} + \begin{bmatrix} 2 \\ 4 \end{bmatrix} = \begin{bmatrix} 2 \\ 4 \end{bmatrix} + \begin{bmatrix} 1 \\ 3 \end{bmatrix} = \begin{bmatrix} 3 \\ 7 \end{bmatrix}$$

3. Vector addition is associative.

$$(\mathbf{a} + \mathbf{b}) + \mathbf{d} = \mathbf{a} + (\mathbf{b} + \mathbf{d})$$

$$\text{Let} \quad \mathbf{d} = \begin{bmatrix} 3 \\ 5 \end{bmatrix} \qquad \text{Then} \quad \overset{\mathbf{a}+\mathbf{b}}{\begin{bmatrix} 3 \\ 7 \end{bmatrix}} + \overset{\mathbf{d}}{\begin{bmatrix} 3 \\ 5 \end{bmatrix}} = \overset{\mathbf{a}}{\begin{bmatrix} 1 \\ 3 \end{bmatrix}} + \overset{\mathbf{b}+\mathbf{d}}{\begin{bmatrix} 5 \\ 9 \end{bmatrix}} = \begin{bmatrix} 6 \\ 12 \end{bmatrix}$$

4. There exists a null or zero vector **0** having the property **a** + **0** = **a** for any vector **a**.

$$\begin{bmatrix} 1 \\ 3 \end{bmatrix} + \begin{bmatrix} 0 \\ 0 \end{bmatrix} = \begin{bmatrix} 1 \\ 3 \end{bmatrix} = \mathbf{a}$$

5. Each vector **a** has a counterpart negative vector –**a** so that **a** + –**a** = **0**.

$$\begin{bmatrix} 1 \\ 3 \end{bmatrix} + \begin{bmatrix} -1 \\ -3 \end{bmatrix} = \begin{bmatrix} 0 \\ 0 \end{bmatrix} = \mathbf{0}$$

We shall have occasion to refer to one or more of these properties quite frequently in subsequent dicussions.

2.3.4 Scalar Multiplication of a Vector

Assume we have some real number k. As pointed out earlier, this is called a scalar in vector algebra. *Scalar multiplication of a vector involves multiplying each component of the vector by the scalar.*

$$k\mathbf{a} = k\begin{bmatrix} a_1 \\ a_2 \\ \vdots \\ a_n \end{bmatrix} = \begin{bmatrix} ka_1 \\ ka_2 \\ \vdots \\ ka_n \end{bmatrix}$$

To illustrate the scalar multiplication of vectors, assume

$$\mathbf{a} = \begin{bmatrix} 1 \\ 2 \\ 3 \end{bmatrix} \quad \text{and} \quad k = 3$$

Then

$$k\mathbf{a} = 3\begin{bmatrix} 1 \\ 2 \\ 3 \end{bmatrix} = \begin{bmatrix} 3 \times 1 \\ 3 \times 2 \\ 3 \times 3 \end{bmatrix} = \begin{bmatrix} 3 \\ 6 \\ 9 \end{bmatrix}$$

Next, let $\mathbf{b}' = (4, 5, 6)$. Then

$$k\mathbf{b}' = 3(4, 5, 6) = (3 \times 4, 3 \times 5, 3 \times 6) = (12, 15, 18)$$

As in the case for vector addition, scalar multiplication of vectors exhibits a number of useful properties.[4]

1. If $\mathbf{a}$ is a vector and k is a scalar, the product $k\mathbf{a}$ is a uniquely defined vector.

$$\text{Let} \quad \mathbf{a} = \begin{bmatrix} 1 \\ 3 \end{bmatrix} \quad \text{and} \quad k = 2$$

$$\text{Then} \quad 2\begin{bmatrix} 1 \\ 3 \end{bmatrix} = \begin{bmatrix} 2 \\ 6 \end{bmatrix} \quad \text{is unique.}$$

2. Scalar multiplication is associative. For example, for two scalars k_1 and k_2, it is the case that $k_1(k_2\mathbf{a}) = (k_1k_2)\mathbf{a}$.

$$\text{Let} \quad k_1 = 2 \quad \text{and} \quad k_2 = 3. \quad \text{Then} \quad 2\underset{k_2\mathbf{a}}{\begin{bmatrix} 3 \\ 9 \end{bmatrix}} = 6\underset{\mathbf{a}}{\begin{bmatrix} 1 \\ 3 \end{bmatrix}} = \begin{bmatrix} 6 \\ 18 \end{bmatrix}$$

[4] These properties and the properties listed in Section 2.3.3 collectively define a *vector space* for all vectors **a**, **b**, **c**, etc., and all scalars (real numbers) k_1, k_2, etc.

3. Scalar multiplication is distributive. For example, $k(\mathbf{a} + \mathbf{b}) = k\mathbf{a} + k\mathbf{b}$. Also, $(k_1 + k_2)\mathbf{a} = k_1\mathbf{a} + k_2\mathbf{a}$.

$$\text{Let} \quad \mathbf{b} = \begin{bmatrix} 2 \\ 4 \end{bmatrix}. \quad \text{Then} \quad 2\overset{\mathbf{a}+\mathbf{b}}{\begin{bmatrix} 3 \\ 7 \end{bmatrix}} = 2\overset{\mathbf{a}}{\begin{bmatrix} 1 \\ 3 \end{bmatrix}} + 2\overset{\mathbf{b}}{\begin{bmatrix} 2 \\ 4 \end{bmatrix}} = \begin{bmatrix} 6 \\ 14 \end{bmatrix}$$

$$\text{Also} \quad 5\overset{\mathbf{a}}{\begin{bmatrix} 1 \\ 3 \end{bmatrix}} = 2\overset{\mathbf{a}}{\begin{bmatrix} 1 \\ 3 \end{bmatrix}} + 3\overset{\mathbf{a}}{\begin{bmatrix} 1 \\ 3 \end{bmatrix}} = \begin{bmatrix} 5 \\ 15 \end{bmatrix}$$

4. For any vector $\mathbf{a}$, we have the products $0 \cdot \mathbf{a} = \mathbf{0}$; $1\mathbf{a} = \mathbf{a}$ and $-1\mathbf{a} = -\mathbf{a}$. For example,

$$0\overset{\mathbf{a}}{\begin{bmatrix} 1 \\ 3 \end{bmatrix}} = \begin{bmatrix} 0 \\ 0 \end{bmatrix}; \quad 1\overset{\mathbf{a}}{\begin{bmatrix} 1 \\ 3 \end{bmatrix}} = \begin{bmatrix} 1 \\ 3 \end{bmatrix}; \quad -1\overset{\mathbf{a}}{\begin{bmatrix} 1 \\ 3 \end{bmatrix}} = \begin{bmatrix} -1 \\ -3 \end{bmatrix}$$

We can now consider an operation that generalizes both vector addition and scalar multiplication of vectors.

2.3.5 Linear Combinations of Vectors

Most of our comments about vector addition and scalar multiplication in Sections 2.3.3 and 2.3.4 can be succinctly summarized in terms of the concept of a linear combination of a set of vectors. Let $\mathbf{a}_1, \mathbf{a}_2, \ldots, \mathbf{a}_m$ denote a set of m vectors (each of order $n \times 1$) and let $k_1, k_2, \ldots, k_m$ denote a set of m scalars. *A linear combination of a set of vectors is defined as*

$$\boxed{\mathbf{v}_1 = k_1\mathbf{a}_1 + k_2\mathbf{a}_2 + \cdots + k_m\mathbf{a}_m}$$

If we take another (arbitrary) linear combination involving another set of m scalars $k_1^*, k_2^*, \ldots, k_m^*$, we have

$$\boxed{\mathbf{v}_2 = k_1^*\mathbf{a}_1 + k_2^*\mathbf{a}_2 + \cdots + k_m^*\mathbf{a}_m}$$

Next, suppose we add the two linear combinations. If so, it will be the case that the following properties hold:

1. $\mathbf{v}_1 + \mathbf{v}_2 = (k_1 + k_1^*)\mathbf{a}_1 + (k_2 + k_2^*)\mathbf{a}_2 + \cdots + (k_m + k_m^*)\mathbf{a}_m$

2. Moreover, if c denotes still another scalar, then $c\mathbf{v}_1 = (ck_1)\mathbf{a}_1 + (ck_2)\mathbf{a}_2 + \ldots + (ck_m)\mathbf{a}_m$ is also a linear combination of $\mathbf{a}_1, \mathbf{a}_2, \ldots, \mathbf{a}_m$.

What this means is that linear combinations of vectors can be added together (such as $\mathbf{v}_1 + \mathbf{v}_2$) or can be multiplied by a scalar (such as $c\mathbf{v}_1$), resulting in new vectors that bear simple relationships to the old.

To illustrate, let

$$\mathbf{a}_1 = \begin{bmatrix} 1 \\ 2 \\ 3 \end{bmatrix}; \qquad \mathbf{a}_2 = \begin{bmatrix} 0 \\ 3 \\ 2 \end{bmatrix}; \qquad \mathbf{a}_3 = \begin{bmatrix} 1 \\ 4 \\ 2 \end{bmatrix}$$

$$k_1 = 2; \quad k_2 = 3; \quad k_3 = 1; \quad k_1{}^* = 0; \quad k_2{}^* = 4; \quad k_3{}^* = 5; \quad c = 2$$

Then, we can first write $\mathbf{v}_1$ and $\mathbf{v}_2$ as

$$\mathbf{v}_1 = k_1\mathbf{a}_1 + k_2\mathbf{a}_2 + k_3\mathbf{a}_3$$

$$= 2\begin{bmatrix} 1 \\ 2 \\ 3 \end{bmatrix} + 3\begin{bmatrix} 0 \\ 3 \\ 2 \end{bmatrix} + 1\begin{bmatrix} 1 \\ 4 \\ 2 \end{bmatrix} = \begin{bmatrix} 3 \\ 17 \\ 14 \end{bmatrix}$$

$$\mathbf{v}_2 = k_1{}^*\mathbf{a}_1 + k_2{}^*\mathbf{a}_2 + k_3{}^*\mathbf{a}_3$$

$$= 0\begin{bmatrix} 1 \\ 2 \\ 3 \end{bmatrix} + 4\begin{bmatrix} 0 \\ 3 \\ 2 \end{bmatrix} + 5\begin{bmatrix} 1 \\ 4 \\ 2 \end{bmatrix} = \begin{bmatrix} 5 \\ 32 \\ 18 \end{bmatrix}$$

The first property above can now be illustrated as

$$\mathbf{v}_1 + \mathbf{v}_2 = \begin{bmatrix} 3 \\ 17 \\ 14 \end{bmatrix} + \begin{bmatrix} 5 \\ 32 \\ 18 \end{bmatrix} = 2\begin{bmatrix} 1 \\ 2 \\ 3 \end{bmatrix} + 7\begin{bmatrix} 0 \\ 3 \\ 2 \end{bmatrix} + 6\begin{bmatrix} 1 \\ 4 \\ 2 \end{bmatrix}$$

$$= \begin{bmatrix} 2 \\ 4 \\ 6 \end{bmatrix} + \begin{bmatrix} 0 \\ 21 \\ 14 \end{bmatrix} + \begin{bmatrix} 6 \\ 24 \\ 12 \end{bmatrix} = \begin{bmatrix} 8 \\ 49 \\ 32 \end{bmatrix}$$

The second property above can now be illustrated as

$$c\mathbf{v}_1 = \begin{bmatrix} 6 \\ 34 \\ 28 \end{bmatrix} = 4\begin{bmatrix} 1 \\ 2 \\ 3 \end{bmatrix} + 6\begin{bmatrix} 0 \\ 3 \\ 2 \end{bmatrix} + 2\begin{bmatrix} 1 \\ 4 \\ 2 \end{bmatrix}$$

$$= \begin{bmatrix} 4 \\ 8 \\ 12 \end{bmatrix} + \begin{bmatrix} 0 \\ 18 \\ 12 \end{bmatrix} + \begin{bmatrix} 2 \\ 8 \\ 4 \end{bmatrix} = \begin{bmatrix} 6 \\ 34 \\ 28 \end{bmatrix}$$

The concept of a linear combination of a set of vectors is one of the most important aspects of vector algebra. We shall return to this topic in the next chapter dealing with the geometric aspects of vectors.

2.3.6 The Scalar Product of Two Vectors

The last operation involving vectors to be discussed in Section 2.3 is that of the scalar product (sometimes called inner product, or dot product) of two vectors. As is well known, when we multiply two numbers (scalars) together, we obtain another element of the same kind, namely, a number that represents their product. However, in vector algebra, multiplication of two vectors need not lead to a vector. For example, one way of multiplying two vectors (of the same order of course) yields a *number* rather than a vector. This number is called their scalar product. To illustrate the scalar product of two vectors, consider the column vectors

$$\mathbf{a} = \begin{bmatrix} a_1 \\ a_2 \\ \vdots \\ a_n \end{bmatrix} \quad \text{and} \quad \mathbf{b} = \begin{bmatrix} b_1 \\ b_2 \\ \vdots \\ b_n \end{bmatrix}$$

Their scalar product is defined as

$$\boxed{\begin{aligned} \mathbf{a}'\mathbf{b} &= a_1 b_1 + a_2 b_2 + \cdots + a_k b_k + \cdots + a_n b_n \\ &= \sum_{k=1}^{n} a_k b_k \end{aligned}}$$

Notice that the first vector is treated as a row vector and the second vector is treated as a column vector. However, either one can serve as the first (row) vector. To illustrate,

$$\mathbf{a} = \begin{bmatrix} 1 \\ 4 \\ 0 \\ 3 \end{bmatrix}; \quad \mathbf{b} = \begin{bmatrix} 0 \\ 2 \\ 7 \\ 4 \end{bmatrix}$$

Hence

$$\begin{aligned} \mathbf{a}'\mathbf{b} = \mathbf{b}'\mathbf{a} &= (1 \times 0) + (4 \times 2) + (0 \times 7) + (3 \times 4) \\ &= (0 \times 1) + (2 \times 4) + (7 \times 0) + (4 \times 3) \\ &= 20 \end{aligned}$$

We might now check to see if the associative and distributive laws are valid for scalar products. As it turns out, the associative law, illustrated by $(\mathbf{a}'\mathbf{b})\mathbf{c}$ is *not* valid because the scalar product of a scalar, that results from $\mathbf{a}'\mathbf{b}$, and a vector $\mathbf{c}$ has *not* been defined. *That is, the scalar product idea is limited to the product of a row and column vector.* Of course, we earlier defined the operation of multiplying a vector by a scalar, but this is not a scalar product.

However, the distributive laws for the scalar product, with respect to addition, *are* valid:

$$a'(b + c) = (a'b) + (a'c)$$

Also,

$$(a + b)'c = (a'c) + (b'c)$$

To illustrate the distributive laws, let,

$$a = \begin{bmatrix} 1 \\ 2 \\ 3 \end{bmatrix}; \qquad b = \begin{bmatrix} 4 \\ 0 \\ 1 \end{bmatrix}; \qquad c = \begin{bmatrix} 3 \\ 2 \\ 2 \end{bmatrix}$$

Then

$$a'(b + c) = (a'b) + (a'c)$$

$$(1, 2, 3)\begin{bmatrix} 7 \\ 2 \\ 3 \end{bmatrix} = (1, 2, 3)\begin{bmatrix} 4 \\ 0 \\ 1 \end{bmatrix} + (1, 2, 3)\begin{bmatrix} 3 \\ 2 \\ 2 \end{bmatrix}$$

$$20 = 7 + 13$$

Also,

$$(a + b)'c = (a'c) + (b'c)$$

$$(5, 2, 4)\begin{bmatrix} 3 \\ 2 \\ 2 \end{bmatrix} = (1, 2, 3)\begin{bmatrix} 3 \\ 2 \\ 2 \end{bmatrix} + (4, 0, 1)\begin{bmatrix} 3 \\ 2 \\ 2 \end{bmatrix}$$

$$27 = 13 + 14$$

2.3.7 Some Special Cases of the Scalar Product

In the definition of scalar product given above, no requirement was made that **a** had to differ from **b**. That is, one can legitimately compute the scalar product of a vector with itself. To illustrate,

$$\text{If} \qquad a = \begin{bmatrix} 1 \\ 2 \\ 3 \end{bmatrix}, \qquad \text{then} \qquad a'a = 1 + 4 + 9 = 14$$

Notice, then, that one obtains a *sum of squares* if one takes the scalar product of a vector with itself. And $a'a > 0$ unless, of course, $a = \mathbf{0}$.

Consider now the unit vector $\mathbf{1}' = (1, 1, 1)$ and the vector

$$\mathbf{a} = \begin{bmatrix} 1 \\ 2 \\ 3 \end{bmatrix}$$

Their scalar product is

$$\mathbf{1}'\mathbf{a} = 1 + 2 + 3 = 6$$

Thus, the scalar product of the unit vector and a given vector results in the *sum* of the entries in the given vector. If the vector is a sign vector, then the algebraic sum is taken. For example,

$$(-1, 1, -1) \begin{bmatrix} 1 \\ 2 \\ 3 \end{bmatrix} = -2$$

Finally, consider the relationships $\mathbf{a}'(k\mathbf{b}) = (k\mathbf{a}')\mathbf{b} = k(\mathbf{a}'\mathbf{b})$, where k is a scalar. To demonstrate that these relations hold, let

$$\mathbf{a} = \begin{bmatrix} 1 \\ 2 \\ 3 \end{bmatrix}; \qquad \mathbf{b} = \begin{bmatrix} 4 \\ 0 \\ 1 \end{bmatrix}; \qquad k = 2$$

Then

$$\mathbf{a}'(k\mathbf{b}) = (k\mathbf{a}')\mathbf{b}$$

$$(1, 2, 3) \begin{bmatrix} 8 \\ 0 \\ 2 \end{bmatrix} = (2, 4, 6) \begin{bmatrix} 4 \\ 0 \\ 1 \end{bmatrix} = 14$$

Also

$$\mathbf{a}'(k\mathbf{b}) = k(\mathbf{a}'\mathbf{b})$$

$$(1, 2, 3) \begin{bmatrix} 8 \\ 0 \\ 2 \end{bmatrix} = 2(1, 2, 3) \begin{bmatrix} 4 \\ 0 \\ 1 \end{bmatrix} = 14$$

We can sum up this part of the discussion by recapitulating the following properties of scalar products, as illustrated above:

1. $\mathbf{a}'(\mathbf{b} + \mathbf{c}) = \mathbf{a}'\mathbf{b} + \mathbf{a}'\mathbf{c}$
2. $(\mathbf{a} + \mathbf{b})'\mathbf{c} = \mathbf{a}'\mathbf{c} + \mathbf{b}'\mathbf{c}$
3. $\mathbf{a}'(k\mathbf{b}) = (k\mathbf{a}')\mathbf{b} = k(\mathbf{a}'\mathbf{b})$

We shall have more to say about the utility of scalar product multiplication in the concluding section of the chapter, which deals with the computation of various matrices derived from statistical data.

Finally, it should be mentioned that two types of vector-by-vector multiplication are *not* defined in matrix algebra. That is,

1. a row vector cannot be multiplied by a row vector;
2. a column vector cannot be multiplied by a column vector.

The remaining case—that of multiplying a column vector by a row vector— is covered in our discussion of matrices since in this instance their product is a matrix, not a scalar.

2.3.8 Some More Examples

To help review the vector operations described in this section, consider the following:

$$\mathbf{a} = \begin{bmatrix} 1 \\ 2 \\ 3 \end{bmatrix}; \qquad \mathbf{a}' = (1, 2, 3); \qquad \mathbf{b} = \begin{bmatrix} 0 \\ 2 \\ 5 \end{bmatrix}; \qquad \mathbf{b}' = (0, 2, 5)$$

$$k_1 = 2; \qquad k_2 = 3$$

We can now illustrate the following operations.

Transpose of Vector We recall that the transpose of the 3 x 1 column vector

$$\mathbf{a} = \begin{bmatrix} 1 \\ 2 \\ 3 \end{bmatrix}$$

is the 1x 3 row vector, written as

$$\mathbf{a}' = (1, 2, 3)$$

Moreover, were we to take the transpose of $\mathbf{a}'$, we would have

$$(\mathbf{a}')' = \mathbf{a} = \begin{bmatrix} 1 \\ 2 \\ 3 \end{bmatrix}$$

That is, the transpose of a transpose equals the original vector. Similarly, we can find

$$((\mathbf{a}')')' = \mathbf{a}' = (1, 2, 3)$$

Addition and Subtraction The sum and difference of $\mathbf{a}'$ and $\mathbf{b}'$, respectively, are simply

$$\mathbf{a}' + \mathbf{b}' = (1 + 0, 2 + 2, 3 + 5) = (1, 4, 8)$$

$$\mathbf{a}' - \mathbf{b}' = (1 - 0, 2 - 2, 3 - 5) = (1, 0, -2)$$

Scalar Multiplication of a Vector Some illustrations of scalar multiplication of a vector are

$$k_1\mathbf{a}' = (2\times 1, 2\times 2, 2\times 3); \qquad 0\mathbf{a} = \begin{bmatrix} 0\times 1 \\ 0\times 2 \\ 0\times 3 \end{bmatrix} = \begin{bmatrix} 0 \\ 0 \\ 0 \end{bmatrix} = \mathbf{0}$$
$$= (2,4,6)$$

$$k_1(k_2\mathbf{a}') = (k_1k_2)\mathbf{a}' \quad ; \qquad k_2\mathbf{b} = 3\begin{bmatrix} 0 \\ 2 \\ 5 \end{bmatrix} = \begin{bmatrix} 0 \\ 6 \\ 15 \end{bmatrix}$$
$$= 6(1,2,3)$$
$$= (6,12,18)$$

Linear Combinations of Vectors Illustrations of linear combinations of vectors are

$$\mathbf{v}_1 = k_1\mathbf{a} + k_2\mathbf{b} = 2\begin{bmatrix} 1 \\ 2 \\ 3 \end{bmatrix} + 3\begin{bmatrix} 0 \\ 2 \\ 5 \end{bmatrix} = \begin{bmatrix} 2 \\ 10 \\ 21 \end{bmatrix} \quad ;$$

$$\mathbf{v}_2' = k_1\mathbf{b}' - k_2\mathbf{a}'$$
$$= 2(0,2,5) - 3(1,2,3)$$
$$= (0,4,10) - (3,6,9)$$
$$= (-3,-2,1)$$

If $c = 4$, then

$$c\mathbf{v}_2' = ck_1\mathbf{b}' - ck_2\mathbf{a}'$$
$$= 8(0,2,5) - 12(1,2,3)$$
$$= (0,16,40) - (12,24,36)$$
$$= (-12,-8,4)$$

The Scalar Product Some scalar products of interest are

$$\mathbf{a}'\mathbf{b} = (1,2,3)\begin{bmatrix} 0 \\ 2 \\ 5 \end{bmatrix}$$
$$= (1\times 0) + (2\times 2) + (3\times 5)$$
$$= 19$$

$$\mathbf{b}'\mathbf{b} = (0,2,5)\begin{bmatrix} 0 \\ 2 \\ 5 \end{bmatrix} = 29; \qquad \mathbf{0}'\mathbf{b} = (0,0,0)\begin{bmatrix} 0 \\ 2 \\ 5 \end{bmatrix} = 0$$

$$\mathbf{a}'\mathbf{1} = (1,2,3)\begin{bmatrix} 1 \\ 1 \\ 1 \end{bmatrix} \quad ; \qquad \mathbf{a}'(\mathbf{1}+\mathbf{b}) = (1,2,3)\begin{bmatrix} 1 \\ 1 \\ 1 \end{bmatrix} + (1,2,3)\begin{bmatrix} 0 \\ 2 \\ 5 \end{bmatrix} = 6 + 19$$
$$= 6 \qquad\qquad = 25$$

Additional examples appear at the end of the chapter.

2.4 MATRIX REPRESENTATION

As in our introduction to vector arithmetic, our purpose here is to describe various operations involving matrices as they relate to subsequent discussion of multivariate procedures. Again, we attempt no definitive treatment of the topic but, rather, select those aspects of particular relevance to subsequent chapters.

We first present a discussion of elementary relations and operations on matrices and then turn to a description of special types of matrices. More advanced topics in matrix algebra are relegated to subsequent chapters and the appendixes.

2.5 BASIC DEFINITIONS AND OPERATIONS ON MATRICES

A matrix **A** *of order m by n, and with general entry* (a_{ij}), *consists of a rectangular array of real numbers (scalars) arranged in m rows and n columns.*

$$\mathbf{A}_{m\times n} = \begin{bmatrix} a_{11} & a_{12} & \cdots & a_{1j} & \cdots & a_{1n} \\ a_{21} & a_{22} & \cdots & a_{2j} & \cdots & a_{2n} \\ \vdots & \vdots & & \vdots & & \vdots \\ a_{i1} & a_{i2} & \cdots & a_{ij} & \cdots & a_{in} \\ \vdots & \vdots & & \vdots & & \vdots \\ a_{m1} & a_{m2} & \cdots & a_{mj} & \cdots & a_{mn} \end{bmatrix} = (a_{ij})_{m\times n}$$

For example, a 4 x 5 matrix would be explicitly written, in brackets, as

$$\mathbf{A}_{4\times 5} = \begin{bmatrix} a_{11} & a_{12} & a_{13} & a_{14} & a_{15} \\ a_{21} & a_{22} & a_{23} & a_{24} & a_{25} \\ a_{31} & a_{32} & a_{33} & a_{34} & a_{35} \\ a_{41} & a_{42} & a_{43} & a_{44} & a_{45} \end{bmatrix}$$

where $i = 1, 2, 3, 4$ and $j = 1, 2, 3, 4, 5$. As is the case for vectors, matrices will appear in boldfaced type, such as **A**, **B**, **C**, etc.

A matrix can exhibit any relation between m, the number of rows, and n, the number of columns. For example, if either $m > n$ or $n > m$, we have a rectangular matrix. (The former is often called a vertical matrix, while the latter is often called a horizontal matrix.)

If $m = n$, the matrix is called square. To illustrate,

$$\mathbf{B}_{3\times 3} = \begin{bmatrix} b_{11} & b_{12} & b_{13} \\ b_{21} & b_{22} & b_{23} \\ b_{31} & b_{32} & b_{33} \end{bmatrix}$$

The set of elements on the diagonal, from upper left to lower right,

$$\{b_{11}, b_{22}, b_{33}\}$$

is called the main or principal diagonal of the square matrix **B**. Square matrices occur quite frequently as derived matrices in multivariate analysis. For example, a correlation matrix, to be described later in the chapter, is a square matrix.

In either the rectangular or square matrix case, the order or "dimensionality" specifies the number of rows and columns of the matrix. Sometimes this order is made explicit in the form of subscripts:

$$\mathbf{C}_{2 \times 3}$$

In other cases, the order is inferred from context, such as

$$\mathbf{D} = \begin{bmatrix} 1 & 2 & 7 & 9 & 3 \\ 0 & 4 & 3 & 1 & 1 \end{bmatrix}$$

While we note that no subscript appears on **D**, it is clear that this matrix is of order 2 x 5.

If $m = 1$, the matrix is equivalent to a row vector. If $n = 1$, the matrix is equivalent to a column vector. If $m = n = 1$, we have a 1 x 1 matrix.[5]

A column vector, written as

$$\begin{bmatrix} a_1 \\ a_2 \\ \vdots \\ a_m \end{bmatrix}$$

can now be viewed as an m by 1 matrix and a row vector $\mathbf{a}' = (a_1, a_2, \ldots, a_n)$ can now be viewed as a 1 by n matrix.

As will be shown later, analogous to earlier discussion of vectors, various kinds of special matrices can be defined. For the moment, however, we define the matrix that is a generalization of the **0** vector. *This matrix, called a null matrix, and denoted* $\boldsymbol{\phi}$, *consists of entries that are all zeros.*

Illustrations of null matrices of various orders are

$$\boldsymbol{\phi} = \begin{bmatrix} 0 & 0 & 0 \\ 0 & 0 & 0 \end{bmatrix}; \qquad \boldsymbol{\phi} = \begin{bmatrix} 0 & 0 \\ 0 & 0 \end{bmatrix}; \qquad \boldsymbol{\phi} = \begin{bmatrix} 0 & 0 \\ 0 & 0 \\ 0 & 0 \\ 0 & 0 \end{bmatrix}$$

Notice that each of the above null matrices is made up entirely of **0** vectors and all entries are 0's.

[5] It is often convenient to consider a scalar as a 1 x 1 matrix.

2.5.1 Matrix Transpose

Consider the 2 by 3 matrix:

$$\mathbf{A} = \begin{bmatrix} -1 & 4 & 3 \\ 0 & 5 & 2 \end{bmatrix}$$

Suppose we write the elements of each row of A as columns and obtain

$$\mathbf{A}' = \begin{bmatrix} -1 & 0 \\ 4 & 5 \\ 3 & 2 \end{bmatrix}$$

Note that this new matrix $\mathbf{A}'$ is a 3 by 2 matrix in which the entries of the first row of $\mathbf{A}'$ denote, in the same order, the first column of $\mathbf{A}$. This is also true of the elements in the second and third rows of $\mathbf{A}'$ compared, respectively, to the second and third columns of $\mathbf{A}$.

The new matrix $\mathbf{A}'$ represents the transpose of the original matrix $\mathbf{A}$. *A transpose of $\mathbf{A}_{m \times n} = (a_{ij})_{m \times n}$ is a matrix obtained from* $\mathbf{A}$ *by interchanging rows and columns so that*

$$\boxed{\mathbf{A}'_{m \times n} = (a_{ij})'_{m \times n} = (a_{ji})_{n \times m}}$$

To illustrate,

$$\text{If} \quad \mathbf{A} = \begin{bmatrix} 1 & 4 \\ 2 & 5 \\ 3 & 6 \end{bmatrix}, \quad \text{then} \quad \mathbf{A}' = \begin{bmatrix} 1 & 2 & 3 \\ 4 & 5 & 6 \end{bmatrix}$$

$$\text{If} \quad \mathbf{B} = \begin{bmatrix} 1 & 4 & 7 & 9 \\ 3 & 1 & 0 & 2 \\ 4 & 2 & 1 & 3 \end{bmatrix}, \quad \text{then} \quad \mathbf{B}' = \begin{bmatrix} 1 & 3 & 4 \\ 4 & 1 & 2 \\ 7 & 0 & 1 \\ 9 & 2 & 3 \end{bmatrix}$$

$$\text{If} \quad \boldsymbol{\phi} = \begin{bmatrix} 0 & 0 \\ 0 & 0 \\ 0 & 0 \end{bmatrix}, \quad \text{then} \quad \boldsymbol{\phi}' = \begin{bmatrix} 0 & 0 & 0 \\ 0 & 0 & 0 \end{bmatrix}$$

Next, we can use the operation of matrix transpose to describe a symmetric matrix. *A square matrix* $\mathbf{A}$ *is called symmetric if*

$$\boxed{\mathbf{A} = (a_{ij}) = \mathbf{A}' = (a_{ji})}$$

That is, a symmetric matrix equals its transpose. To illustrate,

$$\mathbf{A} = \begin{bmatrix} 1 & 2 \\ 2 & 3 \end{bmatrix}; \qquad \mathbf{A}' = \begin{bmatrix} 1 & 2 \\ 2 & 3 \end{bmatrix}$$

Finally, it should also be evident that the transpose of the transpose of a given matrix is the original matrix itself. That is,

$$\boxed{(\mathbf{A}')' = \mathbf{A}}$$

To illustrate,

$$\mathbf{A} = \begin{bmatrix} 1 & 5 \\ 2 & 6 \\ 3 & 7 \end{bmatrix}; \qquad \mathbf{A}' = \begin{bmatrix} 1 & 2 & 3 \\ 5 & 6 & 7 \end{bmatrix}; \qquad (\mathbf{A}')' = \begin{bmatrix} 1 & 5 \\ 2 & 6 \\ 3 & 7 \end{bmatrix}$$

2.5.2 Matrix Equality, Addition, Scalar Multiplication, and Subtraction

Two matrices **A** *and* **B** *are equal if and only if they are of the same order and each entry of the first is equal to the corresponding entry of the second.* That is,

$$\boxed{\begin{gathered} \mathbf{A} = \mathbf{B} \\ \text{if and only if} \\ (a_{ij}) = (b_{ij}) \\ \text{for} \quad i = 1, 2, \ldots, m; \quad j = 1, 2, \ldots, n \end{gathered}}$$

To illustrate,

$$\mathbf{A} = \begin{bmatrix} -1 & 4 & 3 \\ 0 & 5 & 2 \end{bmatrix} = \mathbf{B} = \begin{bmatrix} -1 & 4 & 3 \\ 0 & 5 & 2 \end{bmatrix}$$

since they are of the same order and $a_{ij} = b_{ij}$, entry by entry.

In matrix addition each entry of a sum matrix is the sum of the corresponding entries of the two matrices being added, again assuming they are of the same order.

That is, we define the matrix **C**, denoting the result of adding **A** to **B** as

$$\boxed{\begin{gathered} \mathbf{C} = \mathbf{A} + \mathbf{B} \\ \text{if and only if} \\ (c_{ij}) = (a_{ij}) + (b_{ij}) \\ \text{for} \quad i = 1, 2, \ldots, m; \quad j = 1, 2, \ldots, n \end{gathered}}$$

To illustrate,

$$\mathbf{A} = \begin{bmatrix} -1 & 4 & 3 \\ 0 & 5 & 2 \end{bmatrix}; \quad \mathbf{B} = \begin{bmatrix} 0 & 2 & 1 \\ -1 & 4 & 3 \end{bmatrix}; \quad \mathbf{A} + \mathbf{B} = \mathbf{C} = \begin{bmatrix} -1 & 6 & 4 \\ -1 & 9 & 5 \end{bmatrix}$$

Next, we can consider the case of the transpose of the sum of two matrices.

If **A** *and* **B** *are of common order and if* **C** = **A** + **B**, *then*

$$\boxed{\mathbf{C}' = \mathbf{A}' + \mathbf{B}'}$$

To illustrate,

$$\mathbf{A} = \begin{bmatrix} 1 & 2 & 3 \\ 4 & 5 & 6 \end{bmatrix}; \quad \mathbf{B} = \begin{bmatrix} 0 & 1 & 1 \\ 5 & 1 & 3 \end{bmatrix}$$

$$\mathbf{C} = \begin{bmatrix} 1 & 3 & 4 \\ 9 & 6 & 9 \end{bmatrix}; \quad \mathbf{C}' = \begin{bmatrix} 1 & 9 \\ 3 & 6 \\ 4 & 9 \end{bmatrix} = \overset{\mathbf{A}'}{\begin{bmatrix} 1 & 4 \\ 2 & 5 \\ 3 & 6 \end{bmatrix}} + \overset{\mathbf{B}'}{\begin{bmatrix} 0 & 5 \\ 1 & 1 \\ 1 & 3 \end{bmatrix}}$$

Matrices can also be multiplied by a number (scalar), and this is called scalar multiplication of the matrix. The procedure is simple: One merely multiplies each entry of the matrix by the scalar k. That is,

$$\boxed{\begin{gathered} \mathbf{E} = k\mathbf{A} \\ \text{if and only if} \\ (e_{ij}) = k(a_{ij}) \\ \text{for} \quad i = 1, 2, \ldots, m; \quad j = 1, 2, \ldots, n \end{gathered}}$$

For example, if we wish to multiply **A** by 3, we have

$$3\mathbf{A} = 3\begin{bmatrix} 1 & 2 & 3 \\ 4 & 5 & 6 \end{bmatrix} = \begin{bmatrix} 3 & 6 & 9 \\ 12 & 15 & 18 \end{bmatrix}$$

Subtraction of matrices is now defined as involving the case in which the matrix being subtracted is first multiplied by −1 and then the two matrices are added. That is

$$\boxed{\begin{gathered} C = A - B \\ \text{if and only if} \\ (c_{ij}) = (a_{ij}) - (b_{ij}) \\ \text{for} \quad i = 1, 2, \ldots, m; \quad j = 1, 2, \ldots, n \end{gathered}}$$

To illustrate,

$$\mathbf{A} = \begin{bmatrix} -1 & 4 & 3 \\ 0 & 5 & 2 \end{bmatrix}; \quad \mathbf{B} = \begin{bmatrix} 0 & 2 & 1 \\ -1 & 4 & 3 \end{bmatrix}; \quad \mathbf{A} - \mathbf{B} = \mathbf{C} = \begin{bmatrix} -1 & 2 & 2 \\ 1 & 1 & -1 \end{bmatrix}$$

2.5.3 Properties of Matrix Addition and Scalar Multiplication

Some of the properties exhibited by matrix addition[6] and scalar multiplication are listed below for future use:

1. Matrix addition is commutative:

$$\mathbf{A} + \mathbf{B} = \mathbf{B} + \mathbf{A}$$

2. Matrix addition is associative:

$$\mathbf{A} + (\mathbf{B} + \mathbf{C}) = \mathbf{A} + \mathbf{B} + \mathbf{C} = (\mathbf{A} + \mathbf{B}) + \mathbf{C}$$

3. Scalar multiplication of a matrix is commutative:

$$\mathbf{A}(k\mathbf{B}) = (k\mathbf{A})\mathbf{B}$$

4. Scalar multiplication is associative:

$$k_1(k_2\mathbf{A}) = (k_1k_2)\mathbf{A}$$

5. Scalar multiplication is distributive:

$$(k_1 + k_2)\mathbf{A} = k_1\mathbf{A} + k_2\mathbf{A}$$

6. There exists a null matrix (already defined) ϕ with the property that

$$\mathbf{A} + \phi = \mathbf{A}$$

7. Every matrix $\mathbf{A}$ has a counterpart matrix $-\mathbf{A}$ such that

$$\mathbf{A} + (-\mathbf{A}) = \phi$$

Not surprisingly, the preceding rules are similar to the ones discussed for vector operations in Sections 2.3.3 and 2.3.4 and could be numerically illustrated in similar fashion.

2.5.4 Matrix Multiplication

In discussing the multiplication of two (or more) matrices, *conformability* should first be pointed out. Similar to the previous discussion of matrix equality, addition, and subtraction, in which the matrices were assumed to be of common order before the relation or operation was meaningful, in multiplication the matrices must be conformable. If we wish to multiply $\mathbf{A}$ by $\mathbf{B}$, they must exhibit commonality of *interior* dimensions.

For example, if $\mathbf{A}$ is of order m rows by n columns, it can be written as $\mathbf{A}_{m \times n}$. Next, suppose we have a second matrix $\mathbf{B}$ of order n rows by p columns, written as $\mathbf{B}_{n \times p}$. Using this form, we have

$$\boxed{\mathbf{A}_{m \times n} \quad \mathbf{B}_{n \times p} = \mathbf{C}_{m \times p}}$$

[6] It should be noted, however, that matrix subtraction is neither commutative nor associative.

Note that **A**, called the prefactor, has n columns. This is the "interior" dimension. If **A** has an interior dimension of n columns, then **B**, called the postfactor, must have an interior dimension of n rows. This is the condition of conformability and is necessary for matrix multiplication. Note that their product **C**, then, is of order m by p. These are the "exterior" dimensions of **A** and **B**, respectively.

A simple way to find the order of the matrix product is shown below:

$$\mathbf{A}_{m \times n} \quad \mathbf{B}_{n \times p} = \mathbf{C}_{m \times p}$$

Note that the interior dimensions are the same (n columns of **A** and n rows of **B**) and that the exterior dimensions are obtained from the "outer" dimensions of **A** and **B**, respectively.

This same idea holds for more than two matrices, again assuming that all are conformable. For example, with three matrices, we have

$$\mathbf{A}_{m \times n} \quad \mathbf{B}_{n \times p} \quad \mathbf{C}_{p \times r} = \mathbf{D}_{m \times r}$$

Note further that the interior dimensions of the matrices conform.

Matrix multiplication follows a row-by-column rule, equivalent to the scalar product (Section 2.3.6) of each row of the first matrix with each column of the second. That is, we take the entries of each row of the prefactor, and these are multiplied by the corresponding entries of each column of the postfactor and then summed. If we use the first row of **A** and the first column of **B**, then the first element in **C** (i.e., c_{11}) will be the result of the preceding operation. The second element of **C** (i.e., c_{12}) is found by using the first row of **A** and the second column of **B**, and so on.

This can be summarized as follows:

$$\mathbf{C} = \mathbf{AB}$$

is defined as

$$c_{ij} = \sum_{k=1}^{n} a_{ik} b_{kj}$$

for $i = 1, 2, \ldots, m; \quad j = 1, 2, \ldots, p$

To illustrate, we let

$$\mathbf{A}_{2 \times 3} = \begin{bmatrix} -1 & 3 & 2 \\ 2 & 0 & 1 \end{bmatrix} \quad \text{and} \quad \mathbf{B}_{3 \times 2} = \begin{bmatrix} 2 & 3 \\ 1 & 4 \\ 1 & 2 \end{bmatrix}$$

Then

$$\mathbf{C} = \mathbf{AB}$$

$$= \begin{bmatrix} (-1 \times 2 + 3 \times 1 + 2 \times 1) & (-1 \times 3 + 3 \times 4 + 2 \times 2) \\ (\ 2 \times 2 + 0 \times 1 + 1 \times 1) & (\ 2 \times 3 + 0 \times 4 + 1 \times 2) \end{bmatrix}$$

$$\mathbf{C}_{2 \times 2} = \begin{bmatrix} 3 & 13 \\ 5 & 8 \end{bmatrix}$$

Now, let us reverse the order of multiplication. In this case we can do so since **B** is of order 3 by 2 and **A** is of order 2 by 3. In other instances, however (e.g., if **A** were of order 3 by 3), we could not multiply **B** by **A** since they would then not be conformable.

If we now multiply **BA** = **D**, we obtain the following:

$$\mathbf{D} = \mathbf{B}_{3 \times 2}\ \mathbf{A}_{2 \times 3} = \begin{bmatrix} 2 & 3 \\ 1 & 4 \\ 1 & 2 \end{bmatrix} \begin{bmatrix} -1 & 3 & 2 \\ 2 & 0 & 1 \end{bmatrix}$$

$$\mathbf{D}_{3 \times 3} = \begin{bmatrix} 4 & 6 & 7 \\ 7 & 3 & 6 \\ 3 & 3 & 4 \end{bmatrix}$$

Notice that $\mathbf{D} \neq \mathbf{C}$ and, as a matter of fact, they are not even of the same order. Even if two matrices are conformable, in general, $\mathbf{AB} \neq \mathbf{BA}$. That is, matrix multiplication, in general, is noncommutative. Hence, in discussing matrix multiplication we should *refer explicitly to the order in which they multiply.* For example, the matrix product **AB** can be described as "**A** is postmultiplied by **B**," or "**B** is premultiplied by **A**." Alternatively, we could use the terms "prefactor" and "postfactor" as mentioned earlier.

2.5.4.1 Multiplication of a Vector and a Matrix In some cases of interest we shall want to postmultiply some vector **a** by a matrix **B**. To illustrate, suppose we have the following:

$$\mathbf{a}' = (1, 0, 3); \qquad \mathbf{B} = \begin{bmatrix} 2 & 3 \\ 1 & 2 \\ 0 & 1 \end{bmatrix}$$

where $\mathbf{a}'$ is a row vector of order 1 by 3. Note that $\mathbf{a}'$ and **B** are conformable, and we have

$$(1, 0, 3) \begin{bmatrix} 2 & 3 \\ 1 & 2 \\ 0 & 1 \end{bmatrix} = (2, 6)$$

where their product displays the order of the exterior dimensions of $\mathbf{a}'$ and $\mathbf{B}$, namely, a 1×2 row vector. Alternatively, suppose we have the column vector $\mathbf{c} = \begin{bmatrix} 1 \\ 2 \end{bmatrix}$. Then we can find the product $\mathbf{B}$ as

$$\begin{bmatrix} 2 & 3 \\ 1 & 2 \\ 0 & 1 \end{bmatrix} \begin{bmatrix} 1 \\ 2 \end{bmatrix} = \begin{bmatrix} 8 \\ 5 \\ 2 \end{bmatrix}$$

where their product is a 3×1 column vector. As can be seen, no new rules are involved for either of these cases.

2.5.4.2 Matrix Product of Two Vectors We now might ask what happens when two vectors are multiplied. As shown in Section 2.3.6, the scalar product of two vectors results in a single number (scalar) if row vector times column vector multiplication is performed. However, one might have the case of a column vector multiplying a row vector. In this case the results are quite different, and we consider it next.

To illustrate, assume we have the column vector

$$\begin{bmatrix} 2 \\ 1 \\ 3 \end{bmatrix}$$

and the row vector (1, 1, 2). Their matrix (or outer) product is obtained as

$$\begin{bmatrix} 2 \\ 1 \\ 3 \end{bmatrix} (1, 1, 2) = \begin{bmatrix} (2 \times 1) & (2 \times 1) & (2 \times 2) \\ (1 \times 1) & (1 \times 1) & (1 \times 2) \\ (3 \times 1) & (3 \times 1) & (3 \times 2) \end{bmatrix} = \begin{bmatrix} 2 & 2 & 4 \\ 1 & 1 & 2 \\ 3 & 3 & 6 \end{bmatrix}$$

which is a 3 by 3 *matrix*. Note in this case that each row of the first "matrix" has only a *single* element, as does each column of the second. Thus, the row-by-column rule is not violated in this special case.

2.5.4.3 Triple Product—Vector, Matrix, Vector Multiplication To round out the discussion we might wish to consider the triple product of

$$\mathbf{a}' = (1, 1, 2); \qquad \mathbf{B} = \begin{bmatrix} 1 & 2 & 3 \\ 0 & 1 & 2 \\ 3 & 0 & 1 \end{bmatrix}; \qquad \mathbf{c} = \begin{bmatrix} 2 \\ 1 \\ 3 \end{bmatrix}$$

That is, we desire to find the product $\mathbf{a}'\mathbf{B}\mathbf{c}$. If so, we can proceed in stages. We first find the vector by matrix product:

$$\mathbf{a}'\mathbf{B} = (1, 1, 2) \begin{bmatrix} 1 & 2 & 3 \\ 0 & 1 & 2 \\ 3 & 0 & 1 \end{bmatrix} = (7, 3, 7)$$

Next, we find the scalar product:

$$\mathbf{a}'\mathbf{Bc} = (7, 3, 7)\begin{bmatrix} 2 \\ 1 \\ 3 \end{bmatrix} = 38$$

Hence, we see that the result of all this is a scalar. Note, of course, that it is of "order 1 x 1" and, thus, is in agreement with the order of the exterior dimensions of $\mathbf{a}'$ and $\mathbf{c}$.

A special case of vector, matrix, vector multiplication takes the form of $\mathbf{a}'\mathbf{Ba}$. This can be illustrated by

$$\mathbf{a}' = (1, 1, 2); \qquad \mathbf{B} = \begin{bmatrix} 1 & 2 & 3 \\ 0 & 1 & 2 \\ 3 & 0 & 1 \end{bmatrix}; \qquad \mathbf{a} = \begin{bmatrix} 1 \\ 1 \\ 2 \end{bmatrix}$$

Then

$$\mathbf{a}'\mathbf{B} = (1, 1, 2)\begin{bmatrix} 1 & 2 & 3 \\ 0 & 1 & 2 \\ 3 & 0 & 1 \end{bmatrix} = (7, 3, 7)$$

and

$$\mathbf{a}'\mathbf{Ba} = (7, 3, 7)\begin{bmatrix} 1 \\ 1 \\ 2 \end{bmatrix} = 24$$

As noted, the result is also a scalar. Both of these cases are relevant to multivariate analysis and are discussed in later chapters.

2.5.5 Some Properties of Matrix Multiplication

We have already pointed out that matrix multiplication, in general, is noncommutative. However, matrix multiplication *does* obey certain other properties.

1. Associativity—assuming all matrices to be conformable we can state that

$$(\mathbf{AB})\mathbf{C} = \mathbf{A}(\mathbf{BC})$$

2. Distributivity—again assuming conformable matrices we can state that

$$\mathbf{A}(\mathbf{B} + \mathbf{C}) = \mathbf{AB} + \mathbf{AC}; \qquad \text{left distributive law}$$

$$(\mathbf{B} + \mathbf{C})\mathbf{A} = \mathbf{BA} + \mathbf{CA}; \qquad \text{right distributive law}$$

3. If k is a scalar, then we have the associativity property:

$$k(\mathbf{AB}) = (k\mathbf{A})\mathbf{B}$$

Finally, it is of interest to point out the rule involving the transpose of the product of two (or more) matrices. In the case of two matrices, the rule is

$$\boxed{(\mathbf{AB})' = \mathbf{B}'\mathbf{A}'}$$

That is, the transpose of the product of two (or more) matrices is equal to the product of their respective transposes, multiplied in reverse order. To illustrate,

$$\mathbf{A} = \begin{bmatrix} 1 & 3 \\ 2 & 4 \end{bmatrix}; \quad \mathbf{B} = \begin{bmatrix} 1 & 3 \\ 2 & 2 \end{bmatrix}; \quad \mathbf{A}' = \begin{bmatrix} 1 & 2 \\ 3 & 4 \end{bmatrix}; \quad \mathbf{B}' = \begin{bmatrix} 1 & 2 \\ 3 & 2 \end{bmatrix}$$

Then

$$\mathbf{AB} = \begin{bmatrix} 1 & 3 \\ 2 & 4 \end{bmatrix} \begin{bmatrix} 1 & 3 \\ 2 & 2 \end{bmatrix} = \begin{bmatrix} 7 & 9 \\ 10 & 14 \end{bmatrix}$$

$$(\mathbf{AB})' = \begin{bmatrix} 7 & 10 \\ 9 & 14 \end{bmatrix} = \mathbf{B}'\mathbf{A}' = \begin{bmatrix} 1 & 2 \\ 3 & 2 \end{bmatrix} \begin{bmatrix} 1 & 2 \\ 3 & 4 \end{bmatrix} = \begin{bmatrix} 7 & 10 \\ 9 & 14 \end{bmatrix}$$

2.5.6 Some Differences between Scalar and Matrix Arithmetic

Probably the most difficult temptation to suppress in working with matrix multiplication involves attributing properties to matrices that we associate with ordinary scalars. Table 2.1 shows some of the pitfalls that one should be wary of in doing

TABLE 2.1

Some Differences between Scalar and Matrix Relations

Scalars	Matrices
1. $ab = ba$	1. $\mathbf{AB} \neq \mathbf{BA}$, in general
2. If $ab = ac$ and $a \neq 0$, then $b = c$.	2. If $\mathbf{AB} = \mathbf{AC}$ and $\mathbf{A} \neq \phi$, then it is not necessary that $\mathbf{B}$ equals $\mathbf{C}$.
3. If $ab = 0$, then either $a = 0$, or $b = 0$, or both $a, b = 0$.	3. If $\mathbf{AB} = \phi$, then it is not necessarily the case that either $\mathbf{A} = \phi$, $\mathbf{B} = \phi$, or both $\mathbf{A}, \mathbf{B} = \phi$.
4. If $ab = 0$, then $ba = 0$.	4. If $\mathbf{AB} = \phi$, then $\mathbf{BA}$ does not necessarily equal ϕ.

arithmetic with matrices. As shown in the table, some marked differences exist. Not only does commutativity fail to hold in general for matrix multiplication, but other characteristics involving products equal to zero also do not hold generally. For example, if some matrix product **AB** equals the null matrix ϕ, we note that neither **A**, the prefactor, nor **B**, the postfactor, need to be equal to ϕ.

2.5.7 The Problem of Matrix Division

Up to this point we have discussed addition, subtraction, and multiplication of matrices, but division has been conspicuous by its absence. *And for good reason: division, as we know it in scalar arithmetic, is not defined in matrix algebra.*

What is defined is something more analogous to multiplication by a reciprocal. For example, in ordinary arithmetic, instead of dividing some number by 5, we could multiply the number by the reciprocal of 5:

$$1/5 = (5)^{-1}$$

assuming that the divisor is not equal to zero.

The analogous operation in matrix algebra is called *matrix inversion.* This operation is so special (and considerably more complex) that we defer discussion of it until Chapter 4. What can be said for now is that the inverse of a matrix **A**, if said inverse exists, is analogous to multiplication of **A** by a reciprocal in ordinary algebra. As such, in matrix algebra there is an analogy to the scalar relation:

$$a \times a^{-1} = 1$$

Needless to say, we shall spend a considerable amount of time on the topic of matrix inversion in subsequent chapters.

2.5.8 Some More Examples of Matrix Operations

To facilitate the review of matrix operations described in Section 2.5, consider the following:

$$\mathbf{A} = \begin{bmatrix} 1 & 2 & 3 \\ 2 & 1 & 4 \end{bmatrix}; \qquad \mathbf{B} = \begin{bmatrix} 2 & 4 \\ 2 & 5 \\ 3 & 6 \end{bmatrix}$$

$$k = 3; \qquad \mathrm{a}' = (2, 1); \qquad \mathrm{b} = \begin{bmatrix} 5 \\ 2 \\ 3 \end{bmatrix}$$

Matrix Transpose The transpose of **A** is

$$\mathbf{A}' = \begin{bmatrix} 1 & 2 \\ 2 & 1 \\ 3 & 4 \end{bmatrix}$$

and the transpose of **B** is

$$\mathbf{B}' = \begin{bmatrix} 2 & 2 & 3 \\ 4 & 5 & 6 \end{bmatrix}$$

Addition of Matrices Addition of matrices is illustrated by

$$\mathbf{C} = \mathbf{A} + \mathbf{B}' = \begin{bmatrix} 1 & 2 & 3 \\ 2 & 1 & 4 \end{bmatrix} + \begin{bmatrix} 2 & 2 & 3 \\ 4 & 5 & 6 \end{bmatrix} = \begin{bmatrix} 3 & 4 & 6 \\ 6 & 6 & 10 \end{bmatrix}$$

Subtraction of Matrices Matrix subtraction is illustrated by

$$\mathbf{C} = \mathbf{A} - \mathbf{B}' = \begin{bmatrix} 1 & 2 & 3 \\ 2 & 1 & 4 \end{bmatrix} - \begin{bmatrix} 2 & 2 & 3 \\ 4 & 5 & 6 \end{bmatrix} = \begin{bmatrix} -1 & 0 & 0 \\ -2 & -4 & -2 \end{bmatrix}$$

Scalar Multiplication of a Matrix

$$\mathbf{C} = k\mathbf{A} = 3\begin{bmatrix} 1 & 2 & 3 \\ 2 & 1 & 4 \end{bmatrix} = \begin{bmatrix} 3 & 6 & 9 \\ 6 & 3 & 12 \end{bmatrix}$$

Varieties of Multiplication

$$\mathbf{C} = \mathbf{AB} = \begin{bmatrix} 1 & 2 & 3 \\ 2 & 1 & 4 \end{bmatrix} \begin{bmatrix} 2 & 4 \\ 2 & 5 \\ 3 & 6 \end{bmatrix} = \begin{bmatrix} 15 & 32 \\ 18 & 37 \end{bmatrix}$$

$$\mathbf{C}' = (\mathbf{AB})' = \mathbf{B}'\mathbf{A}' = \begin{bmatrix} 2 & 2 & 3 \\ 4 & 5 & 6 \end{bmatrix} \begin{bmatrix} 1 & 2 \\ 2 & 1 \\ 3 & 4 \end{bmatrix} = \begin{bmatrix} 15 & 18 \\ 32 & 37 \end{bmatrix}$$

$$\mathbf{D} = k\mathbf{AB} = 3\begin{bmatrix} 15 & 32 \\ 18 & 37 \end{bmatrix} = \begin{bmatrix} 45 & 96 \\ 54 & 111 \end{bmatrix}$$

$$\mathbf{E} = \mathbf{a}'\mathbf{Ab} = (2, 1)\begin{bmatrix} 1 & 2 & 3 \\ 2 & 1 & 4 \end{bmatrix} \begin{bmatrix} 5 \\ 2 \\ 3 \end{bmatrix} = 60$$

$$\mathbf{F} = \mathbf{b}'\mathbf{B}(\mathbf{a}')' = (5, 2, 3)\begin{bmatrix} 2 & 4 \\ 2 & 5 \\ 3 & 6 \end{bmatrix} \begin{bmatrix} 2 \\ 1 \end{bmatrix} = 94$$

Additional examples appear at the end of the chapter.

2.6 SOME SPECIAL MATRICES

So far we have been dealing mainly with arbitrary rectangular matrices. In a few cases we have used square matrices for illustrative purposes. In matrix algebra there are a number of special matrices that are encountered in multivariate analysis. We consider some of these here, particularly those that are frequently utilized in multivariate procedures.

2.6.1 Symmetric Matrices

Figure 2.1 shows, in schematic form, various special matrices of interest to multivariate analysis. The first property for categorizing types of matrices concerns whether they are square ($m = n$) or rectangular. In turn, rectangular matrices can be either vertical ($m > n$) or horizontal ($m < n$).

As we shall show in later chapters, square matrices play an important role in multivariate analysis. In particular, the notion of matrix symmetry is important. Earlier, a symmetric matrix was defined as a square matrix that satisfies the relation

$$\boxed{\mathbf{A} = \mathbf{A}' \qquad \text{or, equivalently,} \qquad (a_{ij}) = (a_{ji})}$$

That is, a symmetric matrix is a square matrix that is equal to its transpose. For example,

$$\mathbf{A} = \begin{bmatrix} 3 & 2 & 4 \\ 2 & 0 & -5 \\ 4 & -5 & 1 \end{bmatrix}; \qquad \mathbf{A}' = \begin{bmatrix} 3 & 2 & 4 \\ 2 & 0 & -5 \\ 4 & -5 & 1 \end{bmatrix}$$

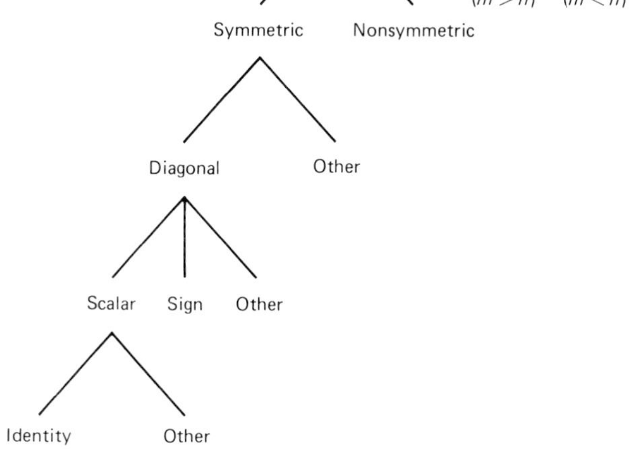

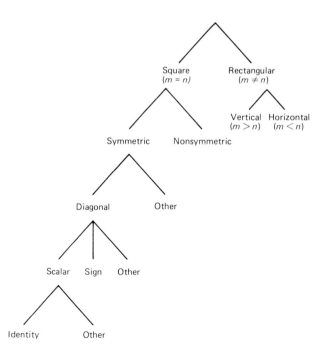

Fig. 2.1 Various types of matrices.

Symmetric matrices, such as correlation matrices and covariance matrices, are quite common in multivariate analysis, and we shall come across them repeatedly in later chapters.[7]

A few properties related to symmetry in matrices are of interest to point out:

1. The product of any (not necessarily symmetric) matrix and its transpose is symmetric; that is, both $\mathbf{AA}'$ and $\mathbf{A}'\mathbf{A}$ are symmetric matrices.
2. If $\mathbf{A}$ is any square (not necessarily symmetric) matrix, then $\mathbf{A} + \mathbf{A}'$ is symmetric.
3. If $\mathbf{A}$ is symmetric and k is a scalar, then $k\mathbf{A}$ is a symmetric matrix.
4. The sum of any number of symmetric matrices is also symmetric.
5. The product of two symmetric matrices is not necessarily symmetric.

Later chapters will discuss still other characteristics of symmetric matrices and the special role that they play in such topics as matrix eigenstructures and quadratic forms.

2.6.2 Diagonal, Scalar, Sign, and Identity Matrices

A special case of a symmetric matrix is a diagonal matrix. *A diagonal matrix is defined as a square matrix in which all off-diagonal entries are zero.* (Note that a diagonal matrix is necessarily symmetric.) Entries on the main diagonal may or may not be zero.

Examples of diagonal matrices are

$$\mathbf{A} = \begin{bmatrix} 2 & 0 & 0 \\ 0 & 3 & 0 \\ 0 & 0 & 1 \end{bmatrix}; \qquad \mathbf{B} = \begin{bmatrix} -1 & 0 & 0 \\ 0 & 0 & 0 \\ 0 & 0 & 3 \end{bmatrix}; \qquad \mathbf{C} = \begin{bmatrix} 1 & 0 \\ 0 & 5 \end{bmatrix}$$

If all entries on the main diagonal are equal scalars, then the diagonal matrix is called a *scalar matrix.*

Examples of scalar matrices are

$$\mathbf{A} = \begin{bmatrix} 3 & 0 & 0 \\ 0 & 3 & 0 \\ 0 & 0 & 3 \end{bmatrix}; \qquad \mathbf{B} = \begin{bmatrix} 2 & 0 \\ 0 & 2 \end{bmatrix}; \qquad \mathbf{C} = \begin{bmatrix} -4 & 0 \\ 0 & -4 \end{bmatrix}$$

If some of the entries on the main diagonal are -1 and the rest are $+1$, the diagonal matrix is called a *sign matrix.* Examples of sign matrices are

$$\mathbf{A} = \begin{bmatrix} -1 & 0 & 0 \\ 0 & 1 & 0 \\ 0 & 0 & 1 \end{bmatrix}; \qquad \mathbf{B} = \begin{bmatrix} 1 & 0 \\ 0 & -1 \end{bmatrix}; \qquad \mathbf{C} = \begin{bmatrix} -1 & 0 \\ 0 & 1 \end{bmatrix}$$

[7] While we do not go into detail here, a *skew symmetric* matrix $\mathbf{A}$ is a square matrix in which all main diagonal elements a_{ii} are zero and $A = -A'$. For example,

$$\text{If } \mathbf{A} = \begin{bmatrix} 0 & -3 & 1 \\ 3 & 0 & 2 \\ 1 & -2 & 0 \end{bmatrix}; \qquad \text{then} \quad -\mathbf{A}' = \begin{bmatrix} 0 & -3 & -1 \\ 3 & 0 & 2 \\ 1 & -2 & 0 \end{bmatrix}$$

is skew symmetric.

If the entries on the diagonal of a scalar matrix are each equal to unity, then this type of scalar matrix is called an *identity matrix,* denoted **I**. Examples are

$$\mathbf{I} = \begin{bmatrix} 1 & 0 & 0 \\ 0 & 1 & 0 \\ 0 & 0 & 1 \end{bmatrix}; \qquad \mathbf{I} = \begin{bmatrix} 1 & 0 \\ 0 & 1 \end{bmatrix}$$

A number of useful properties are associated with diagonal matrices and, hence, the special cases of scalar, sign, and identity matrices.

1. The transpose of a diagonal matrix is equal to the original matrix.
2. Sums and differences of diagonal matrices are also diagonal matrices.
3. Premultiplication of a matrix **A** by a diagonal matrix **D** results in a matrix in which each entry in a given row is the product of the original entry in **A** corresponding to that row and the diagonal element in the corresponding row of the diagonal matrix. To illustrate,

$$\underset{\mathbf{D}}{\begin{bmatrix} 3 & 0 & 0 \\ 0 & 2 & 0 \\ 0 & 0 & 1 \end{bmatrix}} \underset{\mathbf{A}}{\begin{bmatrix} 1 & 2 & 3 & 4 \\ 2 & 1 & 4 & 3 \\ 3 & 2 & 1 & 1 \end{bmatrix}} = \underset{\mathbf{DA}}{\begin{bmatrix} 3 & 6 & 9 & 12 \\ 4 & 2 & 8 & 6 \\ 3 & 2 & 1 & 1 \end{bmatrix}}$$

4. Postmultiplication of a matrix **A** by a diagonal matrix **D** results in a matrix in which each entry in a given column is the product of the original entry in **A** corresponding to that column and the diagonal element in the corresponding column of the diagonal matrix. To illustrate,

$$\underset{\mathbf{A}}{\begin{bmatrix} 1 & 2 & 3 \\ 2 & 1 & 2 \\ 3 & 4 & 1 \\ 4 & 3 & 1 \end{bmatrix}} \underset{\mathbf{D}}{\begin{bmatrix} 3 & 0 & 0 \\ 0 & 2 & 0 \\ 0 & 0 & 1 \end{bmatrix}} = \underset{\mathbf{AD}}{\begin{bmatrix} 3 & 4 & 3 \\ 6 & 2 & 2 \\ 9 & 8 & 1 \\ 12 & 6 & 1 \end{bmatrix}}$$

5. Pre- and postmultiplication of a matrix **A** by diagonal matrices $\mathbf{D}_1$ and $\mathbf{D}_2$ result in a matrix whose *ij*th entry is the product of the *i*th entry in the diagonal of the premultiplier, the *ij*th entry of **A**, and the *j*th entry of the postmultiplier. For example,

$$\underset{\mathbf{D}_1}{\begin{bmatrix} 2 & 0 & 0 \\ 0 & 1 & 0 \\ 0 & 0 & 3 \end{bmatrix}} \underset{\mathbf{A}}{\begin{bmatrix} 1 & 3 \\ 2 & 2 \\ 4 & 1 \end{bmatrix}} \underset{\mathbf{D}_2}{\begin{bmatrix} 1 & 0 \\ 0 & 3 \end{bmatrix}} = \underset{\mathbf{D}_1\mathbf{A}\mathbf{D}_2}{\begin{bmatrix} 2 & 18 \\ 2 & 6 \\ 12 & 9 \end{bmatrix}}$$

6. The product of any number of diagonal matrices is a diagonal matrix, each of whose entries is the product of the corresponding diagonal entries of the matrices. For example,

$$\underset{\mathbf{D}_1}{\begin{bmatrix} 1 & 0 \\ 0 & 3 \end{bmatrix}} \underset{\mathbf{D}_2}{\begin{bmatrix} 2 & 0 \\ 0 & 2 \end{bmatrix}} \underset{\mathbf{D}_3}{\begin{bmatrix} 4 & 0 \\ 0 & -1 \end{bmatrix}} = \underset{\mathbf{D}_1\mathbf{D}_2\mathbf{D}_3}{\begin{bmatrix} 8 & 0 \\ 0 & -6 \end{bmatrix}}$$

7. Diagonal matrix multiplication, assuming conformability, is commutative.

8. Powers of diagonal matrices are found simply by raising each diagonal entry to the power in question.[8] (Roots are found analogously.)

9. Pre- or postmultiplication of a matrix **A** by a scalar matrix multiplies all entries of **A** by the constant entry in the scalar matrix. It is equivalent to scalar multiplication of the matrix, by that scalar appearing on the diagonal.

10. As a special case, pre- or postmultiplication of a matrix **A** by **I**, the identity matrix, leaves the original matrix unchanged.

11. Powers of an identity matrix equal the original matrix.

While the above properties are by no means exhaustive of the characteristics of diagonal matrices, or the special cases of scalar, sign, and identity matrices, they do represent the main properties of interest to applied researchers.

2.6.3 Some Additional Examples

As an aid to integrating some of the discussion of this section, consider the following:

$$\mathbf{A} = \begin{bmatrix} 1 & 4 & 1 \\ 2 & 5 & 1 \\ 3 & 6 & 2 \end{bmatrix}; \quad \mathbf{D}_1 = \begin{bmatrix} 3 & 0 & 0 \\ 0 & 2 & 0 \\ 0 & 0 & 1 \end{bmatrix}; \quad \mathbf{D}_2 = \begin{bmatrix} 2 & 0 & 0 \\ 0 & 2 & 0 \\ 0 & 0 & 2 \end{bmatrix}$$

Premultiplication by a Diagonal

$$\mathbf{D}_1\mathbf{A} = \begin{bmatrix} 3 & 0 & 0 \\ 0 & 2 & 0 \\ 0 & 0 & 1 \end{bmatrix} \begin{bmatrix} 1 & 4 & 1 \\ 2 & 5 & 1 \\ 3 & 6 & 2 \end{bmatrix} = \begin{bmatrix} 3 & 12 & 3 \\ 4 & 10 & 2 \\ 3 & 6 & 2 \end{bmatrix}$$

Postmultiplication by a Diagonal

$$\mathbf{A}\mathbf{D}_1 = \begin{bmatrix} 1 & 4 & 1 \\ 2 & 5 & 1 \\ 3 & 6 & 2 \end{bmatrix} \begin{bmatrix} 3 & 0 & 0 \\ 0 & 2 & 0 \\ 0 & 0 & 1 \end{bmatrix} = \begin{bmatrix} 3 & 8 & 1 \\ 6 & 10 & 1 \\ 9 & 12 & 2 \end{bmatrix}$$

[8] In general, a square matrix **A** can be raised to any power n that is a positive whole number by multiplying it by itself n times, denoted as $\mathbf{A}^n$. Roots can also be found for certain square matrices (not restricted to being diagonal). If a square matrix **A** has an nth root, then the matrix $\mathbf{A}^{1/n}$, when multiplied by itself n times, equals **A**. The topic of powers and roots of (square) matrices is covered in Chapter 5.

Pre- and Postmultiplication by Diagonals

$$\mathbf{D}_2\mathbf{A}\mathbf{D}_1 = \begin{bmatrix} 2 & 0 & 0 \\ 0 & 2 & 0 \\ 0 & 0 & 2 \end{bmatrix} \begin{bmatrix} 1 & 4 & 1 \\ 2 & 5 & 1 \\ 3 & 6 & 2 \end{bmatrix} \begin{bmatrix} 3 & 0 & 0 \\ 0 & 2 & 0 \\ 0 & 0 & 1 \end{bmatrix} = \begin{bmatrix} 6 & 16 & 2 \\ 12 & 20 & 2 \\ 18 & 24 & 4 \end{bmatrix}$$

Scalar Matrix Multiplication

$$3\begin{bmatrix} 1 & 4 & 1 \\ 2 & 5 & 1 \\ 3 & 6 & 2 \end{bmatrix} = \begin{bmatrix} 3 & 0 & 0 \\ 0 & 3 & 0 \\ 0 & 0 & 3 \end{bmatrix} \begin{bmatrix} 1 & 4 & 1 \\ 2 & 5 & 1 \\ 3 & 6 & 2 \end{bmatrix} = \begin{bmatrix} 3 & 12 & 3 \\ 6 & 15 & 3 \\ 9 & 18 & 6 \end{bmatrix}$$

Powers and Roots of a Diagonal with Positive Entries

$$\mathbf{D}_1^{\,2} = \begin{bmatrix} 3 & 0 & 0 \\ 0 & 2 & 0 \\ 0 & 0 & 1 \end{bmatrix} \begin{bmatrix} 3 & 0 & 0 \\ 0 & 2 & 0 \\ 0 & 0 & 1 \end{bmatrix} = \begin{bmatrix} 9 & 0 & 0 \\ 0 & 4 & 0 \\ 0 & 0 & 1 \end{bmatrix}$$

$$\mathbf{D}_2^{\,1/2} = \begin{bmatrix} 2 & 0 & 0 \\ 0 & 2 & 0 \\ 0 & 0 & 2 \end{bmatrix}^{1/2} = \begin{bmatrix} 1.414 & 0 & 0 \\ 0 & 1.414 & 0 \\ 0 & 0 & 1.414 \end{bmatrix}$$

Identity and Sign Matrices

$$\mathbf{IA} = \mathbf{AI} = \begin{bmatrix} 1 & 4 & 1 \\ 2 & 5 & 1 \\ 3 & 6 & 2 \end{bmatrix} \begin{bmatrix} 1 & 0 & 0 \\ 0 & 1 & 0 \\ 0 & 0 & 1 \end{bmatrix} = \begin{bmatrix} 1 & 4 & 1 \\ 2 & 5 & 1 \\ 3 & 6 & 2 \end{bmatrix} = \mathbf{A}$$

Let

$$\mathbf{F} = \begin{bmatrix} -1 & 0 & 0 \\ 0 & -1 & 0 \\ 0 & 0 & 1 \end{bmatrix}$$

Then

$$\mathbf{AF} = \begin{bmatrix} 1 & 4 & 1 \\ 2 & 5 & 1 \\ 3 & 6 & 2 \end{bmatrix} \begin{bmatrix} -1 & 0 & 0 \\ 0 & -1 & 0 \\ 0 & 0 & 1 \end{bmatrix} = \begin{bmatrix} -1 & -4 & 1 \\ -2 & -5 & 1 \\ -3 & -6 & 2 \end{bmatrix}$$

Additional examples appear at the end of the chapter.

2.7 DETERMINANTS OF MATRICES

The determinant of a matrix plays an important role in more advanced matrix concepts such as matrix inversion and matrix rank, as well as in multivariate analysis involving generalized measures of variance. Only square matrices have determinants. The determinant of a square matrix is a scalar function of the entries of the matrix. We denote the determinant of a matrix **A** by the symbol $|\mathbf{A}|$ and reiterate that the value of the determinant is expressed as a single number (scalar).

The early development of determinants was intimately connected with procedures for solving simultaneous equations. As historical background, and motivational interest, consider the two linear equations:

$$ax + by = c$$
$$dx + ey = f$$

These equations could be expressed in matrix-times vector form as

$$\begin{bmatrix} a & b \\ d & e \end{bmatrix} \begin{bmatrix} x \\ y \end{bmatrix} = \begin{bmatrix} c \\ f \end{bmatrix}$$

Note that the left-hand side of the equations is simply the product of a 2 x 2 matrix times a 2 x 1 vector while the right-hand side is another 2 x 1 vector.

As may be recalled from elementary algebra, this system of equations can be solved, for, say, x by the formula

$$x = \frac{ce - fb}{ae - db}$$

assuming that the denominator of the above ratio is not equal to zero.

We can consider the right-hand side of the above equation in the context of determinants by expressing both numerator and denominator of the ratio as

$$x = \frac{\begin{vmatrix} c & b \\ f & e \end{vmatrix}}{\begin{vmatrix} a & b \\ d & e \end{vmatrix}}$$

In the simple case shown here, the determinants of $[\begin{smallmatrix} c & b \\ f & e \end{smallmatrix}]$ and $[\begin{smallmatrix} a & b \\ d & e \end{smallmatrix}]$ are easy to define. That is

$$\begin{vmatrix} c & b \\ f & e \end{vmatrix} = ce - fb \qquad \text{and} \qquad \begin{vmatrix} a & b \\ d & e \end{vmatrix} = ae - db$$

these are called *second-order* determinants. The unknown quantity x is the ratio of two determinants (scalars).

Historically, determinants were employed widely in the solution of simultaneous equations. With the advent of newer solution methods, however, their application in this context has diminished. Still, it is important to have some grasp of the rudiments of determinants, if only as a precursor to other procedures for solving equations that are developed in subsequent chapters.

2.7.1 Operational Definition of a Determinant

The theoretical definition of a determinant for matrices of larger order than 2 x 2 is rather cumbersome and, therefore, as in the 2 x 2 case, we shall define it operationally as a series of computational steps. Assuming a square matrix **A** of order $m \times m$ with general entry (a_{ij}), the determinant of that matrix is found by carrying out the following sequence:

1. Form all possible products of m factors each, such that each factor is an entry of **A** and no two factors are drawn from the same row or column of **A**. There are $m!$ (m factorial) products of this type. For example, if **A** is of order 3 x 3, we have 3! or six products:

(i) $a_{11} \quad a_{22} \quad a_{33}$ (ii) $a_{12} \quad a_{23} \quad a_{31}$ (iii) $a_{13} \quad a_{21} \quad a_{32}$

(iv) $a_{13} \quad a_{22} \quad a_{31}$ (v) $a_{11} \quad a_{23} \quad a_{32}$ (vi) $a_{12} \quad a_{21} \quad a_{33}$

We note that each of the subscripts (1, 2, or 3) appears just once as a row subscript and just once as a column subscript in each of the six triple products. The connections shown below illustrate these six products.

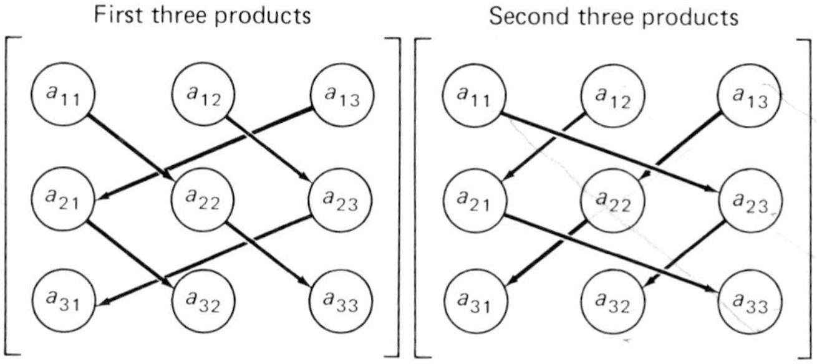

2. Within each separate triple product arrange the factors so that row subscripts are in their natural order; this has been done above. Then count the number of inversions or transpositions involving *column* subscripts. In this case an inversion takes place when a larger column subscript precedes a smaller one. For the six triple products above, we have the following frequencies of inversions:

(i) 0 inversion (ii) 2 inversions (iii) 2 inversions

(iv) 3 inversions (v) 1 inversion (vi) 1 inversion

For example in case (ii), involving the product $a_{12}a_{23}a_{31}$, we note that column subscripts 1 and 2 need to be interchanged, followed by the interchange of column subscripts 3 and 2, in order to obtain the natural order.

3. Having done this for all $m!$ products, multiply each product that has an odd number of inversions by -1. If zero or an even number of inversions is involved, multiply by $+1$; that is, leave the product as is. In the above case the first three products (associated with an incidence of even-type inversions of 0, 2, and 2) will *not* be changed in sign, while the last three products will.

(i) $1(a_{11}a_{22}a_{33})$ (ii) $1(a_{12}a_{23}a_{31})$ (iii) $1(a_{13}a_{21}a_{32})$

(iv) $-1(a_{13}a_{22}a_{31})$ (v) $-1(a_{11}a_{23}a_{32})$ (vi) $-1(a_{12}a_{21}a_{33})$

4. Add the products (observing sign) together. This *sum* is the determinant.

$$|\mathbf{A}| = +(a_{11}a_{22}a_{33}) + (a_{12}a_{23}a_{31}) + (a_{13}a_{21}a_{32})$$
$$-(a_{13}a_{22}a_{31}) - (a_{11}a_{23}a_{32}) - (a_{12}a_{21}a_{33})$$

5. Notice, then, that three steps are involved in finding a determinant. The first step is to form all possible products that can be obtained by taking one element out of one row and column, another out of another row and column, and so on. A matrix of order $m \times m$ will yield $m!$ such products, each composed of m elements. The second step is to affix an algebraic sign to each product via the rule proposed above. The third step is to sum the $m!$ signed products.

6. The procedure can now be formalized by defining the determinant of $\mathbf{A}_{m \times m}$ as the sum of all $m!$ products (each with m factors) in $\mathbf{A}$ of the form

$$(-1)^t a_{1j_1} a_{2j_2} \ldots a_{mj_m}$$

where the sum is understood to be taken over all permutations of the second subscripts. The exponent t denotes the number of inversions required to bring the second subscripts into their natural sequence $(1, 2, \ldots, m)$.

Now let us illustrate the computation of determinants for two simple cases.

The 2 x 2 Case

$$\mathbf{A} = \begin{bmatrix} a_{11} & a_{12} \\ a_{21} & a_{22} \end{bmatrix} = \begin{bmatrix} 1 & 2 \\ 3 & 4 \end{bmatrix}$$

$$|\mathbf{A}| = (-1)^0 a_{11}a_{22} + (-1)^1 a_{12}a_{21} = a_{11}a_{22} - a_{12}a_{21} = (1 \times 4) - (2 \times 3) = -2$$

The 3 x 3 Case

$$\mathbf{A} = \begin{bmatrix} a_{11} & a_{12} & a_{13} \\ a_{21} & a_{22} & a_{23} \\ a_{31} & a_{32} & a_{33} \end{bmatrix} = \begin{bmatrix} 1 & 2 & 3 \\ 2 & -1 & 4 \\ 2 & 1 & 1 \end{bmatrix}$$

$$\begin{aligned} |\mathbf{A}| &= (-1)^0 a_{11}a_{22}a_{33} + (-1)^2 a_{12}a_{23}a_{31} + (-1)^2 a_{13}a_{21}a_{32} \\ &\quad + (-1)^3 a_{13}a_{22}a_{31} + (-1)^1 a_{11}a_{23}a_{32} + (-1)^1 a_{12}a_{21}a_{33} \\ &= (1 \times -1 \times 1) + (2 \times 4 \times 2) + (3 \times 2 \times 1) - (3 \times -1 \times 2) - (1 \times 4 \times 1) - (2 \times 2 \times 1) \\ &= -1 + 16 + 6 + 6 - 4 - 4 \end{aligned}$$

$$|\mathbf{A}| = 19$$

2.7.2 Expansion of Determinants by Cofactors

Even on the basis of the step-by-step demonstration shown above, the evaluation of a determinant (i.e., the process of finding the numerical value of the determinant) is rather complicated and prone to error. Understandably, we might seek some easier procedure in which the arithmetic is simpler and the computations more straightforward. Expansion by cofactors is one such method.

As shown above, a particularly simple determinant can be computed for the case of a 2×2 matrix:

$$\mathbf{A} = \begin{bmatrix} a_{11} & a_{12} \\ a_{21} & a_{22} \end{bmatrix}$$

In this case $|\mathbf{A}| = a_{11}a_{22} - a_{21}a_{12}$.

We can take advantage of this simple 2×2 case in attempting to evaluate high-order determinants (i.e., those of matrices of order 3×3 and higher).

To do this, we first define the minor of an entry (a_{ij}) *of a square matrix* $\mathbf{A} = (a_{ij})$ *as the determinant of a submatrix obtained by deleting the ith row and jth column of* $\mathbf{A}$. For example, the minor of the entry a_{23} in the matrix.

$$\mathbf{A} = \begin{bmatrix} a_{11} & a_{12} & a_{13} \\ a_{21} & a_{22} & a_{23} \\ a_{31} & a_{32} & a_{33} \end{bmatrix}$$

is

$$\text{minor}\ (a_{23}) = \begin{vmatrix} a_{11} & a_{12} \\ a_{31} & a_{32} \end{vmatrix}$$

Notice that this entails omitting those entries in the shaded area:

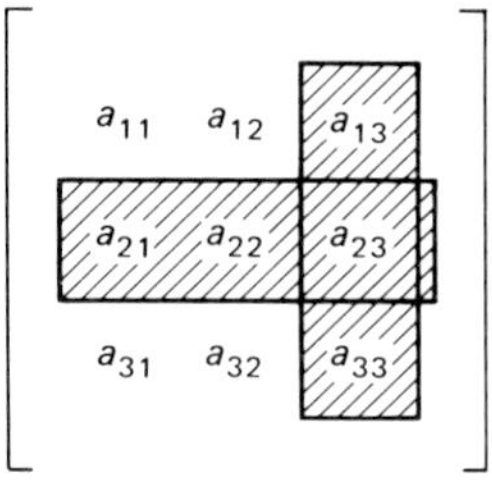

Similarly, we could find the minors of each of the other eight entries in **A**.

The cofactor of an entry a_{ij} *of a square matrix* $\mathbf{A} = (a_{ij})$ *is the product of the minor of* (a_{ij}) *and* $(-1)^{i+j}$. The cofactor is also called a *signed* minor and is denoted by $\mathbf{A}_{ij}$. In the above case,

$$A_{23} = (-1)^{2+3} \begin{vmatrix} a_{11} & a_{12} \\ a_{31} & a_{32} \end{vmatrix} = -[(a_{11}a_{32}) - (a_{31}a_{12})]$$

Notice that the placement of signs follows an alternating pattern:

$$\begin{vmatrix} + & - & + & - & + & \cdots \\ - & + & - & + & - & \cdots \\ + & - & + & - & + & \cdots \\ - & + & - & + & - & \cdots \\ \vdots & \vdots & \vdots & \vdots & \vdots & \vdots \end{vmatrix}$$

Similarly, we could develop a cofactor (or a signed minor) for each of the eight remaining entries in **A**. However, to evaluate |**A**|, it turns out that we need only develop cofactors for any single *row*, or any single *column*, of the original matrix (rather than one each for all nine entries in, for example, the 3 x 3 matrix above).

To see why this is so, we can rearrange the entries in |**A**| so that those in the second row appear first.

$$|\mathbf{A}| = a_{22}(a_{11}a_{33}) + a_{23}(a_{12}a_{31}) + a_{21}(a_{13}a_{32}) - a_{22}(a_{13}a_{31}) - a_{23}(a_{11}a_{32}) - a_{21}(a_{12}a_{33})$$

We then simplify as follows:

$$|\mathbf{A}| = a_{21}(a_{13}a_{32} - a_{12}a_{33}) + a_{22}(a_{11}a_{33} - a_{13}a_{31}) + a_{23}(a_{12}a_{31} - a_{11}a_{32})$$

However, since the cofactor A_{23} has already been defined as

$$A_{23} = (-1)^{2+3} \begin{vmatrix} a_{11} & a_{12} \\ a_{31} & a_{32} \end{vmatrix} = -[(a_{11}a_{32} - a_{12}a_{31})] = (a_{12}a_{31} - a_{11}a_{32})$$

we can substitute the cofactor for the last term on the right in the above expression for |**A**|. Similarly, we can substitute the other two cofactors: A_{21} and A_{22} involving the second row of **A**.

These are

$$A_{21} = (-1)^{2+1} \begin{vmatrix} a_{12} & a_{13} \\ a_{32} & a_{33} \end{vmatrix} = -[(a_{12}a_{33} - u_{13}a_{32})] = (a_{13}a_{32} - a_{12}a_{33})$$

$$A_{22} = (-1)^{2+2} \begin{vmatrix} a_{11} & a_{13} \\ a_{31} & a_{33} \end{vmatrix} = (a_{11}a_{33} - a_{13}a_{31})$$

Having done all this, we can evaluate |**A**| via cofactor expansion as

$$|A| = a_{21}A_{21} + a_{22}A_{22} + a_{23}A_{23}$$

Alternatively, we can evaluate |**A**| by expanding along the first or third rows, or along any of the three columns of **A**.

In summary, in computing a determinant of a matrix of order m, expansion by cofactors transforms the problem into evaluating m determinants of order $m - 1$ and forming a linear combination of these. This procedure is continued along successive stages until second-order determinants are reached. For high-order matrices (e.g., $m \geqslant 4$), expansion by cofactors provides a simple stagewise, if still tedious, way to compute determinants by hand, ultimately arriving at computations involving second-order determinants.

Let us now illustrate the evaluation of determinants through cofactor expansion for the 3 x 3 and 4 x 4 cases:

$$\mathbf{A} = \begin{bmatrix} 1 & 2 & 3 \\ 2 & -1 & 4 \\ 2 & 1 & 1 \end{bmatrix}$$

We can now apply cofactor expansion using, illustratively, the second row of **A**:

$$A_{21} = (-1)^{2+1} \begin{vmatrix} 2 & 3 \\ 1 & 1 \end{vmatrix} = 1$$

$$A_{22} = (-1)^{2+2} \begin{vmatrix} 1 & 3 \\ 2 & 1 \end{vmatrix} = -5$$

$$A_{23} = (-1)^{2+3} \begin{vmatrix} 1 & 2 \\ 2 & 1 \end{vmatrix} = 3$$

Then, by continued expansion we get

$$|A| = 2(1) - 1(-5) + 4(3) = 19$$

The same principle, illustrated above, applies in the case of fourth- or higher-ordered determinants. To illustrate, suppose we expand the determinant around the *first column* of the matrix

$$\mathbf{A} = \begin{bmatrix} 1 & 2 & 3 & 5 \\ 0 & 1 & 3 & 3 \\ 2 & 1 & 0 & 1 \\ 0 & 1 & 2 & 2 \end{bmatrix}$$

As noted, the entries of the first column of **A** are 1, 0, 2, and 0. (By choosing a column with several zeros in it, the computations are simplified.)
We first find

$$|\mathbf{A}| = 1(-1)^{1+1} \begin{vmatrix} 1 & 3 & 3 \\ 1 & 0 & 1 \\ 1 & 2 & 2 \end{vmatrix} + 0 + 2\,(-1)^{3+1} \begin{vmatrix} 2 & 3 & 5 \\ 1 & 3 & 3 \\ 1 & 2 & 2 \end{vmatrix} + 0$$

We now continue to expand around the first column of each of the two 3 x 3 minors of **A**, above.

$$|\mathbf{A}| = 1\left\{1(-1)^{(1+1)} \begin{vmatrix} 0 & 1 \\ 2 & 2 \end{vmatrix} + 1(-1)^{(2+1)} \begin{vmatrix} 3 & 3 \\ 2 & 2 \end{vmatrix} + 1(-1)^{(3+1)} \begin{vmatrix} 3 & 3 \\ 0 & 1 \end{vmatrix}\right\}$$

$$+ 2\left\{2(-1)^{(1+1)} \begin{vmatrix} 3 & 3 \\ 2 & 2 \end{vmatrix} + 1(-1)^{(2+1)} \begin{vmatrix} 3 & 5 \\ 2 & 2 \end{vmatrix} + 1(-1)^{(3+1)} \begin{vmatrix} 3 & 5 \\ 3 & 3 \end{vmatrix}\right\}$$

$$= 1(-2 - 0 + 3) + 2(0 + 4 - 6)$$

$$|\mathbf{A}| = -3$$

While we stop our illustrations with the case of fourth-order determinants, the same principles can be applied to fifth and higher-ordered determinants. Fortunately, the availability of computer programs takes the labor out of finding determinants in problems of realistic size.

2.7.3 Some Properties of Determinants

A number of useful properties are associated with determinants. The most important of these are listed below:

1. If a matrix **B** is formed from a matrix **A** by interchanging a pair of rows (or a pair of columns), then $|\mathbf{A}| = -|\mathbf{B}|$.
2. If all entries of some row or column of **A** are zero, then $|\mathbf{A}| = 0$.
3. If two rows (or two columns) of **A** are equal, then $|\mathbf{A}| = 0$.
4. The determinant of **A** equals that of its transpose **A**′; that is, $|\mathbf{A}| = |\mathbf{A}'|$.
5. The determinant of the product of two (square) matrices of the same order equals the product of the determinants of the two matrices; that is, $|\mathbf{AB}| = |\mathbf{A}|\,|\mathbf{B}|$.
6. If every entry of a row (or column) of **A** is multiplied by a scalar k, then the value of the determinant is $k|\mathbf{A}|$.
7. If the entries of a row (or column) of **A** are multiplied by a scalar and the results added or subtracted from the corresponding entries of another row (or column, respectively), then the determinant is unchanged.

Illustrations of these various properties follow:

Property 1

$$\mathbf{A} = \begin{bmatrix} 3 & 1 \\ 2 & 4 \end{bmatrix}; \qquad \mathbf{B} = \begin{bmatrix} 1 & 3 \\ 4 & 2 \end{bmatrix}$$

$$|\mathbf{A}| = -|\mathbf{B}| = 10$$

Property 2

$$\mathbf{A} = \begin{bmatrix} 0 & 1 \\ 0 & 4 \end{bmatrix}; \qquad |\mathbf{A}| = 0$$

Property 3

$$\mathbf{A} = \begin{bmatrix} 3 & 3 \\ 2 & 2 \end{bmatrix}; \qquad |\mathbf{A}| = 0$$

Property 4

$$\mathbf{A} = \begin{bmatrix} 3 & 1 \\ 2 & 4 \end{bmatrix}; \qquad \mathbf{A}' = \begin{bmatrix} 3 & 2 \\ 1 & 4 \end{bmatrix}$$

$$|\mathbf{A}| = |\mathbf{A}'| = 10$$

Property 5

$$\mathbf{A} = \begin{bmatrix} 3 & 1 \\ 2 & 4 \end{bmatrix}; \qquad \mathbf{B} = \begin{bmatrix} 4 & 2 \\ 3 & 5 \end{bmatrix}; \qquad \mathbf{AB} = \begin{bmatrix} 15 & 11 \\ 20 & 24 \end{bmatrix}$$

$$|\mathbf{A}| = 10; \qquad |\mathbf{B}| = 14; \qquad |\mathbf{AB}| = |\mathbf{A}|\,|\mathbf{B}| = 140$$

Property 6

$$\mathbf{A} = \begin{bmatrix} 3 & 1 \\ 2 & 4 \end{bmatrix}; \qquad k = 3; \qquad \mathbf{B} = \begin{bmatrix} 9 & 1 \\ 6 & 4 \end{bmatrix}$$

$$|\mathbf{A}| = 10; \qquad |\mathbf{B}| = 3|\mathbf{A}| = 30$$

Property 7

$$\mathbf{A} = \begin{bmatrix} 3 & 1 \\ 2 & 4 \end{bmatrix}; \qquad k = 3; \qquad \mathbf{b} = 3\begin{bmatrix} 3 \\ 2 \end{bmatrix} = \begin{bmatrix} 9 \\ 6 \end{bmatrix}$$

$$\mathbf{C} = \begin{bmatrix} 3 & 1 \\ 2 & 4 \end{bmatrix} - \begin{bmatrix} 0 & 9 \\ 0 & 6 \end{bmatrix} = \begin{bmatrix} 3 & -8 \\ 2 & -2 \end{bmatrix}$$

$$|\mathbf{A}| = |\mathbf{C}| = 10$$

In addition to the above (selected) properties of a determinant, we state two very important aspects of determinants that are relevant for discussion in subsequent chapters.

1. *A (square) matrix* **A** *is said to be singular if* $|\mathbf{A}| = 0$. *If* $|\mathbf{A}| \neq 0$, *it is said to be nonsingular.* This aspect of determinants will figure quite prominently in our future discussion of the regular inverse of a square matrix.

2. *The rank of a matrix is the order of the largest square submatrix whose determinant does not equal zero.*

To illustrate the characteristics of these definitions, consider the matrix:

$$\mathbf{A} = \begin{bmatrix} 1 & 2 & 3 \\ 0 & 1 & 2 \\ 2 & 4 & 6 \end{bmatrix}$$

Assume that we wish to find its determinant by cofactor expansion. We expand along the first column.

$$|\mathbf{A}| = 1(-1)^{1+1}\begin{vmatrix} 1 & 2 \\ 4 & 6 \end{vmatrix} + 0 + 2(-1)^{1+3}\begin{vmatrix} 2 & 3 \\ 1 & 2 \end{vmatrix} = -2 + 0 + 2$$

$$|\mathbf{A}| = 0$$

We see that $|\mathbf{A}| = 0$ and, according to the definition above, **A** is singular. In this case we note that the entries of the third row of **A** are precisely twice their counterparts in the first row of **A**. *In general, if a particular row (or column) can be perfectly predicted from a linear combination of the other rows (columns), the matrix is said to be singular.*

Proceeding to the next topic (i.e., the rank of **A**), we check to see if a 2×2 submatrix exists whose determinant does not equal zero.

$$\text{minor}(a_{11}) = \begin{vmatrix} 1 & 2 \\ 4 & 6 \end{vmatrix} = -2$$

Note that we need go no further, since we have found a 2 x 2 submatrix whose determinant does not equal zero.

Notice, also, that even though the matrix is of order 3 x 3, the rank of **A** cannot be 3, since $|\mathbf{A}| = 0$. However, it does turn out that at least one submatrix of order 2 x 2, as illustrated above, has a nonzero determinant. Hence, the rank of **A**, in this case, is 2.

2.7.4 The Pivotal Method of Evaluating Determinants

In relatively large matrices, such as those of fourth and higher order, the evaluation of determinants even by cofactor expansion, becomes time consuming. Over the years mathematicians have developed a wide variety of numerical methods for evaluating determinants. One of these techniques, the pivotal method (Rao, 1952), has been chosen to illustrate this class of procedures. While we illustrate the method in the context of evaluating determinants, much more is obtained, as will be demonstrated in Chapter 4.

The easiest way to describe the pivotal method is by a numerical example. For illustrative purposes let us evaluate the determinant of a fourth-order matrix:

$$\mathbf{A} = \begin{bmatrix} 2 & 3 & 1 & 2 \\ 4 & 2 & 3 & 4 \\ 1 & 4 & 2 & 2 \\ 3 & 1 & 0 & 1 \end{bmatrix}$$

Evaluating the determinant of **A** proceeds in a step-by-step fashion, with the aid of a work sheet similar to that appearing in Table 2.2.

The top, left-hand portion of Table 2.2 shows the original matrix **A**, whose determinant we wish to evaluate. To the right of this matrix is shown an identity matrix of the same order (4 x 4) as the matrix **A**. The last column (column 9) is a check sum column, each entry of which represents the algebraic sum of the specific row of interest. (Other than for arithmetic checking purposes, column 9 plays no role in the computations.)

The objective behind the pivotal method is to reduce the columnar entries in **A** successively so that for each column of interest we have only one entry and this single entry is unity. Specifically, the boxed entry in row 01 (the number 2) serves as the first pivot. Row 10 is obtained from row 01 by dividing each entry in row 01 by 2, the pivot item. Note that *all* entries in row 01 are divided by the pivot, including the entries under the identity matrix and the check sum column. Dividing 2 by itself, of course, produces the desired entry of unity in the first column of row 10.

Row 11 is obtained from the results of two operations. First, we multiply each entry of row 10 by 4, the first entry in row 02. This particular step is not shown in the work sheet, but the nine products are

$$4; \quad 6; \quad 2; \quad 4; \quad 2; \quad 0; \quad 0; \quad 0; \quad 18$$

These are then subtracted from their counterpart entries in row 02, so as to obtain row 11.

TABLE 2.2

Evaluating a Determinant by the Pivotal Method

Row No.	Original matrix				Identity matrix				Check sum column
0	1	2	3	4	5	6	7	8	9
01	[2]	3	1	2	1	0	0	0	9
02	4	2	3	4	0	1	0	0	14
03	1	4	2	2	0	0	1	0	10
04	3	1	0	1	0	0	0	1	6
10	1	1.5	0.5	1	0.5	0	0	0	4.5
11		[−4]	1	0	−2	1	0	0	−4
12		2.5	1.5	1	−0.5	0	1	0	5.5
13		−3.5	−1.5	−2	−1.5	0	0	1	−7.5
20		1	−0.25	0	0.5	−0.25	0	0	1
21			[2.125]	1	−1.75	0.625	1	0	3.0
22			−2.375	−2	0.25	−0.875	0	1	−4.0
30			1	0.471	−0.824	0.294	0.471	0	1.412
31				[−0.881]	−1.707	−0.177	1.119	1	−0.646
40				1	1.938	0.201	−1.270	−1.135	0.733
30*			1		−1.737	0.199	1.069	0.534	
20*		1			0.066	−0.200	0.267	0.134	1.267
10*	1				−0.668	−0.001	0.334	0.667	1.332

$|\mathbf{A}| = (2)(-4)(2.125)(-0.881) = 15$

Note, particularly, that this subtraction has the desired effect of producing a zero (shown as a blank) in the first entry of row 11. Note further that the entries of row 11 add up to −4, the row check sum in the last column; the check sum column is provided for all rows and serves as an arithmetic check on the computations.

Row 12 is obtained in an analogous way; here, since the first element in row 03 is 1, we multiply row 10 by unity and then subtract the row 10 elements from their counterparts in row 03. Row 13 is also obtained in the same way. First, the row 10 entries are multiplied by 3, the first entry in row 04. Then these entries are subtracted from their counterparts in row 04. Finally, we see that in rows 10 through 13, all entries in column 1 are zero (and represented by blanks) except the first entry which is unity.

At the next stage in the computations, the first element in row 11 becomes the pivot. All entries in row 11 are divided by −4, the new pivot, and the results are shown in row

20. Row 20 now becomes the reference row. For example, row 21 is found in a way analogous to row 11. First, we multiply all entries of row 20 by 2.5, the first entry of row 12. Although not shown in the work sheet, these are

$$2.5; \quad -0.625; \quad 0; \quad 1.25; \quad -0.625; \quad 0; \quad 0; \quad 2.5$$

These elements are then subtracted from their counterparts in row 12 and the results appear in row 21.

The procedure is then repeated by multiplying row 20 by −3.5, the leading element in row 13 and subtracting these new entries from their counterparts in row 13. Note that in rows 20 through 22, entries in columns 1 and 2 are all zero, except for the leading element of 1 in row 20.

The third pivot item is the entry 2.125 in row 21. All entries in row 21 are divided by 2.125 and the results listed in row 30. Finally, the entries of row 30 are multiplied by −2.375, the leading entry in row 22. Although not shown in the work sheet, these entries are

$$-2.375; \quad -1.119; \quad 1.957; \quad -0.698; \quad -1.119; \quad 0; \quad -3.353$$

These entries are subtracted from their counterparts in row 22, providing row 31. The last pivot item is −0.881 and appears in row 31.

Finally, the four pivots are multiplied together, leading to the determinant

$$|\mathbf{A}| = (2)(-4)(2.125)(-0.881) = 15$$

At this point the reader may well wonder what is the role played by the various changes being made in the identity matrix as the pivot procedure is applied. Moreover, we have not discussed the various calculations appearing in rows 40 through 10*.

As it turns out, the pivotal method is much more versatile and useful than illustrated here. While the determinant of the matrix is, indeed, obtained, the pivotal method can be employed for three important purposes:

1. computing determinants (as the product of pivot elements);[9]
2. solving a set of simultaneous equations;
3. finding the inverse of a matrix.

Here we have only described the first objective. Later on (in Chapter 4) we review the pivotal method in terms of all three of the above objectives and, in the process, discuss the remaining computations in Table 2.2.

The reader may also have wondered about what happens when a candidate pivot is zero (which, fortunately, did not happen in the preceding example). Clearly, we cannot divide the other entries of that row by zero. It turns out, however, that there is a straightforward way of dealing with this problem. We shall illustrate it in the continued discussion of this method in the context of matrix inversion in Chapter 4.

[9] In general, the determinant of an upper triangular matrix (i.e., a square matrix, all of whose elements below the main diagonal are zero) is given by the product of its main diagonal elements. Similar remarks pertain to the determinant of a lower triangular matrix (i.e., a square matrix, all of whose elements above the main diagonal are zero). The pivot procedure produces a derived triangular matrix via transformation.

In summary, our discussion of determinants does not end here. Since determinants figure quite prominently in other topics such as matrix inversion and matrix rank, we shall return to further discussion of them in subsequent chapters.

2.8 APPLYING MATRIX OPERATIONS TO STATISTICAL DATA

Much of the foregoing discussion has been introduced for a specific purpose, namely, to describe matrix and vector operations that are relevant for multivariate procedures. One of the main virtues of matrix algebra is its conciseness, that is, the succinct way in which many statistical operations can be described.

To illustrate the compactness of matrix formulation, consider the artificial data of Table 2.3. For ease of comparison these are the same data that appeared in the sample problem of Table 1.2 in Chapter 1. That is, Y denotes the employee's number of days

TABLE 2.3

Computing Various Types of Cross-Product Matrices from Sample Data

Employee	Y	Y^2	X_1	X_1^2	X_2	X_2^2	YX_1	YX_2	X_1X_2
a	1	1	1	1	1	1	1	1	1
b	0	0	2	4	1	1	0	0	2
c	1	1	2	4	2	4	2	2	4
d	4	16	3	9	2	4	12	8	6
e	3	9	5	25	4	16	15	12	20
f	2	4	5	25	6	36	10	12	30
g	5	25	6	36	5	25	30	25	30
h	6	36	7	49	4	16	42	24	28
i	9	81	10	100	8	64	90	72	80
j	13	169	11	121	7	49	143	91	77
k	15	255	11	121	9	81	165	135	99
l	16	256	12	144	10	100	192	160	120
	75	823	75	639	59	397	702	542	497

Raw cross-product matrix

$$\mathbf{B} = \begin{matrix} Y \\ X_1 \\ X^2 \end{matrix} \overset{\begin{matrix} Y & X_1 & X_2 \end{matrix}}{\begin{bmatrix} 823 & 702 & 542 \\ 702 & 639 & 497 \\ 542 & 497 & 397 \end{bmatrix}}$$

SSCP matrix

$$\mathbf{S} = \begin{matrix} Y \\ X_1 \\ X_2 \end{matrix} \overset{\begin{matrix} Y & X_1 & X_2 \end{matrix}}{\begin{bmatrix} 354.25 & 233.25 & 173.25 \\ 233.25 & 170.25 & 128.25 \\ 173.25 & 128.25 & 106.92 \end{bmatrix}}$$

Covariance matrix

$$\mathbf{C} = \begin{matrix} Y \\ X_1 \\ X_2 \end{matrix} \overset{\begin{matrix} Y & X_1 & X_2 \end{matrix}}{\begin{bmatrix} 29.52 & 19.44 & 14.44 \\ 19.44 & 14.19 & 10.69 \\ 14.44 & 10.69 & 8.91 \end{bmatrix}}$$

Correlation matrix

$$\mathbf{R} = \begin{matrix} Y \\ X_1 \\ X_2 \end{matrix} \overset{\begin{matrix} Y & X_1 & X_2 \end{matrix}}{\begin{bmatrix} 1.00 & 0.95 & 0.89 \\ 0.95 & 1.00 & 0.95 \\ 0.89 & 0.95 & 1.00 \end{bmatrix}}$$

absent during the past year; X_1 denotes his attitude rating (the higher the score the less favorable his attitude toward the firm); and X_2 denotes the number of years he has been employed by the firm.

As recalled from Chapter 1, this miniature data bank will be used later on in the book to demonstrate several multivariate techniques, including multiple regression, principal components analysis, and multiple discriminant analysis. For the moment, however, let us consider the role of matrix algebra in the development of data summaries *prior* to employing specific analytical techniques.

The computation of means, variances, covariances, correlations, etc., is a necessary preliminary to subsequent multivariate analyses in addition to being useful in its own right as a way to summarize aspects of variation in the data.

2.8.1 Sums, Sums of Squares, and Cross Products

To demonstrate the compactness of matrix notation, suppose we are concerned with computing the usual sums, sums of squares, and sums of cross products of the "raw" scores involving, for example, Y and X_1 in Table 2.3:

$$\Sigma Y; \quad \Sigma X_1; \quad \Sigma Y^2; \quad \Sigma X_1{}^2; \quad \Sigma YX_1$$

In scalar products form, the first two expressions are simply

$$\Sigma Y = \mathbf{1}'\mathbf{y} = 75; \qquad \Sigma X_1 = \mathbf{1}'\mathbf{x}_1 = 75$$

where $\mathbf{1}'$ is a 1×12 unit vector, with all entries unity, and $\mathbf{y}$ and $\mathbf{x}_1$ are the Y and X observations expressed as vectors. Notice in each case that a scalar product of two vectors is involved.

Similarly, the scalar product notion can be employed to compute three other quantities involving Y and X_1:

$$\Sigma Y^2 = \mathbf{y}'\mathbf{y} = 823 \qquad \Sigma X_1{}^2 = \mathbf{x}'_1\mathbf{x}_1 = 639 \qquad \Sigma YX_1 = \mathbf{y}'\mathbf{x}_1 = 702$$

Table 2.3 lists the numerical values for all of these products and, in addition, the products involving X_2 as well.

As a matter of fact, if we designate the matrix $\mathbf{A}$ to be the 12×3 matrix of original data involving variables Y, X_1, and X_2, the following expression

$$\boxed{\mathbf{B} = \mathbf{A}'\mathbf{A}}$$

which is often called the *minor product moment* (of $\mathbf{A}$), will yield a symmetric matrix $\mathbf{B}$ of order 3×3. The diagonal entries of the matrix $\mathbf{B}$ denote the raw sums of squares of each variable, and the off-diagonal elements denote the raw sums of cross products as shown in Table 2.3.

2.8.2 Mean-Corrected (SSCP) Matrix

We can also express the sums of squares and cross products as deviations about the means of Y, X_1, and X_2. The mean-corrected sums of squares and cross-products matrix is often more simply called the SSCP (sums of squares and cross products) matrix and is expressed in matrix notation as

$$\mathbf{S} = \mathbf{A}'\mathbf{A} - \frac{1}{m}(\mathbf{A}'\mathbf{1})(\mathbf{1}'\mathbf{A})$$

where $\mathbf{1}$ denotes a 12 x 1 unit vector and m denotes the number of observations; $m = 12$. The last term on the right-hand side of the equation represents the correction term and is a generalization of the usual scalar formula for computing sums of squares about the mean:

$$\Sigma x^2 = \Sigma X^2 - \frac{(\Sigma X)^2}{m}$$

where $x = X - \bar{X}$; that is, where x denotes deviation-from-mean form. Alternatively, if the columnar means are subtracted out of $\mathbf{A}$ to begin with, yielding the mean-corrected matrix $\mathbf{A}_\text{d}$, then

$$\mathbf{S} = \mathbf{A}_\text{d}'\mathbf{A}_\text{d}$$

For example, the mean-corrected sums of squares and cross products for Y and X_1 are

$$\Sigma y^2 = \Sigma Y^2 - \frac{(\Sigma Y)^2}{m} = 823 - \frac{(75)^2}{12} = 354.25$$

$$\Sigma x_1{}^2 = \Sigma X_1{}^2 - \frac{(\Sigma X_1)^2}{m} = 639 - \frac{(75)^2}{12} = 170.25$$

$$\Sigma yx_1 = \Sigma YX_1 - \frac{(\Sigma Y \Sigma X_1)}{m} = 702 - \frac{(75 \times 75)}{12} = 233.25$$

The SSCP matrix for all three variables appears in Table 2.3.

2.8.3 Covariance and Correlation Matrices

The covariance matrix, shown in Table 2.3, is obtained from the (mean-corrected) SSCP matrix by simply dividing each entry of $\mathbf{S}$ by the scalar m, the sample size. That is,

$$\mathbf{C} = \frac{1}{m}\mathbf{S}$$

In summational form the off-diagonal elements of $\mathbf{C}$ can be illustrated for the variables Y and X_1 by the notation

$$\text{cov}(YX_1) = \Sigma yx_1/m = 233.25/12 = 19.44$$

Note that a covariance, then, is merely an averaged cross product of mean-corrected scores. The diagonals of $\mathbf{C}$ are, of course, variances; for example,

$$s_y{}^2 = \Sigma y^2/m$$

(In some applications we may wish to obtain an unbiased estimate of the population covariance matrix; if so, we use the divisor $m - 1$ instead of m).

The *correlation* between two variables, y and x_1, is often obtained as

$$\boxed{r_{yx} = \frac{\Sigma y x_1}{\sqrt{\Sigma y^2}\,\sqrt{\Sigma x_1{}^2}}}$$

where y and x_1 are each expressed in deviation-from-mean form (as noted above).

Not surprisingly, **R** the correlation matrix is related to **S**, the SSCP matrix, and **C**, the covariance matrix. For example, let us return to **S**. The entries on the main diagonal of **S** represent mean-corrected sums of squares of the three variables Y, X_1, and X_2.

If we take the square roots of these three entries and enter the reciprocals of these square roots in a diagonal matrix, we have

$$\mathbf{D} = \begin{bmatrix} 1/\sqrt{\Sigma y^2} & 0 & 0 \\ 0 & 1/\sqrt{\Sigma x_1{}^2} & 0 \\ 0 & 0 & 1/\sqrt{\Sigma x_2{}^2} \end{bmatrix}$$

Then, by pre- and postmultiplying **S** by **D** we can obtain the correlation matrix **R**.

$$\boxed{\mathbf{R} = \mathbf{DSD}}$$

$$\mathbf{R} = \begin{bmatrix} \dfrac{\Sigma y^2}{\sqrt{\Sigma y^2}\,\sqrt{\Sigma y^2}} & \dfrac{\Sigma y x_1}{\sqrt{\Sigma y^2}\,\sqrt{\Sigma x_1{}^2}} & \dfrac{\Sigma y x_2}{\sqrt{\Sigma y^2}\,\sqrt{\Sigma x_2{}^2}} \\ \dfrac{\Sigma y x_1}{\sqrt{\Sigma y^2}\,\sqrt{\Sigma x_1{}^2}} & \dfrac{\Sigma x_1{}^2}{\sqrt{\Sigma x_1{}^2}\,\sqrt{\Sigma x_1{}^2}} & \dfrac{\Sigma x_1 x_2}{\sqrt{\Sigma x_1{}^2}\,\sqrt{\Sigma x_2{}^2}} \\ \dfrac{\Sigma y x_2}{\sqrt{\Sigma y^2}\,\sqrt{\Sigma x_2{}^2}} & \dfrac{\Sigma x_1 x_2}{\sqrt{\Sigma x_1{}^2}\,\sqrt{\Sigma x_2{}^2}} & \dfrac{\Sigma x_2{}^2}{\sqrt{\Sigma x_2{}^2}\,\sqrt{\Sigma x_2{}^2}} \end{bmatrix}$$

The above matrix is the derived matrix of correlations between each pair of variables and is also shown in Table 2.3

Ordinarily, we could then go on and use **R** in further calculation, for example, to find the regression of Y on X_1 and X_2. Since our purpose here is only to show the conciseness of matrix notation, we defer these additional steps until later. In future chapters we shall have occasion to discuss all four of the preceding matrices: (a) the raw sums and cross-products matrix, (b) the (mean-corrected) SSCP matrix, (c) the covariance matrix, and (d) the correlation matrix.

At this point, however, we should note that they are all variations on a common theme: All involve computing the minor product moment of some matrix.

1. Raw sums of squares and cross-products matrix:

$$\mathbf{B} = \mathbf{A}'\mathbf{A}$$

2. The (mean-corrected) SSCP matrix:

$$\mathbf{S} = \mathbf{A}_\text{d}'\mathbf{A}_\text{d}$$

where $\mathbf{A}_\text{d}$ is the matrix of deviation-from-mean scores; that is, each column of $\mathbf{A}_\text{d}$ sums to zero since each columnar mean has been subtracted from each datum.

3. The covariance matrix

$$\mathbf{C} = 1/m\,\mathbf{A}_\text{d}'\mathbf{A}_\text{d}$$

4. The correlation matrix

$$\mathbf{R} = 1/m\,\mathbf{A}_\text{s}'\mathbf{A}_\text{s}$$

where $\mathbf{A}_\text{s}$ is the matrix of standardized scores.

As can be found from Table 1.2, in which the sample problem data first appear, both deviation-from-mean and standardized scores are shown along with the original scores.

Finally, the matrices $\mathbf{A}_\text{d}$ of mean-corrected scores and $\mathbf{A}_\text{s}$ of standardized scores are derived from $\mathbf{A}$, the matrix of original scores, in the following way. We first find

$$\mathbf{A}_\text{d} = \mathbf{A} - \mathbf{1}\bar{\mathbf{a}}'$$

where $\mathbf{1}$ is a 12×1 unit column vector and $\bar{\mathbf{a}}'$ is a 1×3 row vector of variable means. The vector of means is, itself, obtained from

$$\bar{\mathbf{a}}' = \mathbf{1}'\mathbf{A}/m$$

where $\mathbf{1}'$ is now a 1×12 row vector. Next, we find the matrix of standardized scores from $\mathbf{A}_\text{d}$ as follows:

$$\mathbf{A}_\text{s} = \mathbf{A}_\text{d}\mathbf{D}$$

where $\mathbf{D}$ is a diagonal matrix whose entries along the main diagonal are the reciprocals of the standard deviations of the variables in $\mathbf{A}$.

The standard deviation of any column of $\mathbf{A}_\text{d}$, say $\mathbf{a}_{\text{d}j}$, is simply

$$s_{a_{\text{d}j}} = \sqrt{\mathbf{a}'_{\text{d}j}\mathbf{a}_{\text{d}j}/m}$$

In summary, any of the operations needed to find various cross-product matrices are readily expressible in matrix format. In so doing we arrive at a very compact and graceful way to portray some otherwise cumbersome operations.

2.9 SUMMARY

The purpose of this chapter has been to introduce the reader to relations and operations on vectors and matrices. Our emphasis has been on defining various operations and describing the mechanics by which one manipulates vectors and matrices. Such elementary operations as addition and subtraction, multiplication of vectors and matrices by scalars, the scalar product of two vectors, vector times matrix multiplication, etc., were described and illustrated numerically. Various properties of these operations were also described.

Special kinds of vectors (e.g., null, sign, unit) and special kinds of matrices (e.g., diagonal, scalar, identity) were also defined and illustrated numerically. We then turned to an introductory discussion of determinants of (square) matrices. Evaluation of determinants via expansion by cofactors and the pivotal method was described and illustrated.

We concluded the chapter with a demonstration of how matrix algebra can be used to provide concise descriptions of various statistical operations that are preparatory to specific multivariate analyses. These matrix operations are particularly amenable to computer programming and are used extensively in programs that deal with multivariate procedures.

REVIEW QUESTIONS

1. Write the following equations in matrix form:

a.
$$\begin{aligned} 4x + y - z &= 0 \\ 3x - 4y + 2z &= 1 \\ 5x - y - 2z &= 7 \end{aligned}$$

b.
$$\begin{aligned} 2x + 3y + z &= 11 \\ x + y + 7z &= 24 \\ 3x + 5y + 4z &= 25 \end{aligned}$$

2. Given the matrices

$$\mathbf{A} = \begin{bmatrix} 1 & 2 & -3 \\ 4 & 0 & 1 \end{bmatrix}; \quad \mathbf{B} = \begin{bmatrix} 2 & 3 & 4 \\ -1 & 2 & 0 \end{bmatrix}; \quad \mathbf{C} = \begin{bmatrix} 0 & 1 & 0 \\ 4 & -1 & -2 \end{bmatrix}$$

find

a. $\mathbf{A} + \mathbf{B}$ b. $(\mathbf{A} + \mathbf{C}) + \mathbf{B}$ c. $\mathbf{A} + (\mathbf{B} + \mathbf{C})$

d. $\mathbf{A} - (\mathbf{B} + \mathbf{C})$ e. $-(\mathbf{A} + \mathbf{B})$ f. $(\mathbf{A} - \mathbf{B}) + \mathbf{C}$

3. Given the vectors

$$\mathbf{a} = \begin{bmatrix} 1 \\ 2 \\ 4 \end{bmatrix}; \quad \mathbf{b} = \begin{bmatrix} 1 \\ 3 \\ 4 \end{bmatrix}$$

and the scalars

$$k_1 = 2; \quad k_2 = 5$$

find

a. $\mathbf{b}'\mathbf{b}$ b. $-k_1\mathbf{a}$ c. $k_2\mathbf{b}'$ d. $\mathbf{a}'\mathbf{b}$ e. $k_1k_2(\mathbf{a}'\mathbf{a})$ f. $\frac{1}{k_1}(\mathbf{b}'\mathbf{a})$

4. Given the matrices, vectors, and scalars of Problems 2 and 3, find

a. $\mathbf{a}'\mathbf{A}'$ b. $k_1\mathbf{B}$ c. $(\mathbf{AB}')'$ d. $k_1\mathbf{C}$ e. $k_2\mathbf{BA}'\mathbf{C}$ f. $\mathbf{ab}'$

5. Examine the relationships among $(\mathbf{DE})'$, $\mathbf{D}'\mathbf{E}'$, and $\mathbf{E}'\mathbf{D}'$ under the following two sets of conditions.

Let:

$$\text{a.}\quad \mathbf{D} = \begin{bmatrix} a & b \\ c & d \end{bmatrix} \quad \text{and} \quad \mathbf{E} = \begin{bmatrix} e & f \\ g & h \end{bmatrix}$$

Let:

$$\text{b.}\quad \mathbf{D} = \begin{bmatrix} 1 & 3 \\ 0 & 2 \end{bmatrix} \quad \text{and} \quad \mathbf{E} = \begin{bmatrix} 3 & 4 \\ 0 & 2 \end{bmatrix}$$

6. Given the matrices

$$\mathbf{F} = \begin{bmatrix} -1 & 3 & 5 \\ 1 & -3 & -5 \\ -1 & 3 & 5 \end{bmatrix} \quad \text{and} \quad \mathbf{G} = \begin{bmatrix} 2 & -3 & -5 \\ -1 & 4 & 5 \\ 1 & -3 & -4 \end{bmatrix}$$

demonstrate that

a. $\mathbf{FG} = \boldsymbol{\phi}$ does not imply that either $\mathbf{F} = \boldsymbol{\phi}$ or $\mathbf{G} = \boldsymbol{\phi}$.
b. Find $\mathbf{GF}$. Is this product equal to $\boldsymbol{\phi}$?

7. Given the matrices and vectors in Problems 3 and 6, find the products:

a. $\mathbf{a}'\mathbf{Fb}$ b. $\mathbf{b}'\mathbf{Gb}$ c. $\mathbf{a}'\mathbf{FGa}$ d. $\mathbf{b}'\mathbf{FGa}$

8. Consider the diagonal matrices

$$\mathbf{H}_1 = \begin{bmatrix} 4 & 0 & 0 \\ 0 & 3 & 0 \\ 0 & 0 & 1 \end{bmatrix} \quad \text{and} \quad \mathbf{H}_2 = \begin{bmatrix} -1 & 0 & 0 \\ 0 & 0 & 0 \\ 0 & 0 & 2 \end{bmatrix}$$

and the vectors and matrices of Problems 3 and 6. Find

a. $\mathbf{a}'(\mathbf{H}_1\mathbf{F})$ b. $\mathbf{b}'(\mathbf{H}_1\mathbf{G}\mathbf{H}_2)$ c. $\mathbf{a}'(\mathbf{H}_1\mathbf{H}_2)\mathbf{b}$ d. $\mathbf{a}'(\mathbf{F}\mathbf{G}\mathbf{H}_2)$

9. In ordinary algebra, we have the relationship

$$x^2 - x - 2 = (x + 1)(x - 2)$$

In matrix algebra, if

$$\mathbf{X} = \begin{bmatrix} a & b \\ c & d \end{bmatrix} \quad \text{and} \quad \mathbf{I} = \begin{bmatrix} 1 & 0 \\ 0 & 1 \end{bmatrix}$$

see if the following holds:

$$\mathbf{X}^2 - \mathbf{X} - 2\mathbf{I} = (\mathbf{X} + \mathbf{I})(\mathbf{X} - 2\mathbf{I})$$

10. If

$$\mathbf{J} = \begin{bmatrix} 1 & 2 \\ 2 & 3 \end{bmatrix} \quad \text{and} \quad \mathbf{K} = \begin{bmatrix} 2 & 0 \\ 0 & 3 \end{bmatrix}$$

Find

a. $\mathbf{J}^2$ b. $\mathbf{K}^2$ c. $(\mathbf{JK})^2$ d. $(\mathbf{KJ})^2 + (\mathbf{JK})'$

11. Evaluate the determinants of the following 2 x 2 matrices:

a. $\mathbf{L}_1 = \begin{bmatrix} x^2 & x \\ x^4 & x^3 \end{bmatrix}$ b. $\mathbf{L}_2 = \begin{bmatrix} -1 & 0 \\ 1 & 0 \end{bmatrix}$

c. $\mathbf{L}_3 = \begin{bmatrix} 1/2 & 1/3 \\ 1/4 & 1/6 \end{bmatrix}$ d. $\mathbf{L}_4 = \begin{bmatrix} a & -b \\ b & a \end{bmatrix}$

12. By means of cofactor expansion, evaluate the determinants of

a. $\mathbf{M}_1 = \begin{bmatrix} 4 & -12 & -4 \\ 2 & 1 & 3 \\ -1 & -3 & 2 \end{bmatrix}$

b. $\mathbf{M}_2 = \begin{bmatrix} 0 & 3 & 5 \\ 2 & 6 & 7 \\ 4 & 1 & 1 \end{bmatrix}$

c. $\mathbf{M}_3 = \begin{bmatrix} 1 & 5 & 2 & 1 \\ 3 & 7 & 4 & 5 \\ 2 & 9 & 1 & 2 \\ 4 & 0 & 1 & 3 \end{bmatrix}$

13. Evaluate the determinant of the fourth-order matrix used in Section 2.7.4 via cofactor expansion and check to see that it equals the value of the determinant found from the pivotal method.

14. Apply the pivotal method to matrix $\mathbf{M}_3$ in Problem 12 and check your answer with that found by cofactor expansion.

15. Assume the following data bank:

	Y	X_1	X_2	X_3
a	2	1	0	9
b	4	2	3	8
c	3	5	2	4
d	7	3	4	5
e	8	7	7	2
f	9	8	7	1

Find, via matrix methods,

a. ΣY; $\bar{X}_1$; ΣYX_2; $\Sigma X_3^2 - (\Sigma X_3)^2/m$

b. the 4 x 4 (mean-corrected) SSCP matrix **S**

c. the covariance matrix **C** d. the correlation matrix **R**

e. the matrix of mean-corrected scores

f. show that the sum of the deviations about the mean equals zero for the first column Y.

CHAPTER 3

Vector and Matrix Concepts from a Geometric Viewpoint

3.1 INTRODUCTION

This chapter is, in part, designed to provide conceptual background for many of the vector and matrix operations described in Chapter 2. Here we are interested in "what goes on" when a scalar product, for example, is computed. Since geometry often provides a direct intuitive appeal to one's understanding, liberal use is made of diagrams and geometric reasoning.

To set the stage for the various geometric descriptions to come, we define a Euclidean space—the cornerstone of most multivariate procedures. This provides the setting in which point representations of vectors and such related concepts as vector length and angle are described. The operations of vector addition, subraction, multiplication by a scalar, and scalar product are then portrayed geometrically.

We next turn to a discussion of the meaning of linear independence and the dimensionality of a vector space. The concept of a *basis* of a vector space is described, and the process by which a basis can be changed is also illustrated geometrically. Special kinds of bases—orthogonal and orthonormal—are illustrated, as well as the Gram–Schmidt process of orthonormalizing an arbitrary basis. Some comments are also made regarding general (oblique) Cartesian coordinate systems.

Our discussion then turns to one of the most common types of transformations—orthogonal transformations (i.e., rotations) of axes. These transformations are portrayed in terms of simple geometric figures and also serve as illustrations of matrix multiplication in the context of Chapter 2.

We conclude the chapter with a geometric description of some commonly used association measures, such as covariance and correlation. Moreover, the idea of viewing a determinant of a matrix of association coefficients (e.g., a covariance or a correlation matrix) as a generalized scalar measure of dispersion is also described geometrically and tied in with counterpart material that has already been covered in Chapter 2. In brief, presentation of the material in this chapter covers some of the same ground discussed in Chapter 2. Here, however, our emphasis is on the *geometry* rather than the algebra of vectors.[1]

[1] In this chapter (and succeeding chapters as well) we shall typically present the material in terms of row vectors $\mathbf{a}'$, $\mathbf{b}'$, etc., particularly when *explicit* forms of the vectors are used, such as $\mathbf{a}' = (1, 2, 2)$. This is strictly to conserve on space. The reader should get used to moving back and forth between column vectors (as emphasized in Chapter 2) and row vectors as emphasized here.

3.2 EUCLIDEAN SPACE AND RECTANGULAR CARTESIAN COORDINATES

Before moving right into a discussion of the geometric aspects of vectors, it is useful to establish some preliminaries, even though they may be familiar to many readers. These preliminaries involve the construction of a coordinate system and a description of standard basis vectors.

3.2.1 Coordinate Systems

For illustrative purposes let us review a system that is familiar to most, namely, a three-dimensional coordinate system.[2] To do this, we need three things:

1. A point called the origin of the system, that will be identified by $\mathbf{0}'$, the zero vector.
2. Three lines, called the coordinate axes, that go through the origin. We shall assume for the time being that each line is perpendicular to the other two, and we shall call these lines rectangular Cartesian axes, denoted by x, y, and z
3. One point, other than the origin, on each of the three axes. We need these points to establish scale units and the notion of direction, positive or negative, relative to the origin. Here we assume that the unit of length on each axis is the same.

Figure 3.1 shows a simple illustration of the type of coordinate system that we can set up.

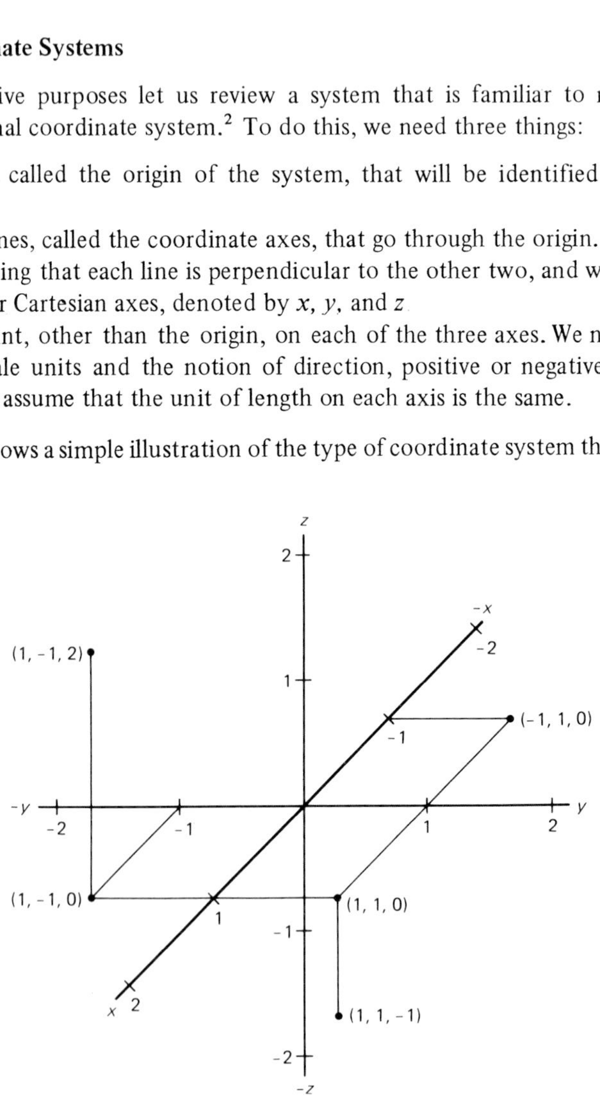

Fig. 3.1 A three-dimensional coordinate system with illustrative points.

[2] Later on we shall refer to two-dimensional as well as to three-dimensional systems. However, this particularization should cause no problems in interpretation.

By convention we have marked the positive directions of x, y, and z. Note further that three coordinate planes are also established:

1. The xy plane containing the x and y axes; this is the plane perpendicular to the z axis and containing the origin.
2. The xz plane containing the x and z axes; this is the plane perpendicular to the y axis and containing the origin.
3. The yz plane containing the y and z axes; this is the plane perpendicular to the x axis and containing the origin.

These planes cut the full space into eight octants. The first octant, for example, is the one above the xy plane in which all coordinates are positive.

Having established a coordinate system and the idea of signed distances along the axes, we can assign to each point in the space an ordered triple of real numbers:

$$\mathbf{a}' = (a_1, a_2, a_3)$$

where a_1 is the coordinate associated with the projection of $\mathbf{a}'$ onto the x axis, a_2 is the coordinate associated with the projection of $\mathbf{a}'$ onto the y axis, and a_3 is the coordinate associated with the projection of $\mathbf{a}'$ onto the z axis.

The (perpendicular) projection of a point onto a line is a vector on the line whose terminus or arrowhead is at the foot of the perpendicular dropped from the given point to the line. With the x, y, and z axes that have been set up in Fig. 3.1, the length of each projection is described on each axis by a single number, its coordinate. The coordinate is a *signed* distance from the origin; the sign is plus if the projection points in the positive direction and minus if the projection points in the negative direction. Figure 3.1 shows a few illustrative cases in different octants of the space.

In Chapter 2 we talked about a vector as a mathematical object having direction and magnitude. We need both characteristics since we can have an infinity of vectors, all having the same direction (but varying in length or magnitude), or all having the same length (but varying in direction). Furthermore, before we can talk meaningfully about direction, we need to fix a set of reference axes so that "direction" is considered relative to some standard.

In one sense vectors can originate and terminate anywhere in the space, as illustrated in Fig. 3.2. However, as also illustrated in Fig. 3.2, we can always move some arbitrary vector in a parallel direction so that the vector's tail starts at the origin. All vectors that start from the origin are called *position* vectors, and we essentially confine our attention to these. *Since we have not changed either the direction or the length of the arbitrary vector by this parallel displacement, any vector can be portrayed as a position vector.*

By concentrating our interest on position vectors, it turns out that any such vector can also be represented by a triple of numbers that we called components of a vector in Chapter 2. In the present context these components are also coordinates. By convention, the ith component of a vector is associated with the ith coordinate axis. This is illustrated in Fig. 3.3, by the projection of the terminus of $\mathbf{a}'$ onto x, y and z, the coordinate axes. Notice that each projection lies along the particular axis of interest. The (signed) length of each of these projections is, of course, described by a single number that is plus or minus, depending upon its direction along each axis relative to the origin.

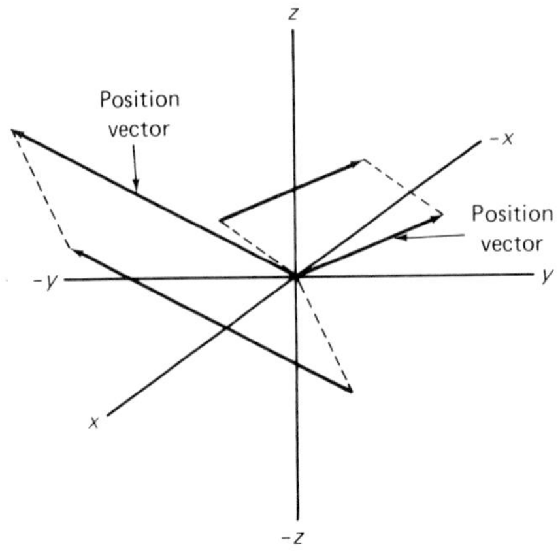

Fig. 3.2

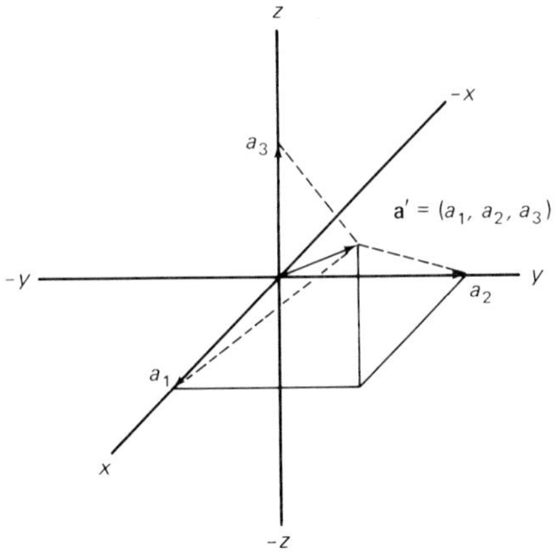

Fig. 3.3

Fig. 3.2 Parallel displacement ot arbitrary vectors.
Fig. 3.3 Vector components shown as signed lengths of projections.

Thus, given a fixed origin that is called $\mathbf{0}'$, we can always make a one-to-one correspondence between position vectors and points. For each point P we can find a corresponding position vector from the origin to P; for each position vector with its tail at $\mathbf{0}'$ we can locate a point P at the vector's terminus.

By restricting our attention to vectors emanating from the origin, any vector is both a geometric object, possessing length and direction, and an n-tuple of numbers (three numbers in this case). Since the vectors that we shall discuss *will* have their tails at the origin, *two vectors are equal if and only if they terminate at the same point.* If it were the case that two vectors had their tails at two different points, then they would be equal if and only if one of the vectors could be moved, without changing its direction or length, so that it coincided with the other.

In summary, then, by making sure that all of the vectors are position vectors (i.e., start at the origin of the coordinate system), we can pass freely back and forth between the *geometric* character of a vector (length and direction) and its *algebraic* character (an ordered n-tuple of scalars). The length of a vector's projection is given by the vector's coordinate on the x, y, and z axes, respectively, and the sign of its projection on x, y, and z depends on where the projection terminates, relative to the origin.

3.2.2 Standard Basis Vectors

Continuing on with the preliminaries, let us next consider Fig. 3.4. This figure shows a three-dimensional space with the vector $\mathbf{a}' = (1, 2, 2)$ appearing as a directed line segment.

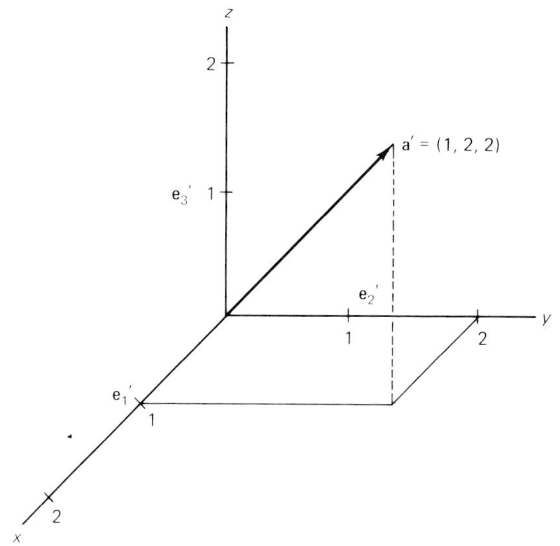

Fig. 3.4 Vector representation in three-dimensional space.

To set up this coordinate space, we define a special set of zero–one coordinate vectors, denoted $\mathbf{e}_i'$, as

1. vector $\mathbf{e}_1'$ of unit length in the positive (by convention) x direction:
$$\mathbf{e}_1' = (1, 0, 0)$$
2. vector $\mathbf{e}_2'$ of unit length in the positive y direction:
$$\mathbf{e}_2' = (0, 1, 0)$$
3. vector $\mathbf{e}_3'$ of unit length in the positive z direction:
$$\mathbf{e}_3' = (0, 0, 1)$$

We shall continue to let $\mathbf{0}'$, the zero vector, denote the origin of the space. As suggested in the discussion of vector addition and scalar multiplication of a vector in Chapter 2, we can now write the vector $\mathbf{a}' = (1, 2, 2)$ as a linear combination of the coordinate vectors:

$$\begin{aligned} 1\mathbf{e}_1' + 2\mathbf{e}_2' + 2\mathbf{e}_3' &= 1(1, 0, 0) + 2(0, 1, 0) + 2(0, 0, 1) \\ &= (1, 0, 0) + (0, 2, 0) + (0, 0, 2) \\ \mathbf{a}' &= (1, 2, 2) \end{aligned}$$

Note that what we have done is to perform scalar multiplication followed by vector addition, relative to the coordinate vectors $\mathbf{e}_i'$. We shall call the $\mathbf{e}_i'$ vectors a *standard basis* and comment later on the meaning of basis vectors, generally.

Note, further, that if we had the oppositely directed vector $-\mathbf{a}'$, this could also be represented in terms of the standard basis vectors as the linear combination:

$$-1\mathbf{e}_1' - 2\mathbf{e}_2' - 2\mathbf{e}_3' = (-1, 0, 0) + (0, -2, 0) + (0, 0, -2) = (-1, -2, -2)$$

In this case $-\mathbf{a}'$ would extend in the negative directions of x, y, and z.

What is shown above in particularized form can be generalized in accordance with the discussion of linear combinations of vectors in Chapter 2. As recalled:

Given p n-component vectors $\mathbf{b}_1', \mathbf{b}_2', \ldots, \mathbf{b}_p'$ *the n-component vector*

$$\mathbf{a}' = \sum_{i=1}^{p} k_i \mathbf{b}_i' = k_1 \mathbf{b}_1' + k_2 \mathbf{b}_2' + \cdots + k_p \mathbf{b}_p'$$

is a linear combination of p vectors, $\mathbf{b}_1', \mathbf{b}_2', \ldots, \mathbf{b}_p'$ *for any set of scalars* k_i $(i = 1, 2, \ldots, p)$.

In the illustration above we have $p = 3$ basis vectors, each containing $n = 3$ components. The components of the vector $\mathbf{a}' = (1, 2, 2)$ involve $p = 3$ scalars. The $\mathbf{b}_i'$ vectors in the more general expression above correspond to the specific $\mathbf{e}_i'$ vectors in the preceding numerical illustration.

The introduction of a set of standard basis vectors allows us to write *any* n-component vector $\mathbf{a}'$, relative to a standard basis of n-component $\mathbf{e}_i'$ vectors, as

$$\boxed{\mathbf{a}' = \sum_{i=1}^{n} a_i \mathbf{e}_i'}$$

where a_i denotes the ith component of $\mathbf{a}'$, and each of the n basis vectors has a 1 appearing in the ith position and zeros elsewhere. In this special case of a linear combination, the number of vectors p equals the number of components in $\mathbf{a}'$, namely, n.

Figure 3.5, incidentally, shows $\mathbf{a}'$ in terms of the triple of numbers (1, 2, 2). This point representation, as we now know, is equally acceptable for representing $\mathbf{a}'$ since the vector is already positioned with its tail at the origin.

The important point to remember is that $\mathbf{a}'$, itself, can be represented as a linear combination of other vectors–in this case, the standard basis vectors $\mathbf{e}_i'$. In a sense the $\mathbf{e}_i'$

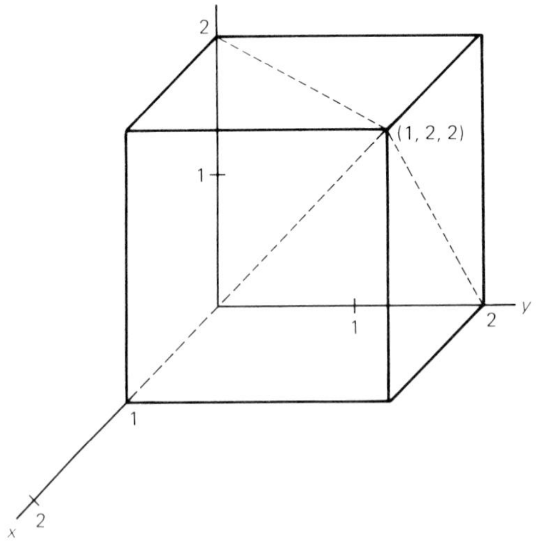

Fig. 3.5 Point representation in three-dimensional space.

represent a standard scale unit across the three axes. Any projection of $\mathbf{a}'$, then, can be considered as involving (signed) multiples of the appropriate $\mathbf{e}_i'$ vector.

With these preliminaries out of the way, we can now introduce the central concept of the chapter, namely, the Euclidean space and the associated idea of the distance between two points, or vector termini, in Euclidean space. This idea, in turn, leads to the concepts of angle and length.

3.2.3 Definition of Euclidean Space

A Euclidean space of n dimensions is the collection of all n-component vectors for which the operations of vector addition and multiplication by a scalar are permissible. Moreover, for any two vectors in the space, there is a nonnegative number, called the Euclidean distance between the two vectors.[3]

The function[4] that produces this nonnegative number is called a Euclidean distance function and is defined as

$$\boxed{\|\mathbf{a}' - \mathbf{b}'\| = [(a_1 - b_1)^2 + (a_2 - b_2)^2 + \cdots + (a_n - b_n)^2]^{1/2}}$$

Alternatively, we can define $\|\mathbf{a}' - \mathbf{b}'\|$ in terms of a function of the now-familiar scalar product of $(\mathbf{a} - \mathbf{b})$ with itself:

$$\boxed{\|\mathbf{a}' - \mathbf{b}'\| = [(\mathbf{a} - \mathbf{b})'(\mathbf{a} - \mathbf{b})]^{1/2}}$$

where the vector $(\mathbf{a} - \mathbf{b})$ is a difference vector.

To get some geometric view of the Euclidean distance between two position vector termini (i.e., between two points), let us first examine Panel I of Fig. 3.6. Here in two dimensions are the two points

$$\mathbf{a}' = (1, 1); \qquad \mathbf{b}' = (1.5, 2)$$

Note that their straight-line distance can be represented by the square root of the hypotenuse of the right triangle, as sketched in the chart. In terms of the distance formula, we have

$$\|\mathbf{a}' - \mathbf{b}'\| = [(1 - 1.5)^2 + (1 - 2)^2]^{1/2} = \sqrt{1.25} = 1.12$$

Panel II of Fig. 3.6 merely extends the same idea to three dimensions for two new points:

$$\mathbf{a}' = (1, 1, -2); \qquad \mathbf{b}' = (2, 1, 2)$$

$$\|\mathbf{a}' - \mathbf{b}'\| = [(1 - 2)^2 + (1 - 1)^2 + (-2 - 2)^2]^{1/2} = \sqrt{17} = 4.12$$

[3] A more formal definition considers a Euclidean space as a finite-dimensional vector space on which a real-valued scalar or inner product is defined.

[4] The Euclidean metric is, itself, a special case of the Minkowski metric. The Minkowski metric also obeys the three distance axioms (positivity, symmetry, and triangle inequality). Since we have used the single bars $|\mathbf{A}|$ to denote the determinant of a matrix in Chapter 2, we use the double bars $\|\mathbf{a}' - \mathbf{b}'\|$ to denote the distance between two vectors, taken here to mean Euclidean distance.

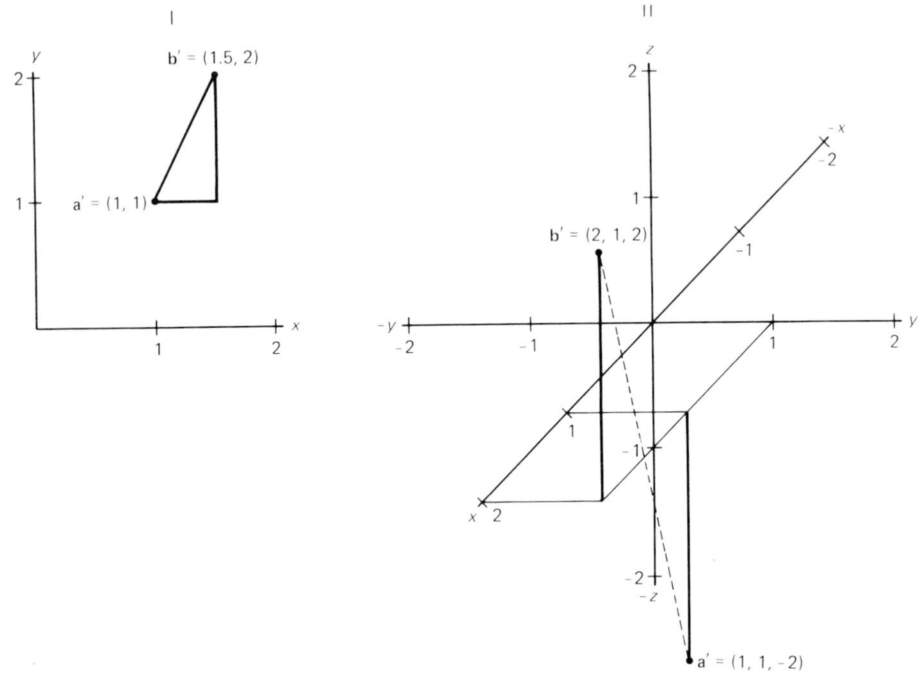

Fig. 3.6 Illustrations of Euclidean distances between pairs of points. Key: I, two dimensions; II, three dimensions.

Euclidean distance, then, entails adding up the squared differences in projections of each point on each axis in turn, and then taking the square root of the sum.

As might be surmised from either panel in Fig. 3.6, the Euclidean distance function possesses the following properties:

$\|\|\mathbf{a}' - \mathbf{b}'\|\| > 0$	unless $\mathbf{a}' - \mathbf{b}' = \mathbf{0}'$;	positivity
$\|\|\mathbf{a}' - \mathbf{b}'\|\| = \|\|\mathbf{b}' - \mathbf{a}'\|\|$;		symmetry
$\|\|\mathbf{a}' - \mathbf{b}'\|\| + \|\|\mathbf{b}' - \mathbf{c}'\|\| \geqslant \|\|\mathbf{a}' - \mathbf{c}'\|\|$;		triangle inequality

The first of the above properties, positivity, precludes the possibility of negative distances. Symmetry, the second property, means that the distance from $\mathbf{a}'$ to $\mathbf{b}'$ is the same as the distance from $\mathbf{b}'$ to $\mathbf{a}'$. The third property, triangle inequality, states that the sum of the distances between $\mathbf{a}'$ and $\mathbf{b}'$ and between $\mathbf{b}'$ and some third point $\mathbf{c}'$ is no less than the direct distance between $\mathbf{a}'$ and $\mathbf{c}'$. If $\mathbf{b}'$ lies on the line connecting $\mathbf{a}'$ and $\mathbf{c}'$, then the sum of the distances of $\mathbf{a}'$ to $\mathbf{b}'$ and $\mathbf{b}'$ to $\mathbf{c}'$ equals the direct distance from $\mathbf{a}'$ to $\mathbf{c}'$.

We next define the concept of vector length or magnitude. The length of a vector $\mathbf{a}' = (a_1, a_2, \ldots, a_n)$ is defined as

$$||\mathbf{a}'|| = \left[\sum_{i=1}^{n} a_i^2\right]^{1/2}$$

Note that this is a special case of the Euclidean distance function in which the second vector is the origin of the space, or the $\mathbf{0}'$ vector. That is,

$$||\mathbf{a}'|| = ||\mathbf{a}' - \mathbf{0}'|| = [(a_1-0)^2 + (a_2-0)^2 + \cdots + (a_n-0)^2]^{1/2}$$

Furthermore, we can also observe that the *squared* vector length equals the scalar product of $\mathbf{a}$ with itself:

$$\boxed{||\mathbf{a}'||^2 = \mathbf{a}'\mathbf{a}}$$

Thus, in the case of $\mathbf{a}' = (1, 2, 2)$, we see that

$$||\mathbf{a}'||^2 = \sum_{i=1}^{3} a_i^2 = (1)^2 + (2)^2 + (2)^2$$

$$= [(a_1-0)^2 + (a_2-0)^2 + (a_3-0)^2] = (1)^2 + (2)^2 + (2)^2$$

$$= \mathbf{a}'\mathbf{a} = (1,2,2)'(1,2,2) = (1)^2 + (2)^2 + (2)^2 = 9$$

are all equivalent ways of finding the squared length of $\mathbf{a}'$. The square root of $||\mathbf{a}'||^2$, that is, $\sqrt{9} = 3$, is, of course, the Euclidean distance or vector length of the vector terminus as measured from the origin $\mathbf{0}'$.

We now discuss some of these notions in more detail. Since it will be intuitively easier to present the concepts in terms of the standard basis vectors $\mathbf{e}_i'$–that is, where rectangular (mutually perpendicular) Cartesian coordinates are used–we discuss this case first and later briefly discuss more general coordinate systems in which the axes are *not* necessarily mutually perpendicular, although the space is still assumed to be Euclidean.

3.3 GEOMETRIC REPRESENTATION OF VECTORS

We have already commented on the fact that a vector can be equally well represented by the directed line segment, starting from the origin (Fig. 3.4), or the triple of point coordinates (Fig. 3.5). In both representations, the coordinate on each axis is the foot of the perpendicular dropped from $\mathbf{a}'$ to each axis.

Our interest now is in expanding some of these geometric notions so as to come up with graphical counterparts to the various algebraic operations on vectors that were described in Chapter 2.

3.3.1 Length and Direction Angles of a Single Vector

Let us again examine vector $\mathbf{a}' = (1, 2, 2)$, represented as the directed line segment in Fig. 3.4. As shown earlier, the length or Euclidean distance, denoted $||\mathbf{a}'||$, of $\mathbf{a}'$ from the origin $\mathbf{0}'$ is

$$||\mathbf{a}'|| = [(a_1-0)^2 + (a_2-0)^2 + (a_3-0)^2]^{1/2}$$

$$= [(1)^2 + (2)^2 + (2)^2]^{1/2} = 3$$

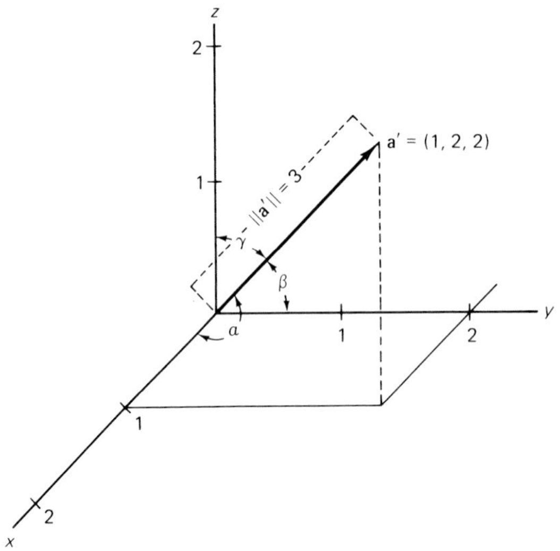

Fig. 3.7 Vector length and direction cosines.

Our interest now focuses on various aspects of length and *direction angles.* First, observe from Fig. 3.7 that **a**′ makes angles of α, β, and γ with the three reference axes x, y, and z. The cosines of these angles are called *direction cosines* relative to the reference axes and are computed as *ratios of each vector component to the vector's length.*

Since the length of **a**′ is 3 and the components of **a**′ are 1, 2, and 2, the cosines of α, β, and γ, respectively, are

$$\cos\alpha = \tfrac{1}{3}; \qquad \cos\beta = \tfrac{2}{3}; \qquad \cos\gamma = \tfrac{2}{3}$$

Notice that these can be written out in full as

$$\cos\alpha = \frac{a_1}{[a_1^2 + a_2^2 + a_3^2]^{1/2}} = \frac{1}{3}$$

$$\cos\beta = \frac{a_2}{[a_1^2 + a_2^2 + a_3^2]^{1/2}} = \frac{2}{3}$$

$$\cos\gamma = \frac{a_3}{[a_1^2 + a_2^2 + a_3^2]^{1/2}} = \frac{2}{3}$$

The angles corresponding to these cosines are

$$\alpha \cong 71^\circ; \qquad \beta \cong 48^\circ; \qquad \gamma \cong 48^\circ$$

Notice that our use of the cosine is in accordance with basic trigonometry. For example, the cosine of the angle α, which the vector **a**′ makes with the x axis, is equal to the length of the adjacent side of the right triangle, formed by projection of **a**′ onto the x axis, divided by the hypotenuse of that right triangle. The adjacent side has length 1, or unit distance from the origin, and the hypotenuse is of length 3. Hence, the cosine of α is $\frac{1}{3}$. By similar reasoning the cosine of β is $\frac{2}{3}$ with respect to the y axis, and that of γ is $\frac{2}{3}$ with respect to the z axis.

We can also discuss some of these notions in somewhat more general terms. Once a coordinate system is chosen, any position vector that emanates from the origin can be represented by

1. the angles α, β, γ, made by the line with respect to the x, y, and z axes, where $0 \leqslant \alpha, \beta, \gamma \leqslant 180°$, and
2. the vector's length or magnitude.

We have already discussed the case of vectors that emanate from locations other than the origin of the space. Therefore, by appropriate parallel displacement to a position vector, any vector in the space can be represented by its direction angles and length.

If we had a vector $-\mathbf{a}' = (-1, -2, -2)$ that was oppositely directed from $\mathbf{a}'$, this would cause no problems since the direction cosines and angles would then be

$$\cos\alpha = -\tfrac{1}{3}; \qquad \alpha = 109° = 180° - 71°$$

$$\cos\beta = -\tfrac{2}{3}; \qquad \beta = 132° = 180° - 48°$$

$$\cos\gamma = -\tfrac{2}{3}; \qquad \gamma = 132° = 180° - 48°$$

It is also useful to examine the sum of the squared cosines of α, β, and γ. Since $a_1^2 + a_2^2 + a_3^2 = \|\mathbf{a}'\|^2$, we have

$$\boxed{\cos^2\alpha + \cos^2\beta + \cos^2\gamma = 1}$$

We can state the above result in words: The sum of the squares of the direction cosines of some vector $\mathbf{a}'$, *originating at the origin, is equal to 1.* This fact holds true in any dimensionality, not just three dimensions.

Furthermore, it is a simple matter to work backward to find the components of a vector if we know its direction angles and length. Continuing with the illustrative vector,

$$\mathbf{a}' = (1, 2, 2)$$

with direction angles and cosines,

$$\alpha = 71°, \quad \cos\alpha = \tfrac{1}{3}; \qquad \beta = 48°, \quad \cos\beta = \tfrac{2}{3}; \qquad \gamma = 48°, \quad \cos\gamma = \tfrac{2}{3}$$

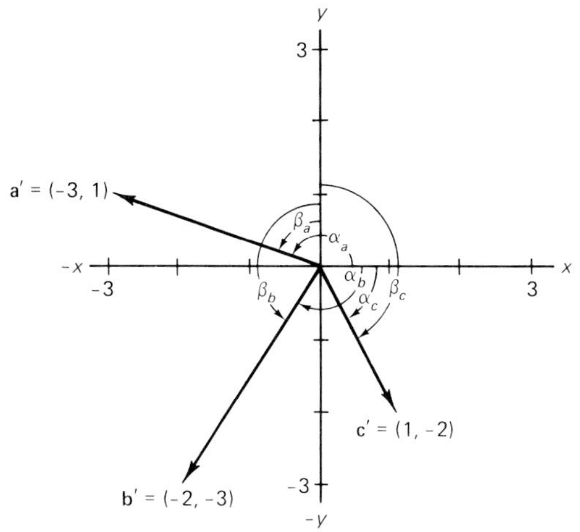

Fig. 3.8 Direction angles and lengths of illustrative vectors.

and length $\|\mathbf{a}'\| = 3$, we have, by simple algebra, the vector components:

$$a_1 = \tfrac{1}{3}(3) = 1; \qquad a_2 = \tfrac{2}{3}(3) = 2; \qquad a_3 = \tfrac{2}{3}(3) = 2$$

In the case of negative coordinates, the cosines, of course, would be negative for those axes involving negative projections.

Working with negative cosines is most easily shown in two dimensions, as illustrated in Fig. 3.8. Here are portrayed three different vectors, terminating in three different quadrants. We first note the smaller angle ($\leqslant 180°$) made by each vector with each axis. Then, by means of the formulas shown earlier, the direction cosines of each vector are computed as follows:

$$\boxed{\mathrm{a}'} \quad \cos\alpha_a = \frac{-3}{[(-3)^2 + (1)^2]^{1/2}} = -0.95; \qquad \cos\beta_a = \frac{1}{[(-3)^2 + (1)^2]^{1/2}} = 0.32$$

$$\boxed{\mathrm{b}'} \quad \cos\alpha_b = \frac{-2}{[(-2)^2 + (-3)^2]^{1/2}} = -0.55; \qquad \cos\beta_b = \frac{-3}{[(-2)^2 + (-3)^2]^{1/2}} = -0.83$$

$$\boxed{\mathrm{c}'} \quad \cos\alpha_c = \frac{1}{[(1)^2 + (-2)^2]^{1/2}} = 0.45; \qquad \cos\beta_c = \frac{-2}{[(1)^2 + (-2)^2]^{1/2}} = -0.89$$

with correspondent direction angles:

$$\alpha_a \cong 161°; \qquad \beta_a \cong 71°$$
$$\alpha_b \cong 123°; \qquad \beta_b \cong 146°$$
$$\alpha_c \cong 63°; \qquad \beta_c \cong 153°$$

as shown in Fig. 3.8.

Notice, in particular, that as any angle becomes obtuse, the formulas for finding direction cosines still hold since changes in the sign of the cosine are taken care of by corresponding changes in the appropriate vector components.

In summary, then, any position vector is uniquely determined by knowledge of its magnitude and direction. In turn, its direction is given by the angles it makes with the reference axes. These angles are obtained from the cosines that are computed by the expression

$$\boxed{\cos\Psi_i = \frac{a_i}{\|\mathbf{a}'\|}}$$

where Ψ_i denotes the angle between the vector and the ith reference axis ($0° \leqslant \Psi_i \leqslant 180°$), a_i denotes the ith component of $\mathbf{a}'$, and $\|\mathbf{a}'\|$ denotes its length.

Note, in particular, that if $\cos\Psi_i = 0$, *then the angle is* 90° *and the vector is said to be orthogonal or perpendicular to the ith reference axis.*

3.3.2 Geometric Aspects of Vector Addition and Multiplication by a Scalar

While we have earlier discussed in Chapter 2 the rules of vector addition and subtraction and multiplication of a vector by a scalar, it is useful now to show these

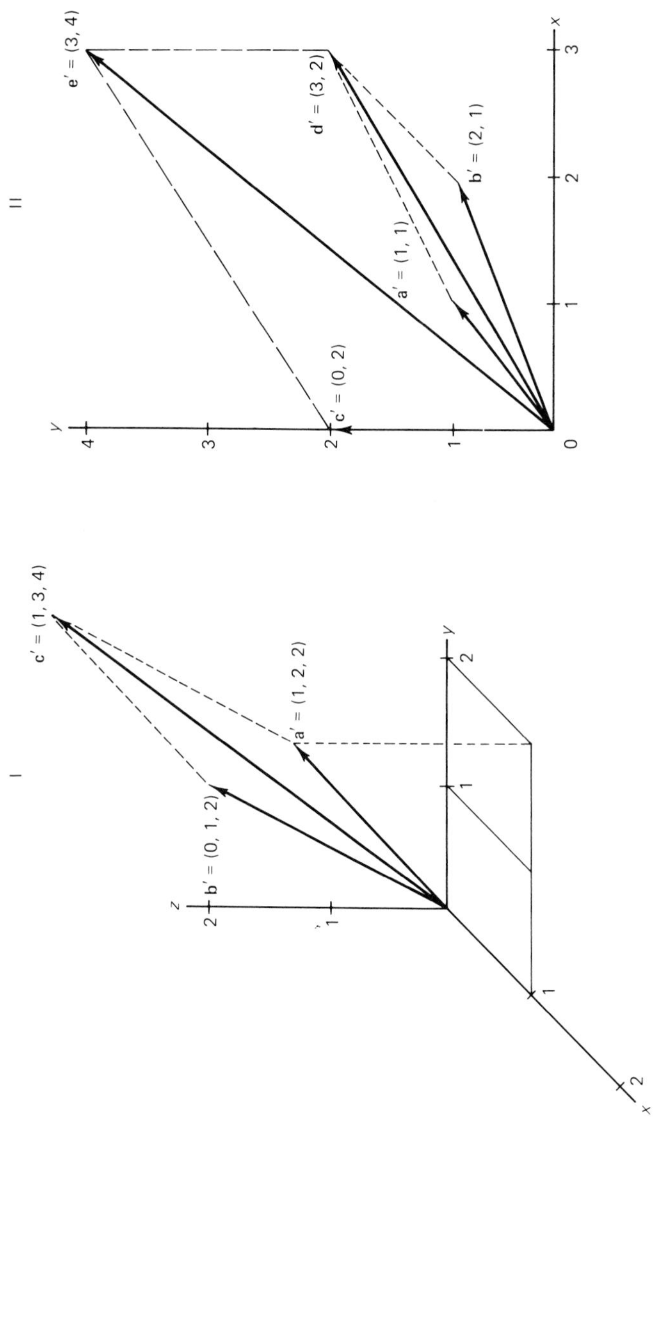

Fig. 3.9 Illustrations of vector addition.

operations geometrically. First consider the two vectors $\mathbf{a}' = (1, 2, 2)$ and $\mathbf{b}' = (0, 1, 2)$ shown in Panel I of Fig. 3.9. As already known from Chapter 2, their vector sum is

$$(1+0, 2+1, 2+2) = (1, 3, 4)$$

We can formalize this by saying that if $\mathbf{a}'$ *and* $\mathbf{b}'$ *are 1* × *n vectors, their vector sum is defined by*

$$\boxed{\mathbf{a}' + \mathbf{b}' = (a_1 + b_1, a_2 + b_2, \ldots, a_i + b_i, \ldots, a_n + b_n)}$$

As noted from Panel I of the figure, vector addition proceeds on a component-by-component basis. Geometrically, $\mathbf{c}' = \mathbf{a}' + \mathbf{b}'$ is represented by the diagonal of a parallelogram determined by $\mathbf{a}'$ and $\mathbf{b}'$.

Panel II of Fig. 3.9 shows a case for three vectors in two dimensions, $\mathbf{a}'$, $\mathbf{b}'$ and $\mathbf{c}'$. When $\mathbf{a}'$ and $\mathbf{b}'$ are added, their sum is represented by $\mathbf{d}'$, the diagonal of a parallelogram. The parallelogram rule also applies as $\mathbf{d}'$ is added to $\mathbf{c}'$, resulting in their vector sum, shown by $\mathbf{e}'$.

Vector subtraction presents no major additional complications. Suppose, for example, that we wish to show the difference

$$\mathbf{d}' = \mathbf{a}' - \mathbf{b}' = (1-0, 2-1, 2-2) = (1, 1, 0)$$

geometrically. Figure 3.10 shows the *difference* vector, denoted by $\mathbf{d}'$, as a vector emanating from the origin with the *same length and direction* as that indicated by the line connecting the arrowheads of $\mathbf{a}'$ and $\mathbf{b}'$. Notice, then, that we maintain the concept of position vector by making a parallel displacement of the difference between $\mathbf{a}'$ and $\mathbf{b}'$ so that $\mathbf{d}'$ starts from the origin.

If we had the vector $\mathbf{a}'$ and another vector $-\mathbf{a}'$, it would, of course, be the case that

$$\mathbf{a}' + (-\mathbf{a}') = \mathbf{0}'$$

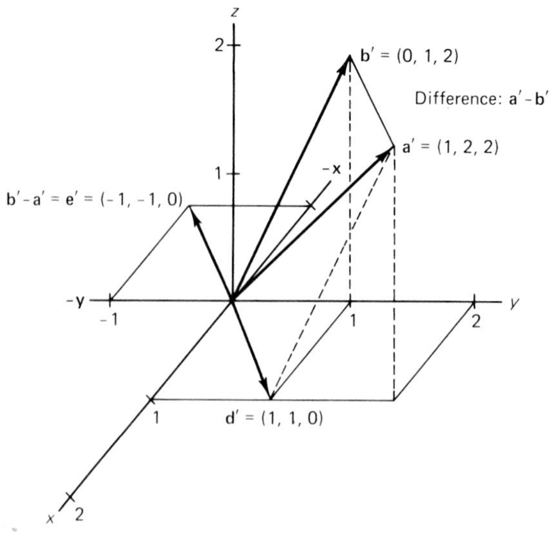

Fig. 3.10 Illustrations of vector subtraction.

Note also that the vector subtraction

$$\mathbf{e}' = \mathbf{b}' - \mathbf{a}' = (0-1, 1-2, 2-2) = (-1, -1, 0)$$

is handled analogously and, furthermore, that $-1\mathbf{e}' = \mathbf{d}'$, as should be the case. We find that $\mathbf{d}'$ and $\mathbf{e}'$ are merely oppositely directed vectors of equal length.[5]

Multiplication of a vector **a** *by a scalar k is formally defined as*

$$\boxed{k(\mathbf{a}') = (ka_1, ka_2, \ldots, ka_i, \ldots, ka_n)}$$

and is also illustrated in Fig. 3.10 for the special case in which $k = -1$. That is,

$$\mathbf{e}' = -1(\mathbf{d}') = -1(1, 1, 0) = (-1, -1, 0)$$

As a more general example, Fig. 3.11 shows the case of multiplying the vector

$$\mathbf{a}' = (2, 3, 2)$$

by $k_1 = -1$, $k_2 = \frac{1}{2}$, $k_3 = 2$. We note that the sign of k determines the direction of $k\mathbf{a}'$ while the magnitude of k determines how far $k\mathbf{a}'$ extends in the appropriate direction from the origin, relative to $\|\mathbf{a}'\|$, the length of $\mathbf{a}'$ when $k = 1$.

As a concluding example we combine the operations of addition and scalar multiplication of a vector by considering the case of a linear combination:

$$\tfrac{1}{2}\mathbf{a}' + 2\mathbf{b}' = \tfrac{1}{2}(1, 2, 2) + 2(0, 1, 2) = (\tfrac{1}{2}, 1, 1) + (0, 2, 4) = (\tfrac{1}{2}, 3, 5)$$

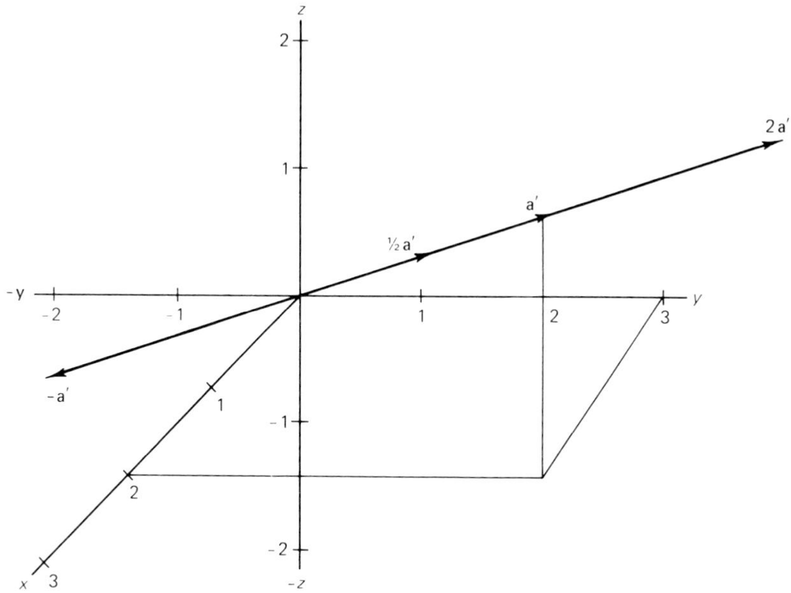

Fig. 3.11 Illustrations of vector multiplication by a scalar.

[5] The vector $\mathbf{e}'$ is used here as an arbitrary vector and is not to be confused with the standard basis vectors $\mathbf{e}_i'$, introduced earlier in the chapter.

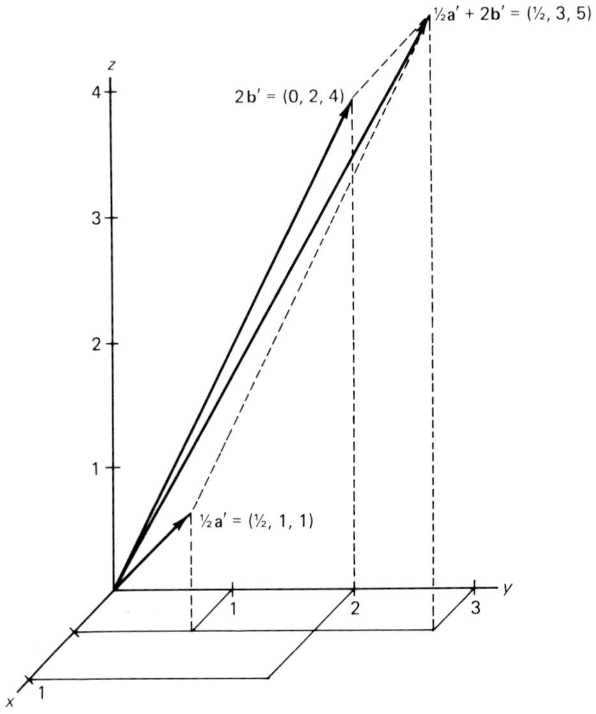

Fig. 3.12 An illustration of the combined operations of scalar multiplication and addition.

The result of these operations appears in Fig. 3.12. This same idea can, of course, be extended to more than two vectors. For example, in three dimensions the sum of three three-component vectors would be represented by the diagonal of a parallelepiped formed from the three contributing vectors. As long as we confine the number of components of each vector to at most three, it becomes quite straightforward to picture the operations of addition, subtraction, scalar multiplication of a vector, and their generalization, a linear combination of vectors.

The properties of addition, subtraction, and multiplication of a vector by a scalar were listed in Chapter 2. These properties, of course, apply here since our current purpose is simply to portray the same vector relations geometrically rather than algebraically.

3.3.3 Distance and Angle between Two Vectors

In Section 3.3.1 we considered the special case of the angle between two vectors when one of those vectors was a coordinate axis. We can now discuss the general situation of the angle between any pair of position vectors in Euclidean space. Suppose we continue to consider the case of the two vectors $\mathbf{a}' = (1, 2, 2)$ and $\mathbf{b}' = (0, 1, 2)$. As shown earlier, vector $\mathbf{a}'$ has length 3. Vector $\mathbf{b}'$ has length

$$\|\mathbf{b}'\| = [(0)^2 + (1)^2 + (2)^2]^{1/2} = \sqrt{5}$$

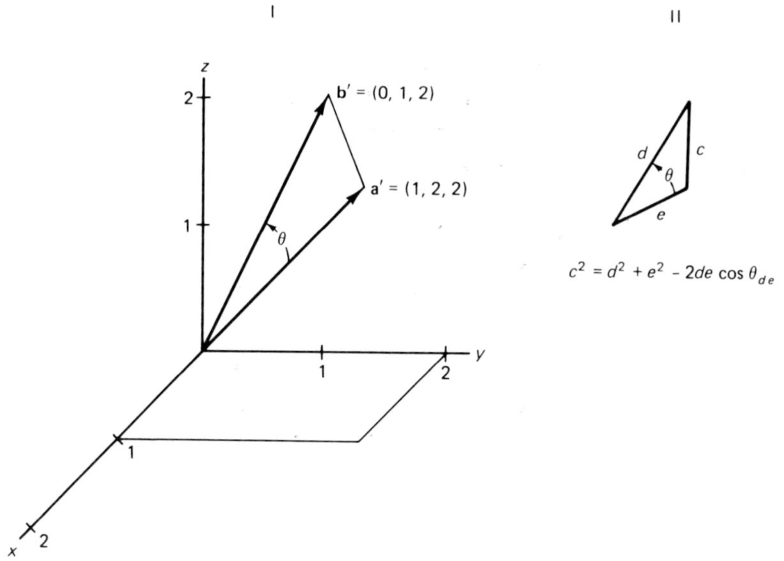

Fig. 3.13 Finding the angle between the two vectors. Key: I, position vectors; II, arbitrary triangle.

with direction cosines and angles, with respect to x, y, and z, of

$$\cos \alpha^* = 0/\sqrt{5} = 0; \qquad \alpha^* = 90^\circ$$

$$\cos \beta^* = 1/\sqrt{5} = 0.45; \qquad \beta^* \cong 63^\circ$$

$$\cos \gamma^* = 2/\sqrt{5} = 0.89; \qquad \gamma^* \cong 27^\circ$$

So far, nothing new. Now we ask: What is the Euclidean distance between $\mathbf{a}'$ and $\mathbf{b}'$?

Again, as we know, the distance between $\mathbf{a}'$ and $\mathbf{b}'$ can be computed as

$$\|\mathbf{a}' - \mathbf{b}'\| = [(1-0)^2 + (2-1)^2 + (2-2)^2]^{1/2} = \sqrt{2} = 1.41$$

That is, we find the difference between the two vectors on a component-by-component basis, square each of these differences, sum the squared differences, and then take the square root of the result. Notice that this is similar to finding a vector's length in which the origin, or zero vector, plays the role of the second vector.

Again, nothing new. However, at this point we can note from Panel I of Fig. 3.13 that $\mathbf{a}'$ and $\mathbf{b}'$ make some angle θ with each other. The problem, now, is to determine what this angle is. That is, analogous to the case of finding the angle that a single vector makes with each of the reference axes, we now wish to find the angle between two different vectors referred to the same set of coordinate axes. To do so, we make use of the *cosine law of trigonometry*.

As the reader may recall from basic trigonometry, the law of cosines states:

For any triangle with sides c, d, and e, the square of any side is equal to the sum of the squares of the other two sides minus twice the product of the other two sides and the cosine of their included angle θ. Or, to illustrate (see arbitrary triangle in Panel II of Fig. 3.13),

$$c^2 = d^2 + e^2 - 2de \cos \theta_{de}$$

Similarly, we could find d^2 or e^2, as the case may be.

Returning to our specific example in Panel I of Fig. 3.13, by simple algebra we can first express the law of cosines in terms of the cosine of

$$\cos \theta_{\mathbf{a}'\mathbf{b}'} = \frac{\|\mathbf{a}'\|^2 + \|\mathbf{b}'\|^2 - \|\mathbf{a}' - \mathbf{b}'\|^2}{2\|\mathbf{a}'\| \cdot \|\mathbf{b}'\|}$$

where the above formula represents the particularized version of

$$\cos \theta_{de} = \frac{d^2 + e^2 - c^2}{2de}$$

as applying to any triangle of interest.

In terms of our specific problem, the cosine of the angle θ between the vectors $\mathbf{a}'$ and $\mathbf{b}'$ is expressed as a ratio in which the numerator is the squared length of $\mathbf{a}'$ plus the squared length of $\mathbf{b}'$ minus the squared length of the difference vector $\mathbf{a}' - \mathbf{b}'$; the denominator of the ratio is simply 2 times the product of the lengths of $\mathbf{a}'$ and $\mathbf{b}'$.

If we then substitute the appropriate numerical quantities, we have

$$\cos \theta_{\mathbf{a}'\mathbf{b}'} = \frac{\|\mathbf{a}'\|^2 + \|\mathbf{b}'\|^2 - \|\mathbf{a}' - \mathbf{b}'\|^2}{2\|\mathbf{a}'\| \cdot \|\mathbf{b}'\|} = \frac{9 + 5 - 2}{2(3)(\sqrt{5})} = \frac{12}{13.416} = 0.894$$

with the correspondent angle

$$\theta_{\mathbf{a}'\mathbf{b}'} \cong 27^\circ$$

Notice further that we can turn this procedure around. If we know the *angle* that two vectors make with each other and their lengths, another way of finding the squared distance between them makes use of a rearrangement of the above formula to

$$\|\mathbf{a}' - \mathbf{b}'\|^2 = \|\mathbf{a}'\|^2 + \|\mathbf{b}'\|^2 - 2 \cos \theta_{\mathbf{a}'\mathbf{b}'} \|\mathbf{a}'\| \cdot \|\mathbf{b}'\|$$

The concepts illustrated here for two dimensions also hold true in higher dimensions since two noncollinear vectors will entail a (plane) triangle embedded in higher dimensionality.[6] The vector lengths, of course, will be based on projections on all axes of the higher-dimensional space.

A few other observations are of interest. First, if the angle θ between two vectors is 90°, then $\cos \theta = 0$, and one has the familiar Pythagorean theorem for a right triangle in which the square of the hypotenuse is equal to the sum of the squares of the sides. In the case where $\cos \theta = 0$, the two vectors are said to be *orthogonal* (as mentioned earlier). If $\cos \theta_{\mathbf{a}'\mathbf{b}'} = 1$, then $\mathbf{a}'$ and $\mathbf{b}'$ are collinear in the same direction, and the sum of the

[6] By noncollinear is meant that the vectors are not superimposed so that all points of one vector fall on the other vector.

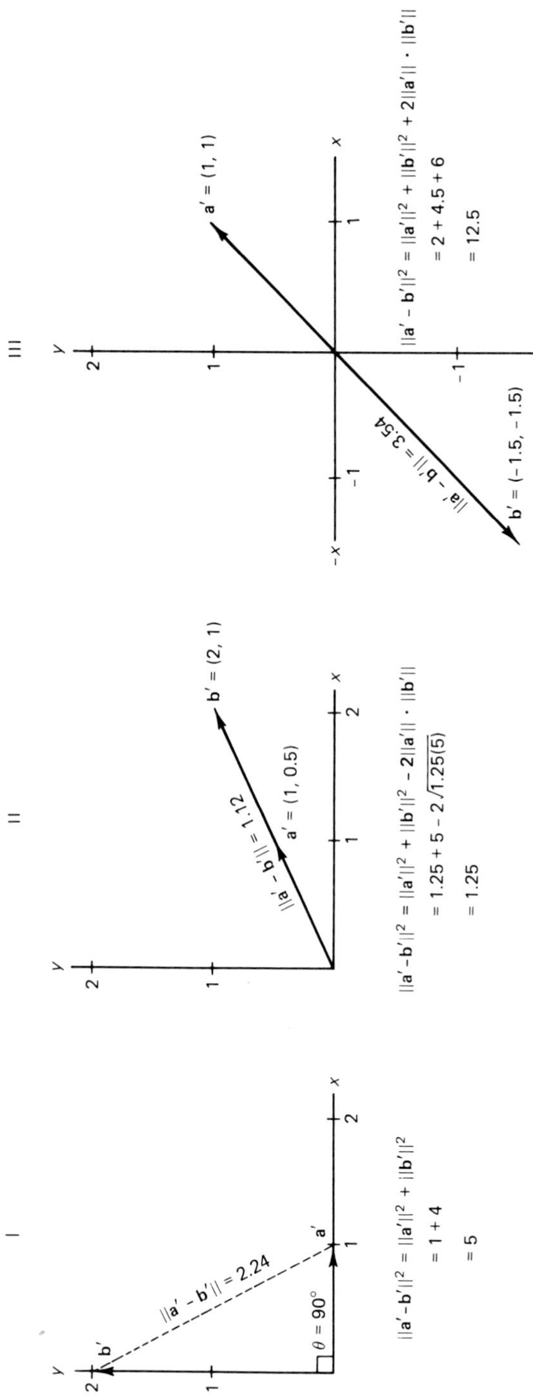

Fig. 3.14 Some special cases of the squared distance between two vectors.

squared lengths of $\mathbf{a}'$ and $\mathbf{b}'$ is appropriately reduced by $2\|\mathbf{a}'\| \cdot \|\mathbf{b}'\|$. If $\cos \theta_{\mathbf{a}'\mathbf{b}'} = -1$, then $\mathbf{a}'$ and $\mathbf{b}'$ are oppositely directed, and the sum of the squared lengths of $\mathbf{a}'$ and $\mathbf{b}'$ is increased by $2\|\mathbf{a}'\| \cdot \|\mathbf{b}'\|$.

These latter two relationships are easily seen by recalling that for scalars we have the identities

$$\boxed{(x-y)^2 \equiv x^2 + y^2 - 2xy}$$

and

$$\boxed{[x-(-y)]^2 = (x+y)^2 \equiv x^2 + y^2 + 2xy}$$

Figure 3.14 shows geometrical examples of all three of the preceding cases.

Later on, when we discuss some of the more common measures of statistical association, we shall find that the above relationships are useful in portraying various statistical measures from a geometric standpoint. At this point, however, we proceed to a geometric description of still another concept of vector algebra, namely, the *scalar product* of two vectors and its relationship to Euclidean distance.

3.3.4 The Scalar Product of Two Vectors

In Chapter 2 we defined the scalar (or inner or dot) product of two vectors $\mathbf{a}$ and $\mathbf{b}$ (of conformable order) as

$$\boxed{\mathbf{a}'\mathbf{b}}$$

in which, if $\mathbf{a}' = (a_1, a_2, \ldots, a_k, \ldots, a_n)$ and $\mathbf{b}' = (b_1, b_2, \ldots, b_k, \ldots, b_n)$, then

$$\boxed{\mathbf{a}'\mathbf{b} = \sum_{k=1}^{n} a_k b_k}$$

and the result was a single number, or scalar.

A geometrically motivated (and more general) definition of scalar product, which takes into consideration the *angle* θ made between the two vectors and their respective lengths, can now be presented. This definition of scalar product is given by the expression

$$\boxed{\mathbf{a}'\mathbf{b} = \|\mathbf{a}\| \cdot \|\mathbf{b}\| \cos \theta_{\mathbf{ab}}}$$

In the above example in which $\mathbf{a}' = (1, 2, 2)$ and $\mathbf{b}' = (0, 1, 2)$, we have

$$\mathbf{a}'\mathbf{b} = 3(\sqrt{5})(0.894) = 6$$

We also recall that $\mathbf{a}'\mathbf{b} = \mathbf{b}'\mathbf{a}$. The geometric counterpart of this is

$$\boxed{\mathbf{b}'\mathbf{a} = \|\mathbf{b}\| \cdot \|\mathbf{a}\| \cos \theta_{\mathbf{ab}}}$$

Moreover, the counterpart to $\mathbf{a}'\mathbf{a}$, the scalar product of a vector with itself, is simply

$$\|\mathbf{a}\| \cdot \|\mathbf{a}\| \cos \theta_{\mathbf{aa}} = \|\mathbf{a}\|^2 \cdot 1$$

This is the vector's squared length, inasmuch as the angle θ that a vector makes with itself is, of course, zero; hence $\cos \theta_{\mathbf{aa}} = 1$.

3.3.5 Vector Projections and Scalar Products

Still another way of looking at the scalar product of two vectors is in terms of the signed length of a projection of one vector along another. At the beginning of this chapter, we talked informally about the projection of a vector onto the coordinate axes x, y, and z. Its projection on some axis, say x, was referred to as the signed distance from the origin, along x, to the foot of a perpendicular dropped from the vector onto x. Similar interpretations pertained to the vector's projections on axes y and z.

However, suppose we have two arbitrary position vectors in the space. Clearly, we could consider the projection of one vector onto the other, in a fashion analogous to coordinate projections. This concept is most simply described in two dimensions. Accordingly, let us select two arbitrary vectors

$$\mathbf{a}' = (1,2) \qquad \text{and} \qquad \mathbf{b}' = (0,2)$$

These vectors are shown in Panel I of Fig. 3.15. We now project $\mathbf{b}'$ onto $\mathbf{a}'$ by dropping a perpendicular from $\mathbf{b}'$'s terminus to $\mathbf{a}'$. *The number*

$$\|\mathbf{b}_p'\| = \left| \|\mathbf{b}'\| \cos \theta_{\mathbf{a}'\mathbf{b}'} \right| = \frac{\mathbf{a}'\mathbf{b}}{\|\mathbf{a}'\|}$$

is defined as the length[7] *of the projection of the vector* $\mathbf{b}'$ *along the vector* $\mathbf{a}'$. The length of the projection is also frequently called the *component* of $\mathbf{b}'$ along $\mathbf{a}'$.

This concept is most easily understood by first recalling that the cosine can be viewed in terms of the length of the projection of a *unit length* vector, in this case one in the direction of $\mathbf{b}'$, onto the adjacent side (vector $\mathbf{a}'$) of a right triangle. Here, the unit length is multiplied by $\|\mathbf{b}'\|$. In this example $\|\mathbf{b}'\| = [(0)^2 + (2)^2]^{1/2} = 2$.

The cosine $\theta_{\mathbf{a}'\mathbf{b}'}$ is next found from the cosine law:

$$\cos \theta_{\mathbf{a}'\mathbf{b}'} = \frac{\|\mathbf{a}'\|^2 + \|\mathbf{b}'\|^2 - \|\mathbf{a}' - \mathbf{b}'\|^2}{2\|\mathbf{a}'\| \cdot \|\mathbf{b}'\|} = \frac{5 + 4 - 1}{2(\sqrt{5})(\sqrt{4})} = 0.89$$

$$\theta_{\mathbf{a}'\mathbf{b}'} \cong 27^\circ$$

[7] Note that in defining $\|\mathbf{b}_p'\|$ we use the *absolute value* of the expression $\|\mathbf{b}'\| \cos \theta_{\mathbf{a}'\mathbf{b}'}$ since lengths are taken to be nonnegative. However, the *signed* distance is in the direction of $\mathbf{a}'$ if $\cos_{\mathbf{a}'\mathbf{b}'}$ is positive (i.e., the angle $\theta_{\mathbf{a}'\mathbf{b}'}$ is acute) and in the direction of $-\mathbf{a}'$ if $\cos \theta_{\mathbf{a}'\mathbf{b}'}$ is negative (i.e., the angle $\theta_{\mathbf{a}'\mathbf{b}'}$ is obtuse).

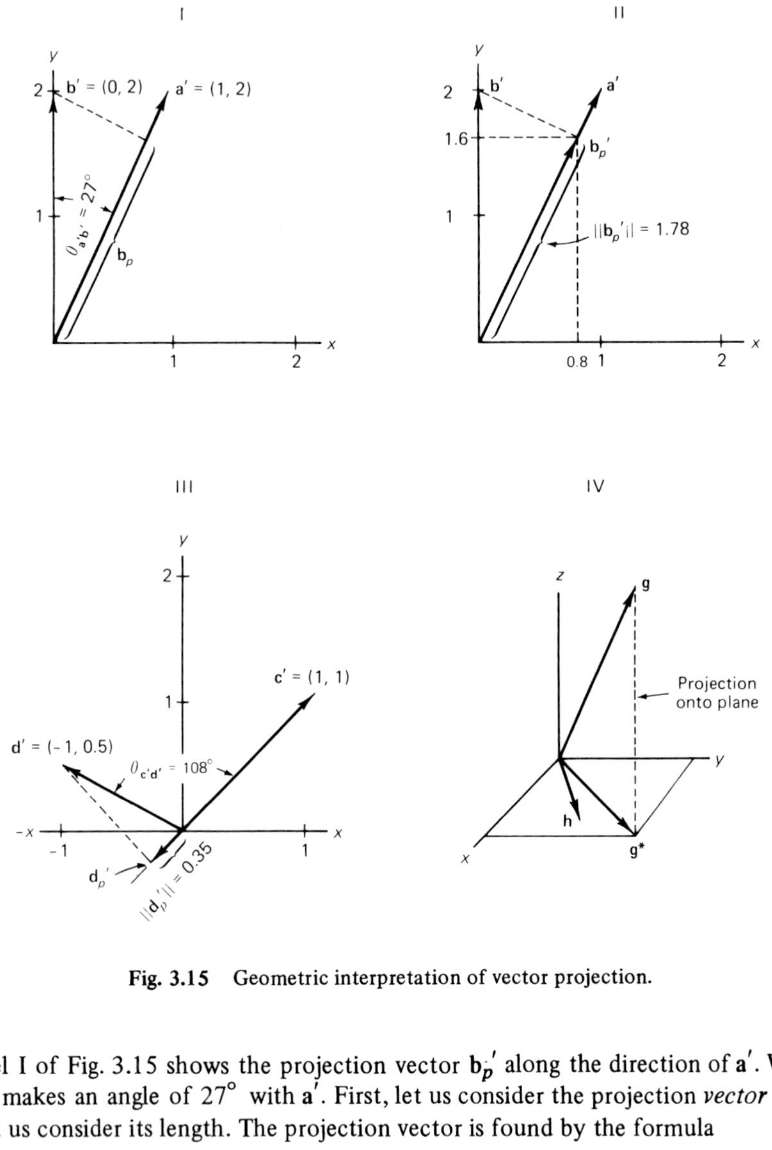

Fig. 3.15 Geometric interpretation of vector projection.

Panel I of Fig. 3.15 shows the projection vector $\mathbf{b}_p'$ along the direction of $\mathbf{a}'$. We note that $\mathbf{b}'$ makes an angle of 27° with $\mathbf{a}'$. First, let us consider the projection *vector* $\mathbf{b}_p'$, and then let us consider its length. The projection vector is found by the formula

$$\boxed{\mathbf{b}_p' = \left[\frac{||\mathbf{b}'||\cos\theta_{\mathbf{a}'\mathbf{b}'}}{||\mathbf{a}'||}\right]\mathbf{a}'}$$

In terms of the problem, we have

$$\mathbf{b}_p' = \left[\frac{2(0.89)}{\sqrt{5}}\right](1, 2) = (0.8, 1.6)$$

Panel II of Fig. 3.15 shows the coordinates of $\mathbf{b}_p' = (0.8, 1.6)$. Since the angle $\theta_{\mathbf{a}'\mathbf{b}'} = 27°$ is acute, $\mathbf{b}_p'$ is in the *same* direction as $\mathbf{a}'$.

The length of $\mathbf{b}_p{}'$ is given by

$$\|\mathbf{b}_p{}'\| = \left|\|\mathbf{b}'\| \cos\theta_{\mathbf{a}'\mathbf{b}'}\right| = 2(0.89) = 1.78$$

and this also appears in Panel II of Fig. 3.15.

Should we desire the length of the projection of a along b, this is obtained analogously as

$$\|\mathbf{a}_p{}'\| = \left|\|\mathbf{a}'\| \cos\theta_{\mathbf{a}'\mathbf{b}'}\right| = \sqrt{5}(0.89) = 2$$

Notice that if $\mathbf{a}'$ and $\mathbf{b}'$ are each of unit length, the length of the projection of $\mathbf{b}'$ along $\mathbf{a}'$ (or $\mathbf{a}'$ along $\mathbf{b}'$) is simply $|\cos\theta_{\mathbf{a}'\mathbf{b}'}|$.

If the two vectors should make an obtuse angle with each other, the procedure remains the same, but the direction of the projection vector is opposite to that of the reference vector. Panel III of Fig. 3.15 shows a case in which $\mathbf{c}' = (1, 1)$ and $\mathbf{d}' = (-1, 0.5)$ make an angle of 108° with each other. The cosine of this angle is -0.316, and we have

$$\|\mathbf{d}_p{}'\| = \left|\|\mathbf{d}'\| \cos\theta_{\mathbf{c}'\mathbf{d}'}\right| = \left|1.12(-0.316)\right| = 0.35$$

as shown in Panel III. However, since $\cos\theta_{\mathbf{c}'\mathbf{d}'}$ is negative, the direction of $\mathbf{d}_p'$ is opposite to that of $\mathbf{c}'$.

The idea of (orthogonal) projection can, of course, be extended to the projection of a vector in three dimensions into a subspace, such as the xy plane in Panel IV of Fig. 3.15. For example, the vector $\mathbf{g}$ can be projected into the xy plane by dropping a perpendicular from the terminus of $\mathbf{g}$ to the xy plane. The distance between the foot of the projection (represented by the terminus of $\mathbf{g}^*$) and $\mathbf{g}$ must be the minimum distance between $\mathbf{g}$ and the xy plane. Any other vector in the xy plane, such as $\mathbf{h}$, must have a terminus that is farther away from $\mathbf{g}$ since the hypotenuse of a right triangle must be longer than either side. Subspace projections are discussed later (in Section 4.6.4).

All of this discussion can be straightforwardly related to the geometric aspects of a scalar product. The scalar product $\mathbf{a}'\mathbf{b}$ was earlier defined in general terms as

$$\boxed{\mathbf{a}'\mathbf{b} = \cos\theta_{\mathbf{ab}} \|\mathbf{a}\| \cdot \|\mathbf{b}\|}$$

which can now be expressed in absolute-value terms as the product of two scalars:

$$|\mathbf{a}'\mathbf{b}| = \|\mathbf{b}_p{}'\| \cdot \|\mathbf{a}'\| = 1.78(\sqrt{5}) = 4$$

Furthermore, the preceding definition of projection length is fully consistent with an informal description presented at the beginning of the chapter. For example, if we have the vector $\mathbf{a}' = (1, 2)$, its projection lengths onto the standard basis vectors $\mathbf{e}_1{}'$ and $\mathbf{e}_2{}'$ are found as follows:

$$\mathbf{a}_{p1}' = \left[\frac{\mathbf{a}'\mathbf{e}_1}{\|\mathbf{e}_1'\| \cdot \|\mathbf{e}_1'\|}\right]\mathbf{e}_1{}' = \left[\frac{\|\mathbf{a}\| \cos\theta_{\mathbf{a}'\mathbf{e}_1'}}{\|\mathbf{e}_1'\|}\right]\mathbf{e}_1{}' = \left[\frac{(1,2)'(1,0)}{1}\right](1,0) = (1,0)$$

It follows that

$$||\mathbf{a}_{p1}'|| = 1$$

$$\mathbf{a}_{p2}' = \left[\frac{\mathbf{a}'\mathbf{e}_2}{||\mathbf{e}_2'|| \cdot ||\mathbf{e}_2'||}\right]\mathbf{e}_2' = \left[\frac{(1,2)'(0,1)}{1}\right](0,1) = (0,2)$$

and

$$||\mathbf{a}_{p2}'|| = 2$$

Incidentally, we shall always take $\theta_{\mathbf{a}'\mathbf{b}'}$ to be the smaller angle between $\mathbf{a}'$ and $\mathbf{b}'$. If the vectors are oppositely directed, the direction of the projection will be the negative of the reference vector's direction since $\cos\theta_{\mathbf{a}'\mathbf{b}'}$ will be negative, as illustrated in Panel III of Fig. 3.15.

3.3.6 Recapitulation

At this point we have provided geometric interpretations of all the various algebraic operations on vectors that were illustrated in Chapter 2. In particular, the addition of two vectors followed a parallelogram rule, as illustrated in Fig. 3.9. Subtraction of two vectors also involved a parallelogram rule, in which the difference vector was displaced so as to start at the origin; this is shown in Fig. 3.10.

Multiplication of a vector by a scalar k involves stretching the vector if $k > 1$ and compressing it if $0 < k < 1$. These cases are illustrated in Fig. 3.11. If $k = 1$, the vector remains unchanged. If $k = 0$ the vector becomes $\mathbf{0}$, the zero vector. If k is negative, the vector is stretched and oppositely directed if $|k| > 1$ and compressed and oppositely directed if $0 < |k| < 1$, as shown in Fig. 3.11.

The operations of sum and difference between two (or more) vectors and multiplication of a vector by a scalar are summarized in terms of the concept of linear combination, as illustrated in Fig. 3.12.

The definition of a Euclidean space enabled us to consider the distance and angle between two vectors. By means of the cosine law, illustrated in Fig. 3.13, the cosine of the angle formed by two vectors and their lengths were related to the (squared) Euclidean distance between them. This concept, in turn, led to the geometric portrayal of the projection of one vector onto another, as illustrated in Fig. 3.15. From here it was a short step toward portraying the scalar product of two vectors as a *signed* distance involving the product of the component (projection length) of one vector along some reference vector and the reference vector's length. In short, all of the algebraic operations of Chapter 2 involving vectors were given geometric interpretations here.

We can summarize the various formulas involving aspects of the scalar product as follows:

1. $||\mathbf{a}-\mathbf{b}||^2 = ||\mathbf{a}||^2 + ||\mathbf{b}||^2 - 2||\mathbf{a}|| \cdot ||\mathbf{b}|| \cos\theta_{\mathbf{ab}} = ||\mathbf{a}||^2 + ||\mathbf{b}||^2 - 2[\mathbf{a}'\mathbf{b}]$

2. $\cos\theta_{\mathbf{ab}} = \dfrac{||\mathbf{a}||^2 + ||\mathbf{b}||^2 - ||\mathbf{a}-\mathbf{b}||^2}{2||\mathbf{a}|| \cdot ||\mathbf{b}||} = \dfrac{\mathbf{a}'\mathbf{b}}{||\mathbf{a}|| \cdot ||\mathbf{b}||}$

3. $$\|\mathbf{b}_p\| = \left|\frac{\mathbf{a}'\mathbf{b}}{\|\mathbf{a}\|}\right| = \left|\|\mathbf{b}\| \cos\theta_{\mathbf{ab}}\right|$$

$$\|\mathbf{a}_p\| = \left|\frac{\mathbf{a}'\mathbf{b}}{\|\mathbf{b}\|}\right| = \left|\|\mathbf{a}\| \cos\theta_{\mathbf{ab}}\right|$$

It is worth noting that the scalar product plays a central role in all of these formulas.

3.4 LINEAR DEPENDENCE OF VECTORS

In the beginning of the chapter we chose a set of reference vectors $\mathbf{e}_i'$, called standard basis vectors, that in three dimensions were defined as follows:

$$\mathbf{e}_1' = (1,0,0); \qquad \mathbf{e}_2' = (0,1,0); \qquad \mathbf{e}_3' = (0,0,1)$$

As we shall see in a moment this set of vectors is linearly independent. The concept of linear independence plays a major role in vector algebra and multivariate analysis. As we know from elementary geometry, a line is one-dimensional, an area is two-dimensional, and a volume is three-dimensional. By analogy, a space of n dimensions entails "hypervolume."

Loosely speaking, linear independence of vectors has to do with the minimum number of vectors in terms of which any given vector in the space can be expressed and, in effect, is related to the "volume" of the space spanned by the vectors. Linearly dependent vectors display a kind of redundancy or superfluity in the sense that at least one vector of a linearly dependent set can be written as a linear combination of the other vectors.

Somewhat more formally, if $\mathbf{a}_1', \mathbf{a}_2', \ldots, \mathbf{a}_p'$ denote a set of p vectors and $k_1, k_2, \ldots, k_p$ denote a set of p scalars, it may be the case that the following linear equation is satisfied:

$$\boxed{k_1\mathbf{a}_1' + k_2\mathbf{a}_2' + \cdots + k_p\mathbf{a}_p' = \mathbf{0}'}$$

where $\mathbf{0}'$ is the zero vector.

For example, if $k_1 = k_2 = \ldots = k_p = 0$, *any* set of p vectors trivially satisfies the above equation. If, however, the equation can be satisfied *without* all k_i being equal to zero, the solution is called "nontrivial."

If a nontrivial solution can be found, then we say that the set of vectors is linearly dependent. If only the trivial solution is satisfied, the set of vectors is said to be linearly independent.

To illustrate the case of nontrivial satisfaction of the above equation, let us assume three four-component vectors:

$$\mathbf{a}_1' = (1, 2, 0, 4); \qquad \mathbf{a}_2' = (-1, 0, 5, 1); \qquad \mathbf{a}_3' = (1, 6, 10, 14)$$

and let

$$k_1 = 3; \qquad k_2 = 2; \qquad k_3 = -1$$

Since

$$\begin{aligned} &3(1, 2, 0, 4) + 2(-1, 0, 5, 1) - 1(1, 6, 10, 14) \\ &\qquad = (3, 6, 0, 12) + (-2, 0, 10, 2) + (-1, -6, -10, -14) = (0, 0, 0, 0) \end{aligned}$$

it is seen that $\mathbf{a}_1'$, $\mathbf{a}_2'$, and $\mathbf{a}_3'$ are linearly dependent and at least one of the vectors is a linear combination of the remaining $p - 1$ vectors. To see that this is so, we note that

$$\begin{aligned} \mathbf{a}_1' &= -\frac{k_2}{k_1}(\mathbf{a}_2') - \frac{k_3}{k_1}(\mathbf{a}_3') \\ &= -\tfrac{2}{3}(-1, 0, 5, 1) + \tfrac{1}{3}(1, 6, 10, 14) \\ &= (\tfrac{2}{3}, 0, -\tfrac{10}{3}, -\tfrac{2}{3}) + (\tfrac{1}{3}, 2, \tfrac{10}{3}, \tfrac{14}{3}) \\ \mathbf{a}_1' &= (1, 2, 0, 4) \end{aligned}$$

and $\mathbf{a}_1'$ is, indeed, a linear combination of $\mathbf{a}_2'$ and $\mathbf{a}_3'$. It is also pertinent to note that any set of p vectors is *always* linearly dependent if $p > n$, where n is the number of vector components in an n by 1 column vector or a 1 by n row vector, as the case may be.

While no proof of this assertion is given, the statement relates to the fact that if one wished to solve n equations for n unknowns, one could take the first n vectors, assuming they are linearly independent, and solve for any of the other vectors as linear combinations of these n linearly independent vectors.

The concept of linear independence is of particular importance to multivariate analysis. A set of linearly independent vectors is said to *span* some Euclidean space of interest. Ultimately the idea of linear independence relates to the *dimensionality* of the space in which the researcher is working. And, as we shall see, once a set of such vectors is found, all other vectors can be expressed as linear combinations of these.

In brief, then, two ideas are involved in the study of linear independence. First, we wish to find a set of *nonredundant* vectors. Second, we wish to make sure that we have *enough* linearly independent vectors to span some space of interest or, as indicated earlier, to contain some hypervolume of interest.

3.4.1 Dimensionality of a Vector Space and the Concept of Basis

In line with our earlier discussions involving geometric analogy, we can now examine the dimensionality of a vector space. *The dimensionality of a vector space is equal to the maximum number of linearly independent vectors in that space.* To illustrate for the case of three dimensions, we return to the $\mathbf{e}_i'$ standard coordinate vectors:

$$\mathbf{e}_1' = (1, 0, 0); \qquad \mathbf{e}_2' = (0, 1, 0); \qquad \mathbf{e}_3' = (0, 0, 1)$$

If we set up the equation

$$k_1\mathbf{e}_1' + k_2\mathbf{e}_2' + k_3\mathbf{e}_3' = \mathbf{0}$$

we find that the above equation is satisfied only if $k_1 = k_2 = k_3 = 0$. Hence, $\mathbf{e}_1'$, $\mathbf{e}_2'$, and $\mathbf{e}_3'$ are linearly independent, and the dimensionality of the space is three. *In general, if* $\mathbf{a}_1'$, $\mathbf{a}_2'$, . . . , $\mathbf{a}_n'$ *denote a set of n linearly independent n-component vectors, then any other vector of that n-space can be written as*

$$\boxed{\mathbf{b}' = k_1\mathbf{a}_1' + k_2\mathbf{a}_2' + \cdots + k_n\mathbf{a}_n'}$$

and the $\mathbf{a}_1'$, $\mathbf{a}_2'$, . . . , $\mathbf{a}_n'$ *vectors are said to constitute a basis of the n-space.* In a space of n dimensions, *any* set of n linearly independent vectors can constitute a basis of the space. Thus the basis vectors $\mathbf{e}_i'$ above represent only one type of basis, one that we have called the standard basis.

The $\mathbf{e}_i'$ standard basis vectors, however, are particularly convenient. *Indeed, unless stated otherwise we shall assume that the particular basis being chosen is the standard basis.* Still, we should indicate that any other set of n linearly independent vectors could qualify as the reference set. Accordingly, we spend some time on the process by which one can change one set of basis vectors to some other set, for example, to a set of standard basis vectors.

3.4.2 Change of Basis Vectors

Up to this point we have emphasized rectangular Cartesian coordinates, where it is natural to view the coordinate vectors $\mathbf{e}_i'$ as both (a) mutually orthogonal (i.e., exhibiting pairwise scalar products of zero) and (b) of unit length. *This type of basis is called orthonormal.* In this intuitively simple case, the vector $\mathbf{a}' = (a_1, a_2, \ldots, a_n)$ can be easily written as

$$\boxed{\mathbf{a}' = a_1\mathbf{e}_1' + a_2\mathbf{e}_2' + \cdots + a_n\mathbf{e}_n'}$$

where $\mathbf{e}_1' = (1, 0, \ldots, 0)$, $\mathbf{e}_2' = (0, 1, 0, \ldots, 0)$, and $\mathbf{e}_n' = (0, 0, \ldots, 1)$. Hence $(a_1, a_2, \ldots, a_n)$ are the coordinates of $\mathbf{a}'$ relative to the orthonormal basis $\mathbf{e}_1'$, $\mathbf{e}_2'$, . . . , $\mathbf{e}_n'$.

An equally satisfactory way of showing this concept is to represent the standard coordinate vectors in columnar form. A given vector $\mathbf{a}$ can then be written as

$$\mathbf{a} = a_1\begin{bmatrix}1\\0\\0\\\vdots\\0\end{bmatrix} + a_2\begin{bmatrix}0\\1\\0\\\vdots\\0\end{bmatrix} + \cdots + a_n\begin{bmatrix}0\\0\\0\\\vdots\\1\end{bmatrix} = \begin{bmatrix}a_1\\a_2\\a_3\\\vdots\\a_n\end{bmatrix}$$

and we have an illustration of a linear combination of standard basis vectors in which the components of $\mathbf{a}$ (i.e., a_1, a_2, etc.) are the scalars of interest.

Orthonormal bases are easy to work with, and we shall usually assume that this type of basis, more specifically, the standard basis vectors $\mathbf{e}_i$, underlies the coordinate representation of interest. However, as indicated above, any set of linearly independent vectors, unit length or not, orthogonal or not, can be used to define a basis. Hence, it is pertinent to point out how one can move from one basis of a space to some other basis of that space.

Accordingly, let us now illustrate the idea of *general* coordinate systems whose basis vectors need not be mutually orthogonal or of unit length. Suppose we start with two sets of basis vectors—first, the more familiar $\mathbf{e}_i'$ standard basis vectors, $\mathbf{e}_1' = (1, 0)$ and $\mathbf{e}_2' = (0, 1)$, and second, another set of basis vectors $\mathbf{f}_1' = c_1\mathbf{e}_1' + c_2\mathbf{e}_2'$ and $\mathbf{f}_2' = d_1\mathbf{e}_1' + d_2\mathbf{e}_2'$.

To be specific, we let $c_1 = 0.707$, $c_2 = 0.707$, $d_1 = 0.940$, and $d_2 = 0.342$. Then

$$\mathbf{f}_1' = 0.707\mathbf{e}_1' + 0.707\mathbf{e}_2' = (0.707, 0.707)$$

$$\mathbf{f}_2' = 0.940\mathbf{e}_1' + 0.342\mathbf{e}_2' = (0.940, 0.342)$$

Note that $\mathbf{f}_1'$ and $\mathbf{f}_2'$ are each of unit length but are not orthogonal; that is, $\mathbf{f}_1'\mathbf{f}_2 \neq 0$. Note further that we can write the preceding equations in columnar form as

$$\mathbf{f}_1 = \begin{bmatrix} f_{11} \\ f_{21} \end{bmatrix} = 0.707 \overset{\mathbf{e}_1}{\begin{bmatrix} 1 \\ 0 \end{bmatrix}} + 0.707 \overset{\mathbf{e}_2}{\begin{bmatrix} 0 \\ 1 \end{bmatrix}} = \begin{bmatrix} 0.707 \\ 0.707 \end{bmatrix}$$

$$\mathbf{f}_2 = \begin{bmatrix} f_{12} \\ f_{22} \end{bmatrix} = 0.940 \overset{\mathbf{e}_1}{\begin{bmatrix} 1 \\ 0 \end{bmatrix}} + 0.342 \overset{\mathbf{e}_2}{\begin{bmatrix} 0 \\ 1 \end{bmatrix}} = \begin{bmatrix} 0.940 \\ 0.342 \end{bmatrix}$$

Figure 3.16 shows a plot of $\mathbf{f}_1$ and $\mathbf{f}_2$ relative to the standard basis $\mathbf{e}_1$ and $\mathbf{e}_2$. By finding their projections on $\mathbf{e}_1$ and $\mathbf{e}_2$, we can note that their coordinates are given by the preceding equations.

Now let us select a new vector $\mathbf{a} = a_1\mathbf{f}_1 + a_2\mathbf{f}_2$. That is, we shall assume that $\mathbf{a}$ is referred to the new (and nonorthogonal) basis, $\mathbf{f}_1$ and $\mathbf{f}_2$. To be specific, we assume that the coordinates of $\mathbf{a}$ relative to $\mathbf{f}_1$ and $\mathbf{f}_2$ are

$$a_1 = 0.5; \qquad a_2 = 0.5$$

We can find these coordinates by extending lines parallel to $\mathbf{f}_1$ and $\mathbf{f}_2$ and noting the coordinates of OQ and OR, respectively, on $\mathbf{f}_1$ and $\mathbf{f}_2$. The basis vectors $\mathbf{f}_1$ and $\mathbf{f}_2$ are often called *oblique* Cartesian axes since the angle that they make with each other is *not* equal to 90°. Notice, however, that $\mathbf{a}$ is still given by the (parallelogram) law for vector addition:

$$\mathbf{a} = a_1\mathbf{f}_1 + a_2\mathbf{f}_2$$

and that $a_1\mathbf{f}_1$ and $a_2\mathbf{f}_2$ are scalar multiples of $\mathbf{f}_1$ and $\mathbf{f}_2$, respectively. Since we have chosen $\mathbf{f}_1$ and $\mathbf{f}_2$ to be of unit length, the coordinates a_1 and a_2 are merely the lengths OQ and OR. Had $\mathbf{f}_1$ and $\mathbf{f}_2$ not been of unit length, a_1 and a_2 would still be regarded as

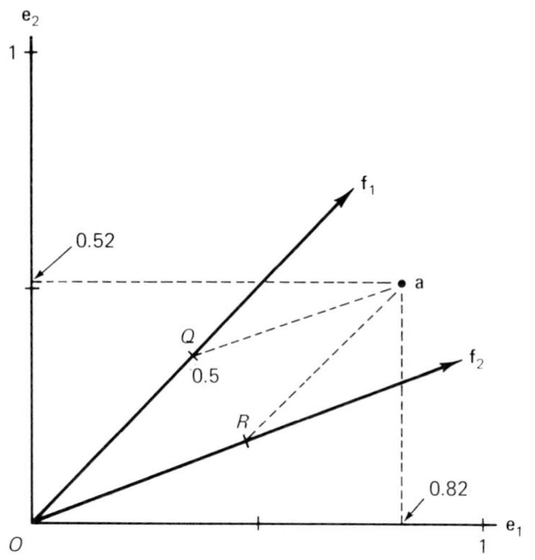

Fig. 3.16 An illustration of generalized coordinates and change of basis.

coordinates, but they would correspond to lengths of $OQ/\|\mathbf{f}_1\|$ and $OR/\|\mathbf{f}_2\|$, respectively.

We now seek a set of coordinates for the vector $\mathbf{a}$, referred to the oblique basis $\mathbf{f}_1$ and $\mathbf{f}_2$, in terms of the *original* (and standard) basis $\mathbf{e}_1$ and $\mathbf{e}_2$. This can be done by the following substitution:

$$\mathbf{a} = a_1\mathbf{f}_1 + a_2\mathbf{f}_2$$

But, since $\mathbf{f}_1$ and $\mathbf{f}_2$ have been defined in terms of $\mathbf{e}_1$ and $\mathbf{e}_2$, we can write

$$\mathbf{a} = a_1(c_1\mathbf{e}_1 + c_2\mathbf{e}_2) + a_2(d_1\mathbf{e}_1 + d_2\mathbf{e}_2)$$
$$= (c_1a_1 + d_1a_2)\mathbf{e}_1 + (c_2a_1 + d_2a_2)\mathbf{e}_2$$

However, since $\mathbf{e}_1$ and $\mathbf{e}_2$ denote a basis, we can also represent $\mathbf{a}$ in terms of $\mathbf{e}_1$ and $\mathbf{e}_2$ as

$$\mathbf{a} = a_1{}^*\mathbf{e}_1 + a_2{}^*\mathbf{e}_2$$

Hence, through substitution of $c_1a_1 + d_1a_2$ for $a_1{}^*$, and $c_2a_2 + d_2a_2$ for $a_2{}^*$, we find

$$a_1{}^* = c_1a_1 + d_1a_2 = 0.707(0.5) + 0.940(0.5) = 0.82$$
$$a_2{}^* = c_2a_1 + d_2a_2 = 0.707(0.5) + 0.342(0.5) = 0.52$$

As can be observed from Fig. 3.16, the length of the projection of $\mathbf{a}$ on $\mathbf{e}_1$ is, indeed, 0.82, and its projection length on $\mathbf{e}_2$ is 0.52.

Thus, one can work "backward" to relate a vector described in terms of one set of basis vectors to a description of that same vector in terms of another set of basis vectors, assuming we know how the basis vectors themselves are connected. And, as a matter of fact, one can *always* find an orthonormal set of axes (mutually orthogonal and of unit length) by which a set of arbitrary basis vectors can be represented, even though the original axes might be oblique and not of unit length. The next section illustrates one procedure for finding an orthonormal basis from an initial set of nonorthonormal basis vectors.

3.4.3 Finding an Orthonormal Basis

As indicated earlier, a special kind of basis in a vector space—one of particular value in multivariate analysis—is an orthonormal basis. This basis is characterized by the facts that (a) the scalar product of any pair of basis vectors is zero and (b) each basis vector is of unit length. As we know, the standard basis vectors $\mathbf{e}_i$ represent one such orthonormal basis.

In multivariate data analysis, it is usually the case that multiple measurements on a set of objects will be associated; for example, weight will be correlated with height. Sometimes we may want to transform the original (and correlated) variables to a set of uncorrelated variables. As will be shown later, this process can be viewed as transforming a set of n nonorthogonal vectors into a set of n orthogonal vectors. In the process we may also want to make all of these vectors unit length; this is the "norming" aspect of the process.

We have already observed that the scalar product is a central concept in vector algebra and is a function that assigns a real number to each pair of vectors in the Euclidean space of interest. *In particular, the concepts of vector length, distance, and cosine can all be expressed in terms of the single idea of a scalar product:*

$$\|\mathbf{a}\| = [\mathbf{a}'\mathbf{a}]^{1/2}$$

$$\|\mathbf{a}-\mathbf{b}\| = [\|\mathbf{a}\|^2 + \|\mathbf{b}\|^2 - 2(\mathbf{a}'\mathbf{b})]^{1/2}$$

$$\cos\theta_{\mathbf{ab}} = \frac{\mathbf{a}'\mathbf{b}}{\|\mathbf{a}\| \cdot \|\mathbf{b}\|}$$

As we shall see in a moment, the scalar product also provides a simple representation of vectors that are mutually orthogonal (perpendicular):

a and **b** are orthogonal

if and only if

$\mathbf{a}'\mathbf{b} = 0$

We can now proceed to construct an orthonormal basis, one whose vectors are mutually orthogonal *and* of unit length.

Any arbitrary basis can be transformed to an orthonormal basis by a procedure known as *Gram–Schmidt orthonormalization.* To illustrate the process, consider the three arbitrary row vectors:

$$\mathbf{a}_1' = (2, 1, 2); \qquad \mathbf{a}_2' = (3, -1, 5); \qquad \mathbf{a}_3' = (0, 1, -1)$$

The Gram–Schmidt process starts out by selecting (arbitrarily) one of the vectors, say $\mathbf{a}_1'$, as the first reference vector.[8] The idea here is to keep this vector fixed and then find other vectors, two other vectors in this case, so that the resultant sets are mutually orthogonal. As a final step each of the orthogonal vectors is normalized to unit length. To start off the process we first set

$$\mathbf{b}_1' = \mathbf{a}_1'$$

and then find

$$\mathbf{b}_2' = \mathbf{a}_2' - \left[\frac{\mathbf{a}_2'\mathbf{b}_1}{\mathbf{b}_1'\mathbf{b}_1}\right]\mathbf{b}_1' = \mathbf{a}_2' - \left[\frac{(3 \times 2) + (-1 \times 1) + (5 \times 2)}{2^2 + 1^2 + 2^2}\right]\mathbf{b}_1'$$

$$= (3, -1, 5) - (15/9)(2, 1, 2) = (-1/3, -8/3, 5/3)$$

$$\mathbf{b}_2' = (-0.33, -2.67, 1.67)$$

Let us now examine the expression

$$\left[\frac{\mathbf{a}_2'\mathbf{b}_1}{\mathbf{b}_1'\mathbf{b}_1}\right]\mathbf{b}_1' = (15/9)(2, 1, 2) = (10/3, 5/3, 10/3)$$

This expression is the orthogonal projection of $\mathbf{a}_2'$ onto $\mathbf{b}_1'$ (as discussed in Section 3.3.5).

The "residual" is then equal to the difference

$$\mathbf{b}_2' = \mathbf{a}_2' - \left[\frac{\mathbf{a}_2'\mathbf{b}_1}{\mathbf{b}_1'\mathbf{b}_1}\right]\mathbf{b}_1' = (3, -1, 5) - (10/3, 5/3, 10/3)$$

$$= (-1/3, -8/3, 5/3)$$

and should be orthogonal to $\mathbf{b}_1'$, as is shown illustratively in *two* dimensions, in Fig. 3.17. That is,

$$\mathbf{b}_2'\mathbf{b}_1 = (-1/3, -8/3, 5/3)\begin{bmatrix}2\\1\\2\end{bmatrix} = 0$$

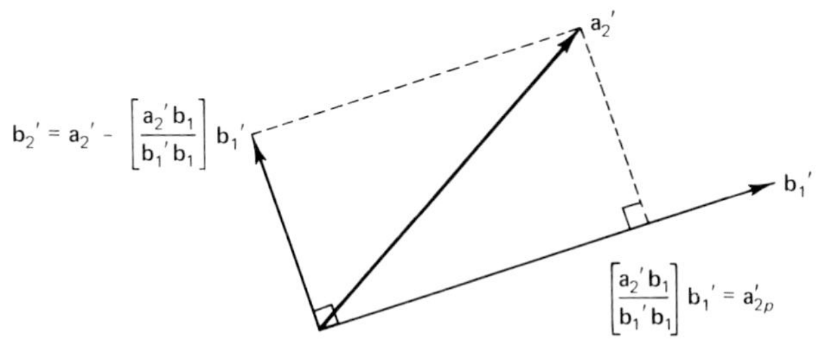

Fig. 3.17 Finding the orthogonal projection of $\mathbf{a}_2'$ onto $\mathbf{b}_1'$ (illustrated in two dimensions).

[8] It should be mentioned that the specific results of the Gram–Schmidt process depend on the order on which the vectors are selected; however, in any case the resulting set will be orthogonal and of unit length.

We encountered orthogonal projections, both in the beginning of the chapter and in Section 3.3.5. Accordingly, there is nothing new here, except for the fact that we are now interested in the *orthogonal complement* or that part of the $\mathbf{a}_2'$ vector that does *not* lie along the reference vector. In this case it is the vector

$$\mathbf{b}_2' = (-0.33, -2.67, 1.67)$$

As we noted above, the scalar product of $\mathbf{b}_2'\mathbf{b}_1$ is indeed zero. Thus, $\mathbf{b}_2'$ is now orthogonal to $\mathbf{b}_1' = \mathbf{a}_1'$. We now have to orthogonalize $\mathbf{a}_3'$ with regard to the two, already orthogonal, vectors $\mathbf{b}_2'$ and $\mathbf{b}_1'$:

$$\begin{aligned}
\mathbf{b}_3' &= \mathbf{a}_3' - \left[\frac{\mathbf{a}_3'\mathbf{b}_2}{\mathbf{b}_2'\mathbf{b}_2}\right]\mathbf{b}_2' - \left[\frac{\mathbf{a}_3'\mathbf{b}_1}{\mathbf{b}_1'\mathbf{b}_1}\right]\mathbf{b}_1' \\
&= (0, 1, -1) + \frac{13}{30}(-1/3, -8/3, 5/3) + \frac{1}{9}(2, 1, 2) \\
&= (0, 1, -1) + (-13/90, -104/90, 65/90) + (2/9, 1/9, 2/9) \\
&= (7/90, -4/90, -5/90) \\
\mathbf{b}_3' &= (0.08, -0.04, -0.06)
\end{aligned}$$

And, in general for r vectors, we would have

$$\mathbf{b}_r' = \mathbf{a}_r' - \left[\frac{\mathbf{a}_r'\mathbf{b}_{r-1}}{\mathbf{b}_{r-1}'\mathbf{b}_{r-1}}\right]\mathbf{b}_{r-1}' - \cdots - \left[\frac{\mathbf{a}_r'\mathbf{b}_1}{\mathbf{b}_1'\mathbf{b}_1}\right]\mathbf{b}_1'$$

After the $\mathbf{b}'$'s are obtained, we would find that they are mutually orthogonal. Each set is then normalized by its respective divisor $\|\mathbf{b}_i'\|$. That is, we find the length of each of the $\mathbf{b}'$'s and divide each vector component by the length of that vector. In the above example, the lengths of $\mathbf{b}_1'$, $\mathbf{b}_2'$, and $\mathbf{b}_3'$, respectively, are 3, 3.17, and 0.108. The normalized vectors then become

$$\begin{aligned}
\mathbf{b}_1^{*\prime} &= (1/3)(2, 1, 2) = (0.67, 0.33, 0.67) \\
\mathbf{b}_2^{*\prime} &= (1/3.17)(-0.33, -2.67, 1.67) = (-0.10, -0.84, 0.53) \\
\mathbf{b}_3^{*\prime} &= (1/0.108)(0.08, -0.04, -0.06) = (0.74, -0.37, -0.56)
\end{aligned}$$

Within rounding error, we first note that all three vectors have unit length. If we then find the scalar product of each pair of vectors, we observe, again within rounding error, that all three scalar products equal zero.

We conclude by saying that the vectors $\mathbf{b}_1^{*\prime}$, $\mathbf{b}_2^{*\prime}$, $\mathbf{b}_3^{*\prime}$ form a three-dimensional *orthonormal* basis—one whose axes are mutually orthogonal and of unit length.

It is rather difficult to show the Gram–Schmidt procedure for the specific vectors utilized in our example. This being the case, Fig. 3.18 shows a more stylized conceptualization of the procedure. The pictures first show orthonormalization of the first two vectors in two dimensions and then orthonormalization of all three in three dimensions. (In the figure the orthonormalized vectors are expressed as column vectors.)

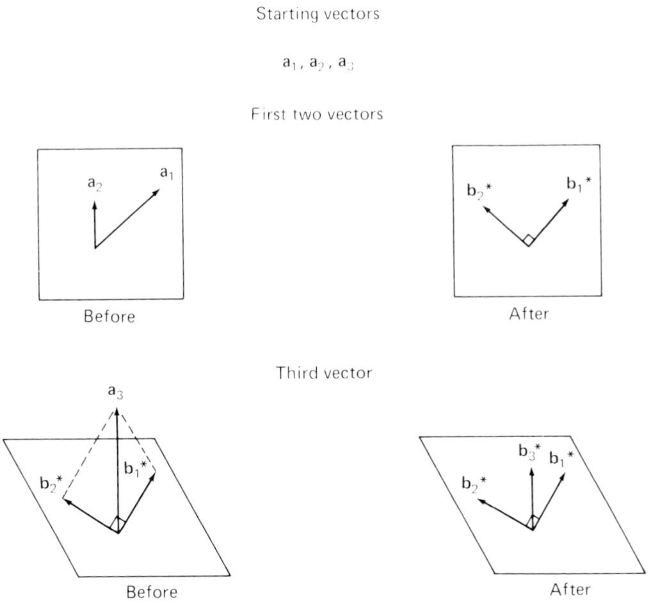

Fig. 3.18 Conceptualization of Gram–Schmidt orthonormalization procedure.

3.4.4 Scalar Products in Oblique Coordinate Systems

Now that we have discussed both oblique and orthonormal bases, it is of interest to point out that vector addition and subtraction as well as multiplication of a vector by a scalar are carried out the same way under either oblique or orthonormal basis conditions. However, this correspondence does *not* hold in the case of the scalar product.

The reason why the usual scalar product formula does not work in the nonorthogonal case is most easily seen by observing the two basis vectors **a** and **b** in the diagram of Fig. 3.19. Note that the angle they make with each other is 45° rather than 90°. If we

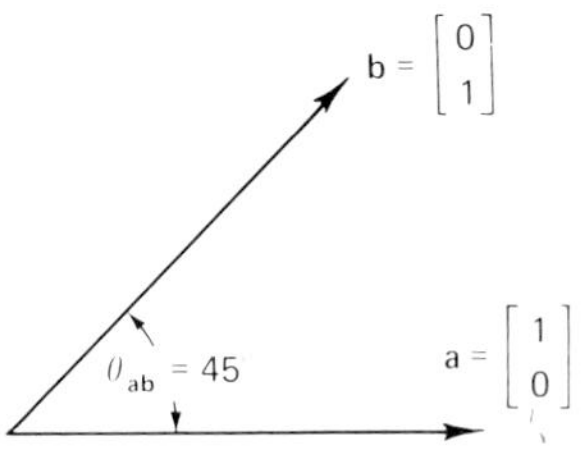

Fig. 3.19 Scalar product in oblique coordinate system.

(incorrectly) assumed that these basis vectors were orthogonal, their scalar product would be zero by application of the special case of a scalar product that entails a′b.[9]

However, using the formula described in Section 3.3.4, we find the following:

$$\mathbf{a}'\mathbf{b} = ||\mathbf{a}|| \cdot ||\mathbf{b}|| \cos \theta_{\mathbf{ab}} = 1(1)(0.707) = 0.707$$

As noted above, this form of computing the scalar product is *not* dependent on referring vectors to an orthonormal basis. Still, as can be observed from this section, orthonormal bases simplify calculations quite a lot.

In more general terms, if we have two column vectors **a** and **b** referred to oblique unit length basis vectors $\mathbf{f}_i$, this situation can be represented as

$$\mathbf{a} = a_1\mathbf{f}_1 + a_2\mathbf{f}_2 + \cdots + a_n\mathbf{f}_n$$

and

$$\mathbf{b} = b_1\mathbf{f}_1 + b_2\mathbf{f}_2 + \cdots + b_n\mathbf{f}_n$$

In this case the scalar product between **a** and **b** is given by

$$\boxed{\mathbf{a}'\mathbf{b} = \sum_{i=1}^{n} \sum_{j=1}^{n} a_i b_j \cos \theta_{ij}}$$

where θ_{ij} is the *angle* between the pair of basis vectors $\mathbf{f}_i$ and $\mathbf{f}_j$ for $i, j = 1, 2, \ldots, n$. However, if $\mathbf{f}_1, \mathbf{f}_2, \ldots, \mathbf{f}_n$ are also orthogonal, $\cos \theta_{ij} = 0$ for all pairs of basis vectors in which $j \neq i$ and, hence, we have the special case, discussed earlier, of

$$\mathbf{a}'\mathbf{b} = \sum_{i=1}^{n} a_i b_i$$

What if the oblique $\mathbf{f}_i$ and $\mathbf{f}_j$ are not of unit length? If this is the case, then the scalar product becomes

$$\boxed{\mathbf{a}'\mathbf{b} = \sum_{i=1}^{n} \sum_{j=1}^{n} a_i b_j \cos \theta_{ij} ||\mathbf{f}_i|| \cdot ||\mathbf{f}_j||}$$

Thus, if we need to account for basis vectors whose lengths are not equal to unity, the more general expression above is applicable.[10]

[9] It is important to note that the definition of scalar product as $\mathbf{a}'\mathbf{b} = \Sigma_{i=1}^{n} a_i b_i$ in Chapter 2 has implicitly assumed that both vectors are referred to standard basis vectors $\mathbf{e}_i$. In general, this will indeed be the case; however, in oblique coordinate systems the geometrically oriented definition $\mathbf{a}'\mathbf{b} = ||\mathbf{a}|| \cdot ||\mathbf{b}|| \cos \theta_{\mathbf{ab}}$ should be used.

[10] The computation of scalar products in an oblique coordinate system entails the concept of a (positive definite) quadratic form, a topic that is discussed in Chapter 5. Throughout the book, however, we shall emphasize the simpler case in which the standard basis vectors $\mathbf{e}_i$ are assumed to be applicable.

3.5 ORTHOGONAL TRANSFORMATIONS

Up to this point we have presented geometric interpretations of all of the principal algebraic operations that were performed on vectors in Chapter 2, such as addition, subtraction, the scalar product of two vectors, and so on. So far, matrices have been ignored for the most part, except in our discussion of basis vectors.

It is now appropriate to discuss some preliminary aspects of a matrix transformation of a vector or set of vectors. In this chapter we limit our discussion to a special but quite important case, namely, orthogonal transformations or rotations.

Most readers probably have an intuitive idea about what is meant by a rigid rotation of a set of points. Often in multivariate analysis we wish to perform a transformation on a set of points that will preserve their angles, lengths, and interpoint distances, while at the same time referring them to a new, perhaps simpler, coordinate system. Since rotations, as a special type of linear transformation, play such a key role in understanding more general kinds of matrix transformations, this concept is introduced here and related to earlier discussions of distance and angle.

The definition of an orthogonal matrix as used in multivariate analysis differs somewhat from researcher to researcher. We shall use the term "orthogonal" to refer to a square matrix $\mathbf{A}$ that exhibits the property

$$\boxed{\mathbf{A}'\mathbf{A} = \mathbf{A}\mathbf{A}' = \mathbf{I}}$$

That is, any two column vectors or any two row vectors in the matrix $\mathbf{A}$ are mutually orthogonal and, furthermore, each vector is of unit length. Some authors call this type of matrix "square, orthonormal," but we shall use the more common term of orthogonal matrix.

3.5.1 Axis and Point Rotations

To motivate the discussion let us consider the column vector $\mathbf{a} = \left[\begin{smallmatrix}2\\1\end{smallmatrix}\right]$ in the diagram of Fig. 3.20. We adopt a set of standard basis vectors $\mathbf{e}_i$ for the space in order to simplify

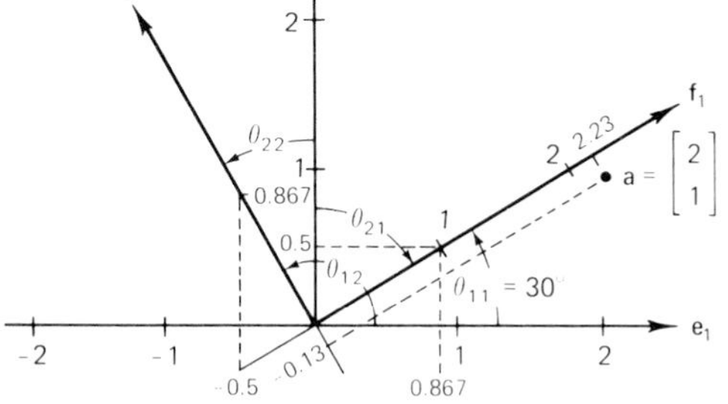

Fig. 3.20 Rotation of reference axes.

subsequent discussion. Now suppose we wish to apply an orthogonal transformation to the vector **a**. Geometrically, this can mean one of two things:

1. we can rigidly rotate the *axes*, either counterclockwise or clockwise, from their original $\mathbf{e}_i$ orientation, while leaving the point fixed, or
2. we can leave the original $\mathbf{e}_i$ axes fixed and rigidly rotate the *vector* $\mathbf{a} = \begin{bmatrix}2\\1\end{bmatrix}$ to a new location.

Let us consider the first case. Suppose we wish to rotate the original axes $\mathbf{e}_1$ and $\mathbf{e}_2$ counterclockwise through an angle of 30°, as shown in Fig. 3.20. To do this we shall need a set of direction cosines for each angle made by the (new) $\mathbf{f}_1$ and $\mathbf{f}_2$ axes with the (original) $\mathbf{e}_1$ and $\mathbf{e}_2$ axes. Let us first find the cosines that we shall need. From basic trigonometry we have

$$\cos 30° = \sqrt{3}/2 = 0.867; \qquad \cos 60° = \tfrac{1}{2} = 0.5; \qquad \cos 120° = -\tfrac{1}{2} = -0.5$$

Next, we shall use the symbol θ_{ij} to denote angles between pairs of axes, where i denotes the original axis and j denotes the new axis. If we examine the four angles $\theta_{11}, \theta_{12}, \theta_{21}, \theta_{22}$, in which the first subscript refers to the old axis and the second to the new axis in Fig. 3.20, we see that

1. θ_{11} involves a 30° counterclockwise rotation with cos 30° = 0.867.
2. θ_{12} involves a 120° counterclockwise rotation with cos 120° = –0.5.
3. θ_{21} involves a 60° clockwise rotation with cos 60° = 0.5.
4. θ_{22} involves a 30° counterclockwise rotation with cos 30° = 0.867.

The angle of 30° that $\mathbf{f}_1$ makes with $\mathbf{e}_1$ involves a cosine that is equal to 0.867. And, since $\mathbf{f}_1$ makes an angle of 60° (with a cosine of 0.5) with $\mathbf{e}_2$, we have the linear combination

$$\mathbf{f}_1 = \cos\theta_{11}\mathbf{e}_1 + \cos\theta_{21}\mathbf{e}_2 = 0.867\begin{bmatrix}1\\0\end{bmatrix} + 0.5\begin{bmatrix}0\\1\end{bmatrix} = \begin{bmatrix}0.867\\0.5\end{bmatrix}$$

as the coordinates of $\mathbf{f}_1$.

Similarly, we can compute the coordinates of $\mathbf{f}_2$ as follows:

$$\mathbf{f}_2 = \cos\theta_{12}\mathbf{e}_1 + \cos\theta_{22}\mathbf{e}_2 = -0.5\begin{bmatrix}1\\0\end{bmatrix} + 0.867\begin{bmatrix}0\\1\end{bmatrix} = \begin{bmatrix}-0.5\\0.867\end{bmatrix}$$

As can be seen from Fig. 3.20, $\mathbf{f}_1$ and $\mathbf{f}_2$ display the coordinates indicated above. We also note that the sum of squares of each set of direction cosines is unity. That is,

$$(0.867)^2 + (0.5)^2 = 1; \qquad (-0.5)^2 + (0.867)^2 = 1$$

At this point we have expressed $\mathbf{f}_1$ and $\mathbf{f}_2$ in terms of $\mathbf{e}_1$ and $\mathbf{e}_2$. We also know that the (assumed fixed) point **a** is expressed in terms of $\mathbf{e}_1$ and $\mathbf{e}_2$, the *original* basis vectors, as

$$\mathbf{a} = 2\begin{bmatrix}1\\0\end{bmatrix} + 1\begin{bmatrix}0\\1\end{bmatrix} = \begin{bmatrix}2\\1\end{bmatrix}$$

Our problem, now, is to find the coordinates of that same point—which we can call $\mathbf{a}^*$—in terms of the *new* basis vectors $\mathbf{f}_i$.

We can express this transformation in the form of a matrix postmultiplied by a vector. That is, we can let $\mathbf{a}^* = \begin{bmatrix} a_1^* \\ a_2^* \end{bmatrix}$ denote the *new* coordinates of the point $\mathbf{a}$ by the following substitution:

$$\begin{bmatrix} a_1^* \\ a_2^* \end{bmatrix} = \begin{bmatrix} \cos\theta_{11} & \cos\theta_{21} \\ \cos\theta_{12} & \cos\theta_{22} \end{bmatrix} \begin{bmatrix} a_1 \\ a_2 \end{bmatrix}$$

$$\begin{bmatrix} 2.23 \\ -0.13 \end{bmatrix} = \begin{bmatrix} 0.867 & 0.5 \\ -0.5 & 0.867 \end{bmatrix} \begin{bmatrix} 2 \\ 1 \end{bmatrix}$$

These are the coordinates of the point with respect to the new basis $\mathbf{f}_i$ in Fig. 3.20.

Let us examine the transformation somewhat more closely in Fig. 3.20. First, as noted above, we see that the unit length portion of $\mathbf{f}_1$ has coordinates of $\mathbf{f}_1 = \begin{bmatrix} 0.867 \\ 0.5 \end{bmatrix}$ with respect to $\mathbf{e}_1$ and $\mathbf{e}_2$, respectively. Similarly, the unit length portion of $\mathbf{f}_2$ has coordinates of $\mathbf{f}_2 = \begin{bmatrix} -0.5 \\ 0.867 \end{bmatrix}$ with respect to $\mathbf{e}_1$ and $\mathbf{e}_2$, respectively.

However, we can turn the coin over and look at the coordinates of $\mathbf{e}_i$ in terms of the new axes $\mathbf{f}_i$. If we project $\mathbf{e}_1$ and $\mathbf{e}_2$ onto $\mathbf{f}_1$ and $\mathbf{f}_2$, we have, from Fig. 3.20,

$$\mathbf{g}_1 = \begin{bmatrix} 0.867 \\ -0.5 \end{bmatrix}; \qquad \mathbf{g}_2 = \begin{bmatrix} 0.5 \\ 0.867 \end{bmatrix}$$

where we use $\mathbf{g}_1$ and $\mathbf{g}_2$ to denote the fact that the reference vectors are now the $\mathbf{f}_i$. Since $\mathbf{a}$ has been defined originally in terms of $\mathbf{e}_i$, and the $\mathbf{e}_i$ have now been represented in terms of $\mathbf{f}_i$, we have

$$\mathbf{a} = 2\mathbf{g}_1 + 1\mathbf{g}_2$$

$$\mathbf{a} = 2\begin{bmatrix} 0.867 \\ -0.5 \end{bmatrix} + 1\begin{bmatrix} 0.5 \\ 0.867 \end{bmatrix} = \begin{bmatrix} 2.23 \\ -0.13 \end{bmatrix}$$

But, as already shown, this can also be written as

$$\begin{bmatrix} a_1^* \\ a_2^* \end{bmatrix} = \begin{bmatrix} \cos\theta_{11} & \cos\theta_{21} \\ \cos\theta_{12} & \cos\theta_{22} \end{bmatrix} \begin{bmatrix} a_1 \\ a_2 \end{bmatrix}$$

$$\begin{bmatrix} 2.23 \\ -0.13 \end{bmatrix} = \begin{bmatrix} 0.867 & 0.5 \\ -0.5 & 0.867 \end{bmatrix} \begin{bmatrix} 2 \\ 1 \end{bmatrix}$$

Thus, while the point remains fixed, its coordinates are determined by the particular basis by which they are expressed.

In summary, we see that the new coordinates of the original point $\mathbf{a} = \begin{bmatrix} 2 \\ 1 \end{bmatrix}$ are now $a_1^* = 2.23$ and $a_2^* = -0.13$ in terms of the $\mathbf{f}_1$ and $\mathbf{f}_2$ axes. However, there is a second way of looking at this transformation. That is, we can make the original $\mathbf{e}_1$, $\mathbf{e}_2$ plane *remain the same* and assume that it is the point $\begin{bmatrix} 2 \\ 1 \end{bmatrix}$ which moves from its old location to the position $\begin{bmatrix} 2.23 \\ -0.13 \end{bmatrix}$. This second way of interpreting things is shown in Fig. 3.21. Notice in

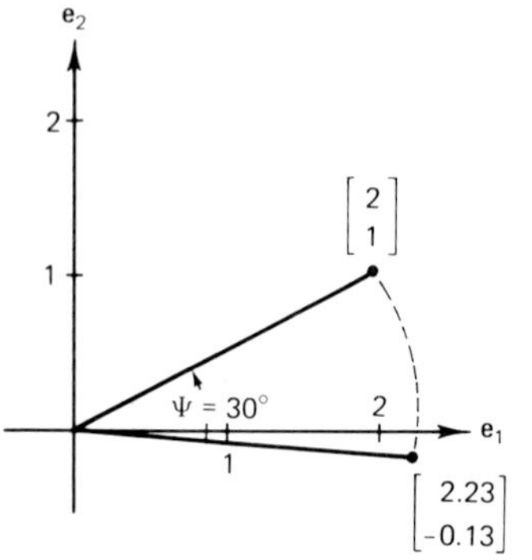

Fig. 3.21 Rotation of point with basis vectors fixed.

this case that the point $\left[\begin{smallmatrix}2\\1\end{smallmatrix}\right]$ is rotated *clockwise* 30°.[11] Either interpretation is equally suitable. The one that is selected will depend on the researcher's purpose since it is only *relative* motion that is indicated.

In Chapter 4 we shall explore basis vector and coordinate transformations much more thoroughly. At this point, however, we wish merely to show that there are two compatible ways of looking at things:

1. One can rotate the basis vectors and refer the unchanged point to the new reference axes.
2. One can rotate the point and refer its new location to the original reference axes.

In each case, under rigid rotations we should note that angles and distances are preserved. Finally, we could simplify the angular representation of the preceding rotation–in the special case of two dimensions–by means of a *single* angle of rotation.

If we let $\Psi = \theta_{11}$, we can note the following:

$$\begin{bmatrix} \cos\theta_{11} & \cos\theta_{21} \\ \cos\theta_{12} & \cos\theta_{22} \end{bmatrix} = \begin{bmatrix} \cos\Psi & \sin\Psi \\ -\sin\Psi & \cos\Psi \end{bmatrix}$$

It is instructive to see how the rotation of a set of basis vectors through the single angle Ψ (in the case of two dimensions) leads to the matrix above.

3.5.2 The Trigonometry of Rotation

The trigonometry of rotation can be shown fairly straightforwardly. Panel I of Fig. 3.22 shows a point A with original coordinates a_1 and a_2 in the $\mathbf{e}_1$, $\mathbf{e}_2$ basis. If the

[11] As will be discussed in more detail in Chapter 4, the *clockwise* rotation of $\mathbf{a} = \left[\begin{smallmatrix}2\\1\end{smallmatrix}\right]$ can also be represented by the product

$$\begin{bmatrix} 2.23 \\ -0.13 \end{bmatrix} = \begin{bmatrix} 0.867 & 0.5 \\ -0.5 & 0.867 \end{bmatrix} \begin{bmatrix} 2 \\ 1 \end{bmatrix}$$

where $\left[\begin{smallmatrix}2\\1\end{smallmatrix}\right]$ and $\left[\begin{smallmatrix}2.23\\-0.13\end{smallmatrix}\right]$ are both expressed in terms of the $\mathbf{e}_i$ basis vectors.

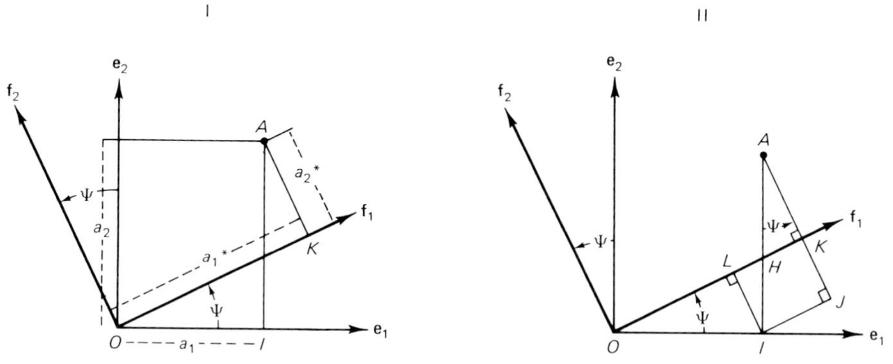

Fig. 3.22 A trigonometric demonstration of basis vector rotation.

original axes are rotated counterclockwise through the angle Ψ, we obtain $a_1{}^*$ and $a_2{}^*$ as the coordinates of A in the $\mathbf{f}_1$, $\mathbf{f}_2$ basis. We now ask: How can $a_1{}^*$ and $a_2{}^*$ be expressed in terms of the old coordinates a_1 and a_2?

The trigonometric argument is simple to describe. Panel II shows the construction of the rectangle *IJKL*. Angle *OHI* is the complement of the angle Ψ and, in turn, equals angle *AHK*. Hence, angle *HAK* is equal to Ψ, the angle of rotation. Given these facts, we can now say

$$a_1{}^* = OK = OH + HK = OL + IJ = OI \cos\Psi + AI \sin\Psi = a_1 \cos\Psi + a_2 \sin\Psi$$

$$a_2{}^* = AK = AJ - JK = AJ - IL = AI \cos\Psi - OI \sin\Psi = a_2 \cos\Psi - a_1 \sin\Psi$$

The coordinates of A in terms of the new basis vectors $\mathbf{f}_1$, $\mathbf{f}_2$ are then given by

$$a_1{}^* = a_1 \cos\Psi + a_2 \sin\Psi; \qquad a_2{}^* = -a_1 \sin\Psi + a_2 \cos\Psi$$

or, in matrix form,

$$\begin{bmatrix} a_1{}^* \\ a_2{}^* \end{bmatrix} = \begin{bmatrix} \cos\Psi & \sin\Psi \\ -\sin\Psi & \cos\Psi \end{bmatrix} \begin{bmatrix} a_1 \\ a_2 \end{bmatrix}$$

as desired.

It should be remembered, however, that expressing a basis vector rotation in terms of a single angle Ψ is restricted to *two* dimensions. On the other hand, the more cumbersome notation involving four angles

$$\theta_{11}; \quad \theta_{12}; \quad \theta_{21}; \quad \theta_{22}$$

is more general, since the concept of direction cosines generalizes to three or more dimensions. Thus, any time that we work with rotations involving three or more dimensions we shall assume that direction cosines are involved throughout.

3.5.3 Higher-Dimensional Rotations

What has been illustrated above for the case of two-dimensional rotations can be extended to three or more dimensions by using the appropriate matrix of direction

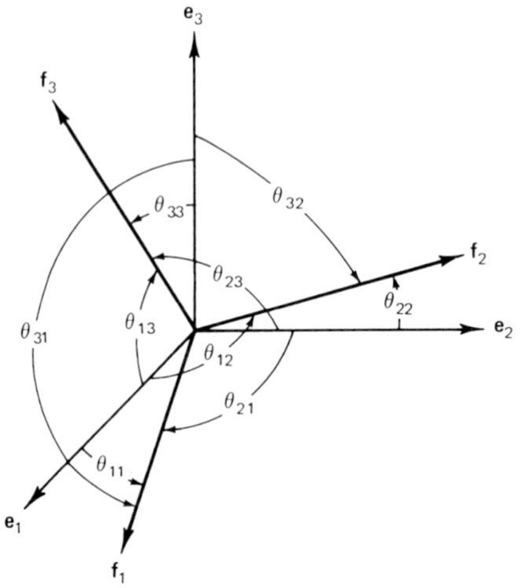

Fig. 3.23 Rotation in three dimensions.

cosines. Figure 3.23 portrays, in general form, a counterclockwise rotation from the $\mathbf{e}_i$ basis to an $\mathbf{f}_i$ basis. The coordinates of some point $\mathbf{a}$ in the original basis can be expressed as $\mathbf{a}^*$ in the new basis by

$$\begin{bmatrix} a_1{}^* \\ a_2{}^* \\ a_3{}^* \end{bmatrix} = \begin{bmatrix} \cos\theta_{11} & \cos\theta_{21} & \cos\theta_{31} \\ \cos\theta_{12} & \cos\theta_{22} & \cos\theta_{32} \\ \cos\theta_{13} & \cos\theta_{23} & \cos\theta_{33} \end{bmatrix} \begin{bmatrix} a_1 \\ a_2 \\ a_3 \end{bmatrix}$$

Fig. 3.23 shows the angles that are considered in this more complex case. However, no new principles are involved.

Orthogonal matrices play a central role in various multivariate procedures, and their special properties should be noted; these are taken up next.

3.5.4 Properties of an Orthogonal Matrix

Now that we have illustrated what goes on when a point, or points, are subjected to a rotation, let us examine some of the properties of the transformation matrix used in Section 3.5.1. We have

$$\mathbf{A} = \begin{bmatrix} \cos\theta_{11} & \cos\theta_{21} \\ \cos\theta_{12} & \cos\theta_{22} \end{bmatrix} = \begin{bmatrix} 0.867 & 0.5 \\ -0.5 & 0.867 \end{bmatrix}$$

First, let us check on the following:

$$\mathbf{A}'\mathbf{A} = \begin{bmatrix} 0.867 & -0.5 \\ 0.5 & 0.867 \end{bmatrix} \begin{bmatrix} 0.867 & 0.5 \\ -0.5 & 0.867 \end{bmatrix} = \begin{bmatrix} 1 & 0 \\ 0 & 1 \end{bmatrix}$$

$$\mathbf{A}\mathbf{A}' = \begin{bmatrix} 0.867 & 0.5 \\ -0.5 & 0.867 \end{bmatrix} \begin{bmatrix} 0.867 & -0.5 \\ 0.5 & 0.867 \end{bmatrix} = \begin{bmatrix} 1 & 0 \\ 0 & 1 \end{bmatrix}$$

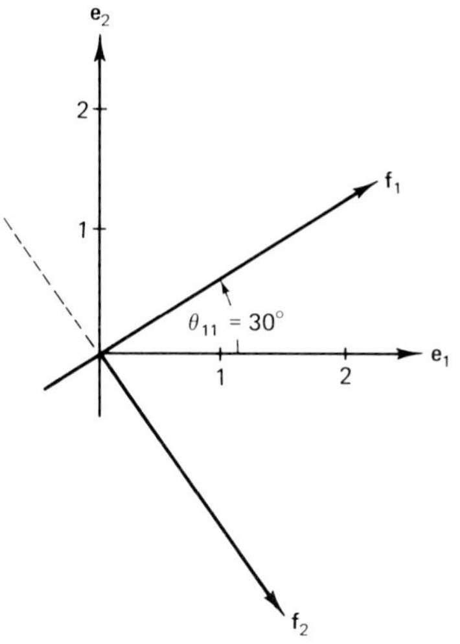

Fig. 3.24 Improper rotation.

As can be seen, within rounding error, we obtain an identity matrix in each case. Hence **A** is observed to be an orthogonal matrix.

However, before concluding that *all* matrices that satisfy the above conditions are rigid rotations, let us consider the following modification of **A**.

$$\mathbf{B} = \begin{bmatrix} \cos\theta_{11} & \cos\theta_{21} \\ -\cos\theta_{12} & -\cos\theta_{22} \end{bmatrix}$$

Note that **B** differs from **A** only in the fact that the second-row entries of **A** have been each multiplied by -1. If we examine the properties of **B**, we see that

$$\mathbf{B}'\mathbf{B} = \mathbf{B}\mathbf{B}' = \mathbf{I}$$

That is, the *same* conditions are met with the **B** matrix as were met with the **A** matrix.[12]

However, what is happening here is something that is a bit different from a rigid rotation. Figure 3.24 illustrates what is going on. In this latter case we have a rigid rotation that leads to a new axis $\mathbf{f}_1$ which is in the same orientation as $\mathbf{f}_1$ in Fig. 3.20 but an axis $\mathbf{f}_2$ which is the *negative* of $\mathbf{f}_2$ in Fig. 3.20.

This new situation represents a case of rotation followed by a *reflection* of the $\mathbf{f}_2$ axis. Alternatively, we could have affixed minus signs to the first row of **A** and, in this case, it would be the $\mathbf{f}_1$ axis that was reflected. However, if all entries of **A** are multiplied by -1, a rigid rotation of $\theta_{11} + 180°$ would result. It is only when an odd number of rows receive minus signs that we have what is known as an "improper" rotation, that is, a rotation followed by reflection.

[12] The reader may verify this numerically or, in more general terms, write out the implied trigonometric relationships.

How do we know before hand whether a proper versus improper rotation is involved? It turns out that this distinction is revealed by examining the determinant of **A**.

1. If the determinant of **A** equals 1, then a proper rotation is involved.
2. If the determinant of **A** equals −1, then an improper rotation is involved.

And, it turns out that any orthogonal matrix will have a determinant that is *either* 1 or −1.

To sum up, if $\mathbf{A}'\mathbf{A} = \mathbf{A}\mathbf{A}' = \mathbf{I}$, we say the matrix is orthogonal. If $|\mathbf{A}| = -1$, it represents a rotation followed by an odd number of reflections, for example, one axis in the 2 x 2 case, one or three in the 3 x 3 case, one or three in the 4 x 4 case, and so on. If $|\mathbf{A}| = 1$, then we are dealing with a proper rotation.[13]

3.6 GEOMETRIC ASPECTS OF CROSS-PRODUCT MATRICES AND DETERMINANTS

In Chapter 2 we defined a determinant as a scalar function of a square matrix. Evaluation of a determinant in terms of both cofactor expansion and the pivotal method was also described and illustrated numerically. At this point attention focuses on the geometric aspects of a determinant and, in particular, its role in portraying *generalized variance* among a set of statistical variables. In the course of describing this relationship, we shall also point out geometric analogies to a number of common statistical concepts.

By way of introduction, we first illustrate some geometric aspects of a determinant at a simple two-dimensional level. We then discuss how determinants can be linked with various statistical measures of interest to multivariate analysis.

3.6.1 The Geometric Interpretation of a Determinant

Certain aspects of the determinant of a matrix can be expressed in geometric format. To illustrate, let us consider the unit square *OIJK*, as shown in Fig. 3.25. The coordinates representing the vertices of the square are

$$O = (0,0); \qquad I = (1,0); \qquad J = (1,1); \qquad K = (0,1)$$

Next, suppose we were to multiply these coordinates by the matrix

$$\mathbf{T} = \begin{bmatrix} 2 & 1 \\ 3 & 4 \end{bmatrix}$$

as follows:

$$\mathbf{U} = \begin{bmatrix} 2 & 1 \\ 3 & 4 \end{bmatrix} \overset{\begin{matrix} O & I & J & K \end{matrix}}{\begin{bmatrix} 0 & 1 & 1 & 0 \\ 0 & 0 & 1 & 1 \end{bmatrix}}; \qquad \mathbf{U} = \overset{\begin{matrix} O & I^* & J^* & K^* \end{matrix}}{\begin{bmatrix} 0 & 2 & 3 & 1 \\ 0 & 3 & 7 & 4 \end{bmatrix}}$$

[13] It should be mentioned that any reflection of two or more axes can itself be represented by a proper rotation followed by just *one* reflection. For example, in the 3 x 3 case, only one axis need be reflected after an appropriate rotation is made.

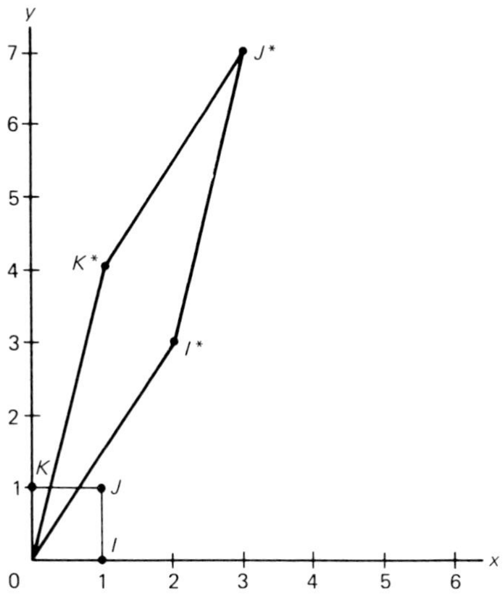

Fig. 3.25 Geometric aspects of a determinant.

These transformed points also appear in Fig. 3.25 as the quadrilateral $OI^*J^*K^*$.

The key aspect of this transformation has to do with the ratio of the area of the quadrilateral to the area of the original unit square. *The ratio of the two areas equals the determinant of the transformation matrix* **T**. That is,

$$|\mathbf{T}| = (2 \times 4) - (3 \times 1) = 5$$

Thus, if one were to measure the area of $OI^*J^*K^*$ and compare it to the area of *OIJK*, one would find that it is exactly five times the latter area. And this would be true for any starting figure that is transformed by **T**.

This concept generalizes to determinants of matrices of order 3 x 3 and higher. In the 3 x 3 case, the determinant measures the ratio of volumes between the original and transformed figures. In the 4 x 4 and higher-order cases, the determinant measures the ratio of hypervolumes between original and transformed figures.

Finally, if the sign of the determinant should be negative, this does not affect the ratio between hypervolumes of original and transformed figures. Rather, the presence of a negative determinant has to do with the orientation of the transformed figure in the space of interest.[14] Hence, it is the *absolute value* of the determinant that indicates the ratio of hypervolumes. Moreover, if that absolute value is less than unity, then the transformed figure's hypervolume is a fraction of that of the original figure.

3.6.2 The Geometry of Statistical Measures

In Chapter 1 we introduced a small and illustrative data bank (Table 1.2), involving only twelve cases and three variables. In Chapter 2 we used this miniature data bank

[14] To illustrate this, the reader should work out the case of $\mathbf{T} = \begin{bmatrix} -2 & -1 \\ 3 & 4 \end{bmatrix}$ with $|\mathbf{T}| = -5$. This entails a reflection of the quadrilateral in Fig. 3.25 across the y axis. However, the ratio of areas is still 5 : 1.

(Table 2.2) to illustrate the application of matrix operations in the computation of various cross-product matrices, such as the SSCP, covariance, and correlation matrices.

We continue to refer to this sample data bank. However, in line with the focus of Chapter 3, the data of the sample problem are now discussed from a geometric viewpoint.

In the course of analyzing multivariate data, it is useful to make various scatter plots for showing relationships among variables. Figures 1.2 and 1.3 are illustrations of the more usual type of plot in which variables are treated as axes, and cases (employees in this example) are treated as points. This more conventional way of portraying data is often called a response surface or point model, since with one criterion variable and two predictors X_1 and X_2, one could visualize the fitting of a response surface to Y, the criterion variable.

Alternatively, however, we could imagine that each of the twelve employees, or cases, represents a dimension, and each of the three variables in the sample problem represents a *vector* embedded in a twelve-dimensional space. (Actually, if the three variables are linearly independent, they will lie in only a three-dimensional subspace of the original twelve dimensions, as is discussed in more detail in later chapters.)

For the moment, let us simplify things even further and consider only two of the variables of the sample problem, namely, Y and X_1. If so, a vector representation of these two variables could be portrayed in only two dimensions, embedded in the full, twelve-dimensional space.

From Table 2.3 we note that the covariance and correlation matrices for only the Y, X_1 pair of variables are

$$\mathbf{C} = \begin{array}{c} \\ Y \\ X_1 \end{array}\begin{array}{c} \begin{array}{cc} Y & X_1 \end{array} \\ \begin{bmatrix} 29.52 & 19.44 \\ 19.44 & 14.19 \end{bmatrix} \end{array}; \qquad \mathbf{R} = \begin{array}{c} \\ Y \\ X_1 \end{array}\begin{array}{c} \begin{array}{cc} Y & X_1 \end{array} \\ \begin{bmatrix} 1.0 & 0.95 \\ 0.95 & 1.0 \end{bmatrix} \end{array}$$

As recalled, both the **C** and **R** matrices are based on mean-corrected variables; as such, the origin of the space will be taken at the centroid of the variables, represented by the **0** vector.[15]

As in Chapter 2, we can define the variance of a variable X_1 as

$s_{x_1}^2 = \dfrac{\Sigma x_{i1}^2}{m}$	where $x_{i1} = X_{i1} - \bar{X}_1$ (i.e., each x is expressed as a deviation about the mean, and m denotes the number of cases[16])

[15] By centroid is meant a vector whose components are the arithmetic means of Y and X_1, respectively. Then, if we allow the centroid to represent the origin or **0** vector, the individual vectors are position vectors whose termini are expressed as deviations from the mean of Y and X_1, respectively.

[16] One could use $m - 1$ in the denominator if one wished to have an unbiased estimate of the population variance. (Such adjustment does not mean that the sample standard deviation is an unbiased estimate of the population standard deviation, however.) Here, for purpose of simplification, we omit the adjustment and use m in the denominator.

Similarly, the correlation of a pair of variables is defined as

$$r_{yx_1} = \frac{\Sigma y_i x_{i1}}{\sqrt{\Sigma y_i^2}\sqrt{\Sigma x_{i1}^2}}$$

where y_i and x_{i1} are each expressed as deviations about their respective means

When Y and X_1 are each expressed in mean-corrected form, their correlation is related to their scalar product as follows:

$$r_{yx_1} = \frac{\mathbf{y}'\mathbf{x}_1}{||\mathbf{y}|| \cdot ||\mathbf{x}_1||} = \cos\theta_{\mathbf{yx}_1}$$

which, we see, is just the cosine of $\theta_{\mathbf{yx}_1}$, their angle of separation.

In terms of the sample problem, the correlation is

$$r_{yx_1} = \cos\theta_{\mathbf{yx}_1} = \frac{233.28}{\sqrt{354.24} \cdot \sqrt{170.28}} = 0.95$$

where the scalar product and squared vector lengths are computed from Y_d and X_{d1} in Table 1.2. The covariance of a pair of variables is defined as

$$\text{cov}_{yx_1} = r_{yx_1} s_y s_{x_1}$$

$$= \frac{\Sigma y_i x_{i1}}{m}$$

where s_y and s_{x_1} are standard deviations of Y and X_1, respectively

Our current objective is to tie in these statistical notions with concepts from vector algebra and, in particular, to show how the *determinant* of a covariance matrix can be used as a generalized scalar measure of dispersion.

To do this, we first imagine a geometric space in which the axes are the m cases (e.g., the employees). The variable Y can then be thought of as a "test" vector $\mathbf{y}$ in a space of m persons. Similarly, the variable X_1 can be thought of as another vector $\mathbf{x}_1$ in the person space. The components of $\mathbf{y}$ and the components of $\mathbf{x}_1$ each sum to zero since each variable is expressed in terms of deviations from its own mean.

Once this has been done, we can see that the length of the vector $\mathbf{y}$ turns out to be proportional to the standard deviation of the variable Y. That is,

$$||\mathbf{y}|| = \sqrt{\sum_{i=1}^{m} y_i^2} = \sqrt{m}\, s_y$$

Similarly, for the vector $\mathbf{x}_1$ we have

$$||\mathbf{x}_1|| = \sqrt{\sum_{i=1}^{m} x_{i1}^2} = \sqrt{m}\, s_{x_1}$$

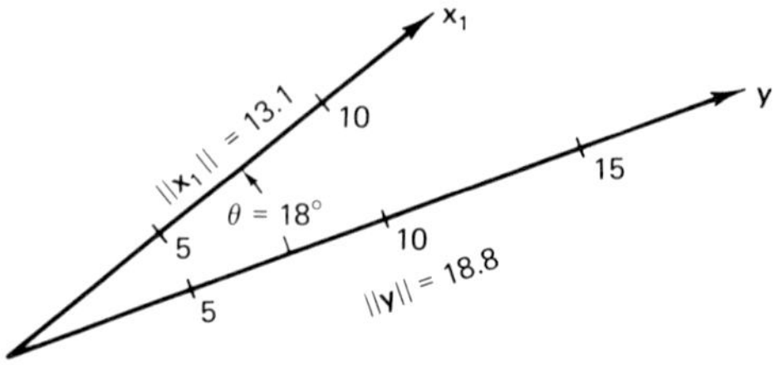

Fig. 3.26 A vector representation of the covariance between Y and X_1 (sample problem).

The correlation r_{yx_1} between the two test vectors, measured as deviations from each variable's mean but based on vector lengths that are each normalized to unity, is equal to the cosine of the angle separating them. Furthermore, the scalar product of these two vectors is proportional to their covariance. That is,

$$\mathbf{y}'\mathbf{x}_1 = \cos\theta_{\mathbf{y}\mathbf{x}_1} ||\mathbf{y}|| \cdot ||\mathbf{x}_1|| = m[\text{cov}_{yx_1}] = m[r_{yx_1} s_y s_{x_1}]$$

Insofar as the sample problem is concerned, Fig. 3.26 shows a plot of the vectors y and $\mathbf{x}_1$. The angle corresponding to a cosine of 0.95 (denoting their correlation) is 18°. Their respective lengths are

$$||\mathbf{y}|| = \sqrt{12} \cdot \sqrt{29.52} = 18.82; \qquad ||\mathbf{x}_1|| = \sqrt{12} \cdot \sqrt{14.19} = 13.05$$

If one is dealing with standardized scores, then vector lengths would, of course, each be equal to $\sqrt{m}$ since s_y and s_{x_1} would each be equal to unity.

In brief, with mean-corrected variables, all three cross-product matrices—the SSCP, covariance, and correlation matrices—can be portrayed in geometric terms. The key concept involves the scalar product between two variables. In all three cases, we have

$$\cos\theta_{\mathbf{y}\mathbf{x}_1} = r_{yx_1}$$

The vector lengths in each case are as follows:

SSCP matrix:	$\sqrt{\sum_{i=1}^{m} y_i^2}$; $\sqrt{\sum_{i=1}^{m} x_{i1}^2}$
Covariance matrix:	$\sqrt{m}\, s_y$; $\sqrt{m}\, s_{x_1}$
Correlation matrix:	$\sqrt{m}$; $\sqrt{m}$

This complementary view of association between variables will serve us in good stead in the interpretation of various aspects of multivariate analysis in later chapters.

3.6.3 A Generalized Variance Measure

Having established the various correspondences shown above, we now wish to illustrate how the determinant of the covariance matrix represents a scalar measure of generalized variance.[17] Given, illustratively, two variables Y and X, the covariance matrix can be written as

$$\mathbf{C} = \begin{bmatrix} s_y{}^2 & r_{yx}s_ys_x \\ r_{yx}s_ys_x & s_x{}^2 \end{bmatrix}$$

where the main diagonal components are variances, and the off-diagonal component is their covariance. If we compute the determinant of this matrix, we get

$$|\mathbf{C}| = s_y{}^2s_x{}^2 - r_{yx}^2s_y{}^2s_x{}^2 = s_y{}^2s_x{}^2(1-r_{yx}^2) = s_y{}^2s_x{}^2(1-\cos^2\theta_{\mathbf{yx}})$$

and, from basic trigonometry, in which we have the identity $\sin^2\theta + \cos^2\theta = 1$, we can write

$$|\mathbf{C}| = s_y{}^2s_x{}^2 \sin^2\theta_{\mathbf{yx}} = (s_ys_x \sin\theta_{\mathbf{yx}})^2$$

where $\theta_{\mathbf{yx}}$ is the angle between the (deviation-score) vectors $\mathbf{y}$ and $\mathbf{x}$.

As pointed out earlier, the standard deviation of a variable is $1/\sqrt{m}$ times the length of its corresponding vector. Hence, we have

$$s_ys_x \sin\theta_{\mathbf{yx}} = \frac{\|\mathbf{y}\|}{\sqrt{m}} \cdot \frac{\|\mathbf{x}\|}{\sqrt{m}} \sin\theta_{\mathbf{yx}}$$

and we can conveniently set up the equivalence

$$h = \frac{\|\mathbf{y}\|}{\sqrt{m}} \sin\theta$$

as the height of a parallelogram with base given by

$$\frac{\|\mathbf{x}\|}{\sqrt{m}}$$

as shown in Fig. 3.27. So, if the vector lengths are each scaled by $1/\sqrt{m}$, we see that the area of the resulting (scaled) parallelogram equals $s_ys_x \sin\theta_{\mathbf{yx}}$. The *square* of this area

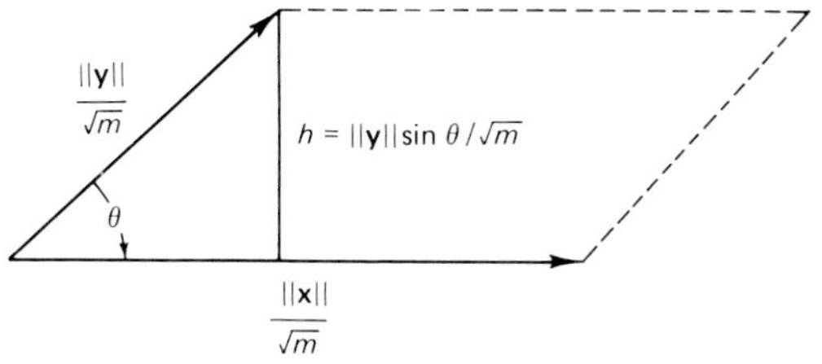

Fig. 3.27 Representing the determinant of a 2 x 2 covariance matrix as the area of a parallelogram.

[17] For example, even in a 2 x 2 covariance matrix we have four dispersionlike entries. Our interest here is on developing a *single* number that represents the four entries in certain multivariate statistical applications.

equals the generalized variance (i.e., determinant of the dispersion matrix). If n variables are involved, the generalized variance equals the square of the volume formed by n such (rescaled) vectors.

Figure 3.27 shows, in general form, the nature of this parallelogram in the two-variable case. We see that each vector appears in scaled (by $1/\sqrt{m}$) form, and the parallelogram is completed as shown.

For a numerical illustration of the correspondence, let us again refer to the sample problem of Table 2.3. We illustrate the equivalence for only the first two variables Y and X_1.

First, from the covariance matrix involving variables Y and X_1,

$$\mathbf{C} = \begin{array}{c} \\ Y \\ X_1 \end{array} \begin{array}{c} \begin{array}{cc} Y & X_1 \end{array} \\ \begin{bmatrix} 29.52 & 19.44 \\ 19.44 & 14.19 \end{bmatrix} \end{array}$$

we obtain

$$\|\mathbf{y}\| = \sqrt{12} \cdot \sqrt{29.52} = 18.82; \qquad \|\mathbf{x}_1\| = \sqrt{12} \cdot \sqrt{14.19} = 13.05$$

$$\cos \theta_{\mathbf{y}\mathbf{x}_1} = 0.95; \qquad \sin \theta_{\mathbf{y}\mathbf{x}_1} = 0.31$$

Hence the area of the parallelogram formed by $\mathbf{y}$ and $\mathbf{x}_1$ is

$$\frac{\|\mathbf{y}\|}{\sqrt{m}} \cdot \frac{\|\mathbf{x}_1\|}{\sqrt{m}} \cdot \sin \theta_{\mathbf{y}\mathbf{x}_1} = \frac{18.82}{\sqrt{12}} \cdot \frac{13.05}{\sqrt{12}} (0.31) = 6.34$$

The square of 6.34 is equal to 40.22. This value, within rounding error, equals the determinant of $\mathbf{C}$, the covariance matrix. Thus we have shown geometrically and numerically how the determinant of $\mathbf{C}$ is equal to the square of the area of the parallelogram in Fig. 3.27.

The concept of generalized variance is quite important in multivariate analysis since it enables us to portray a matrix of variances and covariances *in terms of a single number,* namely, the determinant of the covariance matrix. Just as importantly, we also see that the statistical measures of standard deviation, covariance, and correlation can be portrayed in terms of length and/or angle of test vectors in person and, more generally, object space.

3.7 SUMMARY

The purpose of this chapter has been to describe a number of the vector and matrix operations outlined in Chapter 2 from a geometric standpoint. After setting up a rectangular Cartesian coordinate system and defining the concept of a Euclidean space, we discussed such topics as vector length and angle, vector addition and subtraction, scalar multiplication of a vector, and the scalar product of two vectors from a geometric point of view.

We then described the notion of linear independence. We also illustrated how the Gram-Schmidt process could be employed to find an orthonormal basis starting from any given (arbitrary) basis. Following this we briefly discussed the idea of generalized (nonorthogonal) coordinate systems.

Matrix times vector multiplication was introduced from a geometric viewpoint for the special case of orthogonal (i.e., proper or improper rotation) matrices. The properties of this class of matrices were discussed, and their application was illustrated numerically. We concluded the chapter with a geometric representation of various statistical measures, including the central concept of generalized variance, as applicable to multivariate statistical tests to be considered in later chapters.

REVIEW QUESTIONS

1. Sketch a three-dimensional coordinate system.

a. Plot points with coordinates $(2, 1, 0)$, $(1, -1, -1)$, $(\sqrt{3}, \pi, -2)$. What is the length of each?

b. What is the set of points whose x and y coordinates sum to 1?

c. What is the graph of $z = x^2$?

d. What is the graph of the inequality $x^2 + y^2 + z^2 \leqslant 1$?

2. Let P be the point $(4, 3, -1)$, Q be the point $(1, 0, 2)$, and R be the midpoint of the segment joining P and Q.

a. What are the coordinates of R?

b. Sketch the vectors PR, OR, and PQ, where O denotes the origin.

c. Verify that $PR = PQ/2$ by computing the distance from P to R, R to Q, and P to Q; then show that the first two distances are each half of the last distance.

3. In the context of linear combinations,

a. find a scalar k such that

$$(1, 0, 2) + k(2, 1, 1) = (-1, -1, 1)$$

b. find scalars k_1, k_2, and k_3 such that

$$k_1(5\mathbf{e}_1 + \mathbf{e}_2) + k_2(\mathbf{e}_2 + \mathbf{e}_3) + k_3(\mathbf{e}_3) = 5\mathbf{e}_1 + 3\mathbf{e}_2 + \mathbf{e}_3$$

c. find k_1 and k_2 such that

$$k_1(5\mathbf{e}_1 + \mathbf{e}_2) + k_2(\mathbf{e}_1 - \mathbf{e}_2) = \mathbf{0}$$

4. Let $\mathbf{a}$ and $\mathbf{b}$ be vectors with given lengths and angle θ. Compute their scalar product under the conditions

a. $||\mathbf{a}|| = 0.5$; $||\mathbf{b}|| = 4$; $\theta = 45°$

b. $||\mathbf{a}|| = 4$; $||\mathbf{b}|| = 1$; $\theta = 90°$

c. $||\mathbf{a}|| = 1$; $||\mathbf{b}|| = 1$; $\theta = 120°$

d. What is the possible range of values for the scalar product $\mathbf{a}'\mathbf{b}$ if

$$||\mathbf{a}|| = 2 \quad \text{and} \quad ||\mathbf{b}|| = 3?$$

5. Let a, b, and c be vectors. Let $\|\mathbf{a}_p\|$ be the component of a along c and let $\|\mathbf{b}_p\|$ be the component of b along c. What is the component of a + b along c? Sketch the relationship in two-dimensional space.

6. Let $\mathbf{a}' = (2, -1, 7)$ and $\mathbf{b}' = (-3, 6, 1)$. Find direction cosines for

a. $\mathbf{a}' + \mathbf{b}'$ b. $\mathbf{a}' - \mathbf{b}'$ c. $5\mathbf{a}' + 10\mathbf{b}'$ d. $\frac{1}{2}(\mathbf{a}' - \mathbf{b}')$

Next, find the cosine of the angle between the two vectors obtained in parts c and d.

7. Apply the Gram–Schmidt orthonormalization procedure to the following sets of vectors:

a. $\mathbf{a}' = (1, 2, 3)$; $\mathbf{b}' = (3, 0, 2)$; $\mathbf{c}' = (3, 1, 1)$

b. $\mathbf{a}' = (2, 1)$; $\mathbf{b}' = (1, 2)$; $\mathbf{c}' = (1, 1)$

c. What do you notice about the vectors obtained in part b?

8. Find coordinates of the vector $\mathbf{a}' = (2, 3)$ relative to the basis vectors $\mathbf{f}_1{}' = (1, -1)$ and $\mathbf{f}_2{}' = (3, 5)$.

9. Show that the vectors $\mathbf{a}' = (1, 4, -2)$ and $\mathbf{b}' = (2, 1, 3)$ are orthogonal and find a third vector that is orthogonal to both.

10. Find the equations for the ellipse $4x^2 + y^2 = 4$ and the circle $x^2 + y^2 = 1$ after the xy axes have been rotated counterclockwise through angles of

a. 45° b. 60° c. 120°

11. Find x and y so that the vectors $(4, -2, 1, 7)$ and $(2, -3, x, y)$ are linearly dependent.

12. Express the standard basis vectors $\mathbf{e}_1{}' = (1, 0, 0)$, $\mathbf{e}_2{}' = (0, 1, 0)$, and $\mathbf{e}_3{}' = (0, 0, 1)$ as linear combinations of $\mathbf{f}_1{}' = (1, 2, 4)$, $\mathbf{f}_2{}' = (-2, 1, 5)$, and $\mathbf{f}_3{}' = (-1, -1, 2)$.

13. Rotate the vector $\mathbf{a}' = (1, 2)$ counterclockwise through an angle of 45° while keeping the basis vectors fixed. Rotate $\mathbf{b}' = (3, 2)$ clockwise through an angle of 60°.

a. What is the scalar product $\mathbf{a}'\mathbf{b}$ before and after the two rotations?

b. What are the vector lengths of $\mathbf{a}'$ and $\mathbf{b}'$ after the rotations?

c. Show each of the above steps geometrically.

14. Assume that we have the expression

$$\begin{bmatrix} a_1{}^* \\ a_2{}^* \end{bmatrix} = \begin{bmatrix} \cos 45^\circ & -\sin 45^\circ \\ \sin 45^\circ & \cos 45^\circ \end{bmatrix} \begin{bmatrix} a_1 \\ a_2 \end{bmatrix}$$

and OP is the line joining the origin $\mathbf{0}$ to the point $P = (2, 3)$. Show in diagram form the position of OP^*, the rotated point.

15. Apply the transformation

$$\begin{bmatrix} a_1{}^* \\ a_2{}^* \end{bmatrix} = \begin{bmatrix} 3 & 5 \\ 2 & 4 \end{bmatrix} \begin{bmatrix} a_1 \\ a_2 \end{bmatrix}$$

to a square with vertices of (1, 1), (3, 1), (3, 3), (1, 3) and show geometrically that the ratio of the area of the new figure to the area of the original is 2 to 1.

16. In the sample problem of Table 2.3, consider the full 3 x 3 covariance matrix.

a. Plot the mean-corrected $\mathbf{y}$, $\mathbf{x}_1$, and $\mathbf{x}_2$ in a three-dimensional space.

b. Plot the standardized form of $\mathbf{y}$, $\mathbf{x}_1$, and $\mathbf{x}_2$ in a three-dimensional space.

c. Show how the correlation between $\mathbf{y}$ and $\mathbf{x}_2$ is related to

$$\cos \theta_{\mathbf{y}\mathbf{x}_2} = \frac{\mathbf{y}'\mathbf{x}_2}{\|\mathbf{y}\| \cdot \|\mathbf{x}_2\|}$$

CHAPTER 4

Linear Transformations from a Geometric Viewpoint

4.1 INTRODUCTION

In Chapter 3 the reader was exposed briefly to one type of transformation, namely, that involving the rotation of either a point or a set of basis vectors by an orthogonal matrix. A large part of multivariate analysis is concerned with linear transformations, of which rotation represents only one type—albeit an important special case.

The purpose of this chapter is to describe the geometric aspects of various kinds of matrix transformations. In particular, we shall be interested in how general linear transformations can be viewed as composites of simple kinds of matrices that, individually, are more easily portrayed from a geometric viewpoint.

In the course of describing matrix transformations geometrically, two additional concepts of major importance to multivariate analysis—matrix inversion and the rank of a matrix—are described and illustrated numerically. These concepts are useful in the solution of sets of linear equations and play important roles in multivariate analysis.

The chapter starts out with an overview description of matrix transformations and their representation as sets of linear equations. Then the topic of orthogonal transformations, first introduced in Chapter 3, is reviewed and expanded upon. The distinction between point and basis vector transformations is also emphasized. Discussion then proceeds to more extensive cases involving general linear transformations. In the case of basis vector changes, these kinds of transformations require the use of matrix inverses; hence, the basic ideas of matrix inversion are introduced at this point.

The next major section of the chapter is devoted to the geometric representation of various types of matrix transformations, such as rotations, stretches, central dilations and reflections. The geometric character of combinations of various matrix transformations is also illustrated so that the reader can see how simple geometric changes, when taken in combination, lead to complex representations.

The remainder of the chapter focuses on the solution of simultaneous equations and the central roles that matrix inversion and matrix rank play in this activity. In particular, we discuss the solution of linear equations in multivariate analysis and, in the process, tie in the present topic with material presented in earlier chapters on determinants and the pivotal method for solving sets of linear equations.

4.2 SIMULTANEOUS EQUATIONS AND MATRIX TRANSFORMATIONS

The concept of a function or mapping is fundamental to all mathematics. By a mapping we mean an operation by which elements of one set of mathematical entities are transformed into elements of another. In scalar algebra we recall that functions like the following are often employed:

$$y = f(x) = bx; \qquad y = f(x) = e^x$$

$$y = f(x) = ax^b; \qquad y = f(x) = ab^x$$

For example, for a specific value of x, and a value for the parameter b, we can find a value of y from the linear equation $y = bx$. The possible values that x can assume are called the *domain* of the function. The possible values that y can assume are called the *range* of the function.

In scalar algebra our interest centers on the description of rules (i.e., the functions) by which pairs of numbers are related. In vector algebra we are interested in the rules by which pairs of vectors or points are related.

In our discussion of vector mappings we consider only single-valued, linear mappings of one vector space onto another, which may, of course, be the same space. Thus, we are interested in cases where both y and x are vectors.

By restricting ourselves to linear transformations—illustrated by $y = bx$ above—we can state three conditions of interest:

1. Every vector of a vector space is transformed into a uniquely determined vector of the space.
2. If $\mathbf{a}$ is transformed to $\mathbf{a}^*$ by a linear transformation $\mathbf{T}$, then $k\mathbf{a}$ is transformed (by $\mathbf{T}$) to $k\mathbf{a}^*$ for any scalar k.
3. If $\mathbf{a}$ is transformed to $\mathbf{a}^*$ and $\mathbf{b}$ to $\mathbf{b}^*$ by a linear transformation $\mathbf{T}$, then $\mathbf{a} + \mathbf{b}$ is transformed (by $\mathbf{T}$) to $\mathbf{a}^* + \mathbf{b}^*$.

Linear transformations are those that satisfy the above conditions. Moreover, any linear transformation can be represented in matrix form (e.g., as the multiplication of a vector by a matrix).

Returning to the topic of mappings, values obtained by a mapping are often called *images*, while the values being transformed are often called *preimages* of the transformation. Here we are mainly concerned with mappings that transform vectors into vectors, that is, mappings that represent vector-valued functions. Hence, both the preimages and the images of the mapping are vectors. Moreover, we shall mostly be concerned with linear transformations that involve square matrices as representations of the transformation so that the two vector spaces, before and after the transformation, are of the same dimensionality.[1]

[1] Or, essentially the same space, before and after the transformation.

4.2.1 Simultaneous, Linear Equations Expressed in Matrix Form

The need for matrix transformation arises quite naturally in the solution of linear equations. Consider the following set of linear equations:

$$x_1{}^* = a_{11}x_1 + a_{12}x_2 + \cdots + a_{1n}x_n$$
$$x_2{}^* = a_{21}x_1 + a_{22}x_2 + \cdots + a_{2n}x_n$$
$$\vdots$$
$$x_n{}^* = a_{n1}x_1 + a_{n2}x_2 + \cdots + a_{nn}x_n$$

which can be written, in matrix form, as

$$\mathbf{x}^* = \mathbf{A}\mathbf{x}$$

As a simple case of expressing a matrix transformation as a set of simultaneous equations, consider the point $\mathbf{x} = \left[\begin{smallmatrix} x_1 \\ x_2 \end{smallmatrix}\right]$ in Fig. 4.1. Also consider the second point $\mathbf{x}^* = \left[\begin{smallmatrix} x_1{}^* \\ x_2{}^* \end{smallmatrix}\right]$ in the same figure. A linear mapping of $\mathbf{x}$ onto $\mathbf{x}^*$ can always be expressed by a set of linear equations:

$$x_1{}^* = a_{11}x_1 + a_{12}x_2; \qquad x_2{}^* = a_{21}x_1 + a_{22}x_2$$

which relate the coordinates $x_1{}^*$, $x_2{}^*$ to the coordinates x_1, x_2 in the standard basis $\mathbf{e}_i$.

But, as noted above, these equations can be expressed in matrix form as

$$\begin{bmatrix} x_1{}^* \\ x_2{}^* \end{bmatrix} = \begin{bmatrix} a_{11} & a_{12} \\ a_{21} & a_{22} \end{bmatrix} \begin{bmatrix} x_1 \\ x_2 \end{bmatrix}$$

where the matrix

$$\begin{bmatrix} a_{11} & a_{12} \\ a_{21} & a_{22} \end{bmatrix}$$

represents the mapping of $\mathbf{x}$, the preimage, onto $\mathbf{x}^*$, the image.

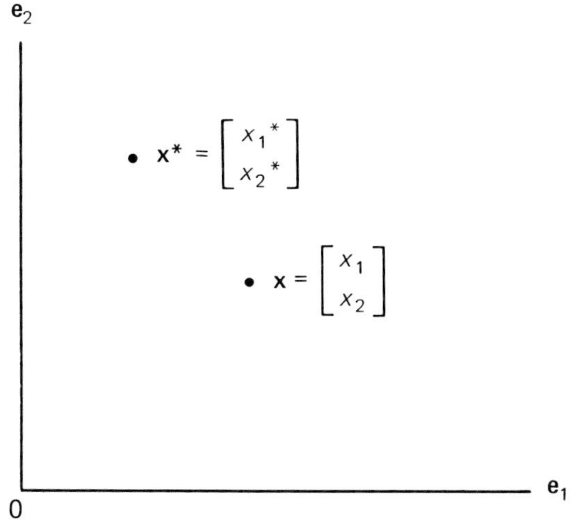

Fig. 4.1 A point transformation.

All of the mappings that we shall be considering are capable of being expressed in terms of a set of linear equations, similar to those illustrated above.

We review some simple examples of matrix transformations, first considered in Chapter 3, and then show how these cases can be extended to more general kinds of linear mappings.

4.2.2 Orthogonal Matrix Transformations

As recalled from Chapter 3, an orthogonal matrix $\mathbf{A}$ is one in which $\mathbf{A}'\mathbf{A} = \mathbf{A}\mathbf{A}' = \mathbf{I}$. That is, rows (and columns) of $\mathbf{A}$ are mutually orthogonal, and each is of unit length. This type of transformation is called a rotation, either proper or improper, depending upon the sign of its determinant.

In the preceding chapter, where we introduced the reader to orthogonal transformations, we recall that rotations were expressed as sets of direction cosines. Although the distinction was not emphasized, we also illustrated two types of rotations:

1. Point rotations, where the original basis vectors remained fixed and the point(s) moved, clockwise or counterclockwise, around the origin.
2. Basis vector rotations, where the original point(s) remained fixed and the basis vectors moved, clockwise or counterclockwise. In this latter case the fixed point was then expressed as a linear combination of the new basis vectors.

Since rotations deal with *relative* motion, either of the above approaches is equally appropriate in interpreting the nature of a rotation.

Figure 4.2 illustrates the four cases that are involved in rotating an arbitrary point $\mathbf{x}$. In Panel I $\mathbf{x}$ undergoes a counterclockwise rotation, mapping onto the point $\mathbf{y}$. Alternatively, we can rotate $\mathbf{x}$ clockwise to map it onto the point $\mathbf{z}$. Note that the standard basis vectors remain fixed throughout both of these rotations.

Panel II of Fig. 4.2 shows the set of basis vector rotations in which the $\mathbf{e}_i$ are mapped onto $\mathbf{f}_i$ via a counterclockwise rotation. Panel III shows the case of mapping the $\mathbf{e}_i$ onto $\mathbf{g}_i$ via a clockwise rotation. In each of these latter two cases, the coordinates of $\mathbf{x}$ are in terms of the new basis vectors. As will be shown, all four types of rotations are variations on a common theme and are related to each other in a straightforward way.

Now, however, we shall want to distinguish more carefully between point transformations and basis vector transformations by adopting a specific notation for each. An image obtained by a point transformation, in which the original basis vectors remain fixed, is denoted by $\mathbf{x}^*$. If obtained by a change in basis vectors—and, hence, the fixed point is referred to the *new* basis—the image is denoted by $\mathbf{x}^\circ$.

Point transformations are much less complicated than basis vector transformations. However, it is important to study both kinds since many multivariate techniques involve situations in which the change of basis vectors simplifies the character of the transformation quite markedly.

As a quick review of the two types of rotations, suppose we choose a standard basis $\mathbf{e}_1 = \left[\begin{smallmatrix}1\\0\end{smallmatrix}\right]$ and $\mathbf{e}_2 = \left[\begin{smallmatrix}0\\1\end{smallmatrix}\right]$ and a point in that space $\mathbf{x} = \left[\begin{smallmatrix}1\\2\end{smallmatrix}\right] = 1\mathbf{e}_1 + 2\mathbf{e}_2$. This linear combination of vectors can also be written as

$$\mathbf{x} = \begin{bmatrix}1\\2\end{bmatrix} = 1\begin{bmatrix}1\\0\end{bmatrix} + 2\begin{bmatrix}0\\1\end{bmatrix}$$

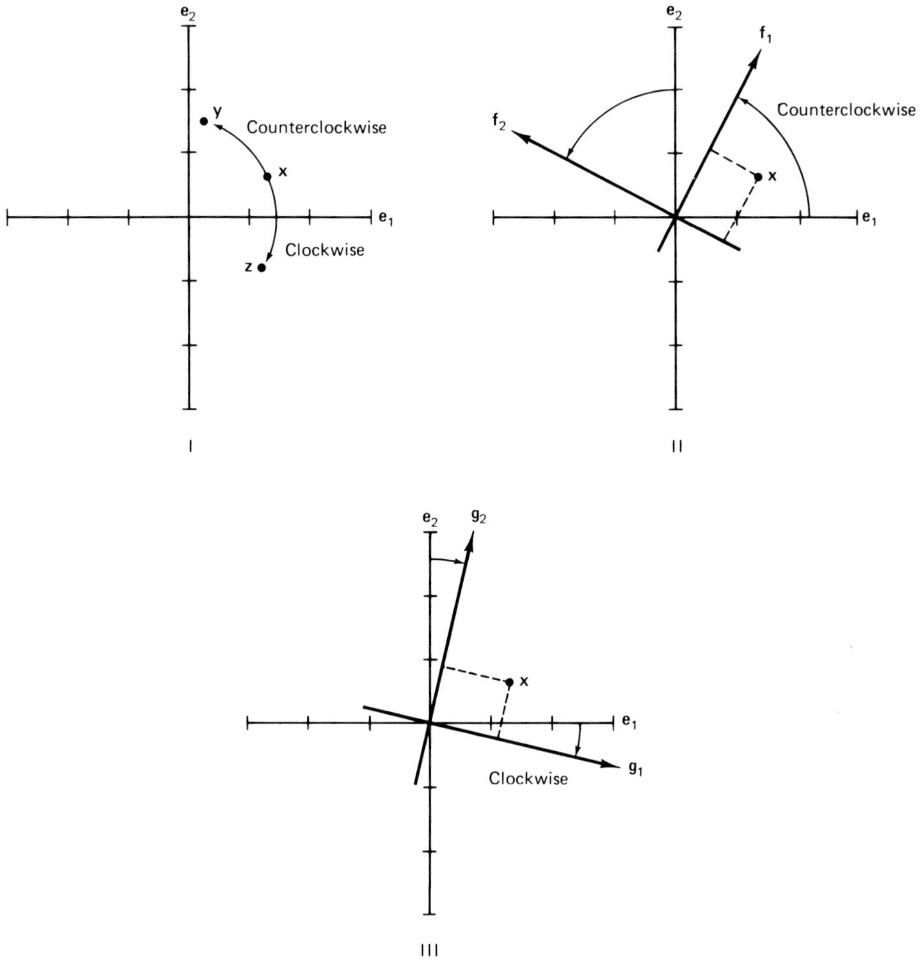

Fig. 4.2 Point and basis vector transformations.

where the standard basis vectors are shown explicitly as column vectors. Let us first consider point rotations and then basis vector rotations.

4.2.2.1 Point Rotations Suppose we examine the rotation of $\mathbf{x} = \left[\begin{smallmatrix}1\\2\end{smallmatrix}\right]$ in a counterclockwise direction through an angle Ψ of 30°. In the case of two dimensions, we have, in general, the equations

$$x_1{}^* = a_{11}x_1 + a_{12}x_2; \qquad x_2{}^* = a_{21}x_1 + c_{22}x_2$$

Here we use $\mathbf{x}^*$ to denote the image of the vector $\mathbf{x}$ under the matrix transformation $\mathbf{A}$.

We assume $\mathbf{A}$ to be orthogonal. The system of equations can be written as

$$\mathbf{x}^* = \mathbf{A}\mathbf{x}$$

Since only two dimensions are involved, we can simplify the problem a bit by relating all direction cosines to the single angle of $\Psi = 30^\circ$, as was illustrated in Chapter 3. Hence for this specific example we have

$$\begin{bmatrix} x_1^* \\ x_2^* \end{bmatrix} = \begin{bmatrix} \cos\Psi & -\sin\Psi \\ \sin\Psi & \cos\Psi \end{bmatrix} \begin{bmatrix} x_1 \\ x_2 \end{bmatrix}$$

$$\begin{bmatrix} -0.13 \\ 2.23 \end{bmatrix} = \begin{bmatrix} 0.87 & -0.50 \\ 0.50 & 0.87 \end{bmatrix} \begin{bmatrix} 1 \\ 2 \end{bmatrix}$$

Notice in this illustration that we made use of the trigonometric facts (involving complementary angles) that

1. $\cos(90^\circ + \Psi) = -\sin\Psi = \cos 120^\circ = -\sin 30^\circ = -0.5$
2. $\cos(90^\circ - \Psi) = \sin\Psi = \cos 60^\circ = \sin 30^\circ = 0.5$

This device allows us to avoid the more complex application of direction cosines involving pairs of angles θ_{ij}, as discussed in Chapter 3, although its use, as recalled, is restricted to two dimensions.

Figure 4.3 shows the results of this mapping. We observe specifically that the new coordinates $\mathbf{x}^* = \left[\begin{smallmatrix} -0.13 \\ 2.23 \end{smallmatrix}\right]$ of the point are still expressed in terms of the old basis, namely, $\mathbf{e}_1$ and $\mathbf{e}_2$. It is the *point*, that is rotated counterclockwise, while the axes maintain their original orientation.

Now let us see what happens when we rotate the point $\mathbf{x} = \left[\begin{smallmatrix} 1 \\ 2 \end{smallmatrix}\right]$ *clockwise* through an angle of $\Psi = 30^\circ$, as given by

$$\begin{bmatrix} x_1^* \\ x_2^* \end{bmatrix} = \begin{bmatrix} \cos\Psi & \sin\Psi \\ -\sin\Psi & \cos\Psi \end{bmatrix} \begin{bmatrix} x_1 \\ x_2 \end{bmatrix}$$

$$\begin{bmatrix} 1.87 \\ 1.23 \end{bmatrix} = \begin{bmatrix} 0.87 & 0.50 \\ -0.50 & 0.87 \end{bmatrix} \begin{bmatrix} 1 \\ 2 \end{bmatrix}$$

In this case the new point coordinates are given, still in terms of the original basis vectors, in Fig. 4.4. Notice, in the case of a clockwise rotation of the point, the matrix $\mathbf{A}'$ is the *transpose* of that ($\mathbf{A}$) used to rotate the point counterclockwise. Finally, if we first rotate the point counterclockwise, given by $\mathbf{A}$, and then rotate it clockwise, given by $\mathbf{A}'$, we end up where we started, since in the case of orthogonal matrices:

$$\mathbf{A}'\mathbf{A} = \mathbf{I}$$

Similarly, had we started with the clockwise rotation and "undone" this by means of the counterclockwise rotation, we would have

$$\mathbf{A}\mathbf{A}' = \mathbf{I}$$

which, again in the special case of an orthogonal matrix, gets us back to where we started.

Point transformations—either orthogonal or more general linear transformations—present relatively few problems and can all be represented simply by

$$\mathbf{x}^* = \mathbf{A}\mathbf{x}$$

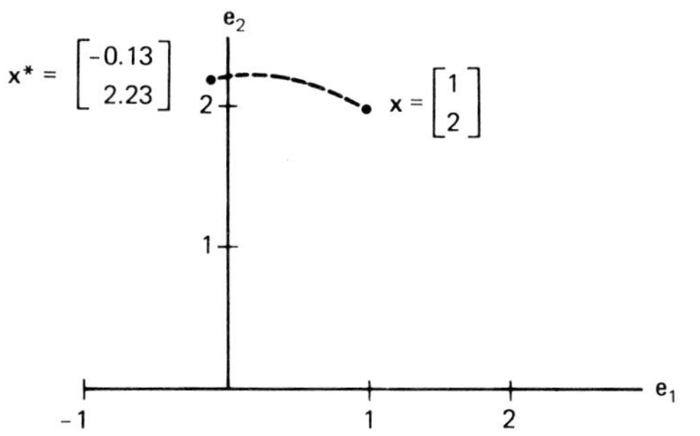

Fig. 4.3

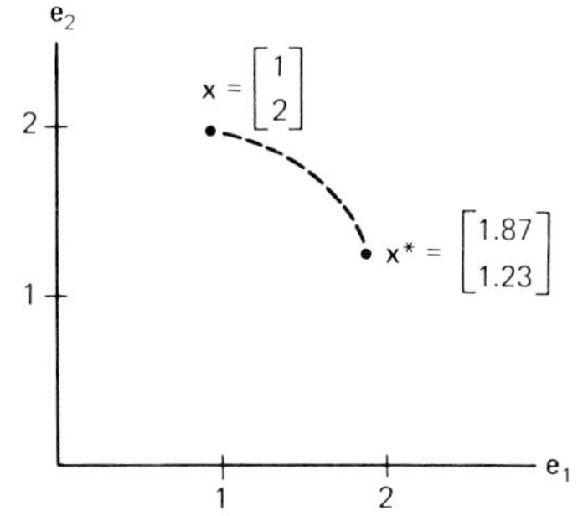

Fig. 4.4

Fig. 4.3 Counterclockwise rotation of point–new coordinates given in terms of old basis vectors.
Fig. 4.4 Clockwise rotation of point–new coordinates given in terms of old basis vectors.

where we should remember, of course, that it is the point(s) or vector(s) that moves relative to a *fixed basis.* Moreover, each row of the transformation matrix represents a linear combination of the *original* point coordinates.

In Chapter 3, however, we also considered the case of rotations which involved *basis vector* changes. In this case the point(s) remains fixed, but is then referred to a set of new (rotated) basis vectors.

4.2.2.2 ***Basis Vector Transformations*** In discussing rotations of basis vectors, we shall want to review a few fundamentals and introduce some new features as well. To be specific, suppose we wish to rotate a set of standard basis vectors clockwise through an angle of 30°. As noted above, the orthogonal matrix that effects this type of rotation is

$$\mathbf{A}' = \begin{bmatrix} \cos\Psi & \sin\Psi \\ -\sin\Psi & \cos\Psi \end{bmatrix} = \begin{bmatrix} 0.87 & 0.50 \\ -0.50 & 0.87 \end{bmatrix}$$

Let us start out with the standard basis vectors $\mathbf{e}_1$ and $\mathbf{e}_2$. Based on our earlier remarks—and as verified by Fig. 4.5—we see that a clockwise rotation of the basis vectors results in a new basis in which the fixed point $\mathbf{x} = \left[\begin{smallmatrix}1\\2\end{smallmatrix}\right]$ is now $\mathbf{x}^\circ = \left[\begin{smallmatrix}-0.13\\2.23\end{smallmatrix}\right]$ in terms of the new basis.[2]

To distinguish the two types of transformations, we let $\mathbf{x}^\circ$ denote the image of $\mathbf{x}$ under a basis vector transformation, while $\mathbf{x}^*$ denotes its image under a point transformation. Note further that the coordinates of $\mathbf{x}^\circ$ are the same as those found when the point was rotated counterclockwise. This is as it should be inasmuch as we are concerned with only relative motion.

The new basis vectors $\mathbf{f}_1$ and $\mathbf{f}_2$ can be expressed in terms of the original as

$$\mathbf{F} = \overset{\mathbf{f}_1 \quad \mathbf{f}_2}{\begin{bmatrix} 0.87 & 0.50 \\ -0.50 & 0.87 \end{bmatrix}} = \overset{\mathbf{E}}{\begin{bmatrix} 1 & 0 \\ 0 & 1 \end{bmatrix}} \overset{\mathbf{A}'}{\begin{bmatrix} 0.87 & 0.50 \\ -0.50 & 0.87 \end{bmatrix}}$$

[2] It is well to remember that the vector $\mathbf{x}$ exists independently of its coordinates. However, the specific *coordinate* values that it assumes *are* dependent upon the basis vectors to which it is referred.

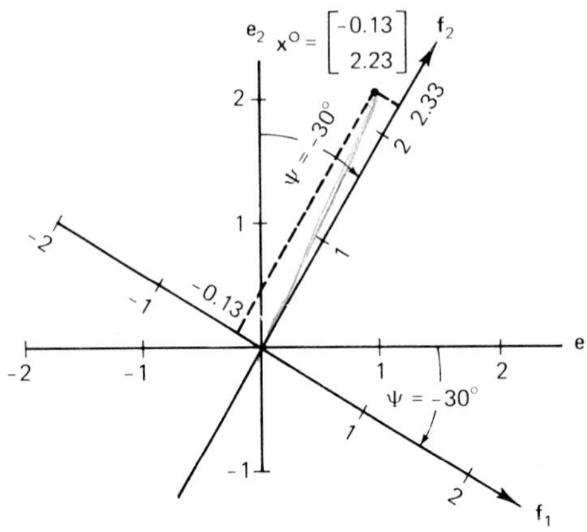

Fig. 4.5

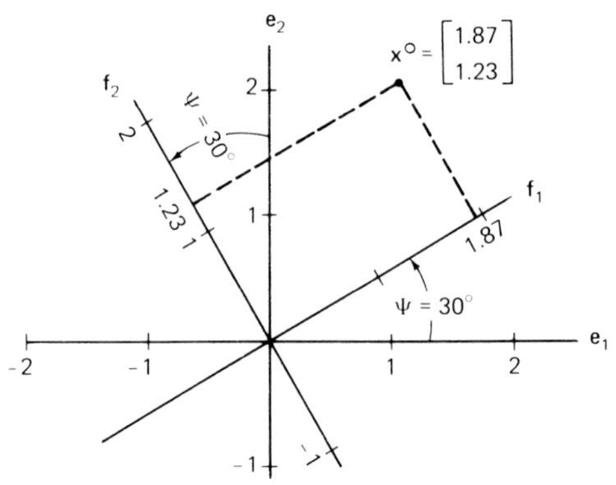

Fig. 4.6

Fig. 4.5 Clockwise rotation of axes—old coordinates given in terms of new basis vectors.

Fig. 4.6 Counterclockwise rotation of axes—old coordinates given in terms of new basis vectors.

Here we employ $\mathbf{E}$ to denote the matrix of the original (standard) basis and $\mathbf{F}$ to denote the matrix of the new basis. Also, note that the columns of $\mathbf{F}$ represent linear combinations of the (column) basis vectors of $\mathbf{E}$.

As can be seen in Fig. 4.5, $\mathbf{f}_1$, the first column of $\mathbf{F}$, passes through the point (0.87, –0.5) positioned in the original $\mathbf{e}_i$ basis, while $\mathbf{f}_2$ passes through the point (0.5, 0.87) in the original $\mathbf{e}_i$ basis.

In this case the original vector $\mathbf{x} = \begin{bmatrix}1\\2\end{bmatrix}$ is now $\mathbf{x}^\circ = \begin{bmatrix}-0.13\\2.23\end{bmatrix}$ when referred to the $\mathbf{f}_1$, $\mathbf{f}_2$ basis. Recall that these are the coordinates found by a 30° *counterclockwise* rotation of the point in the original basis.

Similarly, if the original basis vectors are rotated counterclockwise, this gives us the same result as found by rotating the point clockwise. A picture of this change in basis vectors is shown in Fig. 4.6.

At this point it might sound confusing and redundant to have two ways, point and basis vector changes, for expressing linear transformations. As we shall try to show later, however, there are advantages to considering basis vector as well as point transformations. This is particularly true when the transformation matrix is not orthogonal; that is, when the transformation does not entail a simple rotation.

To sum up, if we wish to find the coordinates of transformed $\mathbf{x}^*$ relative to the standard $\mathbf{e}_1$, $\mathbf{e}_2$ basis (i.e., to move the point relative to the fixed basis), we use the point transformation

$$\mathbf{x}^* = \mathbf{A}\mathbf{x}$$

where $\mathbf{x}$ is originally referred to the $\mathbf{e}_1$, $\mathbf{e}_2$ basis. In this case the coordinates of the point $\mathbf{x}^*$ are also expressed directly in terms of $\mathbf{e}_1$, $\mathbf{e}_2$. (See Figs. 4.3 and 4.4.)

Alternatively,we may care to transform the $\mathbf{e}_i$ *basis itself* to a new set of basis vectors $\mathbf{f}_i$. In this case it is the new basis vectors that are expressed in terms of $\mathbf{e}_i$, while the old coordinates of the point $\mathbf{x}$ are now expressed as $\mathbf{x}^\circ$ in terms of the new basis $\mathbf{f}_i$. (See Figs. 4.5 and 4.6.) Notice, then, that the values of the new coordinates *depend on which method we use to carry out the transformation.*

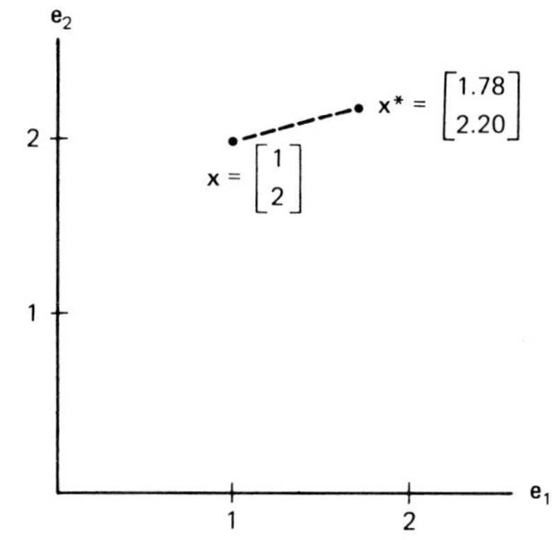

Fig. 4.7 Point transformation of $\mathbf{x}$ with fixed basis vectors.

4.2.3 Generalizing the Results

It is now of interest to describe what goes on when we do *not* restrict ourselves to rotations. Consider the more general transformation matrix

$$\mathbf{T} = \begin{bmatrix} 0.90 & 0.44 \\ 0.60 & 0.80 \end{bmatrix}$$

and the vector $\mathbf{x} = \left[\begin{smallmatrix}1\\2\end{smallmatrix}\right]$, again relative to the standard basis vectors

$$\mathbf{e}_1 = \begin{bmatrix} 1 \\ 0 \end{bmatrix} \quad \text{and} \quad \mathbf{e}_2 = \begin{bmatrix} 0 \\ 1 \end{bmatrix}$$

Application of a point transformation to $\mathbf{x}$ (retaining the original basis) is quite straightforward and is shown in Fig. 4.7. Here we see that $\mathbf{x}$ moves to the position

$$\mathbf{x}^* = \begin{bmatrix} x_1^* \\ x_2^* \end{bmatrix} = \begin{bmatrix} 1.78 \\ 2.20 \end{bmatrix} = \begin{bmatrix} 0.90 & 0.44 \\ 0.60 & 0.80 \end{bmatrix} \begin{bmatrix} 1 \\ 2 \end{bmatrix}$$

However, notice that this point movement does *not* involve a simple rotation. Still, the process of finding the image vector $\mathbf{x}^*$ presents no new problems. That is, we still have a straightforward task of premultiplying the original point by the matrix of the transformation $\mathbf{T}$. Moreover, $\mathbf{x}^*$ is still referred to the original $\mathbf{e}_i$ basis.

Unfortunately, things are not so simple when we attempt to construct the counterpart basis vector transformation. Unlike the special case of rotation, the present transformation matrix $\mathbf{T}$ presents complications in referring $\mathbf{x}$ to a new basis $\mathbf{f}_i$. Rather than try to solve this problem now, we can address ourselves to a related problem in point transformations. Solution of this related problem will pave the way for handling the basis vector transformation in the case of the linear transformation matrix $\mathbf{T}$.

The related problem can be expressed as follows: Suppose we were given the point $\mathbf{x}^* = \left[\begin{smallmatrix}1.78\\2.20\end{smallmatrix}\right]$ in Fig. 4.7 to begin with and wanted to find $\mathbf{x}$, knowing only the transformation matrix $\mathbf{T}$ and the original basis $\mathbf{E}$. In other words, we now wish to find a transformation, call it $\mathbf{T}^{-1}$, that will get us *back* to $\mathbf{x}$, given the transformed coordinates

$\mathbf{x}^*$. For this task we shall need to describe the nature of $\mathbf{T}^{-1}$, the *inverse* of $\mathbf{T}$. Discussion of matrix inversion will require a brief digression, after which we can return to the particular problem at hand.

4.3 MATRIX INVERSION

In Chapter 2 we briefly described a special diagonal matrix, referred to as the identity matrix $\mathbf{I}$. As discussed there, $\mathbf{I}$ consists of a square matrix with 1's along the main diagonal and zeros elsewhere. As will be shown, the identity matrix– and the concept of matrix inverse–play special roles in the matrix algebra "equivalent" of division (or, more appropriately, multiplication by a reciprocal). We recall from Chapters 2 and 3 that although we have discussed addition, subtraction, and multiplication as operations in matrix algebra, we have not discussed, as yet, the companion operation of division. As we shall see, matrix inversion in linear algebra is analogous to the operation of division in scalar algebra.

The identity matrix $\mathbf{I}$ plays a special role in matrix algebra. The effect of pre- or postmultiplying any conformable matrix by $\mathbf{I}$ is to leave the original matrix unchanged:

$$\mathbf{IA} = \mathbf{AI} = \mathbf{A}$$

That is, $\mathbf{I}$ in matrix algebra plays the role of the number 1 in ordinary arithmetic. We now ask if there is a type of matrix that is analogous to the reciprocal of a number, that is, a number, $a \neq 0$, for which the relation $ax = 1$ is true. If so, we should be able to develop the concept of "division" in the context of matrix algebra.

In the case of *square matrices* there is an analogue, in some instances, to the notion of a reciprocal in scalar arithmetic. This is called an *inverse.* As a matter of fact, matrix inverses–in a very generalized sense–can be obtained for rectangular matrices too. We discuss this more advanced topic in Appendix B while here we confine ourselves to *regular* inverses of *square* matrices.

A regular inverse of the (square) matrix $\mathbf{A}$ *is denoted by* $\mathbf{A}^{-1}$ *and, when it exists, it is unique and satisfies the following relations:*

$$\boxed{\mathbf{AA}^{-1} = \mathbf{A}^{-1}\mathbf{A} = \mathbf{I}}$$

For example, let us take the matrix

$$\mathbf{A} = \begin{bmatrix} 1 & 2 & 0 \\ 0 & 3 & 1 \\ 2 & 2 & 4 \end{bmatrix}$$

Assume, now, that we have gone on to solve for $\mathbf{A}^{-1}$ and have found it to be

$$\mathbf{A}^{-1} = \begin{bmatrix} 10/14 & -8/14 & 2/14 \\ 2/14 & 4/14 & -1/14 \\ -6/14 & 2/14 & 3/14 \end{bmatrix}$$

Then the relation $\mathbf{AA}^{-1}$ (or $\mathbf{A}^{-1}\mathbf{A}$) = **I** should hold. The reader can verify that it does:

$$\underset{\mathbf{A}}{\begin{bmatrix} 1 & 2 & 0 \\ 0 & 3 & 1 \\ 2 & 2 & 4 \end{bmatrix}} \underset{\mathbf{A}^{-1}}{\begin{bmatrix} 10/14 & -8/14 & 2/14 \\ 2/14 & 4/14 & -1/14 \\ -6/14 & 2/14 & 3/14 \end{bmatrix}} = \underset{\mathbf{I}}{\begin{bmatrix} 1 & 0 & 0 \\ 0 & 1 & 0 \\ 0 & 0 & 1 \end{bmatrix}}$$

The problem, of course, is to find $\mathbf{A}^{-1}$ when it exists.[3] As mentioned above, if $\mathbf{A}^{-1}$ exists, it will be unique for a given **A**.

4.3.1 The Determinant and the Adjoint of a Matrix

In Chapter 2 we also discussed the concept of the determinant of a (square) matrix. The determinant of **A**, denoted as $|\mathbf{A}|$, is merely a scalar or number that is computed in a certain way. We also discussed cofactors of a matrix and defined them to be determinants of order $n-1$ by $n-1$ obtained from a matrix **A** of order n by n by omitting the ith row and $j$the column of **A** and affixing the sign $(-1)^{i+j}$ to the determinant of the $n-1$ by $n-1$ submatrix. We further recall that a matrix **A** will have as many cofactors as there are entries in **A**.

The cofactors themselves can be, in turn, transformed in a way that possesses some special characteristics relative to **A** and the identity matrix. This matrix, called the adjoint of **A**, possesses the useful property that[4]

$$\boxed{\mathbf{A}\left[\frac{\operatorname{adj}(\mathbf{A})}{|\mathbf{A}|}\right] = \mathbf{I}}$$

We now need to define the adjoint of **A** in terms of the cofactors of **A**.

The adjoint of a (square) matrix **A**, *denoted as* adj(**A**), *is defined as the transpose of the matrix of cofactors obtained from* **A**.

[3] If each column of $\mathbf{A}^{-1}$ is considered originally as a set of unknowns, then all that is involved is solving a set of linear equations. For example,

$$\begin{aligned} 1x_1 + 2x_2 + 0x_3 &= 1 \\ 0x_1 + 3x_2 + 1x_3 &= 0 \\ 2x_1 + 2x_2 + 4x_3 &= 0 \end{aligned}$$

has the solution $x_1 = 10/14$, $x_2 = 2/14$, $x_3 = -6/14$, which is the first column of $\mathbf{A}^{-1}$. Similar procedures lead to the second and third columns of $\mathbf{A}^{-1}$. That is, $\mathbf{A}^{-1}$ could be found by solving three sets of linear equations each, in which the right-hand side of the equation is, respectively, the column vectors

$$\begin{bmatrix} 1 \\ 0 \\ 0 \end{bmatrix}, \quad \begin{bmatrix} 0 \\ 1 \\ 0 \end{bmatrix}, \quad \text{and} \quad \begin{bmatrix} 0 \\ 0 \\ 1 \end{bmatrix}$$

Details of the method appear in Section 4.7.5.

[4] As the reader might surmise, this property is also displayed by $\mathbf{A}^{-1}$; as will be shown, the inverse $\mathbf{A}^{-1}$ *does* equal the adjoint of **A** divided by the determinant of **A**.

In the simple case of a 2 x 2 matrix

$$\mathbf{A} = \begin{bmatrix} a_{11} & a_{12} \\ a_{21} & a_{22} \end{bmatrix}$$

we recall that the determinant $|\mathbf{A}|$ is

$$a_{11}a_{22} - a_{12}a_{21}$$

and the cofactors—consisting of single elements—are

$$A_{11} = a_{22}; \qquad A_{12} = -a_{21}; \qquad A_{21} = -a_{12}; \qquad A_{22} = a_{11}$$

Let us now place these cofactors in a square (2 x 2) matrix. Next, let us take the transpose of this matrix.

Then, the adjoint of **A**, in the 2 x 2 case, is just

$$\boxed{\operatorname{adj}(\mathbf{A}) = \begin{bmatrix} a_{22} & -a_{12} \\ -a_{21} & a_{11} \end{bmatrix}}$$

In words, this says that in the 2 x 2 case, the adjoint of the matrix **A** involves (a) switching the entries along the main diagonal of the original matrix and (b) changing signs of the off-diagonal entries.

If the $n \times n$ matrix **A** is of higher order than 2 x 2, computation of adj(**A**) is a bit more complicated but proceeds in the same manner as stated above:

1. Find the minors of **A** via the procedure described in Chapter 2.
2. Find the cofactors, or signed minors of **A**, again via the procedure described in Chapter 2.
3. Place these cofactors in an $n \times n$ matrix.
4. Find the transpose of this matrix and call this transpose adj(**A**).

From here it is but a short step to finding the inverse $\mathbf{A}^{-1}$

4.3.2 The Matrix Inverse

Both the determinant and adjoint of **A** figure prominently in the computation of its inverse. *If the square matrix* **A** *has an inverse* $\mathbf{A}^{-1}$, *this inverse, defined such that* $\mathbf{A}\mathbf{A}^{-1} = \mathbf{A}^{-1}\mathbf{A} = \mathbf{I}$, *can be computed from*

$$\boxed{\mathbf{A}^{-1} = \frac{1}{|\mathbf{A}|}\operatorname{adj}(\mathbf{A}); \qquad |\mathbf{A}| \neq 0}$$

That is, the inverse is found by scalar multiplication of a matrix. In the simple 2 x 2 case described above, this is

$$\mathbf{A}^{-1} = \frac{1}{|\mathbf{A}|}\begin{bmatrix} a_{22} & -a_{12} \\ -a_{21} & a_{11} \end{bmatrix}; \qquad |\mathbf{A}| \neq 0$$

We now have an operational way, based on the transposed matrix of cofactors divided by the determinant of the matrix, to find the inverse of a square matrix **A**. We next consider this problem in the earlier context of point transformations.

4.3.3 Applying the Concept of an Inverse to Point Transformations

Suppose we now return to the problem of finding the preimage vector **x**, given the image vector **x*** and the transformation matrix

$$\mathbf{T} = \begin{bmatrix} 0.90 & 0.44 \\ 0.60 & 0.80 \end{bmatrix}$$

As recalled from Section 4.2.3, $\mathbf{x}^* = \begin{bmatrix} 1.78 \\ 2.20 \end{bmatrix}$ is shown in Fig. 4.7. Now we wish to find **x** which, of course, we already know to be $\mathbf{x} = \begin{bmatrix} 1 \\ 2 \end{bmatrix}$. The starting equation is $\mathbf{x}^* = \mathbf{Tx}$, and we wish to solve for **x**.

By taking advantage of the fact that $\mathbf{T}^{-1}\mathbf{T} = \mathbf{I}$, we can find **x** from a matrix transformation that premultiplies both sides of $\mathbf{x}^* = \mathbf{Tx}$ by $\mathbf{T}^{-1}$. Thus

$$\mathbf{T}^{-1}\mathbf{x}^* = \mathbf{T}^{-1}\mathbf{Tx} = \mathbf{Ix} = \mathbf{x}$$

In terms of the specific problem of interest, we need to perform the following calculations. First, we obtain the determinant of **T**:

$$|\mathbf{T}| = 0.9(0.8) - 0.6(0.44) = 0.46$$

Based on the simple definition of the adjoint in the 2 x 2 case, we find

$$\operatorname{adj}(\mathbf{T}) = \begin{bmatrix} t_{22} & -t_{12} \\ -t_{21} & t_{11} \end{bmatrix} = \begin{bmatrix} 0.80 & -0.44 \\ -0.60 & 0.90 \end{bmatrix}$$

Having found both $|\mathbf{T}|$ and adj(**T**), we compute $\mathbf{T}^{-1}$ as follows:

$$\mathbf{T}^{-1} = \frac{1}{0.46}\begin{bmatrix} 0.80 & -0.44 \\ -0.60 & 0.90 \end{bmatrix} = \begin{bmatrix} 1.74 & -0.96 \\ -1.30 & 1.96 \end{bmatrix}$$

It now remains to show that

$$\mathbf{x} = \begin{bmatrix} 1 \\ 2 \end{bmatrix} = \overset{\mathbf{T}^{-1}}{\begin{bmatrix} 1.74 & -0.96 \\ -1.30 & 1.96 \end{bmatrix}} \overset{\mathbf{x}^*}{\begin{bmatrix} 1.78 \\ 2.20 \end{bmatrix}}$$

which is, indeed, the case.

We can also verify that, within rounding error, $\mathbf{TT}^{-1} = \mathbf{T}^{-1}\mathbf{T} = \mathbf{I}$. Finally, we should state that if $|\mathbf{T}|$ is zero, then $1/|\mathbf{T}|$ is not defined, and the (regular) inverse of **T** does not exist. In this case the matrix **T** is said to be *singular.* Otherwise, as is the case here, it is called nonsingular.

A nonsingular matrix **A**, *then, is one in which*

$$\boxed{|\mathbf{A}| \neq 0}$$

Nonsingularity is very important to the topic of matrix inversion since every nonsingular matrix has an inverse; moreover, only nonsingular matrices have (regular) inverses.

Now that we have found out how to compute a matrix inverse and solve the equation

$$\mathbf{x} = \mathbf{T}^{-1}\mathbf{x}^*$$

we should also state a property involving the inverse of the product of two (or more) matrices.

Given the product of two or more conformable matrices, $\mathbf{T}_1\mathbf{T}_2 \cdots \mathbf{T}_s$, *the inverse of that product equals the product of the separate inverses in reverse order:*

$$\boxed{(\mathbf{T}_1\mathbf{T}_2 \cdots \mathbf{T}_s)^{-1} = \mathbf{T}_s^{-1} \cdots \mathbf{T}_2^{-1}\mathbf{T}_1^{-1}}$$

Notice that this property is similar to the property involving the transpose of the product of two or more matrices.

Having discussed some introductory aspects of matrix inversion, we return to the topic of vector transformation, but now in the context of changing basis vectors for the case of general linear transformations. As it turns out, the concept of matrix inverse is also needed here.

4.3.4 Transformation by Basis Vector Changes

As recalled in our earlier discussion of basis vector changes in the context of orthogonal transformations, a second way to examine transformations is in terms of referring some vector $\mathbf{x}$ to a new set of basis vectors $\mathbf{f}_i$. Let us return to the discussion involving transformations using the matrix $\mathbf{T}$.

$$\mathbf{T} = \begin{bmatrix} 0.9 & 0.44 \\ 0.6 & 0.8 \end{bmatrix}$$

Our interest now centers on the case of transformation via changed basis vectors where, as we know, $\mathbf{T}$ is *not* orthogonal. To find the new basis vectors $\mathbf{f}_i$ in the current problem of interest, we make note of the fact that linear combinations of the standard basis vectors $\mathbf{e}_1$ and $\mathbf{e}_2$ are found by

$$\mathbf{f}_1 = \begin{bmatrix} 0.90 \\ 0.44 \end{bmatrix} = 0.9\begin{bmatrix} 1 \\ 0 \end{bmatrix} + 0.44\begin{bmatrix} 0 \\ 1 \end{bmatrix}; \qquad \mathbf{f}_2 = \begin{bmatrix} 0.6 \\ 0.8 \end{bmatrix} = 0.6\begin{bmatrix} 1 \\ 0 \end{bmatrix} + 0.8\begin{bmatrix} 0 \\ 1 \end{bmatrix}$$

To express this in matrix form, where the $\mathbf{f}_i$ also appear as column vectors, we have

$$\mathbf{F} = \begin{bmatrix} 0.90 & 0.60 \\ 0.44 & 0.80 \end{bmatrix} = \overset{\mathbf{E}}{\begin{bmatrix} 1 & 0 \\ 0 & 1 \end{bmatrix}} \overset{\mathbf{T}'}{\begin{bmatrix} 0.90 & 0.60 \\ 0.44 & 0.80 \end{bmatrix}}$$

Notice, then, that the new basis vectors are given by the transpose of the matrix $\mathbf{T}$, where $\mathbf{T}$ itself was used in the point transformation in which $\mathbf{x}$ moved relative to the fixed $\mathbf{e}_i$ basis.

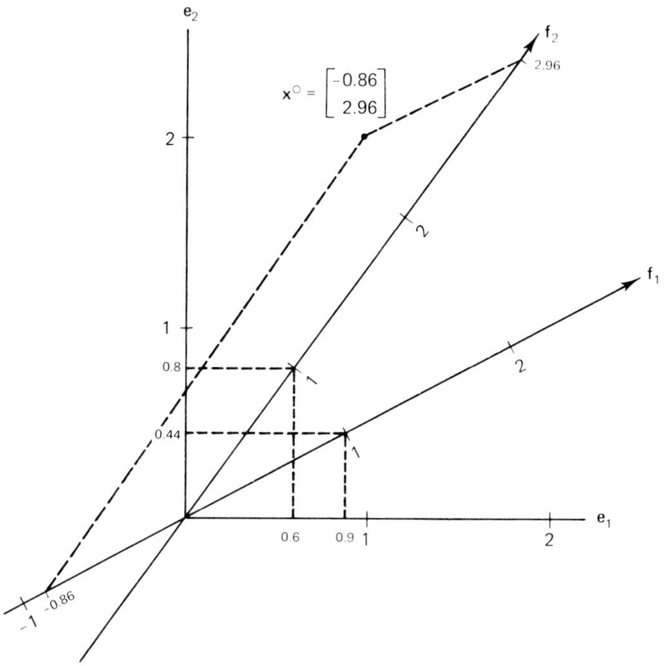

Fig. 4.8 Transformation of x via change in basis vectors.

Fig. 4.8 shows the new basis vectors $\mathbf{f}_1$ and $\mathbf{f}_2$ plotted in terms of the old basis. We note from the figure that the new basis vectors $\mathbf{f}_1, \mathbf{f}_2$ are oblique. Our problem now is to refer the original $\mathbf{x} = \begin{bmatrix} 1 \\ 2 \end{bmatrix}$, as portrayed in $\mathbf{E}$, to the new (oblique) basis in $\mathbf{F}$. As we know, the point transformation $\mathbf{T}$ yields the point $\mathbf{x}^* = \begin{bmatrix} 1.78 \\ 2.20 \end{bmatrix}$, shown in the original $\mathbf{e}_i$ basis of Fig. 4.7.

But we want to find the coordinates of the points, as referred to $\mathbf{F}$. To do this we note that a point can either be expressed as $\mathbf{x}$ in $\mathbf{E}$ or as $\mathbf{x}^\circ$ in $\mathbf{F}$.

$$\mathbf{F}\mathbf{x}^\circ = \mathbf{E}\mathbf{x}$$

But, as we know from the basis change shown above,

$$\mathbf{F} = \mathbf{E}\mathbf{T}'$$

Hence, to solve for $\mathbf{x}^\circ$ in $\mathbf{F}\mathbf{x}^\circ = \mathbf{E}\mathbf{x}$, we can premultiply both sides by $\mathbf{F}^{-1}$ to get

$$\mathbf{F}^{-1}\mathbf{F}\mathbf{x}^\circ = \mathbf{F}^{-1}\mathbf{E}\mathbf{x}$$

Noting that $\mathbf{F}^{-1}\mathbf{F} = \mathbf{I}$ on the left-hand side and substituting $\mathbf{E}\mathbf{T}'$ for $\mathbf{F}$ on the right-hand side, we recall the relationship involving the inverse of the product of two or more matrices shown in the preceding section to get

$$\mathbf{x}^\circ = (\mathbf{E}\mathbf{T}')^{-1}\mathbf{E}\mathbf{x} = (\mathbf{T}')^{-1}\mathbf{E}^{-1}\mathbf{E}\mathbf{x}$$

But, $\mathbf{E}^{-1}\mathbf{E} = \mathbf{I}$, so that

$$\boxed{\mathbf{x}^\circ = (\mathbf{T}')^{-1}\mathbf{x}}$$

What this all says is that to find $\mathbf{x}^\circ$ in terms of the new basis vectors $\mathbf{f}_i$ we are going to have to find $(\mathbf{T}')^{-1}$, the *inverse* of $\mathbf{T}'$.

We have already found $\mathbf{T}^{-1}$ in the case of moving $\mathbf{x}^* = \begin{bmatrix} 1.78 \\ 2.20 \end{bmatrix}$ via a point transformation back to $\mathbf{x} = \begin{bmatrix} 1 \\ 2 \end{bmatrix}$. We use the same type of matrix inversion procedure to find $(\mathbf{T}')^{-1}$, and this turns out to be

$$(\mathbf{T}')^{-1} = \begin{bmatrix} 1.74 & -1.30 \\ -0.96 & 1.96 \end{bmatrix}$$

Having found $(\mathbf{T}')^{-1}$, we then solve for $\mathbf{x}^\circ$ as

$$\mathbf{x}^\circ = \begin{bmatrix} -0.86 \\ 2.96 \end{bmatrix} = \begin{bmatrix} 1.74 & -1.30 \\ -0.96 & 1.96 \end{bmatrix} \begin{bmatrix} 1 \\ 2 \end{bmatrix}$$

Figure 4.8 shows the transformed coordinates, $\mathbf{x}^\circ = \begin{bmatrix} -0.86 \\ 2.96 \end{bmatrix}$ of the original and fixed point $\mathbf{x}$, now expressed in terms of the new (oblique) basis vectors $\mathbf{f}_i$.

How does this result relate to our earlier discussion involving basis vector changes that are given by orthogonal matrices? *In this case it turns out that for orthogonal matrices the following relationship holds:*

$$\boxed{\mathbf{A}^{-1} = \mathbf{A}'}$$

Hence, in the special case of orthogonal matrices we find the relationship

$$\mathbf{x}^\circ = (\mathbf{A}')^{-1}\,\mathbf{x} = \mathbf{A}\mathbf{x}$$

in solving the general equation

$$\mathbf{F}\mathbf{x}^\circ = \mathbf{E}\mathbf{x}$$

for $\mathbf{x}^\circ$. This, of course, is what we expect to find in the case of rotations. In other cases involving basis vector changes that employ general linear transformations such as $\mathbf{T}$, the appropriate expression, as we now know, is

$$\mathbf{x}^\circ = (\mathbf{T}')^{-1}\,\mathbf{x}$$

So, in summary, we can recapitulate the following points:

1. For point transformations, the simpler type of transformation, we have

   a. $\mathbf{x}^* = \mathbf{T}\mathbf{x}$ with $\mathbf{x}^*$ and $\mathbf{x}$ in terms of $\mathbf{E}$

   b. $\mathbf{x} = \mathbf{T}^{-1}\mathbf{x}^*$ with $\mathbf{x}$ and $\mathbf{x}^*$ in terms of $\mathbf{E}$

2. For transformations involving basis vector changes, the new basis vectors are given in terms of the old by $\mathbf{F} = \mathbf{E}\mathbf{T}'$. We then have the cases

   a. $\mathbf{x}^\circ = (\mathbf{T}')^{-1}\mathbf{x}$ with $\mathbf{x}^\circ$ in terms of $\mathbf{F}$ and $\mathbf{x}$ in terms of $\mathbf{E}$

   b. $\mathbf{x} = \mathbf{T}'\mathbf{x}^\circ$ with $\mathbf{x}$ in terms of $\mathbf{E}$ and $\mathbf{x}^\circ$ in terms of $\mathbf{F}$

Finally, in the special case of orthogonal matrices, denoted by **A**, we have

1. For point transformations:

   a. $\mathbf{x}^* = \mathbf{A}\mathbf{x}$ with $\mathbf{x}^*$ and $\mathbf{x}$ in terms of **E**

   b. $\mathbf{x} = \mathbf{A}'\mathbf{x}^*$ with $\mathbf{x}$ and $\mathbf{x}^*$ in terms of **E**

2. For transformations involving basis vector changes, the new basis vectors are still given by the matrix $\mathbf{F} = \mathbf{E}\mathbf{A}'$. We then have the special cases:

   a. $\mathbf{x}^\circ = \mathbf{A}\mathbf{x}$ with $\mathbf{x}^\circ$ in terms of **F** and $\mathbf{x}$ in terms of **E**

   b. $\mathbf{x} = \mathbf{A}'\mathbf{x}^\circ$ with $\mathbf{x}$ in terms of **E** and $\mathbf{x}^\circ$ in terms of **F**

As the reader has probably gathered by now, point transformation, in which the basis vectors remain fixed, is considerably easier to follow intuitively. However, instances arise in multivariate analyses where the selection of an appropriate basis to which the vectors can be referred results in a significant degree of simplification in characterizing the nature of the transformation that the researcher is employing.

For this reason, we carry our analysis one more step, albeit the most complex one so far. We can pose the problem as one of starting with a point transformation of $\mathbf{x}$, as given by the matrix **T**, relative to the standard basis vectors $\mathbf{e}_i$.

However, suppose the $\mathbf{e}_i$ basis is related, in turn, to a new basis $\mathbf{f}_i$ via a matrix **L**. If so, how is the point transformation given by **T** in the context of $\mathbf{e}_i$ represented in the new basis $\mathbf{f}_i$? We shall call the new transformation matrix $\mathbf{T}^\circ$ to take into consideration the fact that it transforms points in the new basis $\mathbf{f}_i$.

The practical import of all of this is that in applied multivariate problems we often seek special sets of basis vectors in which the transformation matrix $\mathbf{T}^\circ$ displays a particularly simple character. This pragmatic aspect of basis vector transformations is deferred until Chapter 5. However, here we can at least go through the mechanics of the process of relating a transformation represented by **T** in the standard basis $\mathbf{e}_i$ to its counterpart $\mathbf{T}^\circ$ in the derived basis $\mathbf{f}_i$.

4.3.5 Transformations under Arbitraιy Changes of Basis Vectors

Suppose we continue to consider the matrix

$$\mathbf{T} = \begin{bmatrix} 0.90 & 0.44 \\ 0.60 & 0.80 \end{bmatrix}$$

but let us also consider a second matrix

$$\mathbf{L} = \begin{bmatrix} 0.83 & 0.55 \\ 0.20 & 0.98 \end{bmatrix}$$

that can be used to transform the vectors in the standard basis $\mathbf{e}_i$ to a new basis $\mathbf{f}_i$. Note that the sums of squares of the row elements of **L** equal unity. Hence, the rows of **L** can be considered as direction cosines. However, since the scalar product of row 1 with row 2 does not equal zero, the new $\mathbf{f}_i$ basis is oblique.

As before, our first job is to find the oblique basis $\mathbf{f}_i$ in terms of $\mathbf{e}_i$, the standard basis. Again, we define the column vectors $\mathbf{f}_i$ as follows:

$$\mathbf{f}_1 = 0.83\mathbf{e}_1 + 0.55\mathbf{e}_2; \qquad \mathbf{f}_2 = 0.20\mathbf{e}_1 + 0.98\mathbf{e}_2$$

which, in matrix multiplication form, are obtained as column vectors from

$$\mathbf{F} = \begin{bmatrix} 0.83 & 0.20 \\ 0.55 & 0.98 \end{bmatrix} = \overset{\mathbf{E}}{\begin{bmatrix} 1 & 0 \\ 0 & 1 \end{bmatrix}} \overset{\mathbf{L}'}{\begin{bmatrix} 0.83 & 0.20 \\ 0.55 & 0.98 \end{bmatrix}}$$

where, of course, we have the linear combinations

$$\mathbf{f}_1 = 0.83\begin{bmatrix} 1 \\ 0 \end{bmatrix} + 0.55\begin{bmatrix} 0 \\ 1 \end{bmatrix}; \qquad \mathbf{f}_2 = 0.20\begin{bmatrix} 1 \\ 0 \end{bmatrix} + 0.98\begin{bmatrix} 0 \\ 1 \end{bmatrix}$$

Panel II of Fig. 4.9 shows the oblique basis vectors $\mathbf{f}_1$ and $\mathbf{f}_2$ plotted in terms of the original basis $\mathbf{e}_i$.

From earlier discussion we know how to find the point transformation:

$$\mathbf{x}^* = \mathbf{T}\mathbf{x}$$

From Fig. 4.7 we note that $\mathbf{x}^* = \left[\begin{smallmatrix} 1.78 \\ 2.20 \end{smallmatrix}\right]$. For ease of comparison this reappears in Panel I of Fig. 4 9 relative to the original basis $\mathbf{E}$.

Our objective now is to find the counterpart point transformation of $\mathbf{x}^\circ = \left[\begin{smallmatrix} 1 \\ 2 \end{smallmatrix}\right]$, referred now to the oblique basis $\mathbf{f}_i$ in Panel II of Fig. 4.9.

Let us first present the solution to this problem and then examine it, piece by piece. First, we have the relationship

$$\boxed{\mathbf{x}^{*\circ} = \mathbf{T}^\circ\mathbf{x}^\circ}$$

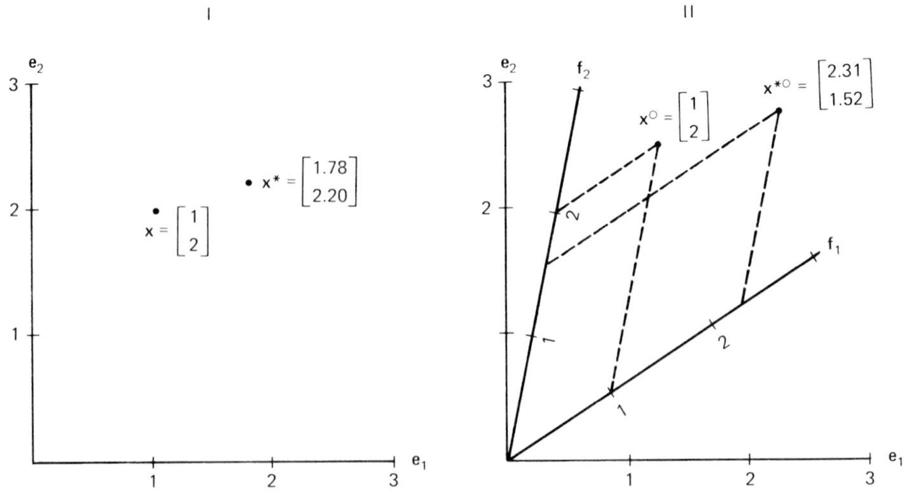

Fig. 4.9 Point transformations relative to different bases.

but $\mathbf{T}^\circ$ is found from the product

$$\boxed{\mathbf{T}^\circ = (\mathbf{L}')^{-1}\mathbf{T}\mathbf{L}'}$$

Since we already have $\mathbf{L}'$ and $\mathbf{T}$ above, the problem now involves the computation of $(\mathbf{L}')^{-1}$. The determinant of $\mathbf{L}'$ is

$$|\mathbf{L}'| = (0.83 \times 0.98) - (0.20 \times 0.55) = 0.703$$

We then need to find adj($\mathbf{L}'$) and, finally, $(\mathbf{L}')^{-1}$. This is carried out as follows:

$$(\mathbf{L}')^{-1} = \frac{1}{0.703}\underbrace{\begin{bmatrix} 0.98 & -0.20 \\ -0.55 & 0.83 \end{bmatrix}}_{\text{adj}(\mathbf{L}')} = \begin{bmatrix} 1.39 & -0.28 \\ -0.78 & 1.18 \end{bmatrix}$$

Hence

$$\mathbf{T}^\circ = \underset{(\mathbf{L}')^{-1}}{\begin{bmatrix} 1.39 & -0.28 \\ -0.78 & 1.18 \end{bmatrix}} \underset{\mathbf{T}}{\begin{bmatrix} 0.90 & 0.44 \\ 0.60 & 0.80 \end{bmatrix}} \underset{\mathbf{L}'}{\begin{bmatrix} 0.83 & 0.20 \\ 0.55 & 0.98 \end{bmatrix}} = \begin{bmatrix} 1.11 & 0.60 \\ 0.34 & 0.59 \end{bmatrix}$$

Returning to the vector $\mathbf{x} = \left[\begin{smallmatrix}1\\2\end{smallmatrix}\right]$, we recall under the fixed basis vectors $\mathbf{e}_1, \mathbf{e}_2$, that $\mathbf{x}$ is transformed by $\mathbf{T}$ onto $\mathbf{x}^* = \left[\begin{smallmatrix}1.78\\2.20\end{smallmatrix}\right]$, as noted in Panel I of Fig. 4.9. However, if we use the basis $\mathbf{f}_1, \mathbf{f}_2$, the point transformation of $\mathbf{x}^\circ = \left[\begin{smallmatrix}1\\2\end{smallmatrix}\right]$ in the basis $\mathbf{F}$ utilizes $\mathbf{T}^\circ$ and is

$$\mathbf{x}^{*\circ} = \begin{bmatrix} 2.31 \\ 1.52 \end{bmatrix} = \begin{bmatrix} 1.11 & 0.60 \\ 0.34 & 0.59 \end{bmatrix}\begin{bmatrix} 1 \\ 2 \end{bmatrix}$$

Notice, then, that $\mathbf{x}^{*\circ}$ is a *point* transformation of $\mathbf{x}^\circ = \left[\begin{smallmatrix}1\\2\end{smallmatrix}\right]$ in the $\mathbf{F}$ basis.

Panel II of Fig. 4.9 shows the nature of the transformation. First, we plot $\mathbf{x}^\circ = \left[\begin{smallmatrix}1\\2\end{smallmatrix}\right]$ in terms of the oblique basis $\mathbf{F}$. Note that $\mathbf{f}_1 = \left[\begin{smallmatrix}0.83\\0.55\end{smallmatrix}\right]$ and $\mathbf{f}_2 = \left[\begin{smallmatrix}0.20\\0.98\end{smallmatrix}\right]$ are, in turn, plotted in terms of the original $\mathbf{E}$ basis, while $\mathbf{x}^\circ = \left[\begin{smallmatrix}1\\2\end{smallmatrix}\right]$ is positioned with respect to the new basis $\mathbf{F}$. The process of finding the point transformation $\mathbf{x}^{*\circ} = \left[\begin{smallmatrix}2.31\\1.52\end{smallmatrix}\right]$ in $\mathbf{F}$ proceeds by decomposing

$$\mathbf{T}^\circ = (\mathbf{L}')^{-1}\mathbf{T}\mathbf{L}'$$

as follows, starting from the far right of the expression to the right of the equals sign:

1. $\mathbf{L}'$ maps $\mathbf{x}^\circ$ onto $\mathbf{x}$; that is,

$$\underset{\mathbf{L}'}{\begin{bmatrix} 0.83 & 0.20 \\ 0.55 & 0.98 \end{bmatrix}} \underset{\mathbf{x}^\circ}{\begin{bmatrix} 1 \\ 2 \end{bmatrix}} = \underset{\mathbf{x}}{\begin{bmatrix} 1.23 \\ 2.51 \end{bmatrix}}$$

2. $\mathbf{T}$ maps $\mathbf{x} = \left[\begin{smallmatrix}1.23\\2.51\end{smallmatrix}\right]$ onto $\mathbf{x}^*$

$$\underset{\mathbf{T}}{\begin{bmatrix} 0.90 & 0.44 \\ 0.60 & 0.80 \end{bmatrix}} \underset{\mathbf{x}}{\begin{bmatrix} 1.23 \\ 2.51 \end{bmatrix}} = \underset{\mathbf{x}^*}{\begin{bmatrix} 2.21 \\ 2.75 \end{bmatrix}}$$

3. $(\mathbf{L}')^{-1}$ maps $\mathbf{x}^*$ onto $\mathbf{x}^{*\circ}$

$$\overset{(\mathbf{L}')^{-1}}{\begin{bmatrix} 1.39 & -0.28 \\ -0.78 & 1.18 \end{bmatrix}} \overset{\mathbf{x}^*}{\begin{bmatrix} 2.21 \\ 2.75 \end{bmatrix}} = \overset{\mathbf{x}^{*\circ}}{\begin{bmatrix} 2.31 \\ 1.52 \end{bmatrix}}$$

We can follow through each of these steps from Panel II of Fig. 4.9. First, in step one we note that the coordinates of $\mathbf{x}^\circ$ after mapping onto $\mathbf{x}$ are, indeed, $\mathbf{x} = \left[\begin{smallmatrix}1.23\\2.51\end{smallmatrix}\right]$ with respect to the $\mathbf{E}$ basis. In step two $\mathbf{x}$ is mapped onto $\mathbf{x}^* = \left[\begin{smallmatrix}2.21\\2.75\end{smallmatrix}\right]$, again with respect to the $\mathbf{E}$ basis. However, to refer the point to the $\mathbf{F}$ basis we employ step three, giving us $\mathbf{x}^{*\circ} = \left[\begin{smallmatrix}2.31\\1.52\end{smallmatrix}\right]$ with respect to the oblique basis $\mathbf{F}$.

As will be pointed out in Chapter 5, the practical matter is to find a suitable basis $\mathbf{F}$ such that the matrix of the transformation with respect to this new basis takes on a particularly simple form, such as a diagonal matrix. Finally, it is worth noting that if $\mathbf{L}$ is orthogonal we have the simplification

$$\mathbf{T}^\circ = \mathbf{LTL}'$$

since, as noted earlier, if $\mathbf{L}$ is orthogonal, then

$$(\mathbf{L}')^{-1} = \mathbf{L}$$

4.3.6 Recapitulation

The concept of vector transformation is central to matrix algebra and multivariate analysis. The simplest type of transformation is represented by a point transformation, relative to a fixed basis:

$$\mathbf{x}^* = \mathbf{Tx}$$

Usually, we choose the fixed basis to be $\mathbf{E}$, the standard basis. Then the axes are mutually orthogonal and of unit length. In point transformations $\mathbf{x}$ moves according to $\mathbf{T}$ while the basis stays fixed. Figure 4.7 shows the geometric character of this type of transformation.

Alternatively, we can allow the point to stay fixed and, instead, transform the basis vectors to some new, and possibly oblique, orientation. This type of transformation, called a basis vector transformation, is exemplified by

$$\mathbf{x}^\circ = (\mathbf{T}')^{-1}\,\mathbf{x}$$

and is illustrated in Fig. 4.8.

As also pointed out, if $\mathbf{T}$ is orthogonal, various simplifications result that make the geometric interpretation easier. Finally, the specific character of some particular mapping, denoted generally by τ, depends on the reference basis. We showed how one matrix of the transformation, represented by $\mathbf{T}$ with respect to $\mathbf{e}_i$, the original basis, can be represented by $\mathbf{T}^\circ$ if we know the transformation that connects $\mathbf{f}_i$, the basis for $\mathbf{T}^\circ$ with $\mathbf{e}_i$, the basis for $\mathbf{T}$.

While not illustrated in the cases that were covered, it should be mentioned in passing that $|\mathbf{T}| = |\mathbf{T}^\circ|$. That is, the determinant of a linear transformation is independent of the basis to which the transformation is referred.

Inasmuch as matrix transformations are so central to the subject, we continue our discussion of the geometric character of various types of special matrices. While we have described transformations represented by orthogonal matrices (i.e., rotations), it turns out that many other kinds of matrices have intuitively simple geometric representations as well.

4.4 GEOMETRIC RELATIONSHIPS INVOLVING MATRIX TRANSFORMATIONS

As we have illustrated, many matrix operations can be usefully represented geometrically if two or three dimensions are involved. At this point it seems useful to extend this line of geometric reasoning to other aspects of matrix transformations. In so doing the reader may get some intuitive understanding of what is involved in higher dimensionalities where geometrical representation is no longer feasible.

In order to motivate the discussion, let us consider the small 9 x 2 matrix of synthetic data shown in Table 4.1. This matrix consists of nine points, positioned in two dimensions, as diagrammed in Panel I of Fig. 4.10. Note that the points involve a square lattice arrangement.

TABLE 4.1

Original Matrix **X** *of Nine Points in Two Dimensions*

Point (code letter)	Dimension 1	Dimension 2
a	1	0
b	1	1
c	1	2
d	2	0
e	2	1
f	2	2
g	3	0
h	3	1
i	3	2
$\bar{X}_j$	2	1

We shall now describe a variety of operations on this synthetic data matrix and show their effects geometrically. Whereas our earlier discussions of matrix transformations involved transforming a single vector $\mathbf{x}$ into another vector $\mathbf{x}^*$, here we shall transform a *set* of vectors, which can be represented as a matrix $\mathbf{X}$. The basic principles remain the same.

In all examples of this section we deal with the simpler case of point transformations, rather than basis vector transformations. However, so as to show the flexibility of matrix transformations, in this section we *postmultiply* the original matrix $\mathbf{X}$ by the transformation matrix, in order to obtain $\mathbf{X}^*$. No new principles are involved in this

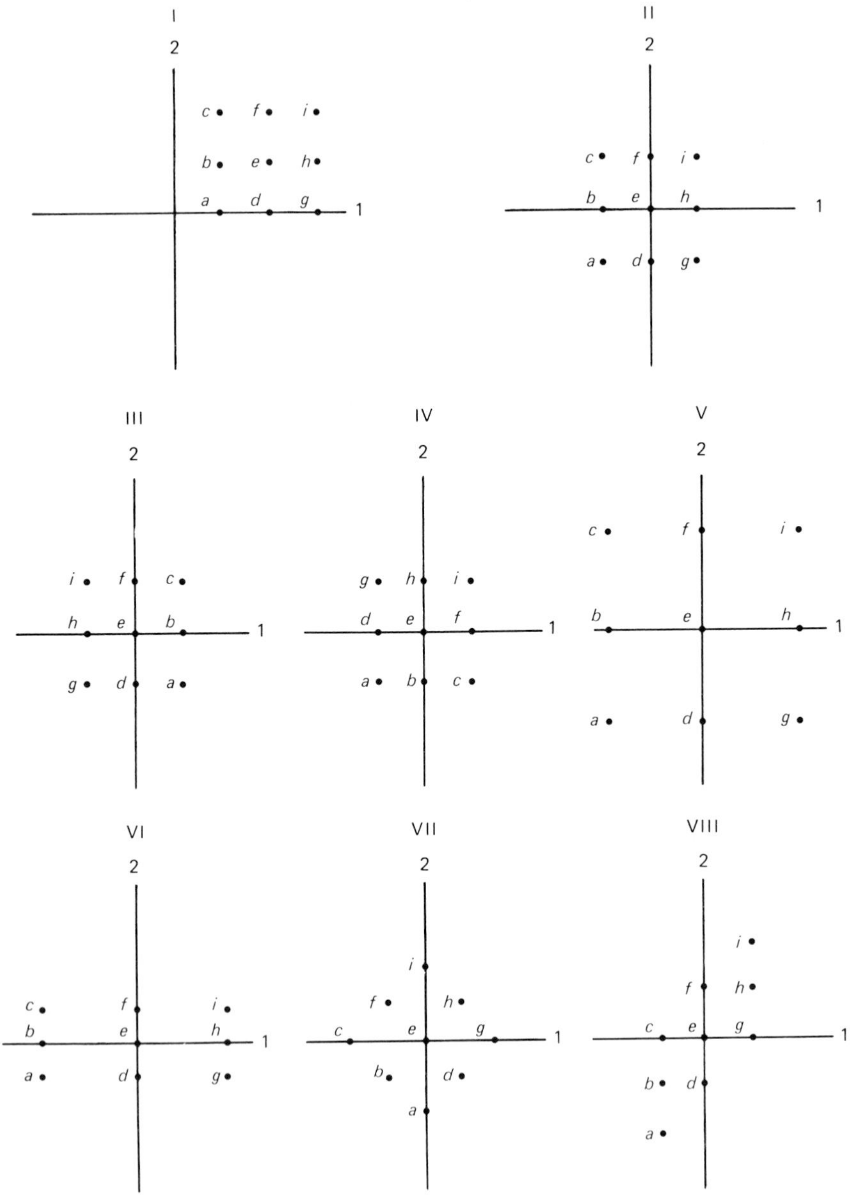

Fig. 4.10 Geometric effect of various simple transformations. Key: I, original data; II, translation; III, reflection; IV, permutation; V, central dilation; VI, stretch; VII, counterclockwise rotation of 45°; VIII, shear.

change, as the reader will note in due course. Thus, the basic format to be followed consists of the transformation

$$\boxed{\mathbf{X}^* = \mathbf{XA}}$$

where **A** assumes various special forms that exhibit simple geometric patterns. These special forms are of particular relevance to multivariate analysis.

Notice, however, that each *row* vector in **X** is being postmultiplied by **A** to obtain each row vector in **X***. As we shall see later, this viewpoint modifies the specific entries of the matrices, although all basic concepts remain unchanged from our earlier discussion in which **A** premultiplied column vectors.

4.4.1 Translation

Matrix translation has to do with the problem of relating a set of points to a particular *origin* in the space. Suppose, for example, that we wished to refer the nine points of Panel I of Fig. 4.10 to a centroid-centered origin. By "centroid" we mean, of course, the arithmetic mean of each set of coordinates on each dimension. As shown in Table 4.1, the mean of the points on the first dimension is 2; their mean on the second dimension is 1. By expressing each original value as a *deviation* from its mean, we arrive at matrix $\mathbf{X}_d$ as shown in Table 4.2. Note that this simply involves a subtraction of $\bar{\mathbf{X}}$, whose entries represent the mean of the points on each dimension, from the original matrix **X**. Panel II of Fig. 4.10 shows the effect of the translation geometrically.

TABLE 4.2

Translation of Matrix **X** *to Origin at Centroid*

$$\begin{array}{ccccc} \mathbf{X} & & \bar{\mathbf{X}} & & \mathbf{X}_d \\ \begin{bmatrix} 1 & 0 \\ 1 & 1 \\ 1 & 2 \\ 2 & 0 \\ 2 & 1 \\ 2 & 2 \\ 3 & 0 \\ 3 & 1 \\ 3 & 2 \end{bmatrix} & - & \begin{bmatrix} 2 & 1 \\ 2 & 1 \\ 2 & 1 \\ 2 & 1 \\ 2 & 1 \\ 2 & 1 \\ 2 & 1 \\ 2 & 1 \\ 2 & 1 \end{bmatrix} & = & \begin{bmatrix} -1 & -1 \\ -1 & 0 \\ -1 & 1 \\ 0 & -1 \\ 0 & 0 \\ 0 & 1 \\ 1 & -1 \\ 1 & 0 \\ 1 & 1 \end{bmatrix} \end{array}$$

A translation, then, involves a parallel displacement of every point to some new origin of interest. In this case the centroid of the points is the origin of interest. The particular nature of the matrix **X**, consisting of the column means of **X**, is

$$\boxed{\bar{\mathbf{X}} = 1/m\mathbf{1}\mathbf{1}'\mathbf{X}}$$

where **1** is the unit column vector and $m = 9$. We can then find $\mathbf{X}_d$ as

$$\boxed{\begin{aligned} \mathbf{X}_d &= \mathbf{X} - 1/m\mathbf{1}\mathbf{1}'\mathbf{X} \\ &= (\mathbf{I} - 1/m\mathbf{1}\mathbf{1}')\mathbf{X} \end{aligned}}$$

In previous sections of the chapter, all of our discussion of matrix transformations, involved a *multiplicative* form, such as

$$\mathbf{x}^* = \mathbf{A}\mathbf{x}$$

or, in the present format,

$$\mathbf{X}^* = \mathbf{X}\mathbf{A}$$

In translating a set of points in two dimensions, denoted by the matrix **X**, we see that for each transformed point we have the coordinates

$$x_{i1}^* = x_{i1} + h; \qquad x_{i2}^* = x_{i2} + k$$

where h and k are constants. Hence, translation departs from the usual matrix multiplication format by involving the sum or difference of two matrices.

However, by using the following device:

$$(x_{i1}^*, x_{i2}^*, 1) = (x_{i1}, x_{i2}, 1) \begin{bmatrix} 1 & 0 & 0 \\ 0 & 1 & 0 \\ h & k & 1 \end{bmatrix}$$

we can obtain

$$x_{i1}^* = x_{i1} + h; \qquad x_{i2}^* = x_{i2} + k; \qquad 1 = 1$$

Note that the last equation (1 = 1) is trivial, but does enable us to express a translation in the multiplicative format used earlier.[5]

Translations are frequently used in multivariate analysis. In particular, the SSCP, covariance, and correlation matrices all utilize mean-corrected scores and involve, among other things, a translation of raw scores into deviation scores.

4.4.2 Reflection

Reflection of a set of points, as noted in the discussion of improper rotation in Chapter 3, entails multiplication of the coordinates of each point to be reflected by −1. For example, suppose we wished to reflect the nine points of Panel II in Fig. 4.10 "across" axis 2. This can be accomplished by multiplying each of the coordinates on axis 1 by a −1, as shown in Table 4.3 and illustrated graphically in Panel III of Fig. 4.10.

The matrix used for this purpose is represented by

$$\begin{bmatrix} -1 & 0 \\ 0 & 1 \end{bmatrix}$$

Similarly, if desired, one could reflect the nine points across axis 1. It should be reiterated, however, that reflection typically involves an odd number of dimensions. If we

[5] This particular computational trick can be useful in the preparation of computer routines for translating the origin of a set of points.

TABLE 4.3

Reflection of Matrix $\mathbf{X}_d$ *across Axis 2*

$$\overset{\mathbf{X}_d}{\begin{bmatrix} -1 & -1 \\ -1 & 0 \\ -1 & 1 \\ 0 & -1 \\ 0 & 0 \\ 0 & 1 \\ 1 & -1 \\ 1 & 0 \\ 1 & 1 \end{bmatrix}} \overset{\text{Reflection}}{\begin{bmatrix} -1 & 0 \\ 0 & 1 \end{bmatrix}} = \begin{bmatrix} 1 & -1 \\ 1 & 0 \\ 1 & 1 \\ 0 & -1 \\ 0 & 0 \\ 0 & 1 \\ -1 & -1 \\ -1 & 0 \\ -1 & 1 \end{bmatrix}$$

were to reflect the points across *both* axis 1 and axis 2, the overall effect–called a reflection through the origin–is equivalent to a rotation of the points about an angle of 180° from their original orientation.

As discussed earlier, rotation followed by reflection is often referrred to as an "improper" rotation in the case of orthogonal matrices whose entries consist of direction cosines. It is relevant to point out that improper rotation satisfies the same conditions as proper rotation:

1. All transformation vectors are of unit length.
2. All transformation vectors are mutually orthogonal.

The differentiating feature of improper rotation is that the determinant of this type of orthogonal matrix is −1, while the determinant of an orthogonal matrix constituting a proper rotation is +1.

Perhaps all of this can be summarized by saying:

1. A reflection can always be described as a proper rotation followed by reflection of *one* dimension.
2. Reflection of an even number of dimensions (e.g., two, four, six, etc.) is equivalent to a proper rotation in the 1–2 plane, the 3–4 plane, and so on.
3. Reflection of an odd number of dimensions is equivalent to a proper rotation followed by reflection of one dimension.

4.4.3 Axis Permutation

Permutation of a set of points, as the name suggests, involves a matrix transformation that carries each coordinate value on axis 1 into a corresponding coordinate on axis 2, and vice versa. This is illustrated in Table 4.4, and the effect is shown graphically in Panel IV of Fig. 4.10.

The permutation matrix

$$\begin{bmatrix} 0 & 1 \\ 1 & 0 \end{bmatrix}$$

TABLE 4.4

Axis Permutation of Matrix $\mathbf{X}_d$

$$\overset{\mathbf{X}_d}{\begin{bmatrix} -1 & -1 \\ -1 & 0 \\ -1 & 1 \\ 0 & -1 \\ 0 & 0 \\ 0 & 1 \\ 1 & -1 \\ 1 & 0 \\ 1 & 1 \end{bmatrix}} \; \overset{\text{Permutation}}{\begin{bmatrix} 0 & 1 \\ 1 & 0 \end{bmatrix}} = \begin{bmatrix} -1 & -1 \\ 0 & -1 \\ 1 & -1 \\ -1 & 0 \\ 0 & 0 \\ 1 & 0 \\ -1 & 1 \\ 0 & 1 \\ 1 & 1 \end{bmatrix}$$

TABLE 4.5

Central Dilation of Matrix $\mathbf{X}_d$

$$\overset{\mathbf{X}_d}{\begin{bmatrix} -1 & -1 \\ -1 & 0 \\ -1 & 1 \\ 0 & -1 \\ 0 & 0 \\ 0 & 1 \\ 1 & -1 \\ 1 & 0 \\ 1 & 1 \end{bmatrix}} \; \overset{\text{Central dilation}}{\begin{bmatrix} 2 & 0 \\ 0 & 2 \end{bmatrix}} = \begin{bmatrix} -2 & -2 \\ -2 & 0 \\ -2 & 2 \\ 0 & -2 \\ 0 & 0 \\ 0 & 2 \\ 2 & -2 \\ 2 & 0 \\ 2 & 2 \end{bmatrix}$$

that postmultiplies $\mathbf{X}_d$ produces an interchange of columns. However, if the permutation matrix premultiplies a matrix of preimages, then the rows of the preimages are interchanged.

In Table 4.4, however, we see that the first and second columns of $\mathbf{X}_d$ are interchanged by means of postmultiplication of $\mathbf{X}_d$ by the permutation matrix.

4.4.4 Central Dilation

Central dilation of a set of points entails scalar multiplication of the matrix of coordinates, which is equivalent to multiplication by a scalar matrix; that is, a diagonal matrix in which each diagonal entry involves the same positive constant λ. Central dilation leads to a uniform expansion, if $\lambda > 1$, or a uniform contraction, if $\lambda < 1$, of each dimension. If $\lambda = 1$, then the scalar matrix becomes an identity matrix, and the point positions remain as originally expressed.

Table 4.5 shows application of a central dilation where $\lambda = 2$. Panel V of Fig. 4.10 shows the results graphically. Scalar matrix transformations are particularly simple from a geometric standpoint since we see that *uniform* stretching or compressing of the dimensions takes place along the original axes of orientation. According to the present case, the scalar matrix is

$$\begin{bmatrix} 2 & 0 \\ 0 & 2 \end{bmatrix}$$

where each of the original axes is dilated to twice its original length.

4.4.5 Stretch

A stretch transformation of a set of points involves application of a *diagonal* matrix where, in general, the diagonal entries are such that $\lambda_{ii} \neq \lambda_{jj}$. In contrast to central dilation, a stretch involves *differential* stretching or contraction (rescaling) of points corresponding, again, to directions along the original axes.

TABLE 4.6

A Stretch of Matrix $\mathbf{X}_d$

$$\overset{\mathbf{X}_d}{\begin{bmatrix} -1 & -1 \\ -1 & 0 \\ -1 & 1 \\ 0 & -1 \\ 0 & 0 \\ 0 & 1 \\ 1 & -1 \\ 1 & 0 \\ 1 & 1 \end{bmatrix}} \overset{\text{Stretch}}{\begin{bmatrix} 2 & 0 \\ 0 & 1 \end{bmatrix}} = \begin{bmatrix} -2 & -1 \\ -2 & 0 \\ -2 & 1 \\ 0 & -1 \\ 0 & 0 \\ 0 & 1 \\ 2 & -1 \\ 2 & 0 \\ 2 & 1 \end{bmatrix}$$

For example, if $\lambda_{11} = 2$ and $\lambda_{22} = 1$, the effect of this transformation is to stretch axis 1 to twice its original length, thus producing a latticelike rectangle out of the original latticelike square. Table 4.6 illustrates the computations involved, and Panel VI of Fig. 4.10 shows the graphical results of the transformation

$$\begin{bmatrix} 2 & 0 \\ 0 & 1 \end{bmatrix}$$

Here we shall restrict the term "stretch" to the case in which all $\lambda_{ii} > 0$. If some λ_{ii} were zero, those dimensions would be annihilated. A $\lambda_{ii} < 0$ would correspond to a stretching (or contraction) followed by reflection.

4.4.6 Rotation

As described earlier in the chapter, axis rotation involves application of a rather special kind of matrix—an orthogonal matrix. An orthogonal matrix is distinguished by the properties: (a) the sum of squares of each column (row) equals 1, and (b) the scalar product of each pair of columns (rows) equals zero. To illustrate, the matrix corresponding to a 45° rotation

$$\begin{bmatrix} 0.707 & 0.707 \\ -0.707 & 0.707 \end{bmatrix}$$

meets these conditions, since, by columns, for example,

$$\text{(a)} \quad (0.707)^2 + (-0.707)^2 = 1; \qquad (0.707)^2 + (0.707)^2 = 1$$

$$\text{(b)} \quad (0.707, -0.707)\begin{bmatrix} 0.707 \\ 0.707 \end{bmatrix} = 0$$

As we know, in two dimensions an orthogonal matrix entails a rigid rotation of the original configuration of points about some angle Ψ. In the preceding example we rotate the points counterclockwise about an angle, $\Psi = 45°$, and this involves the following:

$$\begin{bmatrix} \cos\Psi & \sin\Psi \\ -\sin\Psi & \cos\Psi \end{bmatrix}$$

That is, $\cos 45° = \sin 45° = 0.707$, while $-\sin 45° = -0.707$.

TABLE 4.7

Counterclockwise Rotation of Matrix $\mathbf{X}_d$ *through 45° Angle*

$$\overset{\mathbf{X}_d}{\begin{bmatrix} -1 & -1 \\ -1 & 0 \\ -1 & 1 \\ 0 & -1 \\ 0 & 0 \\ 0 & 1 \\ 1 & -1 \\ 1 & 0 \\ 1 & 1 \end{bmatrix}} \overset{\text{Rotation}}{\begin{bmatrix} 0.707 & 0.707 \\ -0.707 & 0.707 \end{bmatrix}} = \begin{bmatrix} 0 & -1.414 \\ -0.707 & -0.707 \\ -1.414 & 0 \\ 0.707 & -0.707 \\ 0 & 0 \\ -0.707 & 0.707 \\ 1.414 & 0 \\ 0.707 & 0.707 \\ 0 & 1.414 \end{bmatrix}$$

The reader should note that the above rotation matrix is postmultiplying each row vector of the matrix $\mathbf{X}_d$. This is in direct contrast to the rotation depicted in Fig. 4.3, in which the matrix of this transformation is *premultiplying* the column vector of coordinate values $\mathbf{x}$. As such, the orthogonal matrix, whose effect is depicted in Fig. 4.3, still represents a counterclockwise point rotation but now has the form

$$\begin{bmatrix} \cos \Psi & -\sin \Psi \\ \sin \Psi & \cos \Psi \end{bmatrix}$$

This is the transpose of the matrix form shown above.

Table 4.7 shows the application of the 45° counterclockwise point rotation, and Panel VII of Fig. 4.10 summarizes the results graphically. Note that the zero point (intersection of the axes) can be viewed as the hub of a wheel and remains fixed during the rotation.

As first discussed in Chapter 3, entries of the rotation matrix can all be expressed as direction cosines of the angles θ_{11}, θ_{12}, θ_{21}, θ_{22} made between old and new axes. The first subscript refers to the old axis, while the second refers to the new axis. And, as stated earlier, this type of generalization is important when more than two dimensions are involved.

A second point of interest is that the same kind of transformation noted above can be made by rotating the axes, rather than the points, around an angle of −45°, relative to the original orientation. This alternative view, of course, was covered earlier in the chapter.

4.4.7 Shear

A shear transformation is characterized by the following form:

$$\begin{bmatrix} 1 & 1 \\ 0 & 1 \end{bmatrix}$$

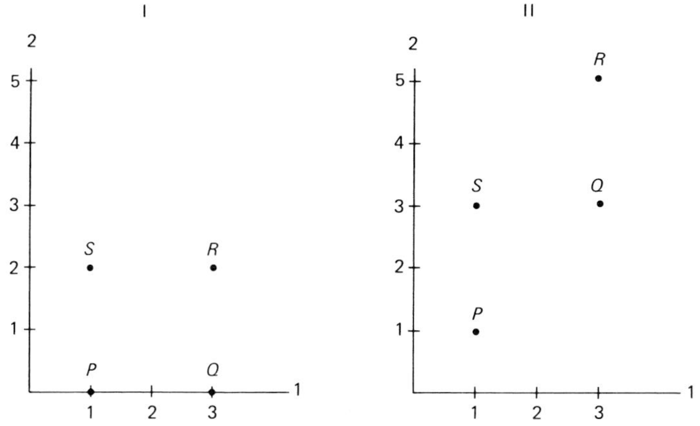

Fig. 4.11 A shear transformation. Key: I, before; II, after.

so that postmultiplication of some matrix by a shear has the effect of adding columns. For example,

$$\begin{bmatrix} a & b \\ c & d \end{bmatrix} \overset{\text{Shear}}{\begin{bmatrix} 1 & 1 \\ 0 & 1 \end{bmatrix}} = \begin{bmatrix} a & a+b \\ c & c+d \end{bmatrix}$$

while premultiplication has the effect of adding rows. For example,

$$\overset{\text{Shear}}{\begin{bmatrix} 1 & 1 \\ 0 & 1 \end{bmatrix}} \begin{bmatrix} a & b \\ c & d \end{bmatrix} = \begin{bmatrix} a+c & b+d \\ c & d \end{bmatrix}$$

The geometric effect of a shear is shown, illustratively, in Fig. 4.11 for the simple case involving the rectangle

$$P = (1,0); \qquad Q = (3,0); \qquad R = (3,2); \qquad S = (1,2)$$

When the shear transformation postmultiplies the vertices of the rectangle, we obtain:

$$\begin{matrix} P \\ Q \\ R \\ S \end{matrix} \begin{bmatrix} 1 & 1 \\ 3 & 3 \\ 3 & 5 \\ 1 & 3 \end{bmatrix} = \begin{bmatrix} 1 & 0 \\ 3 & 0 \\ 3 & 2 \\ 1 & 2 \end{bmatrix} \begin{bmatrix} 1 & 1 \\ 0 & 1 \end{bmatrix}$$

as shown in Panel II of Fig. 4.11.

TABLE 4.8

A Shear Transformation of Matrix $\mathbf{X}_d$

$$\overset{\mathbf{X}_d}{\begin{bmatrix} -1 & -1 \\ -1 & 0 \\ -1 & 1 \\ 0 & -1 \\ 0 & 0 \\ 0 & 1 \\ 1 & -1 \\ 1 & 0 \\ 1 & 1 \end{bmatrix}} \overset{\text{Shear}}{\begin{bmatrix} 1 & 1 \\ 0 & 1 \end{bmatrix}} = \begin{bmatrix} -1 & -2 \\ -1 & -1 \\ -1 & 0 \\ 0 & -1 \\ 0 & 0 \\ 0 & 1 \\ 1 & 0 \\ 1 & 1 \\ 1 & 2 \end{bmatrix}$$

If we apply the shear to the matrix $\mathbf{X}_d$, we obtain the coordinates shown in Table 4.8 and plotted in Panel VIII of Fig. 4.10.

4.5 COMPOSITE TRANSFORMATIONS

The transformations described in the preceding section have each been applied singly. It is instructive to see what happens when some of these transformations are applied on a composite basis. For illustrative purposes we consider the following:

1. a rotation followed by reflection of the first axis
2. a stretch followed by a rotation
3. a rotation followed by a stretch
4. a rotation followed by a stretch followed by another rotation
5. an arbitrary linear transformation.

The theoretical rationale for applying successive matrix transformation is based on the idea that if $\mathbf{T}$ is the matrix of one linear transformation and $\mathbf{S}$ is the matrix of a transformation that maps images obtained from $\mathbf{T}$, then the matrix to be transformed can be mapped by a *composite* transformation that involes the matrix product $\mathbf{TS}$.[6] This same idea can be extended in the same manner, to more than two transformation matrices.

The *order* in which successive matrix transformations are applied is quite important. That is, in general the results of the composite mapping involving $\mathbf{TS}$ are not the same as the images that would be obtained from the composite mapping $\mathbf{ST}$, as will be demonstrated shortly.

Moreover, it should be reiterated that the matrix of a transformation is uniquely defined only relative to a set of *specific* basis vectors.[7] All along we have been using the standard basis vectors $\mathbf{e}_i$, and we shall continue to do so here.[8]

[6] We are continuing to assume that $\mathbf{T}$ (and $\mathbf{S}$) are *postmultiplying* (say) $\mathbf{X}_d$, the initial configuration of points.

[7] Here we continue to distinguish between τ, the transformation (e.g., a stretch or a rotation), and $\mathbf{T}$, its characterization with respect to a specific set of basis vectors.

[8] We shall continue to refer to the transformation of $\mathbf{X}_d$, the configuration of points in Panel II of Fig. 4.10.

4.5.1 Rotation Followed by Reflection

If we multiply the following matrices:

$$\begin{bmatrix} 0.707 & 0.707 \\ -0.707 & 0.707 \end{bmatrix} \begin{bmatrix} -1 & 0 \\ 0 & 1 \end{bmatrix}$$

we obtain

$$\begin{bmatrix} -0.707 & 0.707 \\ 0.707 & 0.707 \end{bmatrix}$$

As can be easily verified, the resultant matrix has a determinant of -1. And as indicated earlier, this type of matrix is called an improper rotation. Although it meets the conditions of an orthogonal matrix, its application actually involves a "proper" rotation, in which the determinant is +1, *followed by a reflection* of axis 1. Table 4.9 shows the

TABLE 4.9

An Improper Rotation of Matrix $\mathbf{X}_d$

$$\overset{\mathbf{X}_d}{\begin{bmatrix} -1 & -1 \\ -1 & 0 \\ -1 & 1 \\ 0 & -1 \\ 0 & 0 \\ 0 & 1 \\ 1 & -1 \\ 1 & 0 \\ 1 & 1 \end{bmatrix}} \overset{\text{Improper rotation}}{\begin{bmatrix} -0.707 & 0.707 \\ 0.707 & 0.707 \end{bmatrix}} = \begin{bmatrix} 0 & -1.414 \\ 0.707 & -0.707 \\ 1.414 & 0 \\ -0.707 & -0.707 \\ 0 & 0 \\ 0.707 & 0.707 \\ -1.414 & 0 \\ -0.707 & 0.707 \\ 0 & 1.414 \end{bmatrix}$$

results of this composite transformation, while Panel I of Fig. 4.12 shows the geometric results. That is, the points in Panel II of Fig. 4.10 are first rotated counterclockwise about an angle of 45°, and then the (implied) $\mathbf{e}_1{}^*$ axis is reflected, as observed in Panel I of Fig. 4.12.

4.5.2 Stretch Followed by Rotation

If we multiply the following matrices:

$$\begin{bmatrix} 2 & 0 \\ 0 & 1 \end{bmatrix} \begin{bmatrix} 0.707 & 0.707 \\ -0.707 & 0.707 \end{bmatrix}$$

we obtain

$$\begin{bmatrix} 1.414 & 1.414 \\ 0.707 & 0.707 \end{bmatrix}$$

This latter composite matrix first involves a stretch of the configuration followed by a counterclockwise rotation of 45°. Table 4.10 shows the computations, while Panel II of

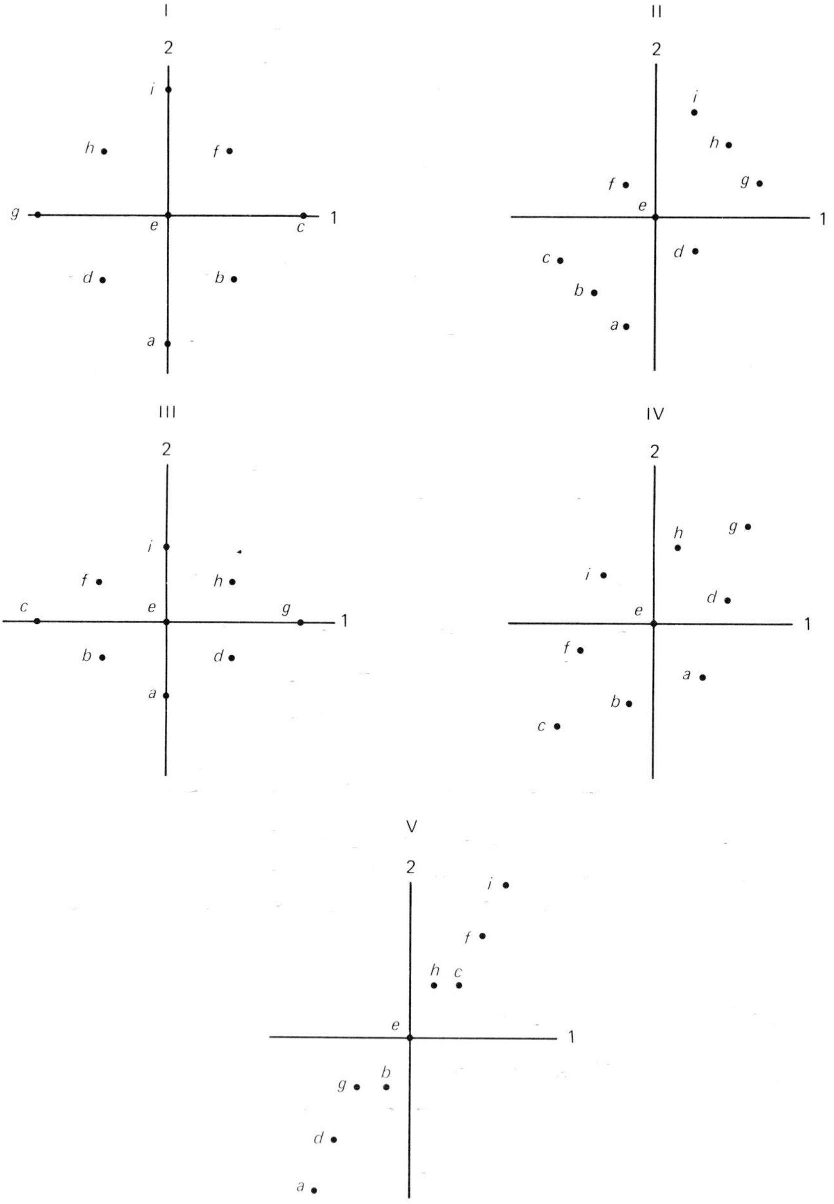

Fig. 4.12 Geometric effect of various composite transformations. Key: I, rotation followed by reflection; II, stretch followed by rotation; III, rotation followed by stretch; IV, rotation–stretch–rotation; V, arbitrary linear transformation.

TABLE 4.10

A Stretch of Matrix $\mathbf{X}_d$ *Followed by Counterclockwise Rotation through an Angle of 45°*

$$\overset{\mathbf{X}_d}{\begin{bmatrix} -1 & -1 \\ -1 & 0 \\ -1 & 1 \\ 0 & -1 \\ 0 & 0 \\ 0 & 1 \\ 1 & -1 \\ 1 & 0 \\ 1 & 1 \end{bmatrix}} \overset{\text{Stretch–rotation}}{\begin{bmatrix} 1.414 & 1.414 \\ -0.707 & 0.707 \end{bmatrix}} = \begin{bmatrix} -0.707 & -2.121 \\ -1.414 & -1.414 \\ -2.121 & -0.707 \\ 0.707 & -0.707 \\ 0 & 0 \\ -0.707 & 0.707 \\ -2.121 & 0.707 \\ 1.414 & 1.414 \\ 0.707 & 2.121 \end{bmatrix}$$

Fig. 4.12 shows the geometric results. The first matrix maps the square lattice into a rectangular lattice, while the second transformation rotates this rectangular lattice 45°, counterclockwise.

4.5.3 Rotation Followed by Stretch

In general, the matrix product $\mathbf{AB} \neq \mathbf{BA}$. That is, usually the multiplication of matrices (even if conformable) is not commutative. This can be illustrated rather dramatically by considering the following matrix product:

$$\begin{bmatrix} 0.707 & 0.707 \\ -0.707 & 0.707 \end{bmatrix} \begin{bmatrix} 2 & 0 \\ 0 & 1 \end{bmatrix} = \begin{bmatrix} 1.414 & 0.707 \\ -1.414 & 0.707 \end{bmatrix}$$

While the same two matrices as those used in the preceding section are employed here, we see that their matrix product differs markedly. In the present case we have a counterclockwise rotation of $\mathbf{X}_d$ through an angle of 45° *followed* by a stretch. The result of this is that even the original "shape" of the points (a square lattice) is deformed

TABLE 4.11

A Counterclockwise Rotation of Matrix $\mathbf{X}_d$ *through an Angle of 45° Followed by a Stretch*

$$\overset{\mathbf{X}_d}{\begin{bmatrix} -1 & -1 \\ -1 & 0 \\ -1 & 1 \\ 0 & -1 \\ 0 & 0 \\ 0 & 1 \\ 1 & -1 \\ 1 & 0 \\ 1 & 1 \end{bmatrix}} \overset{\text{Rotation–stretch}}{\begin{bmatrix} 1.414 & 0.707 \\ -1.414 & 0.707 \end{bmatrix}} = \begin{bmatrix} 0 & -1.414 \\ -1.414 & -0.707 \\ -2.829 & 0 \\ 1.414 & -0.707 \\ 0 & 0 \\ -1.414 & 0.707 \\ 2.829 & 0 \\ 1.414 & 0.707 \\ 0 & 1.414 \end{bmatrix}$$

TABLE 4.12

Rotation–Stretch–Rotation Composite Transformation of Matrix $\mathbf{X}_d$

$$\overset{\mathbf{X}_d}{\begin{bmatrix} -1 & -1 \\ -1 & 0 \\ -1 & 1 \\ 0 & -1 \\ 0 & 0 \\ 0 & 1 \\ 1 & -1 \\ 1 & 0 \\ 1 & 1 \end{bmatrix}} \overset{\text{Rotation–stretch–rotation}}{\begin{bmatrix} 0.87 & 1.32 \\ -1.58 & -0.10 \end{bmatrix}} = \begin{bmatrix} 0.71 & -1.22 \\ -0.87 & -1.32 \\ -2.45 & -1.42 \\ 1.58 & 0.10 \\ 0 & 0 \\ -1.58 & -0.10 \\ 2.45 & 1.42 \\ 0.87 & 1.32 \\ -0.71 & 1.22 \end{bmatrix}$$

TABLE 4.13

Application of an Arbitrary Linear Transformation to Matrix $\mathbf{X}_d$

$$\overset{\mathbf{X}_d}{\begin{bmatrix} -1 & -1 \\ -1 & 0 \\ -1 & 1 \\ 0 & -1 \\ 0 & 0 \\ 0 & 1 \\ 1 & -1 \\ 1 & 0 \\ 1 & 1 \end{bmatrix}} \overset{\mathbf{V}}{\begin{bmatrix} 1 & 2 \\ 3 & 4 \end{bmatrix}} = \begin{bmatrix} -4 & -6 \\ -1 & -2 \\ 2 & 2 \\ -3 & -4 \\ 0 & 0 \\ 3 & 4 \\ -2 & -2 \\ 1 & 2 \\ 4 & 6 \end{bmatrix}$$

into a rhomboidlike figure. The computations appear in Table 4.11, and the geometric results appear in Panel III of Fig. 4.12.

4.5.4 A Rotation–Stretch–Rotation Composite

As an extended case, let us now consider a 45° counterclockwise rotation followed by a stretch followed by a 30° counterclockwise rotation. This combination can be illustrated by the following matrix product:

$$\begin{bmatrix} 0.707 & 0.707 \\ -0.707 & 0.707 \end{bmatrix} \begin{bmatrix} 2 & 0 \\ 0 & 1 \end{bmatrix} \begin{bmatrix} 0.866 & 0.500 \\ -0.500 & 0.866 \end{bmatrix} = \begin{bmatrix} 0.87 & 1.32 \\ -1.58 & -0.10 \end{bmatrix}$$

Table 4.12 shows the results of applying this composite transformation, while Panel IV of Fig. 4.12 portrays the results graphically.

4.5.5 An Arbitrary Linear Transformation

To round out discussion, assume that we had the arbitrarily selected linear transformation

$$\mathbf{V} = \begin{bmatrix} 1 & 2 \\ 3 & 4 \end{bmatrix}$$

and wished to find out what would happen if this transformation were applied to the matrix $\mathbf{X}_d$. If $\mathbf{X}_d$ is postmultiplied by $\mathbf{V}$, we obtain the product shown in Table 4.13.

Panel V of Fig. 4.12 shows the effect graphically. Compared to the preceding cases the pattern of the transformed points may look a bit strange. As we shall show in the next chapter, however, even the arbitrary linear transformation $\mathbf{V}$ can be represented as a composite of more simple transformations, of the types illustrated in Fig. 4.10.

Anticipating material to be described in Chapter 5, Fig. 4.13 shows three configurations that successively portray the movement of the points of $\mathbf{X}_d$ from their original positions in Panel II of Fig. 4.10 to their positions shown in Panel V of Fig. 4.12. For the moment we shall do a bit of "hand waving" and present the results of decomposing $\mathbf{V}$ into the product of simpler transformations.

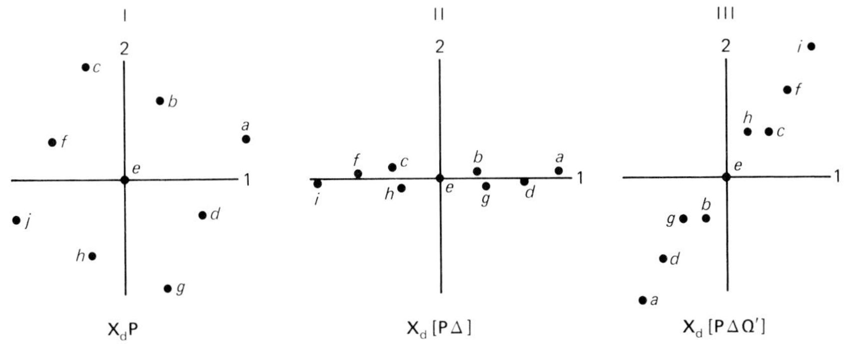

Fig. 4.13 Decomposition of general linear transformation $\mathbf{V} = \mathbf{P\Delta Q}'$.

First, suppose we postmultiply $\mathbf{X}_d$ by the orthogonal transformation

$$\mathbf{P} = \begin{bmatrix} -0.41 & -0.91 \\ -0.91 & 0.41 \end{bmatrix}$$

In this case $\mathbf{P}$ produces a 66° clockwise rotation of points in $\mathbf{X}_d$, followed by a reflection of the first axis. This preliminary result appears in Panel I of Fig. 4.13.

Next, let us assume that the configuration in Panel I is stretched in accordance with the diagonal matrix

$$\Delta = \begin{bmatrix} 5.47 & 0 \\ 0 & 0.37 \end{bmatrix}$$

This operation is shown in Panel II of Fig. 4.13

Finally, let us assume that the configuration in Panel II is further rotated by the orthogonal matrix

$$\mathbf{Q}' = \begin{bmatrix} -0.58 & -0.82 \\ 0.82 & -0.58 \end{bmatrix}$$

This rotation involves a clockwise movement of 125° from the orientation in Panel II. The results appear in Panel III of Fig. 4.13. Table 4.14 shows the accompanying numerical results.

The upshot of all of this is that the arbitrary linear transformation

$$\mathbf{V} = \begin{bmatrix} 1 & 2 \\ 3 & 4 \end{bmatrix}$$

has been decomposed into the product of an improper rotation, followed by a stretch, followed by another rotation. This triple product can be represented as

$$\boxed{\mathbf{V} = \mathbf{P\Delta Q}'}$$

TABLE 4.14

Decomposition of an Arbitrary Linear Transformation

$\mathbf{X}_d$		P Rotation–reflection			$\mathbf{X}_d\mathbf{P}$		Δ Stretch			$\mathbf{X}_d[\mathbf{P}\,\Delta]$		Q′ Rotation			$\mathbf{X}_d[\mathbf{P}\,\Delta\,\mathbf{Q}']$	
−1	−1	−0.41	−0.91	⇒	1.32	0.50	5.47	0	⇒	7.22	0.19	−0.58	−0.82	⇒	−4	−6
−1	0	−0.91	0.41		0.41	0.91	0	0.37		2.24	0.34	0.82	−0.58		−1	−2
−1	1				−0.50	1.32				−2.74	0.49				2	2
0	−1				0.91	−0.41				4.98	−0.15				−3	−4
0	0				0	0				0	0				0	0
0	1				−0.91	0.41				−4.98	0.15				3	4
1	−1				0.50	−1.32				2.74	−0.49				−2	−2
1	0				−0.41	−0.91				−2.24	−3.34				1	2
1	1				−1.32	−0.50				−7.22	−0.19				4	6

where both **P** and **Q** are orthogonal and Δ is diagonal. This rather remarkable result is of extreme importance to multivariate analysis.[9] A major part of Chapter 5 is devoted to discussing this class of decompositions in some detail.

4.5.6 Observations on Composite Transformations

Clearly we could continue with various other kinds of composite transformations, although sufficient variety has been shown for the reader to get some idea of their geometric effect. As noted above, each simple matrix, such as a stretch, rotation, or reflection, is associated with a geometric analogue. It is when these operations are considered in a composite way that the overall transformation appears complex.

As it turns out, however, the value of this approach lies precisely in looking at the other side of the coin. That is, by *decomposing* seemingly complex-appearing matrices into the product of simpler ones, we can gain a geometric understanding of the transformation in a direct, intuitive way.

And, so it turns out, *any nonsingular matrix transformation with real-valued entries can be uniquely decomposed into the product of either (a) a rotation, followed by a stretch, followed by another rotation or (b) a rotation, followed by a reflection, followed by a stretch, followed by another rotation.*

This important and useful result will stand us in good stead in examining the geometric aspects of various multivariate techniques in Chapters 5 and 6. Its value lies in contributing to our understanding of what goes on under various matrix transformations. Indeed, still further generalizations are possible in cases where the matrix transformation is singular, as will be examined in Chapter 5.

4.6 INVERTIBLE TRANSFORMATIONS AND MATRIX RANK

As pointed out at the beginning of the chapter, all matrix transformations involve sets of linear equations; conversely, sets of linear equations can be compactly displayed in matrix form. The purpose of this section is to pull together material briefly presented earlier on the topics of matrix inversion and determinants, along with additional concepts as related to the general objective of solving sets of simultaneous linear equations.

As the reader may recall from basic algebra, in the general problem of attempting to solve m linear equations in n unknowns, three possibilities can arise:

1. The set of equations may have no solution; that is, they may form an inconsistent system.
2. The set of equations, while consistent, may have an infinite number of solutions.
3. The set of equations may be both consistent and have exactly one solution.

Solutions of simultaneous linear equations based on the application of inversion assume that the number of equations equals the number of unknowns. In this instance, the matrix of coefficients is square. If other conditions (to be described) are met, the matrix

[9] The representation $\mathbf{V} = \mathbf{P}\Delta\mathbf{Q}'$ is variously called decomposition to basic structure or singular value decomposition.

of coefficients has a *unique inverse,* and exactly one solution exists for the set of equations. Accordingly, we emphasize the last of the three cases above since this is the one that involves matrix inversion and, furthermore, is most relevant for multivariate analysis.

Appendix B discusses the topic of simultaneous linear equations from a much broader point of view, one that encompasses all three of the preceding cases and describes their characteristics in detail.

As recalled, a set of n linear equations in n variables can be compactly written in matrix form as

$$\mathbf{A}_{n \times n}\mathbf{x}_{n \times 1} = \mathbf{b}_{n \times 1}$$

where $\mathbf{A}$ is the $n \times n$ matrix of coefficients, $\mathbf{x}$ is the $n \times 1$ column vector of unknowns, and $\mathbf{b}$ is the $n \times 1$ column vector of constants. For example, suppose we had the following equations:

$$4x_1 - 10x_2 = -2$$

$$3x_1 + 7x_2 = 13$$

These can be written in matrix form as

$$\underset{\mathbf{A}}{\begin{bmatrix} 4 & -10 \\ 3 & 7 \end{bmatrix}} \underset{\mathbf{x}}{\begin{bmatrix} x_1 \\ x_2 \end{bmatrix}} = \underset{\mathbf{b}}{\begin{bmatrix} -2 \\ 13 \end{bmatrix}}$$

If the inverse $\mathbf{A}^{-1}$ exists, then we also know that the following relationship holds:

$$\mathbf{AA}^{-1} = \mathbf{A}^{-1}\mathbf{A} = \mathbf{I}$$

where $\mathbf{A}^{-1}$ is unique. We can solve for $\mathbf{x}$, the vector of unknowns, as follows:

$$\mathbf{Ax} = \mathbf{b}$$

$$\mathbf{A}^{-1}\mathbf{Ax} = \mathbf{A}^{-1}\mathbf{b}$$

$$\mathbf{Ix} = \mathbf{A}^{-1}\mathbf{b}$$

$$\mathbf{x} = \mathbf{A}^{-1}\mathbf{b}$$

Furthermore, from previous discussion we also know that $\mathbf{A}^{-1}$ exists, provided that $|\mathbf{A}| \neq 0$. In the current example, $|\mathbf{A}| = 58$. From Section 4.3.2, we can find $\mathbf{A}^{-1}$ by dividing each entry of the adjoint of $\mathbf{A}$, adj($\mathbf{A}$), by the determinant of $\mathbf{A}$:

$$\mathbf{A}^{-1} = \frac{1}{58} \underset{\text{adj}(\mathbf{A})}{\begin{bmatrix} 7 & 10 \\ -3 & 4 \end{bmatrix}} = \begin{bmatrix} 7/58 & 5/29 \\ -3/58 & 2/29 \end{bmatrix}$$

If $\mathbf{b}$ is then premultiplied by $\mathbf{A}^{-1}$, we obtain the solution vector:

$$\mathbf{x} = \begin{bmatrix} 2 \\ 1 \end{bmatrix}$$

In short, if the determinant of $|\mathbf{A}|$, the coefficient matrix, is not equal to zero, computing the inverse $\mathbf{A}^{-1}$ is a useful way to solve sets of linear equations in which the number of equations equals the number of unknowns.

Two major questions crop up in the discussion of general solution methods for sets of linear equations:

1. If the conditions are such that matrix inversion methods can be employed, what are some of the properties of matrix inverses?
2. Suppose $\mathbf{A}^{-1}$ does not exist, but we still want to say something about those aspects of the space that are preserved under the linear transformation $\mathbf{A}$. What is the connection between the number of linearly independent dimensions in the transformation and the number of dimensions that are preserved under that transformation?

Discussion of these two questions constitutes the primary focus of this section of the chapter.

4.6.1 Properties of Matrix Inverses

In Section 4.3.3 we described one important property of matrix inverses, namely, that the inverse of the product of two or more (conformable) matrices equals the product of the separate inverses in reverse order:

$$\boxed{(\mathbf{T}_1\mathbf{T}_2 \cdots \mathbf{T}_s)^{-1} = \mathbf{T}_s^{-1} \cdots \mathbf{T}_2^{-1}\mathbf{T}_1^{-1}}$$

Moreover, in Section 4.3.4 we made note of the fact that if a matrix $\mathbf{B}$ is orthogonal, then

$$\mathbf{B}' = \mathbf{B}^{-1}$$

Some other aspects of matrix inverses are also useful to point out:

1. The inverse of an inverse is the original matrix:

$$\boxed{(\mathbf{A}^{-1})^{-1} = \mathbf{A}}$$

2. The inverse of a scalar times a matrix equals the reciprocal of the scalar times the matrix inverse:

$$\boxed{(k\mathbf{A})^{-1} = 1/k\mathbf{A}^{-1}}$$

3. The inverse of the transpose of a matrix, equals the transpose of the inverse:

$$\boxed{(\mathbf{A}')^{-1} = (\mathbf{A}^{-1})'}$$

4. The inverse of the diagonal matrix $\mathbf{D}$ is obtained by simply finding the reciprocals of the entries on the main diagonal:

$$\boxed{(\mathbf{D}^{-1}) = \mathrm{diag}(d_{ii}^{-1})}$$

5. If **A**, **B**, and **C** are each square and of order n by n, and if **A** is nonsingular, then

$$\boxed{\mathbf{AB} = \mathbf{AC} \qquad \text{implies} \qquad \mathbf{B} = \mathbf{C}}$$

Note, then, that the cancellation law of scalar algebra holds over the set of nonsingular matrix transformations. We shall use several of the above properties in Chapters 5 and 6 that employ inverses in various types of multivariate computations. In the present geometrically oriented context, however, we note that inverses relate to *invertible* functions in which for every vector in one space we have one and only one uniquely paired vector in another space. This "other" space may, of course, be the original space. The image vector is then another point in the same space as the preimage vector.

4.6.2 Characteristics of Invertible Transformations

If we consider the following transformation:

$$\mathbf{x}^* = \begin{bmatrix} 2.23 \\ 2.20 \end{bmatrix} = \overset{\mathbf{T}}{\begin{bmatrix} 0.45 & 0.89 \\ 0.60 & 0.80 \end{bmatrix}} \overset{\mathbf{x}}{\begin{bmatrix} 1 \\ 2 \end{bmatrix}}$$

we note that a point $\mathbf{x} = \left[\begin{smallmatrix} 1 \\ 2 \end{smallmatrix}\right]$ in two dimensions is mapped by **T** onto $\mathbf{x}^*$, another point in that same space. Furthermore, by methods discussed earlier, we could find $\mathbf{T}^{-1}$ and observe the following:

$$\mathbf{x} = \begin{bmatrix} 1 \\ 2 \end{bmatrix} = \overset{\mathbf{T}^{-1}}{\begin{bmatrix} -4.60 & 5.11 \\ 3.45 & -2.59 \end{bmatrix}} \overset{\mathbf{x}^*}{\begin{bmatrix} 2.23 \\ 2.20 \end{bmatrix}}$$

Here, the inverse $\mathbf{T}^{-1}$ maps $\mathbf{x}^*$ in the plane onto **x** in the plane. This is an illustration of a one-to-one, or *invertible,* transformation in which to each point in the (x_1, x_2) plane, we have one and only one point in the $(x_1{}^*, x_2{}^*)$ plane. Most of our discussion in Sections 4.4 and 4.5 centered around invertible transformations.

If $\mathbf{T}^{-1}$ exists, every **x** has a unique $\mathbf{x}^*$ and vice versa. Invertible transformations exhibit the important property of being nonsingular and, hence, square. Furthermore, as long as *all* (square) transformations $\mathbf{T}_1, \mathbf{T}_2, \ldots, \mathbf{T}_p$ are each nonsingular, their product is also nonsingular, and all information about **x** is preserved in the mapping in the sense that $(\mathbf{T}_1\mathbf{T}_2 \cdots \mathbf{T}_p)^{-1}$ could undo the original composite transformation by transforming $\mathbf{x}^*$ back to **x**.

What happens if **T** is singular and, hence, $\mathbf{T}^{-1}$ does not exist? As might be surmised, if such is the case, we lose the one-to-one correspondence between **x** and $\mathbf{x}^*$. Moreover, it may be the case that additional information about **x** is irretrievably lost in the mapping process. What can be said about the dimensionality of the image space under these circumstances? A discussion of this question involves a topic of central importance to matrix algebra, namely, the *rank* of a matrix.

4.6.3 The Rank of a Matrix

The concept of matrix rank is related to two topics that have already been discussed in earlier chapters:

1. linear dependence of a set of row or column vectors,
2. the determinant of a matrix.

There are two basic, and compatible, ways of defining the rank of a matrix.

One definition takes a dimensional, or geometric, viewpoint. Assume that we have a matrix **A** that is not necessarily square. *The rank of* **A**, *denoted by* $r(\mathbf{A})$, *is defined as the maximum number of linearly independent rows (columns) of* **A**. While it may seem strange to say that the row rank of a matrix is always equal to its column rank, such is the case, as we shall illustrate subsequently. Any matrix, square or rectangular, has a unique rank, one that equals the maximum number of linearly independent vectors.

In Chapter 3 we discussed the concept of linear independence of row or column vectors. In fact, the Gram–Schmidt process was illustrated as a way of finding an orthonormal basis from a set of arbitrary basis vectors.

At that time we pointed out that n linearly independent vectors, each of n components, are sufficient to define a basis. If more that n vectors are present, then the set cannot be linearly independent. However, if less than n vectors are present, they are insufficient to span a space of n dimensions. That is, either the number of components or the number of vectors is sufficient to constrain the dimensionality.

What this boils down to is that if a matrix **A** is rectangular, its rank cannot exceed its smaller dimension (rows or columns as the case may be). Moreover, its row rank equals its column rank so that we can say without ambiguity that $r(\mathbf{A}) = k$, where k is some nonnegative integer. Of course, this in itself does not say that $r(\mathbf{A})$ must equal the lesser of m or n; its rank k may be *less* than the minimum of m or n. All that is being said is that one can talk just as appropriately about m points in n dimensions as n points in m dimensions and that the lesser of the two numbers denotes the maximum subspace in which the points are contained.

Notice that the definition of matrix rank refers to the *maximum* number of rows (columns) that are linearly independent. As suggested above, if a matrix **A** is of order $m \times n$, and if the matrix has rank $r(\mathbf{A}) = k$, then there exist k rows and k columns, where $k \leqslant \min(m, n)$ that are linearly independent. Furthermore, any set of $k + 1$ rows (columns) is linearly dependent.

The reader will recall that we also discussed determinants in Chapter 2 and elsewhere. It turns out that a fully compatible definition of rank can be developed from the foundation of determinants.

The rank of a matrix **A**, *denoted* $r(\mathbf{A})$, *is the order of the largest square submatrix of* **A** *whose determinant is not zero.*

To illustrate the nature of the latter definition, suppose we have the following square matrix:

$$\mathbf{A} = \begin{bmatrix} 1 & 2 & 4 \\ 2 & 1 & 2 \\ 3 & 2 & 4 \end{bmatrix}$$

By inspection we note that the third column is a multiple of the second; clearly $|\mathbf{A}| = 0$ and $r(\mathbf{A})$ is not 3. However, consider one of the 2 x 2 submatrices, for example,

$$\mathbf{B} = \begin{bmatrix} 1 & 2 \\ 2 & 1 \end{bmatrix}$$

We find that $|\mathbf{B}| = -3$; hence, $r(\mathbf{A}) = 2$.

Next, suppose we have the following rectangular matrix:

$$\mathbf{A} = \begin{bmatrix} 1 & 2 & 3 \\ 1 & 1 & 2 \end{bmatrix}$$

Clearly, there are no square submatrices of order 3 x 3 since the matrix is only 2 x 3; hence, $r(\mathbf{A})$ is at most 2. If we take one of the 2 x 2 submatrices

$$\mathbf{B} = \begin{bmatrix} 2 & 3 \\ 1 & 2 \end{bmatrix}$$

we find that $|\mathbf{B}| = 1$; hence $r(\mathbf{A}) = 2$.

It should be stated that we have to find only one square submatrix (of the desired order) with a nonzero determinant. Once we have done so, we can stop the search.

By the systematic examination of determinants of various submatrices of **A**, we have a way to go about finding $r(\mathbf{A})$ in either the square or rectangular matrix cases. Also, we should note that the lowest rank of any matrix must be zero, and this would happen only if $\mathbf{A} = \phi$; that is, the matrix consisted of all zeros. Otherwise, there would be some nonzero element in the single-element minors, and $r(\mathbf{A})$ would at least be 1.

How do we relate the concept of matrix rank to the topics of matrix inverse and invertible transformations described earlier? Perhaps the most direct way is to state that if we have a square ($n \times n$) transformation matrix **A**, whose inverse $\mathbf{A}^{-1}$ exists, the following statements are all equivalent:

1. **A** is nonsingular; that is, $|\mathbf{A}| \neq 0$.
2. The rank $r(\mathbf{A}) = n$; that is, **A** is of full rank in which its rank equals its order.
3. The row vectors of **A** are linearly independent.
4. The column vectors of **A** are linearly independent.
5. The image space obtained from **A** fully preserves the preimage space in a one-to-one fashion.
6. One can obtain the unique preimages transformed by **A** from their counterpart images by means of $\mathbf{A}^{-1}$, the inverse.
7. The specific image points and preimage points are in one-to-one correspondence.

Of course, not all transformations of interest to multivariate analysis will involve cases in which the inverse $\mathbf{A}^{-1}$ exists. Accordingly, means are needed to find out what happens when the transformation is *not* fully invertible. Accordingly, we now consider some of the difficulties that arise when the transformation matrix is singular.

4.6.4 The Relationship of Rank to Linear Transformations

As might be surmised at this point, the rank of a matrix transformation is quite important in matrix algebra. Consider the transformation matrix

$$\mathbf{T} = \begin{bmatrix} 1 & 0 \\ 0 & 1 \\ 0 & 1 \end{bmatrix}$$

and the vector $\mathbf{x} = \left[\begin{smallmatrix}1\\2\end{smallmatrix}\right]$. Via matrix multiplication, we can find the point transformation $\mathbf{x}^* = \mathbf{T}\mathbf{x}$ as follows:

$$\mathbf{x}^* = \begin{bmatrix} 1 & 0 \\ 0 & 1 \\ 0 & 1 \end{bmatrix} \begin{bmatrix} 1 \\ 2 \end{bmatrix} = \begin{bmatrix} 1 \\ 2 \\ 2 \end{bmatrix}$$

and, presumably $\mathbf{T}$ has taken $\mathbf{x}$ into three dimensions. However, as can be seen in Fig. 4.14, the transformation rotates the $\mathbf{e}_1$, $\mathbf{e}_2$ plane through a 45° angle. All of the points in the $\mathbf{e}_1$, $\mathbf{e}_2$ plane, including $\mathbf{x} = \left[\begin{smallmatrix}1\\2\end{smallmatrix}\right]$, undergo this rotation. The range of the transformation *is still a plane.* What should be remembered is that the range of any transformation that takes a point in n dimensions into a point in m dimensions (where $m > n$) cannot exceed n; that is, the higher space *cannot* be filled from a space of lower dimensionality.

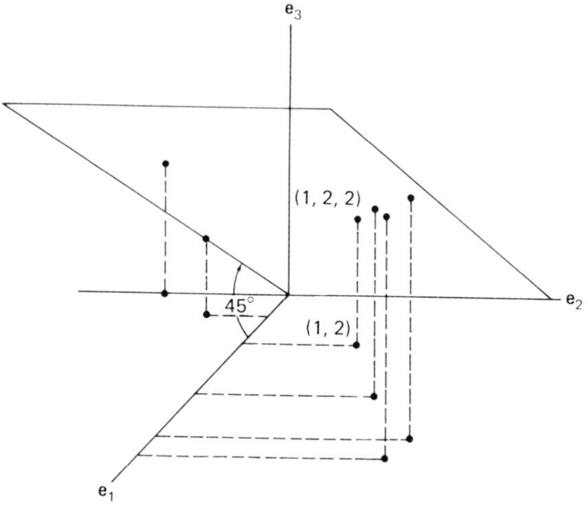

Fig. 4.14

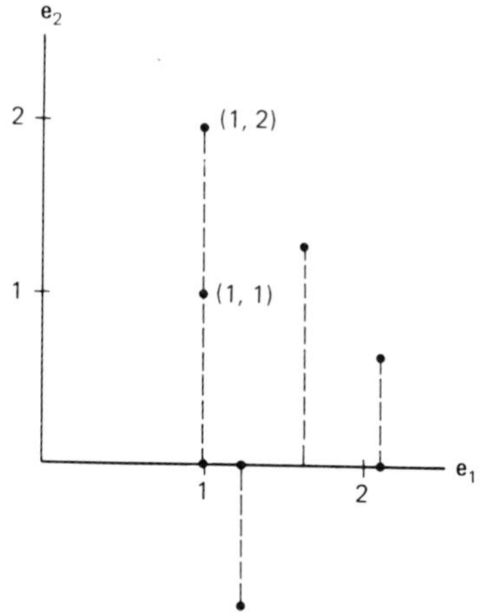

Fig. 4.15

Fig. 4.14 Geometric effect of a linear transformation involving a point in a lower dimensionality.
Fig. 4.15 Geometric effect of a linear transformation involving a point in a higher dimensionality.

Notice that **T** has only two columns, and $r(\mathbf{T})$ is at most equal to 2. In this case $r(\mathbf{T})$ *is* equal to 2, and we observe that the third row of **T** equals the second. Thus, the rank of **T** has placed restrictions on the dimensionality of the transformation.

Now consider the matrix

$$\mathbf{S} = \begin{bmatrix} 1 & 0 \\ 0 & 0 \end{bmatrix}$$

In this case if we desire to find $\mathbf{x}^* = \mathbf{Sx}$, we have

$$\mathbf{x}^* = \begin{bmatrix} 1 & 0 \\ 0 & 0 \end{bmatrix} \begin{bmatrix} 1 \\ 2 \end{bmatrix} = \begin{bmatrix} 1 \\ 0 \end{bmatrix}$$

and, indeed, all points in two dimensions will be mapped onto the (single) $\mathbf{e}_1$ axis. That is, $r(\mathbf{S}) = 1$ and, hence, the transformation maps all vectors, $\mathbf{x}_1$, $\mathbf{x}_2$, $\mathbf{x}_3$, etc., onto a line, regardless of the original dimensionality of $\mathbf{x}_1$, $\mathbf{x}_2$, and $\mathbf{x}_3$. Fig. 4.15 shows this effect geometrically. Note in particular that points (1, 2) and (1, 1) are not distinguished after this transformation. Thus, the transformation preserves the first component of the vector but not the second; these all become zero, and this information is irretrievably lost.

The fact that the rank of a matrix and the number of linearly independent vectors of a matrix are equal is important in the understanding of matrix transformations generally. As noted above, knowledge of the rank of a transformation matrix provides information about the characteristics of the original dimensionality that are "preserved" under the mapping.

To round out the preceding comments, we can extend the discussion of Section 3.3.5 on vector projection to the more general case of projecting a vector in n dimensions onto some hyperplane of dimension k $(k < n)$ passing through the origin of the space (see Panel IV of Fig. 3.15 for an illustrative case).

For example, suppose we have a three-dimensional space and a plane passing through the origin of that space and through the two points

$$\mathbf{b}_1 = \begin{bmatrix} 1 \\ 1 \\ 0 \end{bmatrix}; \qquad \mathbf{b}_2 = \begin{bmatrix} 1 \\ 0 \\ 1 \end{bmatrix}$$

Let us also assume that the vector of interest is represented by

$$\mathbf{a} = \begin{bmatrix} 2 \\ 3 \\ 1 \end{bmatrix}$$

in the full space of three dimensions. If so, the projection $\mathbf{a}^*$ of $\mathbf{a}$ onto the plane defined by $\mathbf{0}$, $\mathbf{b}_1$, and $\mathbf{b}_2$ is given by

$$\boxed{\mathbf{a}^* = \mathbf{B}(\mathbf{B}'\mathbf{B})^{-1}\,\mathbf{B}'\mathbf{a}}$$

where **B** is the matrix whose columns are the vectors $\mathbf{b}_1$ and $\mathbf{b}_2$ in the plane of interest. In terms of this numerical example, we have

$$\mathbf{a}^* = \overset{\mathbf{B}}{\begin{bmatrix} 1 & 1 \\ 1 & 0 \\ 0 & 1 \end{bmatrix}} \overset{(\mathbf{B}'\mathbf{B})^{-1}}{\frac{1}{3}\begin{bmatrix} 2 & -1 \\ -1 & 2 \end{bmatrix}} \overset{\mathbf{B}'}{\begin{bmatrix} 1 & 1 & 0 \\ 1 & 0 & 1 \end{bmatrix}} \overset{\mathbf{a}}{\begin{bmatrix} 2 \\ 3 \\ 1 \end{bmatrix}} = \begin{bmatrix} 8/3 \\ 7/3 \\ 2/3 \end{bmatrix}$$

and $\mathbf{a}^*$ is the vector representing the projection of **a** onto the plane defined by **0**, the origin, and $\mathbf{b}_1$ and $\mathbf{b}_2$.

Since the line through **0** and $\mathbf{b}_1$ is in the plane, an orthogonal projection implies that $(\mathbf{a}^* - \mathbf{a})$ be perpendicular to $\mathbf{b}_1$; similarly so for the line through **0** and $\mathbf{b}_2$, giving the equations

$$(\mathbf{a}^* - \mathbf{a})'\mathbf{b}_1 = 0; \qquad (\mathbf{a}^* - \mathbf{a})'\mathbf{b}_2 = 0$$

A brief sketch of the derivation of $\mathbf{a}^*$ may be in order. In more general terms, if the dimensionality of the full space is n and k denotes the dimensionality of the hyperplane defined by **0** and the $\mathbf{b}_i$, then the $n \times k$ matrix **B** has rank k, and we have

$$(\mathbf{a}^* - \mathbf{a})'\mathbf{b}_1 = 0; \quad (\mathbf{a}^* - \mathbf{a})'\mathbf{b}_2 = 0; \ldots; \quad (\mathbf{a}^* - \mathbf{a})'\mathbf{b}_k = 0$$

or

$$\mathbf{a}^{*\prime}\mathbf{b}_i = \mathbf{a}'\mathbf{b}_i \qquad \text{for} \quad i = 1, 2, \ldots, k$$

In matrix notation, this can be written as

$$\mathbf{a}^{*\prime}\mathbf{B} = \mathbf{a}'\mathbf{B}$$

However, since $\mathbf{a}^*$ is itself in the hyperplane, it represents a linear combination of the vectors $\mathbf{b}_i$:

$$\mathbf{a}^* = \sum_{i=1}^{k} p_i \mathbf{b}_i$$

or, equivalently,

$$\mathbf{a}^* = \mathbf{B}\mathbf{p}$$

where **p** is a vector of arbitrary scalars defining the linear combination. Substituting for $\mathbf{a}^{*\prime}$ above, we have

$$\mathbf{p}'\mathbf{B}'\mathbf{B} = \mathbf{a}'\mathbf{B}$$

B is $n \times k$ and of rank k. $\mathbf{B}'\mathbf{B}$ is $k \times k$ and nonsingular, and we can postmultiply both sides by $(\mathbf{B}'\mathbf{B})^{-1}$ to get

$$\mathbf{p}' = \mathbf{a}'\mathbf{B}(\mathbf{B}'\mathbf{B})^{-1}$$

Next, we can find the transpose of $\mathbf{p}'$:

$$\mathbf{p} = (\mathbf{B}'\mathbf{B})^{-1}\mathbf{B}'\mathbf{a}$$

and recall that $(\mathbf{B}'\mathbf{B})^{-1}$ is symmetric. Finally, we substitute $\mathbf{B}^{-1}\mathbf{a}^*$ for $\mathbf{p}$ and then premultiply both sides by $\mathbf{B}$ to get

$$\mathbf{a}^* = \mathbf{B}(\mathbf{B}'\mathbf{B})^{-1}\mathbf{B}'\mathbf{a}$$

as desired.

The reason for introducing this generalization of vector projection at all is that the idea of (orthogonal) projection plays a central role in least squares, one of the cornerstone methods in multivariate analysis.

4.6.5 Transformations Involving Singular Matrices

To round out our discussion of matrix rank and its relationship to determinants and linear independence, let us consider the geometric effect of a singular matrix from another viewpoint. In Section 3.6.1 we showed how the determinant of a 2×2 transformation matrix $\mathbf{T}$ measures the ratio of areas of the transformed to original figure. As illustrated, in the case of vertices of the unit square

$$O = (0,0); \qquad I = (1,0); \qquad J = (1,1); \qquad K = (0,1)$$

Transformed by

$$\mathbf{T} = \begin{bmatrix} 2 & 1 \\ 3 & 4 \end{bmatrix}$$

we obtained the quadrilateral, reproduced in Panel I of Fig. 4.16. The ratio of areas between transformed and original figure is given by $|\mathbf{T}| = 5$.

Now suppose we transformed the vertices of the unit square by

$$\mathbf{U} = \begin{bmatrix} 2 & 4 \\ 3 & 6 \end{bmatrix}$$

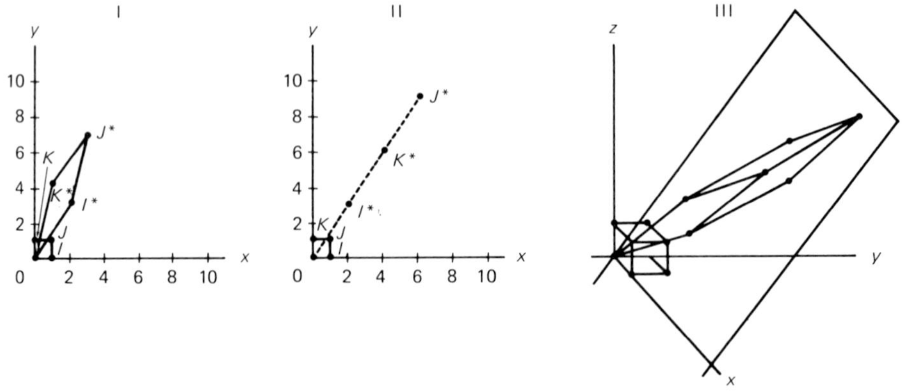

Fig. 4.16 Geometric relationships involving nonsingular and singular transformations.

leading to

$$\mathbf{V} = \begin{bmatrix} 2 & 4 \\ 3 & 6 \end{bmatrix} \overset{\begin{matrix} O & I & J & K \end{matrix}}{\begin{bmatrix} 0 & 1 & 1 & 0 \\ 0 & 0 & 1 & 1 \end{bmatrix}} = \overset{\begin{matrix} O^* & I^* & J^* & K^* \end{matrix}}{\begin{bmatrix} 0 & 2 & 6 & 4 \\ 0 & 3 & 9 & 6 \end{bmatrix}}$$

When we plot the new points of **V** in Panel II of Fig. 4.16, the disconcerting outcome is that **V** becomes a straight line of zero area.

This result, of course, is not hard to understand when we note that $|\mathbf{U}| = 0$ and that the entries in the second column of **U** are twice those in the first. Under $\mathbf{T}^{-1}$, the inverse of **T**, we could reverse the mapping and get back to the unit square in Panel I of the figure. This is not possible in the case of the second matrix **U** since $\mathbf{U}^{-1}$ does not exist. *Hence, whenever a transformation matrix* U *is singular, that mapping is not invertible.*

Panel III shows another case, this one involving a 3 x 3 transformation matrix:

$$\mathbf{W} = \begin{bmatrix} 2 & 1 & 1 \\ 1 & 2 & 2 \\ 2 & 4 & 4 \end{bmatrix}$$

as applied to the vertices of the unit cube. Here, the entries in the third row are twice those of the second, and all points of the unit cube lie on a plane through the origin that has the equation $z = 2y$.

Hence, the parallelepiped that would have been obtained had **W** been nonsingular has collapsed into a parallelogram of only two dimensions. Again we have the case in which the original mapping obtained from **W** is not invertible, and the rows (columns) of **W** are not linearly independent.[10]

4.6.6 Finding the Rank of a Matrix via Determinants

A number of methods are available for finding the rank of an arbitrary matrix. Perhaps the most straightforward, if tedious, precedure is by means of determinants. If **A** is square and of order $n \times n$, we first see if $|\mathbf{A}| \neq 0$. If so, then $r(\mathbf{A}) = n$.

If $|\mathbf{A}|$ is zero, we then examine square submatrices of order $(n-1) \times (n-1)$. If one of these has a nonzero determinant, then we stop and state that $r(\mathbf{A}) = n - 1$. If all determinants are zero, we continue with square submatrices of order $(n-2) \times (n-2)$, and so on.

If **A** is rectangular of order $m \times n$, we know at the outset that $r(\mathbf{A}) \leqslant \min(m, n)$. Having established which order is smaller—suppose it is m—we examine square

[10] Note, however, that the x, y dimensions *are* retained; it is only the third dimension that collapses. The rank of the transformation indicates how many dimensions will be retained (two out of three in this case).

submatrices of order $m \times m$ to see if one can be found whose determinant is nonzero. If so, then $r(\mathbf{A}) = m$. If not, we examine square submatrices of order $(m-1) \times (m-1)$, and so on, as indicated above.

To illustrate how one might go about finding the rank of a matrix via these procedures, we can examine a few examples. First, consider the 2×2 matrix

$$\mathbf{A} = \begin{bmatrix} 1 & 2 \\ 4 & 3 \end{bmatrix}$$

We compute the determinant of **A** and find that $|\mathbf{A}| = -5$. Since $|\mathbf{A}| \neq 0$, we conclude that **A** has rank 2.

On the other hand, consider the following case:

$$\mathbf{B} = \begin{bmatrix} 2 & 3 \\ 1 & 1.5 \end{bmatrix}$$

where we first find that $|\mathbf{B}| = 0$. Since the determinant of the original matrix has vanished, we must examine single-element minors. Since there are first-order determinants, such as $|2|$ or $|3|$, that are not equal to zero, **B** has rank 1.

Now let us examine the 3×4 matrix:

$$\mathbf{C} = \begin{bmatrix} 1 & 1 & 2 & 5 \\ 2 & 2 & 4 & 10 \\ 1 & 0 & 3 & 1 \end{bmatrix}$$

At the outset we know that the rank of **C** cannot exceed 3, the smaller order. However, we see immediately that the second row is twice the first row; hence **C** cannot be of rank 3 since the row vectors are not linearly independent. But, since the first and third rows are linearly independent (i.e., neither is a multiple of the other), **C** has rank 2. This could be checked by observing that any of the six square 2×2 submatrices:

$$\begin{bmatrix} 1 & 1 \\ 1 & 0 \end{bmatrix}; \quad \begin{bmatrix} 1 & 2 \\ 1 & 3 \end{bmatrix}; \quad \begin{bmatrix} 1 & 2 \\ 0 & 3 \end{bmatrix}; \quad \begin{bmatrix} 1 & 5 \\ 1 & 1 \end{bmatrix}; \quad \begin{bmatrix} 1 & 5 \\ 0 & 1 \end{bmatrix}; \quad \begin{bmatrix} 2 & 5 \\ 3 & 1 \end{bmatrix}$$

made up of elements from rows 1 and 3 each has a nonzero determinant. [Of course, only one such submatrix would be needed to establish that $r(\mathbf{C}) = 2$.]

If the order of the matrix is large, the procedure outlined above becomes rather tedious. Fortunately, other ways of finding the rank of a matrix exist. Some of these are described in Chapter 5, and one of these—based on the *echelon matrix* procedure—is discussed in the last main section of this chapter.

4.6.7 The Uses of Rank in Matrix Algebra

The rank of a matrix plays several important roles in matrix algebra. For example, in solving a set of simultaneous linear equations, it is the case that when (and only when) the rank of the matrix of coefficients equals the rank of the augmented matrix, the set of equations has at least one solution. By "augmented matrix" is meant a matrix consisting of the coefficients to which has been appended an additional column made up of the constants. For example, in the equations described earlier:

$$4x_1 - 10x_2 = -2$$

$$3x_1 + 7x_2 = 13$$

the augmented matrix is

$$\begin{bmatrix} 4 & -10 & -2 \\ 3 & 7 & 13 \end{bmatrix}$$

If the rank of this matrix equals the rank of the matrix of coefficients:

$$\begin{bmatrix} 4 & -10 \\ 3 & 7 \end{bmatrix}$$

the system is consistent and has at least one solution.

Second, if the set of equations is consistent, their solution is unique if (and only if) the rank of the coefficients matrix and, hence, that of the augmented matrix equals the number of unknowns. In this case the sytem can be solved by inversion of the coefficients matrix. If the rank is less than the number of unknowns, then an infinite number of solutions exist. While we do not delve into detailed discussion here of these two major uses of matrix rank (see Appendix B for this), enough has been said to show the importance of the concept in the solution of simultaneous equations.

A second important role that is played by matrix rank concerns those aspects of a configuration of points that are preserved under matrix transformations. As we know, two linearly independent vectors are needed to span a plane, three to span a three-dimensional space, and so on. And, as illustrated in Figs. 4.14 and 4.15, the rank of a matrix determines what aspects of the configuration will be retained after transformation.

Furthermore, as illustrated in Fig. 4.16, if the transformation is not of full rank, the image of a two-dimensional unit square could collapse to a line or to the origin; the image of a three-dimensional unit cube could collapse to an area or to a line, or to the origin. *In general, then, the rank of a transformation matrix determines the dimensionality of the image space.* Perhaps more than anything else, this is the essential aspect of matrix rank in multivariate analysis.

At this point we have discussed matrix inversion as a way to solve simultaneous equations in which the number of equations equals the number of unknowns and, hence, where it is possible that an inverse of the matrix of coefficients exists. We have also examined what happens when a square matrix is singular and the effect that this has on the transformation.

However, what has not been discussed in detail as yet are three related topics:

1. What other procedures are available for finding the rank of a matrix?
2. What procedures, other than computation of the adjoint matrix, are available for finding the inverse of a matrix?
3. How can these numerical methods for finding inverses be applied to problems in multivariate analysis?

The last main section of the chapter takes up these questions.

4.7 METHODS FOR RANK DETERMINATION AND MATRIX INVERSION

In multivariate analysis numerous occasions arise in which we wish to solve a set of simultaneous equations. Often the system will be consistent, and the matrix of coefficients will be square and of rank equal to its order; if so, a solution based on matrix inversion can be found.

Sometimes, however, the matrix of coefficients will either be rectangular or, even if square, singular. In the latter case its rank will not equal its order, and we may wish to find out—by means other than the tedious examination of determinants of square submatrices—what the rank of the transformation matrix is. A highly general approach to determining the rank of a matrix makes use of what are called elementary operations and the associated construction of echelon matrices. We first discuss this alternative approach to the determination of rank and its relationship to the solution of simultaneous linear equations.

We then return to the pivotal method, first used in Chapter 2, to compute determinants. As mentioned there, the pivotal method is also applicable to solving simultaneous equations and computing inverses. We discuss these extensions and illustrate their application to the 4 x 4 matrix that was described in Table 2.2 and to statistical data drawn from the sample problem of Table 1.2.

4.7.1 Elementary Operations

Elementary operations play an essential role in the solution of sets of simultaneous equations. Illustratively taking the case of the rows of a transformation matrix, there are three basic operations—called elementary row operations—that can be used to transform one matrix into another. We may

1. interchange any two rows;
2. multiply any row by a nonzero scalar;
3. add to any given row a scalar multiple of some other row.

If we change some matrix **A** *into another matrix* **B** *by the use of elementary row operations, we say that* **B** *is row equivalent to* **A**.

Elementary row operations involve the multiplication of **A** by special kinds of matrices that effect the above transformations. However, we could just as easily talk about elementary column operations—the same kinds as these shown above—that are applied to the columns of **A**. A matrix so transformed would be called column equivalent to the original matrix. To simplify our discussion, we illustrate the ideas via elementary row operations. The reader should bear in mind, however, that the same approach is applicable to the columns of **A**.

To illustrate how elementary row operations can be applied to the general problem of solving a set of simultaneous equations, let us again consider the two equations described earlier:

$$\text{I}\begin{cases} 4x_1 - 10x_2 = -2 \\ 3x_1 + 7x_2 = 13 \end{cases}$$

Suppose we first multiply all members of the first equation by $-\frac{3}{4}$ and then add the result to those of the second equation:

$$\text{II}\begin{cases} 4x_1 - 10x_2 = -2 \\ \frac{29}{2}x_2 = \frac{29}{2} \end{cases}$$

Next, let us multiply the second equation by $\frac{2}{29}$. If so, we obtain

$$\text{III}\begin{cases} 4x_1 - 10x_2 = -2 \\ x_2 = 1 \end{cases}$$

All these sets of equations are row equivalent in the sense that all three sets have the same solution:

$$x_1 = 2$$

$$x_2 = 1$$

and each can be transformed to either of the others via elementary row operations. However, it is clear that the solution to the third set of equations is most apparent and easily found.

While elementary row operations are useful in the general task of solving sets of simultaneous equations, this is not their only desirable feature. A second, and major, attraction is the fact that elementary operations (row or column), as applied to a matrix **A**, *do not change its rank.* Moreover, as will be shown, elementary operations transform the given matrix in such a way as to make its rank easy to determine by inspection. As it turns out, all three sets of equations above have the same rank (rank 2) since they are all equivalent in terms of elementary row operations.

Elementary row operations are performed by a special set of square, nonsingular matrices called elementary matrices. *An elementary matrix is a nonsingular matrix that can be obtained from the identity matrix by an elementary row operation.* For example, if we wanted to interchange two rows of a matrix, we could do so by means of the permutation matrix

$$\begin{bmatrix} 0 & 1 \\ 1 & 0 \end{bmatrix}$$

For example, if we have the point $\mathbf{x} = \begin{bmatrix} 1 \\ 2 \end{bmatrix}$ in two dimensions, the premultiplication of **x** by the permutation matrix above would yield:[11]

$$\mathbf{x}^* = \begin{bmatrix} 0 & 1 \\ 1 & 0 \end{bmatrix}\begin{bmatrix} 1 \\ 2 \end{bmatrix} = \begin{bmatrix} 2 \\ 1 \end{bmatrix}$$

We note that the coordinates of **x** have, indeed, been permuted.

[11] In this section of the chapter we use the general format $\mathbf{x}^* = \mathbf{A}\mathbf{x}$, in which the transformation matrix premultiplies the vector (or matrix) of interest. The reader should become comfortable in using either (pre- or postmultiplication) mode.

As mentioned above, elementary matrices are nonsingular. In the 2 x 2 matrix case, the set of elementary matrices consists of the following:

I. *Permutation*

$$\begin{bmatrix} 0 & 1 \\ 1 & 0 \end{bmatrix}$$

II. *Stretches*

$\begin{bmatrix} k & 0 \\ 0 & 1 \end{bmatrix}$ A stretch or compression of the plane that is parallel to the x axis; $\begin{bmatrix} 1 & 0 \\ 0 & k \end{bmatrix}$ A stretch or compression of the plane that is parallel to the y axis

III. *Shears*

$\begin{bmatrix} 1 & c \\ 0 & 1 \end{bmatrix}$ A shear parallel to the x axis; $\begin{bmatrix} 1 & 0 \\ c & 1 \end{bmatrix}$ A shear parallel to the y axis

Continuing with the numerical example involving $\mathbf{x} = [^1_2]$, we have

Stretches

$$\begin{bmatrix} k & 0 \\ 0 & 1 \end{bmatrix}\begin{bmatrix} 1 \\ 2 \end{bmatrix} = \begin{bmatrix} k \\ 2 \end{bmatrix}; \quad \begin{bmatrix} 1 & 0 \\ 0 & k \end{bmatrix}\begin{bmatrix} 1 \\ 2 \end{bmatrix} = \begin{bmatrix} 1 \\ 2k \end{bmatrix}$$

Shears

$$\begin{bmatrix} 1 & c \\ 0 & 1 \end{bmatrix}\begin{bmatrix} 1 \\ 2 \end{bmatrix} = \begin{bmatrix} 1 + 2c \\ 2 \end{bmatrix}; \quad \begin{bmatrix} 1 & 0 \\ c & 1 \end{bmatrix}\begin{bmatrix} 1 \\ 2 \end{bmatrix} = \begin{bmatrix} 1 \\ c + 2 \end{bmatrix}$$

But, there is nothing stopping us from applying, in some prescribed order, a *series* of premultiplications by elementary matrices. Furthermore, the product of a set of nonsingular matrices will itself be nonsingular.

The geometric character of permutations, stretches, and shears has already been illustrated in Section 4.5. Here we are interested in two major applications of elementary row operations and the matrices that represent them:

1. determining the rank of a matrix, and
2. finding the inverse of a matrix, when such inverse exists. Each application is described in turn.

4.7.2 Elementary Operations and Matrix Rank

Three properties of matrix rank are of general interest to matrix algebra:

1. The rank of an $n \times n$ identity matrix $\mathbf{I}_{n \times n}$, is equal to n.
2. The rank of a matrix is not changed by its premultiplication (or postmultiplication) by a nonsingular matrix. In particular, elementary row operations involve nonsingular matrices and, hence, do not change the rank of the matrix being transformed.

3. If two matrices **A** and **B** have ranks that are denoted by $r(\mathbf{A})$ and $r(\mathbf{B})$, and their product **AB** is possible, then the rank of their product, denoted $r(\mathbf{AB})$, is less than or equal to the smaller rank of the two matrices:

$$\boxed{r(\mathbf{AB}) \leqslant \min [r(\mathbf{A}), r(\mathbf{B})]}$$

Suppose we first look at the implications of the second and third of the above properties. The second property indicates that if some matrix **A** has rank k, then multiplication by an elementary matrix will *not* change its rank. This is a special case of the third property, in which both matrices in the matrix product are square and of rank k.

The third property is important to our earlier discussion of invertible transformations. If the transformation matrix has a rank that is less than the matrix being transformed, all information in the preimages will not be preserved in the image space (as was illustrated in Fig. 4.15).

Suppose, now that we wish to find the rank of some arbitrary matrix **A**. Assume that we operate on **A** via a series of elementary row operations. As noted, application of a sequence of elementary matrices, each of which is nonsingular, will not change the rank of **A** but could transform **A** to a structure in which its rank can be determined by inspection. This particular type of matrix—one that is obtained by a series of elementary row operations—is called an *echelon* matrix. To illustrate, consider the following rectangular matrix:

$$\mathbf{H} = \begin{bmatrix} 0 & 1 & 3 & 0 & 4 \\ 0 & 0 & 1 & -2 & 6 \\ 0 & 0 & 0 & 1 & 3 \\ 0 & 0 & 0 & 0 & 0 \end{bmatrix}$$

This matrix is an example of an echelon matrix. *An echelon matrix is any matrix, square or rectangular, that exhibits the following structure:*

1. Each of the first k rows ($k \geqslant 0$) of **H** has one or more nonzero elements.
2. For each such row, the first nonzero element, as one reads from left to right, is unity.
3. The arrangement of the first k rows is such that the first nonzero element in a given row is always to the right of the first nonzero element of any row that precedes (or lies above) the given row.
4. After the first k rows, the elements of all remaining rows, if any, are all zero.

The importance of echelon matrices relates to the facts that

1. any matrix **A** can be transformed by a sequence of elementary row operations into echelon form;
2. the rank of the matrix is not altered in the process of changing it to echelon form;
3. the number of nonzero rows in **H**, the echelon form of **A**, equals the rank of **A**.

A couple of caveats are in order, however. First, it should be pointed out that **H**, the echelon form of **A**, does not "equal" **A** but, rather, can be *derived* from **A** by elementary operations. Second, **H** is, in general, not a unique representation of **A**; that is, there is no unique echelon form for a given matrix, **A**. However, neither of these caveats weakens the general usefulness of echelon matrices in rank determination and solving sets of simultaneous equations.

Transforming a given matrix to echelon form represents a relatively straightforward procedure. To illustrate, let us take the matrix

$$\mathbf{A} = \begin{bmatrix} 1 & 2 & 3 & 4 \\ 1 & 3 & 2 & 2 \\ 2 & 4 & 1 & 0 \end{bmatrix}$$

and apply elementary row operations to it.

1. Subtract row 1 from row 2; subtract twice row 1 from row 3.

$$\begin{bmatrix} 1 & 2 & 3 & 4 \\ 0 & 1 & -1 & -2 \\ 0 & 0 & -5 & -8 \end{bmatrix}$$

2. Multiply row 3 by $-\frac{1}{5}$.

$$\begin{bmatrix} 1 & 2 & 3 & 4 \\ 0 & 1 & -1 & -2 \\ 0 & 0 & 1 & 8/5 \end{bmatrix}$$

We note that there are three nonzero rows remaining and, hence, the rank of the echelon form of **A**—and the rank of **A** as well—is 3.

Finding the echelon matrix, as indicated by the preceding operations, involves concentrating on one row of the matrix at a time in order to obtain (a) a leading entry of unity in that row and (b) zeros in all lower rows of the column containing the leading entry of unity.

While elementary row operations are useful in finding the echelon form of a matrix and, hence, determining its rank, they are also of value in matrix inversion and the solution of simultaneous linear equations.

4.7.3 Elementary Operations and Simultaneous Equations

Let us now examine how the preceding approach to obtaining echelon matrices can be adapted to solving simultaneous equations. Since we know that the rank of the matrix illustrated above is 3, let us make up a new problem by setting down only the first three columns of the matrix used in the preceding example. Next, assume that the following simultaneous equations represent a system that we would like to solve:

$$x_1 + 2x_2 + 3x_3 = 14$$

$$x_1 + 3x_2 + 2x_3 = 13$$

$$2x_1 + 4x_2 + x_3 = 13$$

As illustrated earlier, this set of equations can be written as

$$\begin{bmatrix} 1 & 2 & 3 \\ 1 & 3 & 2 \\ 2 & 4 & 1 \end{bmatrix} \begin{bmatrix} x_1 \\ x_2 \\ x_3 \end{bmatrix} = \begin{bmatrix} 14 \\ 13 \\ 13 \end{bmatrix}$$

We see that the first three columns are the same as those appearing in the preceding example. Now let us go through the echelon procedure once more, but this time apply each elementary row operation to the full, *augmented* matrix consisting of coefficients *and* the vector of constants:

$$\left[\begin{array}{ccc:c} 1 & 2 & 3 & 14 \\ 1 & 3 & 2 & 13 \\ 2 & 4 & 1 & 13 \end{array}\right]$$

1. Subtract row 1 from row 2; subtract twice row 1 from row 3.

$$\left[\begin{array}{ccc:c} 1 & 2 & 3 & 14 \\ 0 & 1 & -1 & -1 \\ 0 & 0 & -5 & -15 \end{array}\right]$$

2. Multiply row 3 by $-\frac{1}{5}$.

$$\left[\begin{array}{ccc:c} 1 & 2 & 3 & 14 \\ 0 & 1 & -1 & -1 \\ 0 & 0 & 1 & 3 \end{array}\right]$$

If we examine the 3 x 3 matrix to the left of the dotted line, we see that the leading entry of each row is unity. Moreover, in echelon form the rank of the matrix is seen, by inspection, to be 3.

If we refer to the original set of equations, it is apparent that we can now obtain this solution quite readily. From the third row of the echelon form, immediately above, we have

$$x_3 = 3$$

If $x_3 = 3$ is then substituted in the second row, we have

$$x_2 - 3 = -1$$

$$x_2 = 2$$

If $x_2 = 2$ and $x_3 = 3$ are substituted in the first row, we have

$$x_1 + 4 + 9 = 14$$

$$x_1 = 1$$

The process of finding the value of x_3 in the third equation first, and then substituting it in the second equation to find x_2, and so on, is called back substitution. (The whole process is just an application of the pivotal method described, in the context of determinants, in Chapter 2, as will be shown later.)

4.7.4 Finding the Inverse

Now that we have seen how a set of simultaneous equations can be solved by means of elementary row operations, let us carry the same general procedure one step further to find the inverse of the matrix of coefficients. For ease of illustration we continue with the same example.

First of all, it should be clear that the set of simultaneous equations shown above can be written in the following form:

$$\begin{bmatrix} 1 & 2 & 3 \\ 1 & 3 & 2 \\ 2 & 4 & 1 \end{bmatrix} \begin{bmatrix} x_1 \\ x_2 \\ x_3 \end{bmatrix} = \begin{bmatrix} 1 & 0 & 0 \\ 0 & 1 & 0 \\ 0 & 0 & 1 \end{bmatrix} \begin{bmatrix} 14 \\ 13 \\ 13 \end{bmatrix}$$

which, in turn, can be expressed as

$$\mathbf{Ax} = \mathbf{Ib}$$

We could then perform a set of elementary row operations on *both* **A** and **I**. In particular, we shall try to reduce **A** to an identity matrix.

1. Subtract row 1 from row 2; subtract twice row 1 from row 3.

$$\left[\begin{array}{ccc:ccc} 1 & 2 & 3 & 1 & 0 & 0 \\ 0 & 1 & -1 & -1 & 1 & 0 \\ 0 & 0 & -5 & -2 & 0 & 1 \end{array}\right]$$

2. Subtract twice row 2 from row 1. We include this elementary row operation in order to make sure that column 2 has only a *single* nonzero entry (in row 2).

$$\left[\begin{array}{ccc:ccc} 1 & 0 & 5 & 3 & -2 & 0 \\ 0 & 1 & -1 & -1 & 1 & 0 \\ 0 & 0 & -5 & -2 & 0 & 1 \end{array}\right]$$

3. Multiply row 3 by $-\frac{1}{5}$.

$$\left[\begin{array}{ccc:ccc} 1 & 0 & 5 & 3 & -2 & 0 \\ 0 & 1 & -1 & -1 & 1 & 0 \\ 0 & 0 & 1 & 2/5 & 0 & -1/5 \end{array}\right]$$

The 3 x 3 submatrix on the left is still not an identity matrix. Accordingly, we can apply the following additional row operations:

4. Subtract 5 times row 3 from row 1; add row 3 to row 2.

$$\left[\begin{array}{ccc|ccc} 1 & 0 & 0 & 1 & -2 & 1 \\ 0 & 1 & 0 & -3/5 & 1 & -1/5 \\ 0 & 0 & 1 & 2/5 & 0 & -1/5 \end{array}\right]$$

When we do this, we observe that the left side of the matrix is, indeed, an identity matrix.

However, while we have been transforming **A** to **I** via elementary row operations, we have, at the same time, been transforming **I** on the right side of the dotted line. Recalling that

$$\mathbf{Ax} = \mathbf{b} \qquad \mathbf{A}^{-1}\mathbf{Ax} = \mathbf{A}^{-1}\mathbf{b} \qquad \mathbf{Ix} = \mathbf{A}^{-1}\mathbf{b}$$

it seems reasonable to suppose that we have been obtaining the inverse of **A**. That is

$$\mathbf{A}^{-1} = \begin{bmatrix} 1 & -2 & 1 \\ -3/5 & 1 & -1/5 \\ 2/5 & 0 & -1/5 \end{bmatrix}$$

First, we observe that

$$\underset{\mathbf{A}^{-1}}{\begin{bmatrix} 1 & -2 & 1 \\ -3/5 & 1 & -1/5 \\ 2/5 & 0 & -1/5 \end{bmatrix}} \underset{\mathbf{b}}{\begin{bmatrix} 14 \\ 13 \\ 13 \end{bmatrix}} = \underset{\mathbf{x}}{\begin{bmatrix} 1 \\ 2 \\ 3 \end{bmatrix}}$$

as we know it should. Second, we check to see that

$$\underset{\mathbf{I}}{\begin{bmatrix} 1 & 0 & 0 \\ 0 & 1 & 0 \\ 0 & 0 & 1 \end{bmatrix}} = \underset{\mathbf{A}}{\begin{bmatrix} 1 & 2 & 3 \\ 1 & 3 & 2 \\ 2 & 4 & 1 \end{bmatrix}} \underset{\mathbf{A}^{-1}}{\begin{bmatrix} 1 & -2 & 1 \\ -3/5 & 1 & -1/5 \\ 2/5 & 0 & -1/5 \end{bmatrix}}$$

and the solution is complete.

In summary, application of elementary row operations has served three main purposes:

1. By reducing the matrix to echelon form we were able to discern its rank by simply counting the number of rows containing a nonzero entry.

2. If an explicit set of simultaneous equations is involved—here we assume that a unique solution exists—we can apply elementary row operations to the augmented matrix and obtain, via back substitution, the desired values of the unknowns.

3. If the inverse of the given transformation matrix is also desired, we can apply elementary row operations to both **A** and **I**, transforming the former into **I** and the latter into $\mathbf{A}^{-1}$.

In actuality, elementary row operations (or elementary column operations) have applicability to solving sets of simultaneous equations in more general settings where an inverse of the coefficients matrix may not exist. These matters are taken up in Appendix B.

In summary, from a somewhat more theoretical viewpoint, we can use the echelon approach to reduce any matrix to the following form :

$$\begin{bmatrix} \mathbf{I}_{k\times k} & \vdots & \phi \\ \cdots & \cdots & \cdots \\ \phi & \vdots & \phi \end{bmatrix}$$

where the first k rows and k columns represent a $k \times k$ identity matrix (of rank k, of course), and the remaining entries (if any) all consist of zeros.

In the preceding example involving the nonsingular matrix **A**, no zeros appeared since **A** itself was of full rank. In other instances zeros will be found. In general, the above form is found in two steps. Given an arbitrary matrix **B**, we first apply a set of elementary row operations to get the echelon form:

$$\mathbf{H} = \mathbf{FB}$$

where **F** denotes the (nonsingular) matrix product of *all* of the separate elementary matrices used to carry out the reduction of **B** to echelon form. This step has been illustrated above.

Next, we could apply elementary column operations in order to get the matrix product:

$$\mathbf{HG} = \begin{bmatrix} \mathbf{I}_{k\times k} & \vdots & \phi \\ \cdots & \cdots & \cdots \\ \phi & \vdots & \phi \end{bmatrix} = \mathbf{FBG}$$

where $r(\mathbf{B}) = k$. In the preceding numerical example it was not necessary to apply elementary column operations after the elementary row operations were performed. In other cases, it may be more efficient to reduce the matrix to echelon form via row operations and then to obtain the form above via elementary column operations that entail postmultiplication of the echelon matrix **H**. At any rate we include the general approach that involves both row and column operations.

The complete transformation that entails row and column operations is also called an equivalence transformation. *More formally, if* **F** *and* **G** *are nonsingular, an equivalence transformation of an arbitrary matrix* **B** *is defined as*

$$\boxed{\mathbf{C} = \mathbf{FBG}}$$

and **C** is defined to be equivalent, via elementary row and column operations, to the given matrix **B**.

4.7.5 The Pivotal Method

Having seen how elementary operations are used to (a) determine rank, (b) solve a set of simultaneous equations, and (c) find the inverse of the coefficients matrix, our discussion of the pivotal method, first described in Chapter 2, can now be completed. For ease of reference, Table 2.2 is reproduced here as Table 4.15.

TABLE 4.15

Evaluating a Determinant by the Pivotal Method

Row no.	Original matrix				Identity matrix				Check sum column
0	1	2	3	4	5	6	7	8	9
01	**[2]**	3	1	2	1	0	0	0	9
02	4	2	3	4	0	1	0	0	14
03	1	4	2	2	0	0	1	0	10
04	3	1	0	1	0	0	0	1	6
10	1	1.5	0.5	1	0.5	0	0	0	4.5
11		**[−4]**	1	0	−2	1	0	0	−4
12		2.5	1.5	1	−0.5	0	1	0	5.5
13		−3.5	−1.5	−2	−1.5	0	0	1	7.5
20		1	−0.25	0	0.5	−0.25	0	0	1
21			**[2.125]**	1	−1.75	0.625	1	0	3.0
22			−2.375	−2	0.25	−0.875	0	1	−4.0
30			1	0.471	−0.824	0.294	0.471	0	1.412
31				**[−0.881]**	−1.707	−0.177	1.119	1	−0.646
40				1	1.938	0.201	−1.270	−1.135	0.733
30*			1		−1.737	0.199	1.069	0.534	1.067
20*		1			0.066	−0.200	0.267	0.134	1.267
10*	1				−0.668	−0.001	0.334	0.667	1.332

$|\mathbf{A}| = (2)(-4)(2.125)(-0.881) = 15$

In Chapter 2 we illustrated how the pivotal (boxed) entries in Table 4.15 were obtained and how $|\mathbf{A}|$ was found as the product of the four pivotal items. From what we have now learned, we see that the pivotal method is just a systematic way of applying elementary row operations, leading in the example of Table 4.15 to the echelon matrix:

$$\mathbf{H} = \begin{bmatrix} 1 & 1.5 & 0.5 & 1 \\ 0 & 1 & -0.25 & 0 \\ 0 & 0 & 1 & 0.471 \\ 0 & 0 & 0 & 1 \end{bmatrix}$$

which, we note, is of full rank where $r(\mathbf{H})$ equals n, its order. Our task now is to explain the back substitution procedure that leads to the inverse of the original 4 x 4 matrix (call it $\mathbf{A}$) in the top left-hand corner of Table 4.15.

First, if we treat column 5 as a vector of constants, it should be apparent from earlier discussion how we can obtain the following values of x_4, x_3, x_2, x_1 via back substitution:

$$x_4 = 1.938 \qquad \text{(from row 40 and column 5)}$$

$$x_3 + 0.471(1.938) = -0.824 \qquad \text{(from row 30 and column 5)}$$

$$x_3 = -1.737$$

$$x_2 - 0.25(-1.737) + 0(1.938) = 0.5 \qquad \text{(from row 20 and column 5)}$$

$$x_2 = 0.066$$

$$x_1 + 1.5(0.066) + 0.5(-1.737) + 1(1.938) = 0.5 \qquad \text{(from row 10 and column 5)}$$

$$x_1 = -0.668$$

These entries are shown as a column vector in rows 40, 30*, 20*, and 10*, and column 5.

What we are doing here is building up the inverse of **A** one column at a time. Then we repeat the whole procedure, using the entries of column 6, followed by column 7, and finally by column 8. The result of all of this is the inverse $\mathbf{A}^{-1}$, shown in rows 40, 30*, 20*, and 10*, and columns 5, 6, 7, and 8.

However, it is important to note that the inverse $\mathbf{A}^{-1}$ is shown in Table 4.15 in *reverse row order*. That is, the *first* row of $\mathbf{A}^{-1}$ is given by

$$(-0.668 \quad -0.001 \quad 0.334 \quad 0.667)$$

as designated by the row 10*. Hence, if one uses the row order 10*, 20*, 30*, and 40, the inverse $\mathbf{A}^{-1}$ will be in correct row order.

The pivotal procedure proceeds somewhat differently from that used in Section 4.7.4, but the results are the same. Rather than applying additional elementary operations to transform the echelon matrix **H** into identity form, in the pivotal method we simply treat each of the columns—columns 5 through 8—as a separate set of constants and solve for $\mathbf{A}^{-1}$ in this manner. Either way we have the relation

$$\mathbf{A}\mathbf{A}^{-1} = \mathbf{I}$$

as we should.

Furthermore, had an explicit set of constants been present, in the usual form of a set of simultaneous equations, the same procedure could have been applied to this column as well. In brief, the pivotal method can be used for the same purposes noted earlier:

1. finding the determinant of **A**,
2. finding the rank of **A**,
3. solving an explicit set of simultaneous equations,
4. finding the inverse of **A** (if it exists).

In applying the pivotal method one should be on the lookout for cases in which a zero appears in the leading diagonal position. If such does occur, we cannot, of course, divide that row by the pivot. If the matrix is nonsingular—which will usually be the case in data-based applications—the presence of a leading zero suggests that we want to move to the next row and select it as the pivot (i.e., permute rows) before proceeding. This has no effect on the inverse, but will reverse the sign of the determinant (if an odd number of such transpositions occur).

If the matrix is singular, the situation will be revealed by the presence of *all zeros* in some row to be pivoted. This is the type of situation that we encountered earlier in transforming a matrix to echelon form.

In summary, the pivotal method is now seen as just a specific step-by-step way to go about applying elementary row operations. While other approaches are available, the pivotal method does exhibit the virtues of simplicity and directness. To round out discussion, we consider its use in finding the inverse of a correlation matrix in the context of multiple regression.

4.7.6 Matrix Inversion in Multiple Regression

The sample data introduced in Table 1.2 involved three variables:

Y number of days absent

X_1 attitude toward the company

X_2 number of years employed by the company

As discussed in Section 1.6.2, the least-squares principle entails finding a linear equation that minimizes the sum of the squared deviations

$$\sum_{i=1}^{12} e_i^2 = \sum_{i=1}^{12} (Y_i - \hat{Y}_i)^2$$

between the original criterion Y_i and the predicted criterion $\hat{Y}_i$.

The linear equation is represented by the form

$$\hat{Y}_i = b_0 + b_1 X_1 + b_2 X_2$$

where b_0, b_1, and b_2 are parameters to be estimated (by least squares) from the data.

The derivation of the normal equations underlying the least-squares procedure is described in Appendix A.

If we work with the original data from the sample problem, the normal equations can be represented in matrix form as

$$(\mathbf{X}'\mathbf{X})\mathbf{b} = \mathbf{X}'\mathbf{y}$$

where $\mathbf{X}$ is the matrix of predictor variable scores (to which has been appended a column vector of unities); $\mathbf{b}$ is the vector of parameters:

$$\mathbf{b} = \begin{bmatrix} b_0 \\ b_1 \\ b_2 \end{bmatrix}$$

to be solved for, and $\mathbf{y}$ is the criterion vector.

The intercept b_0 denotes the value of Y when X_1 and X_2 are each zero; b_1 measures the change in Y per unit change in X_1 and b_2 measures the change in Y per unit change in X_2.

TABLE 4.16

Computing the Matrix Product Required for Solving the Normal Equations in the Sample Regression Problem

$$\mathbf{X}'\mathbf{X} = \begin{bmatrix} 1,1,1,\ldots, & 1 \\ 1,2,2,\ldots, & 12 \\ 1,1,2,\ldots, & 10 \end{bmatrix} \begin{bmatrix} 1 & 1 & 1 \\ 1 & 2 & 1 \\ 1 & 2 & 2 \\ \vdots & \vdots & \vdots \\ 1 & 12 & 10 \end{bmatrix} = \begin{bmatrix} 12 & 75 & 59 \\ 75 & 639 & 497 \\ 59 & 497 & 397 \end{bmatrix}$$

$$\mathbf{X}'\mathbf{y} = \begin{bmatrix} 1,1,1,\ldots, & 1 \\ 1,2,2,\ldots, & 12 \\ 1,1,2,\ldots, & 10 \end{bmatrix} \begin{bmatrix} 1 \\ 0 \\ 1 \\ \vdots \\ 16 \end{bmatrix} = \begin{bmatrix} 75 \\ 702 \\ 542 \end{bmatrix}$$

Table 4.16 illustrates the computations required to obtain $\mathbf{X}'\mathbf{X}$ and $\mathbf{X}'\mathbf{y}$. To solve for $\mathbf{b}$, the vector of regression parameters, we express the normal equations in the form

$$\boxed{\mathbf{b} = (\mathbf{X}'\mathbf{X})^{-1}\mathbf{X}'\mathbf{y}}$$

and proceed to solve for $\mathbf{b}$.

Table 4.17 shows the pivotal method applied to the sample problem. In this case the original matrix is only 3×3 so the procedure involves fewer steps; otherwise, the approach is the same as that followed in Table 4.15. After $(\mathbf{X}'\mathbf{X})^{-1}$ is found by the pivotal procedure, this is postmultiplied by $\mathbf{X}'\mathbf{y}$. The solution vector $\mathbf{b}$ is shown in the lower portion of Table 4.17. The vector of parameters $\mathbf{b}$ was actually obtained via computer and so differs somewhat from that found by carrying only three decimal places in Table 4.17.

The desired linear equation is

$$\hat{Y}_i = -2.263 + 1.550X_1 - 0.239X_2$$

In Chapter 6 we discuss this sample problem in considerably more detail. At this point, however, we simply wish to show how the pivotal procedure can be used to find a matrix inverse in the context of multiple regression analysis.

Matrix inversion represents a central concept in multivariate analysis. While illustrated here in the context of multiple regression, matrix inversion goes well beyond this type of application. It is used in a variety of multivariate techniques, including analysis of variance and covariance, discriminant analysis and canonical correlation, to name some of the procedures. Along with matrix eigenstructures, to be discussed in Chapter 5, matrix inversion represents one of the most important and commonly applied operations in all of multivariate analysis.

TABLE 4.17

Applying the Pivotal Method to the Sample Regression Problem

Row no. 0	Original matrix 1	2	3	Identity matrix 4	5	6	Check sum column 7
01	12	75	59	1	0	0	147
02	75	639	497	0	1	0	1212
03	59	497	397	0	0	1	954
10	1	6.250	4.917	0.083	0	0	12.250
11		170.250	128.225	−6.225	1	0	293.250
12		128.250	106.897	−4.897	0	1	231.250
20		1	0.753	−0.037	0.006	0	1.722
21			10.325	−0.152	−0.770	1	10.403
30			1	−0.015	−0.075	0.097	1.007
20*		1		−0.026	0.062	−0.073	0.963
10*	1			0.320	−0.019	−0.021	1.280

$$\mathbf{b} = \overset{(\mathbf{X}'\mathbf{X})^{-1}}{\begin{bmatrix} 0.320 & -0.019 & -0.021 \\ -0.026 & 0.062 & -0.073 \\ -0.015 & -0.075 & 0.097 \end{bmatrix}} \overset{\mathbf{X}'\mathbf{y}}{\begin{bmatrix} 75 \\ 702 \\ 542 \end{bmatrix}}; \qquad \mathbf{b} = \begin{bmatrix} -2.263 \\ 1.550 \\ -0.239 \end{bmatrix}$$

4.8 SUMMARY

This chapter has focused on linear transformations, a key concept in multivariate analysis. As indicated at the outset of the chapter, all matrix transformations are linear transformations. Furthermore, in two or three dimensions various kinds of matrix transformations can be portrayed geometrically.

We first reviewed transformations involving orthogonal matrices, that is, rotations. We described how one can move a point (or points) relative to a fixed basis or, alternatively, leave the point fixed and rotate the basis vectors. This idea was then extended to linear transformations generally. Numerical examples were used to illustrate the concepts.

We next discussed matrix inversion and its role in solving for the original vector, given some transformed vector and the matrix of the transformation. Allied concepts involving the adjoint of a matrix and the use of the matrix inverse in finding transformations relative to a new set of basis vectors were also discussed. The concept of basis vector transformation was then extended to the case of showing the effect of a given transformation on points referred to two different sets of basis vectors if one knows (a) the matrix of the transformation with respect to one basis and (b) the linear transformation that relates the two sets of basis vectors. Some discussion was also presented on general (oblique) coordinate systems.

The next principal section of the chapter attempted to integrate many of our earlier comments, in this and preceding chapters, by showing what happens geometrically as various kinds of matrix transformations are applied. An important extension of this concept involves the idea of composite mappings and, in particular, the observation that any nonsingular matrix transformation with real-valued entries can be decomposed into the unique product of either a rotation followed by a stretch followed by another rotation, or a rotation followed first by a reflection and then by a stretch that is followed by another rotation. Explanation of this assertion represents one of the main topics of the next chapter.

The next major area of interest concerned the rank of a matrix and its relationship to matrix inversion and linear independence. Various properties of matrix rank were listed and illustrated in the context of linear transformations. As was pointed out, the rank of a transformation matrix is important in determining what characteristics of the original space are preserved under a matrix transformation.

A topic that is related to the foregoing involves the application of elementary operations to determine matrix rank, solving sets of simultaneous equations, and matrix inversion. This subject was taken up next, and various examples of applying elementary operations were worked through and the results integrated with earlier material.

The concluding section dealt with numerical procedures for matrix inversion and emphasized the pivotal method, first introduced in Chapter 2. This example was continued here and, in addition, the pivotal method was applied to inverting a cross-products matrix in the context of multiple regression.

REVIEW QUESTIONS

1. Let g_1 denote a mapping of E^3 (Euclidean 3-space) that represents the projection of a point onto the xy plane; let g_2 denote a mapping of E^3 that represents the projection of a point onto the z axis. Describe the following mappings geometrically:

a. $g_1 + g_2$ b. $-g_2$ c. $g_1 - g_2$ d. $3g_1 + g_2$

2. Use the transformation

$$\mathbf{x}^* = \begin{bmatrix} 1 & 0 \\ 1/2 & 1 \end{bmatrix} \begin{bmatrix} x_1 \\ x_2 \end{bmatrix}$$

to carry out the following transformations:

a. a triangle with vertices (1, 1), (4, 2), (5, 4)
b. a parallelogram with vertices (1, 1), (4, 2), (5, 4), (2, 3)
c. a rectangle with vertices (1, 2), (2, 2), (2, 5), (1, 5)

By means of diagrams show the original and transformed figures. What aspects—lengths, angles, areas—are invariant over the transformation?

3. Using the same figures as those of the preceding question,

a. show the effect of a stretch with $k_{11} = 2$ and $k_{22} = 3$;
b. show the effect of a shear parallel to the x axis with $c = 2$.

4. Use 2 x 2 matrices to represent the following transformations:

a. A counterclockwise rotation of points through 45° followed by the translation $x_1{}^* = x_1 + 2; x_2{}^* = x_2 - 1$.

b. A counterclockwise rotation of points through 30° followed by the stretch $x_1{}^* = 3x_1; x_2{}^* = x_2/2$.

c. A stretch $x_1{}^* = 3x_1; x_2{}^* = x_2/2$, followed by a counterclockwise rotation of points through a 30° angle.

5. Consider the following transformations:

$$\mathbf{A}_1 = \begin{bmatrix} 1 & 2 \\ 0 & 1 \end{bmatrix}; \qquad \mathbf{A}_2 = \begin{bmatrix} 1/2 & 0 \\ 0 & 1/2 \end{bmatrix}; \qquad \mathbf{A}_3 = \begin{bmatrix} 0.707 & 0.707 \\ -0.707 & 0.707 \end{bmatrix}$$

Express in matrix form, and show geometrically, the following composite transformations:

a. $\mathbf{A}_2\mathbf{A}_1$ b. $\mathbf{A}_3\mathbf{A}_1\mathbf{A}_2$ c. $\mathbf{A}_2\mathbf{A}_3$ d. $\mathbf{A}_3\mathbf{A}_2\mathbf{A}_1$

6. Find $\mathbf{A}^{-1}$, if it exists, and verify that $\mathbf{A}\mathbf{A}^{-1} = \mathbf{A}^{-1}\mathbf{A} = \mathbf{I}$ for the following:

a. $\mathbf{A} = \begin{bmatrix} 5 & 2 \\ 7 & 5 \end{bmatrix}$ b. $\mathbf{A} = \begin{bmatrix} 8 & 4 \\ -4 & -2 \end{bmatrix}$

c. $\mathbf{A} = \begin{bmatrix} 1 & 0 & -1 \\ -1 & 1 & 0 \\ 0 & -1 & 1 \end{bmatrix}$ d. $\mathbf{A} = \begin{bmatrix} a & -b \\ -c & -d \end{bmatrix}$

7. What is the rank of the following matrices:

a. $\mathbf{A} = \begin{bmatrix} 2 & -1 \\ 3 & 4 \end{bmatrix}$ b. $\mathbf{A} = \begin{bmatrix} 3 & 2 & 4 \\ 2 & 0 & 2 \\ 4 & 2 & 3 \end{bmatrix}$

c. $\mathbf{A} = \begin{bmatrix} 4 & 3 & 1 & 18 \\ 2 & 1 & 3 & 10 \\ 5 & 7 & -2 & 29 \end{bmatrix}$

8. Use elementary row operations to find the echelon form of

a. $\mathbf{A} = \begin{bmatrix} 0 & 5 & 6 & 2 & 0 \\ 1 & 2 & 3 & 9 & 2 \\ 0 & 1 & 2 & 1 & 4 \\ 2 & 0 & 4 & 0 & 1 \end{bmatrix}$ b. $\mathbf{A} = \begin{bmatrix} 1 & 4 & 3 \\ 2 & 6 & 8 \\ 3 & 2 & 1 \end{bmatrix}$

c. $\mathbf{A} = \begin{bmatrix} 2 & 3 & 8 \\ 1 & 1 & 1 \\ 6 & 2 & 3 \\ 4 & 7 & 0 \\ 3 & 2 & 2 \end{bmatrix}$

9. For each of the following sets of equations, determine the rank of the coefficients matrix and the rank of the augmented matrix. Calculate the inverse (if it exists) by elementary row operations.

a.
$$\begin{aligned} 2x_1 + x_2 + 4x_3 &= 13 \\ x_1 - x_2 + x_3 &= 2 \\ x_1 + 2x_2 + 3x_3 &= 10 \end{aligned}$$

b.
$$\begin{aligned} x_1 + 2x_2 + 6x_3 &= 16 \\ x_1 + 3x_2 - x_3 &= 12 \\ x_1 + 2x_2 + 0x_3 &= 10 \end{aligned}$$

c.
$$\begin{aligned} x_1 - 2x_2 + x_3 &= 7 \\ 3x_1 - 2x_2 + 5x_3 &= 43 \\ x_1 + 2x_2 + 3x_3 &= 29 \end{aligned}$$

10. Use the pivotal method to solve the following set of equations; invert the matrix of coefficients and find its determinant:

$$\begin{aligned} 2x_1 + 5x_2 + 3x_3 &= 1 \\ 3x_1 + x_2 + 2x_3 &= 1 \\ x_1 + 2x_2 + x_3 &= 0 \end{aligned}$$

11. Use the adjoint method to invert the 2×2 correlation matrix of predictors in the sample problem of Table 1.2 and then perform the indicated multiplication:

$$\mathbf{R} = \begin{matrix} & X_1 & X_2 \\ X_1 & 1.00 & 0.95 \\ X_2 & 0.95 & 1.00 \end{matrix}; \qquad \mathbf{b}^* = \mathbf{R}^{-1}\mathbf{r}(y)$$

where $\mathbf{r}(y)$ is the vector of simple correlations between the criterion Y and the predictors X_1 and, X_2, respectively:

$$\mathbf{r}(y) = \begin{bmatrix} 0.95 \\ 0.89 \end{bmatrix}$$

The vector $\mathbf{b}^*$ denotes the standardized regression coefficients $b_1{}^*$ and $b_2{}^*$ (often called beta weights) that measure the change in Y per unit change in X_1 and X_2, respectively, when all variables are expressed in terms of mean zero and unit standard deviation.

12. Let

$$\mathbf{T} = \begin{bmatrix} 1 & 2 \\ 3 & 4 \end{bmatrix}$$

denote the matrix of a point transformation in the basis $\mathbf{E}$. Let the basis vector transformation relating $\mathbf{F}$ to $\mathbf{E}$ be

$$\begin{aligned} \mathbf{f}_1 &= 11\mathbf{e}_1 + 7\mathbf{e}_2 \\ \mathbf{f}_2 &= 3\mathbf{e}_1 + 2\mathbf{e}_2 \end{aligned}$$

Find the point transformation $\mathbf{T}^\circ$ corresponding to $\mathbf{T}$ relative to the basis $\mathbf{F}$.

a. If $\mathbf{x} = \left[\begin{smallmatrix} 2 \\ 3 \end{smallmatrix}\right]$ in $\mathbf{E}$, what is it after transformation in $\mathbf{F}$?
b. If $\mathbf{x}^* = \left[\begin{smallmatrix} 3 \\ 1 \end{smallmatrix}\right]$ in $\mathbf{E}$, what is its preimage in $\mathbf{E}$?
What is the determinant of $\mathbf{T}$? of $\mathbf{T}^\circ$?

13. A linear transformation relative to the **E** basis has been defined as

$$\begin{aligned} x_1^* &= x_1 + x_2 \\ x_2^* &= x_1 \qquad + x_3 \\ x_3^* &= \qquad x_2 + x_3 \end{aligned}$$

The transformation linking **F**, some new basis, with the original **E** basis is

$$\mathbf{L} = \begin{bmatrix} 1 & 2 & 1 \\ 2 & 1 & 1 \\ 1 & 3 & 1 \end{bmatrix}$$

Find the point transformation defined above in terms of the new basis **F**.

14. Relative to the **E** basis we have the linear transformation:

$$\begin{aligned} x_1^* &= 7x_1 - 2x_2 \\ x_2^* &= -2x_1 + 6x_2 - 2x_3 \\ x_3^* &= \qquad - 2x_2 + 5x_3 \end{aligned}$$

A transformation of **E** to **F** has been found and is represented by

$$\begin{aligned} \mathrm{f}_1 &= \mathrm{e}_1/3 + 2\mathrm{e}_2/3 + 2\mathrm{e}_3/3 \\ \mathrm{f}_2 &= 2\mathrm{e}_1/3 + \mathrm{e}_2/3 - 2\mathrm{e}_3/3 \\ \mathrm{f}_3 &= -2\mathrm{e}_1/3 + 2\mathrm{e}_2/3 - \mathrm{e}_3/3 \end{aligned}$$

What (particularly simple) form does the mapping of x onto x^* take with respect to the new basis **F**?

CHAPTER 5

Decomposition of Matrix Transformations: Eigenstructures and Quadratic Forms

5.1 INTRODUCTION

In the preceding chapter we discussed various special cases of matrix transformations, such as rotations, reflections, and stretches, and portrayed their effects geometrically. We also pointed out the geometric effect of various *composite* transformations, such as a rotation followed by a stretch.

The motivation for this chapter is, however, just the opposite of that in Chapter 4. Here we start out with a more or less arbitrary matrix transformation and consider ways of *decomposing* it into the product of matrices that are simpler from a geometric standpoint. As such, our objective is to provide, in part, a set of complementary approaches to those illustrated in Chapter 4.

Adopting this reverse viewpoint enables us to introduce a number of important concepts in multivariate analysis—matrix eigenvalues and eigenvectors, the eigenstructure properties of symmetric and nonsymmetric matrices, the basic structure of a matrix and quadratic forms. This new material, along with that of the preceding three chapters, should provide most of the background for understanding vector and matrix operations in multivariate analysis. Moreover, we shall examine concepts covered earlier, such as matrix rank, matrix inverse, and matrix singularity, from another perspective—one drawn from the context of eigenstructures.

Finding the eigenstructure of a matrix, like finding its inverse, is almost a routine matter in the current age of computers. Nevertheless, it seems useful to discuss the kinds of computations involved even though we limit ourselves to small matrices of order 2 x 2 or 3 x 3. In this way we can illustrate many of these concepts geometrically as well as numerically.

Since the topic of eigenstructures can get rather complex, we start off the chapter with an overview discussion of eigenstructures in which the eigenvalues and eigenvectors can be found simply and quickly. Emphasis here is on describing the *geometric* aspects of eigenstructures as related to special kinds of basis vector changes that render the nature of the mapping as simple as possible, for example, as a stretch relative to the appropriate set of basis vectors.

This simple and descriptive treatment also enables us to tie in the present material on eigenstructures with the discussion in Chapter 4 that centered on point and basis vector transformations. In so doing, we return to the numerical example shown in Section 4.3 and obtain the eigenstructure of the transformation matrix described there.

The next main section of the chapter continues the discussion of eigenstructures, but now in the context of multivariate analysis. To introduce this complementary approach–one based on finding a linear composite such that the variance of point projections onto it is maximal–we return to the small numerical problem drawn from the sample data of Table 1.2. We assume that we have a set of mean-corrected scores of twelve employees on X_1 (attitude toward the company) and X_2 (number of years employed by the company). The problem is to find a linear composite of the two separate scores that exhibits maximum variance across individuals. This motivation leads to a discussion of matrix eigenstructures involving *symmetric* matrices and the multivariate technique of principal components analysis.

The next main section of the chapter deals with various properties of matrix eigenstructures. The more common case of symmetric matrices (with real-valued entries) is discussed in some detail, while the more complex case involving eigenstructures of nonsymmetric matrices is described more briefly. The relationship of eigenstructure to matrix rank is also described here.

The basic structure of a matrix and its relationship to matrix decomposition is another central concept in multivariate procedures. Accordingly, attention is centered on this topic, and the discussion is also related to material covered in Chapter 4. Here, however, we focus on the *decomposition* of matrices into the product of other matrices that individually exhibit rather simple geometric interpretations.

Quadratic forms are next taken up and related to the preceding material. Moreover, additional discussion about the eigenstructure of nonsymmetric matrices, as related to such multivariate techniques as multiple discriminant analysis and canonical correlation, is presented in the context of the third sample problem in Chapter 1.

Thus, if matrix inversion and matrix rank are important in linear regression and related procedures for studying single criterion, multiple predictor association, matrix eigenstructures and quadratic forms are the essential concepts in dealing with multiple criterion, multiple predictor relationships.

5.2 AN OVERVIEW OF MATRIX EIGENSTRUCTURES

In Chapter 4 we spent a fair amount of time discussing point and basis vector transformations. In particular, in Section 4.3.5 we discussed the problem of finding the transformation matrix $\mathbf{T}^\circ$, relative to some basis $\mathbf{F}$, if we know the transformation matrix $\mathbf{T}$, denoting the same mapping relative to the $\mathbf{E}$ basis. As shown, to find $\mathbf{T}^\circ$ requires that we know $\mathbf{L}$ the transformation that connects $\mathbf{F}$ with $\mathbf{E}$. We can then find $\mathbf{T}^\circ$ from the equation

$$\boxed{\mathbf{T}^\circ = (\mathbf{L}')^{-1}\mathbf{T}\mathbf{L}'}$$

While the discussion at that point may have seemed rather complex, it was pointed out that this procedure for changing basis vectors has practical utility in cases where we are able to find some *special* basis $\mathbf{F}$ in which the matrix (analogous to $\mathbf{T}^\circ$ above) of the linear transformation takes on some particularly simple form, such as a stretch or a stretch followed by a reflection.

The development of a special basis, in which a linear transformation assumes a simple (i.e., diagonal) form, is the motivation for this section of the chapter. As it turns out, if such a basis exists, it will be found from the *eigenstructure* of a matrix that is analogous to **T** above.[1] Moreover, the (diagonal) matrix that represents the same transformation relative to the new basis will also be found at the same time. In all cases we assume that the original matrix of the transformation is square (with real-valued entries, of course).

By way of introduction to matrix eigenstructures, let us first take up an even simpler situation than that covered in Section 4.3.5. Assume that we have a 2 x 2 transformation matrix:

$$\mathbf{A} = \begin{bmatrix} -3 & 5 \\ 4 & -2 \end{bmatrix}$$

Next, suppose we wished to find an image vector

$$\mathbf{x}^* = \begin{bmatrix} x_1^* \\ x_2^* \end{bmatrix}$$

that has the same (or, possibly, precisely the opposite) direction as the preimage vector

$$\mathbf{x} = \begin{bmatrix} x_1 \\ x_2 \end{bmatrix}$$

If we are concerned only with maintaining direction, then $\mathbf{x}^*$ the image vector can be represented by

$$\mathbf{x}^* = \begin{bmatrix} \lambda x_1 \\ \lambda x_2 \end{bmatrix} = \lambda \begin{bmatrix} x_1 \\ x_2 \end{bmatrix}$$

where λ denotes a scalar. That is, we can stretch or compress $\mathbf{x}$, the preimage, in any way we wish as long as $\mathbf{x}^*$ is in the same (or precisely the opposite) direction as $\mathbf{x}$.

If $\mathbf{x}$ is transformed by $\mathbf{A}$ into $\mathbf{x}^* = \lambda\mathbf{x}$, we state the following:

Vectors, which under a given transformation map into themselves or multiples of themselves, are called invariant vectors under that transformation.

It follows, then, that such vectors obey the relation

$$\boxed{\mathbf{A}\mathbf{x} = \lambda\mathbf{x}}$$

where, as noted, λ is a scalar.

[1] What we shall call eigenvalues (and eigenvectors) some authors call characteristic roots (vectors) or latent roots (vectors).

To illustrate, suppose we try out the vector $\mathbf{x}_1 = \begin{bmatrix} 2 \\ 3 \end{bmatrix}$ to see whether this is invariant under $\mathbf{A}$:

$$\overset{\mathbf{A}}{\begin{bmatrix} -3 & 5 \\ 4 & -2 \end{bmatrix}} \overset{\mathbf{x}_1}{\begin{bmatrix} 2 \\ 3 \end{bmatrix}} = \overset{\mathbf{x}_1{}^*}{\begin{bmatrix} 9 \\ 2 \end{bmatrix}}$$

Such is not the case. We see that the relationship for an invariant vector does not hold, since the components of the vector $\mathbf{x}_1{}^* = \begin{bmatrix} 9 \\ 2 \end{bmatrix}$ are not constant multiples of the vector $\mathbf{x}_1 = \begin{bmatrix} 2 \\ 3 \end{bmatrix}$. However, let us next try the vector $\mathbf{x}_2 = \begin{bmatrix} 3 \\ 3 \end{bmatrix}$.

$$\overset{\mathbf{A}}{\begin{bmatrix} -3 & 5 \\ 4 & -2 \end{bmatrix}} \overset{\mathbf{x}_2}{\begin{bmatrix} 3 \\ 3 \end{bmatrix}} = \overset{\mathbf{x}_2{}^*}{\begin{bmatrix} 6 \\ 6 \end{bmatrix}} = 2 \begin{bmatrix} 3 \\ 3 \end{bmatrix}$$

In the case of $\mathbf{x}_2 = \begin{bmatrix} 3 \\ 3 \end{bmatrix}$, we *do* have an invariant vector. Moreover, if we try *any* vector in which the components are in the ratio 1 : 1, we would find that the relation is also satisfied. For example,

$$\begin{bmatrix} -3 & 5 \\ 4 & -2 \end{bmatrix} \begin{bmatrix} 4 \\ 4 \end{bmatrix} = \begin{bmatrix} 8 \\ 8 \end{bmatrix} = 2 \begin{bmatrix} 4 \\ 4 \end{bmatrix}$$

where $\lambda = 2$ is the constant of proportionality.

Is it the case that only preimage vectors of the form $\mathbf{x}_i = \begin{bmatrix} k \\ k \end{bmatrix}$ satisfy the relation? Let us try another vector, namely, $\mathbf{x}_3 = \begin{bmatrix} 5 \\ -4 \end{bmatrix}$:

$$\overset{\mathbf{A}}{\begin{bmatrix} -3 & 5 \\ 4 & -2 \end{bmatrix}} \overset{\mathbf{x}_3}{\begin{bmatrix} 5 \\ -4 \end{bmatrix}} = \overset{\mathbf{x}_3{}^*}{\begin{bmatrix} -35 \\ 28 \end{bmatrix}} = -7 \begin{bmatrix} 5 \\ -4 \end{bmatrix}$$

We see that this form $\mathbf{x}_j = \begin{bmatrix} 5k \\ -4k \end{bmatrix}$, works also. But are there others? As we shall see, there are no others that are not of the form of either

$$\begin{bmatrix} k \\ k \end{bmatrix} \quad \text{or} \quad \begin{bmatrix} 5k \\ -4k \end{bmatrix}$$

To delve somewhat more deeply into the problem, let us return to the matrix equation

$$\boxed{\mathbf{Ax} = \lambda\mathbf{x}}$$

which can be rearranged (by subtracting $\lambda\mathbf{x}$ from both sides) as follows:

$$\mathbf{Ax} - \lambda\mathbf{x} = \mathbf{0}$$

Or equivalently,

$$\mathbf{Ax} - \lambda\mathbf{Ix} = \mathbf{0}$$

where **I** is an identity matrix. Next, we can factor out **x** to get

$$(\mathbf{A}-\lambda\mathbf{I})\mathbf{x}=\mathbf{0}$$

As can now be seen, the problem of finding an invariant vector **x** is reduced to the problem of solving the equation

$$\boxed{(\mathbf{A}-\lambda\mathbf{I})\mathbf{x}=\mathbf{0}}$$

One trivial solution is, of course, to let $\mathbf{x}=\left[\begin{smallmatrix}0\\0\end{smallmatrix}\right]$. Generally, however, we would be interested in nontrivial solutions; that is, solutions in which $\mathbf{x}\neq\left[\begin{smallmatrix}0\\0\end{smallmatrix}\right]$.

For the moment, let us set $\mathbf{A}-\lambda\mathbf{I}$ equal to **B** and examine what is implied about **B** if **x** is to be nontrivial (i.e., contain nonzero elements). The above expression can then be written as

$$\mathbf{Bx}=\begin{bmatrix}a & b\\ c & d\end{bmatrix}\begin{bmatrix}x_1\\ x_2\end{bmatrix}=\begin{bmatrix}0\\ 0\end{bmatrix}$$

which, in turn, can be written as the set of simultaneous linear equations:

$$ax_1+bx_2=0$$

$$cx_1+dx_2=0$$

After multiplying the first equation by d, the second by $-b$, and adding the two, we have

$$(ad-bc)x_1=0$$

We then repeat the process by multiplying the first equation by $-c$, the second by a, and adding the two, to get

$$(ad-bc)x_2=0$$

So, if *either* $x_1\neq 0$ or $x_2\neq 0$, we must have the situation in which

$$\begin{vmatrix}a & b\\ c & d\end{vmatrix}=|\mathbf{B}|=|(\mathbf{A}-\lambda\mathbf{I})|=0$$

What this all says is that the determinant of $\mathbf{A}-\lambda\mathbf{I}$ *must be zero* if we wish to allow x_1 and x_2 to be nonzero.

5.2.1 The Characteristic Equation

Returning to the original expression of $\mathbf{A}-\lambda\mathbf{I}$, the above reasoning says that we want the determinant of this matrix to be zero. We can write out the above matrix explicitly as

$$\mathbf{A}-\lambda\mathbf{I}=\begin{bmatrix}a_{11} & a_{12}\\ a_{21} & a_{22}\end{bmatrix}-\lambda\begin{bmatrix}1 & 0\\ 0 & 1\end{bmatrix}$$

$$=\begin{bmatrix}a_{11} & a_{12}\\ a_{21} & a_{22}\end{bmatrix}-\begin{bmatrix}\lambda & 0\\ 0 & \lambda\end{bmatrix}=\begin{bmatrix}a_{11}-\lambda & a_{12}\\ a_{21} & a_{22}-\lambda\end{bmatrix}$$

Then, we can find the determinant and set it equal to zero:[2]

$$\begin{vmatrix} a_{11}-\lambda & a_{12} \\ a_{21} & a_{22}-\lambda \end{vmatrix} = \lambda^2 - \lambda(a_{11} + a_{22}) + a_{11}a_{22} - a_{12}a_{21} = 0$$

This last expression is called the *characteristic equation* of the transformation matrix **A**. The roots of this equation, which shall be denoted by λ_i, are called *eigenvalues*, and their associated vectors $\mathbf{x}_i$ are obtained by substituting the roots in

$$(\mathbf{A} - \lambda_i \mathbf{I})\mathbf{x}_i = \mathbf{0}$$

and solving for $\mathbf{x}_i$. These vectors $\mathbf{x}_i$ are called *eigenvectors*. They are the vectors that are invariant under transformation by the matrix **A**. That is, by setting up the format of the characteristic equation and then solving for its roots and associated vectors, we have an operational procedure for finding the invariant vectors of interest. We obtain two central results from the process:

1. the eigenvalues λ_i that indicate the magnitude of the stretch (or stretch followed by reflection), and
2. the eigenvectors $\mathbf{x}_i$ that indicate the new directions (basis vectors) along which the stretching or compressing takes place.

In the case of a 2 x 2 matrix, not more than two values of λ_i are possible. We can see this from the fact that the characteristic equation is quadratic, and a quadratic equation has two solutions, or roots. In general, if **A** is $n \times n$, n roots are possible, since a polynomial of degree n is involved.

As indicated above, the characteristic equation of **A** is defined as

$$\boxed{|\mathbf{A} - \lambda_i \mathbf{I}| = 0}$$

The determinant itself is defined as

$$\boxed{|\mathbf{A} - \lambda_i \mathbf{I}|}$$

and is called the *characteristic function* of **A**.

It should be clear, then, that only square matrices have eigenstructures, since we know already that only square matrices have determinants. Moreover, since $\mathbf{Ax} = \lambda\mathbf{x}$, **A** must be square.

5.2.2 A Numerical and Geometric Illustration

Now that we have concerned ourselves with the rationale for finding the eigenstructure (i.e., the eigenvalues and eigenvectors) of a square matrix, let us apply the procedure to the illustrative matrix shown earlier:

$$\mathbf{A} = \begin{bmatrix} -3 & 5 \\ 4 & -2 \end{bmatrix}$$

[2] The reader should note that the characteristic equation is a polynomial of degree n (given that **A** is $n \times n$). It will have, in general, n roots, not all of which may be either real or, even if real, distinct. We consider these possibilities in due course.

First we write

$$(\mathbf{A}-\lambda_i \mathbf{I})\mathbf{x}_i = \mathbf{0}$$

$$\left\{\begin{bmatrix} -3 & 5 \\ 4 & -2 \end{bmatrix} - \lambda_i \begin{bmatrix} 1 & 0 \\ 0 & 1 \end{bmatrix}\right\} \begin{bmatrix} x_{1i} \\ x_{2i} \end{bmatrix} = \begin{bmatrix} 0 \\ 0 \end{bmatrix}$$

Next, we set up the characteristic equation

$$\begin{vmatrix} -3-\lambda_i & 5 \\ 4 & -2-\lambda_i \end{vmatrix} = 0$$

and expand the determinant to get

$$\lambda_i^2 + 5\lambda_i - 14 = 0$$

We then find the roots of this quadratic equation by simple factoring:

$$(\lambda_i + 7)(\lambda_i - 2) = 0$$

$$\lambda_1 = -7$$

$$\lambda_2 = 2$$

Next, let us substitute $\lambda_1 = -7$ in the equation $(\mathbf{A}-\lambda_i \mathbf{I})\mathbf{x}_i = \mathbf{0}$:

$$\left\{\begin{bmatrix} -3 & 5 \\ 4 & -2 \end{bmatrix} - \begin{bmatrix} -7 & 0 \\ 0 & -7 \end{bmatrix}\right\} \begin{bmatrix} x_{11} \\ x_{21} \end{bmatrix} = \begin{bmatrix} 0 \\ 0 \end{bmatrix}$$

$$\begin{bmatrix} 4 & 5 \\ 4 & 5 \end{bmatrix} \begin{bmatrix} x_{11} \\ x_{21} \end{bmatrix} = \begin{bmatrix} 0 \\ 0 \end{bmatrix}$$

The obvious solution to the two equations, each of which is

$$4x_{11} + 5x_{21} = 0$$

is the vector

$$\mathbf{x}_1 = \begin{bmatrix} 5 \\ -4 \end{bmatrix}$$

or, as illustrated earlier, more generally,

$$\mathbf{x}_1 = \begin{bmatrix} 5k \\ -4k \end{bmatrix}$$

Next, let substitute $\lambda_2 = 2$ in the same way, so as to find

$$\left\{\begin{bmatrix} -3 & 5 \\ 4 & -2 \end{bmatrix} - \begin{bmatrix} 2 & 0 \\ 0 & 2 \end{bmatrix}\right\} \begin{bmatrix} x_{12} \\ x_{22} \end{bmatrix} = \begin{bmatrix} 0 \\ 0 \end{bmatrix}$$

$$\begin{bmatrix} -5 & 5 \\ 4 & -4 \end{bmatrix} \begin{bmatrix} x_{12} \\ x_{22} \end{bmatrix} = \begin{bmatrix} 0 \\ 0 \end{bmatrix}$$

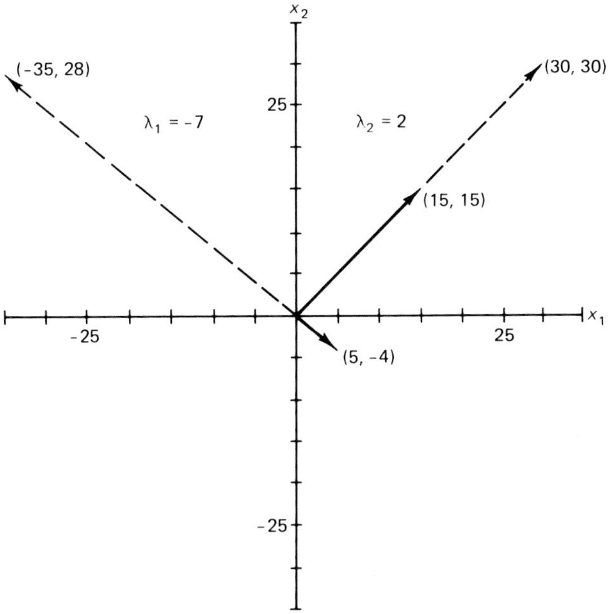

Fig. 5.1 Vectors that are invariant under the transformation matrix **A**.

A solution to these two equations:

$$-5x_{12} + 5x_{22} = 0$$
$$4x_{12} - 4x_{22} = 0$$

is evidently the vector

$$\mathbf{x}_2 = \begin{bmatrix} 1 \\ 1 \end{bmatrix}$$

or, again more generally,

$$\mathbf{x}_2 = \begin{bmatrix} 1k \\ 1k \end{bmatrix}$$

Hence, insofar as $\mathbf{x}_1$ and $\mathbf{x}_2$ are concerned, any vector whose components are in the ratio of either

$$5 : -4 \qquad \text{or} \qquad 1 : 1$$

represents an eigenvector of the transformation given by the matrix **A**.

Figure 5.1 shows the geometric aspects of the preceding computations.[3] If we consider

[3] For ease of presentation, in Fig. 5.1 we let

$$\mathbf{x}_2 = \begin{bmatrix} k \times 1 \\ k \times 1 \end{bmatrix} = \begin{bmatrix} 15 \times 1 \\ 15 \times 1 \end{bmatrix}$$

so that the stretch $\lambda\mathbf{x}_2 = 2\left[\begin{smallmatrix} 15 \\ 15 \end{smallmatrix}\right] = \left[\begin{smallmatrix} 30 \\ 30 \end{smallmatrix}\right]$ is prominent on the figure.

the eigenvector $\mathbf{x}_1 = \left[\begin{smallmatrix}5\\-4\end{smallmatrix}\right]$, we see that this is mapped onto

$$\lambda_1\begin{bmatrix}x_{11}\\x_{21}\end{bmatrix} = -7\begin{bmatrix}5\\-4\end{bmatrix} = \begin{bmatrix}-35\\28\end{bmatrix}$$

while the second eigenvector $\mathbf{x}_2 = \left[\begin{smallmatrix}k \times 1\\k \times 1\end{smallmatrix}\right] = \left[\begin{smallmatrix}15 \times 1\\15 \times 1\end{smallmatrix}\right]$ is mapped onto

$$\lambda_2\begin{bmatrix}x_{12}\\x_{22}\end{bmatrix} = 2\begin{bmatrix}15\\15\end{bmatrix} = \begin{bmatrix}30\\30\end{bmatrix}$$

Furthermore, the eigenvalues $\lambda_1 = -7$ and $\lambda_2 = 2$ represent stretch (or stretch followed by reflection) constants.

5.2.3 Diagonalizing the Transformation Matrix

Let us return to the two eigenvectors found above and next place them in a matrix, denoted by **U**:

$$\mathbf{U} = \begin{bmatrix}5 & 1\\-4 & 1\end{bmatrix}$$

As noted, the two *column* vectors above are the invariant vectors of **A**. We now ask the question: *How would* **A** *behave if one chose as a basis for the space the two eigenvectors, now denoted by* $\mathbf{u}_1$ *and* $\mathbf{u}_2$*, the columns of* **U**?

As we shall show numerically, if **U** is chosen as a new basis of the transformation, originally represented by the matrix **A** relative to the standard **E** basis, then the new transformation matrix is a stretch. This is represented by the diagonal matrix **D**, given by the expression

$$\boxed{\mathbf{D} = \mathbf{U}^{-1}\mathbf{A}\mathbf{U}}$$

If **U** and **D** can be found, we say that **A** is *diagonalizable* via **U**. The matrix **U** consists of the eigenvectors of **A**, and the matrix **D** is a diagonal matrix whose entries are the eigenvalues of **A**. Note, then, that **U** must be nonsingular, and we must find its inverse $\mathbf{U}^{-1}$.

Recalling material from Chapter 4, we know that we can find the inverse of **U** in the 2 x 2 case simply from the determinant and the adjoint of **U**:

$$\mathbf{U}^{-1} = \frac{1}{|\mathbf{U}|}\,\mathrm{adj}(\mathbf{U}) = \frac{1}{9}\begin{bmatrix}1 & -1\\4 & 5\end{bmatrix} = \begin{bmatrix}1/9 & -1/9\\4/9 & 5/9\end{bmatrix}$$

Next, we form the triple product

$$\mathbf{D} = \mathbf{U}^{-1}\mathbf{A}\mathbf{U} = \begin{bmatrix}1/9 & -1/9\\4/9 & 5/9\end{bmatrix}\begin{bmatrix}-3 & 5\\4 & -2\end{bmatrix}\begin{bmatrix}5 & 1\\-4 & 1\end{bmatrix} = \begin{bmatrix}-7 & 0\\0 & 2\end{bmatrix}$$

We see that the transformation matrix **A**, when premultiplied by $\mathbf{U}^{-1}$ and postmultiplied by **U**, the matrix whose columns represent its eigenvectors, *has* been transformed into a diagonal matrix **D** with entries given by the eigenvalues of **A**. *That is, if a set of basis vectors given by* **U** *is employed, the transformation, represented by* **A**, *behaves as a stretch, or possibly as a stretch followed by a reflection, relative to this special basis of eigenvectors.*

We shall be coming back to this central result several times in the course of elaborating upon matrix eigenstructures. The point to remember here is that we have found an instance where, by appropriate choice of basis vectors, a given linear transformation takes on a particularly simple form. This search for a basis, in which the nature of the transformation is particularly simple, represents the primary motivation for presenting the material of this section.

Next, let us look at the expression

$$\boxed{\mathbf{D} = \mathbf{U}^{-1}\mathbf{A}\mathbf{U}}$$

somewhat more closely. First of all, we are struck by the resemblance of this triple product to the triple product

$$\boxed{\mathbf{T}^{\circ} = (\mathbf{L}')^{-1}\mathbf{T}\mathbf{L}'}$$

described in Section 4.3.5. There we found that $\mathbf{T}^{\circ}$ denoted the point transformation of a vector $\mathbf{x}^{\circ}$, referred to a basis $\mathbf{f}_i$, onto a vector $\mathbf{x}^{*\circ}$, also referred to **F**. $\mathbf{T}^{\circ}$ can be found if we know **T**, the matrix of the same linear mapping with respect to the original basis $\mathbf{e}_i$, and **L**, the matrix of the transformation linking the $\mathbf{f}_i$ basis to the $\mathbf{e}_i$ basis.

Note in the present case that **D** plays the role of $\mathbf{T}^{\circ}$, **A** plays the role of **T**, and **U** plays the role of $\mathbf{L}'$. As such, the analogy is complete. Since **U** is the transpose of the matrix used to find the two linear combinations with respect to the standard basis $\mathbf{e}_i$:

$$\mathbf{f}_1 = 5\mathbf{e}_1 - 4\mathbf{e}_2; \qquad \mathbf{f}_1 = 5\begin{bmatrix}1\\0\end{bmatrix} - 4\begin{bmatrix}0\\1\end{bmatrix}$$

$$\mathbf{f}_2 = 1\mathbf{e}_1 + 1\mathbf{e}_2; \qquad \mathbf{f}_2 = 1\begin{bmatrix}1\\0\end{bmatrix} + 1\begin{bmatrix}0\\1\end{bmatrix}$$

we see that the analogy does, indeed, hold. The current material thus provides some motivation for recapitulating, and extending, the discussion of point and basis vector transformations in Chapter 4.

5.2.4 Point and Basis Vector Transformations Revisited

Suppose we now tie in directly the current material on the special basis vectors (eigenvectors) obtained by finding the eigenstructure of a matrix to the material covered in Section 4.3.5. There we set up the transformation matrix

$$\mathbf{T} = \begin{bmatrix}0.9 & 0.44\\0.6 & 0.8\end{bmatrix}$$

and the vector $\mathbf{x} = \left[\begin{smallmatrix}1\\2\end{smallmatrix}\right]$ with respect to the original $\mathbf{e}_i$ basis.

We also considered the transformation matrix

$$\mathbf{L} = \begin{bmatrix} 0.83 & 0.55 \\ 0.20 & 0.98 \end{bmatrix}$$

which denoted point transformations with respect to the $\mathbf{e}_i$ basis according to the transformation $\mathbf{L}$. As recalled, when we wish to find some new basis $\mathbf{f}_i$ with respect to $\mathbf{e}_i$, we use the transpose of $\mathbf{L}$:

$$\mathbf{L}' = \begin{bmatrix} 0.83 & 0.20 \\ 0.55 & 0.98 \end{bmatrix}$$

Note, then, that we must keep in mind the distinction between a linear transformation τ and its matrix representation with respect to a *particular* basis. By way of review, Fig. 5.2 shows the geometric aspects of the mapping:

$$\mathbf{x}^{*\circ} = \mathbf{T}^\circ \mathbf{x}^\circ = (\mathbf{L}')^{-1}\, \mathbf{T}\mathbf{L}'\mathbf{x}^\circ = \begin{bmatrix} 1.11 & 0.60 \\ 0.34 & 0.59 \end{bmatrix} \begin{bmatrix} 1 \\ 2 \end{bmatrix} = \begin{bmatrix} 2.31 \\ 1.52 \end{bmatrix}$$

where

$$\mathbf{L}' = \begin{bmatrix} 0.83 & 0.20 \\ 0.55 & 0.98 \end{bmatrix}; \qquad (\mathbf{L}')^{-1} = \begin{bmatrix} 1.39 & -0.28 \\ -0.78 & 1.18 \end{bmatrix}$$

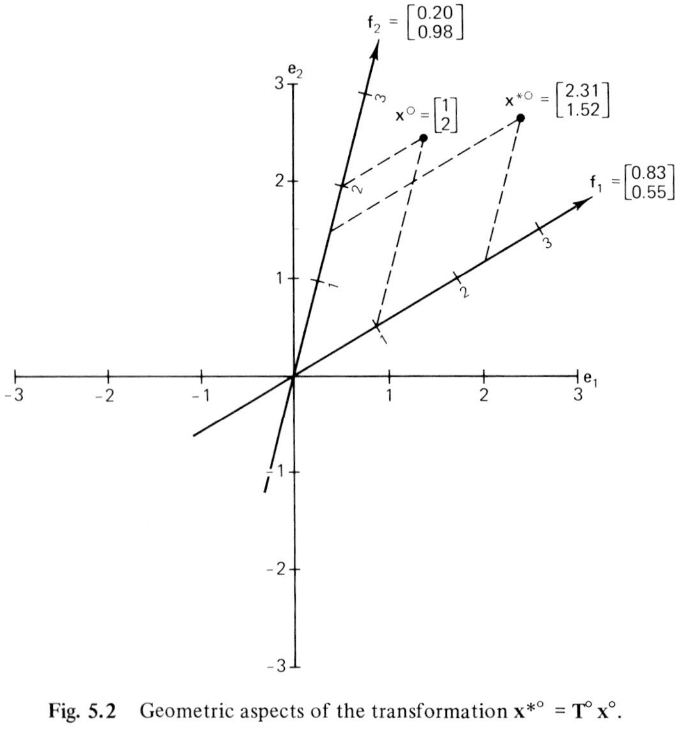

Fig. 5.2 Geometric aspects of the transformation $\mathbf{x}^{*\circ} = \mathbf{T}^\circ \mathbf{x}^\circ$.

in which the vector $\mathbf{x}^\circ$ is mapped onto $\mathbf{x}^{*\circ}$ by the point transformation $\mathbf{T}^\circ$. As recalled $\mathbf{T}^\circ$ is the matrix of the transformation with respect to the $\mathbf{f}_i$ basis (given knowledge of $\mathbf{T}$, the transformation matrix with respect to the $\mathbf{e}_i$ basis), and $\mathbf{L}'$ is the matrix connecting basis vectors in $\mathbf{F}$ with those in $\mathbf{E}$.

In the present context, $\mathbf{U}$ plays the role of $\mathbf{L}'$. Hence, to bring in the new material, we shall want to find the eigenstructure of $\mathbf{T}$. Without delving into computational details, we simply state that the eigenstructure of $\mathbf{T}$ is found in just the same way as already illustrated for the matrix $\mathbf{A}$. In the case of $\mathbf{T}$, the decomposition, as derived from its eigenstructure, is

$$\mathbf{D} = \mathbf{U}^{-1}\mathbf{T}\mathbf{U}$$

$$\underset{\mathbf{D}}{\begin{bmatrix} 1.37 & 0 \\ 0 & 0.66 \end{bmatrix}} = \underset{\mathbf{U}^{-1}}{\begin{bmatrix} -0.80 & -0.62 \\ 0.74 & -0.70 \end{bmatrix}} \underset{\mathbf{T}}{\begin{bmatrix} 0.9 & 0.44 \\ 0.6 & 0.8 \end{bmatrix}} \underset{\mathbf{U}}{\begin{bmatrix} -0.69 & 0.61 \\ -0.73 & -0.79 \end{bmatrix}}$$

Next, in line with the recapitulation in Fig. 5.2, we find the transformation

$$\mathbf{x}^{*\circ} = \mathbf{U}^{-1}\,\mathbf{T}\mathbf{U}\mathbf{x}^\circ = \mathbf{D}\mathbf{x}^\circ = \begin{bmatrix} 1.37 & 0 \\ 0 & 0.33 \end{bmatrix} \begin{bmatrix} 1 \\ 2 \end{bmatrix} = \begin{bmatrix} 1.37 \\ 0.66 \end{bmatrix}$$

Figure 5.3 shows the pertinent results from a geometric standpoint. First, we note that the columns of $\mathbf{U}$ appear as the new basis vectors denoted $\mathbf{f}_1$ and $\mathbf{f}_2$, respectively, so as to maintain the analogy with the column vectors of $\mathbf{L}'$ in Fig. 5.2.

First, $\mathbf{U}$ takes $\mathbf{x}^\circ$ onto $\mathbf{x}$ with respect to $\mathbf{E}$. Then, the transformation $\mathbf{T}$ takes $\mathbf{x}$ onto $\mathbf{x}^*$ with respect to $\mathbf{E}$. Finally, $\mathbf{U}^{-1}$ takes $\mathbf{x}^*$ onto $\mathbf{x}^{*\circ}$ with respect to $\mathbf{F}$, the matrix of the new basis. The interesting aspect of the exercise, however, is that the $\mathbf{f}_i$ basis (given by $\mathbf{U}$ in the present context) is not just any old basis; rather, it is one in which the mapping of $\mathbf{x}^\circ$ onto $\mathbf{x}^{*\circ}$ involves a stretch as given by the transformation

$$\mathbf{x}^{*\circ} = \mathbf{D}\mathbf{x}^\circ = \underset{\mathbf{D}}{\begin{bmatrix} 1.37 & 0 \\ 0 & 0.33 \end{bmatrix}} \underset{\mathbf{x}^\circ}{\begin{bmatrix} 1 \\ 2 \end{bmatrix}} = \underset{\mathbf{x}^{*\circ}}{\begin{bmatrix} 1.37 \\ 0.66 \end{bmatrix}}$$

Figure 5.3 shows the point transformation from $\mathbf{x}^\circ$ to $\mathbf{x}^{*\circ}$ with respect to the $\mathbf{f}_i$ basis.[4] In one sense, then, the eigenstructure problem is precisely analogous to finding the nature of a transformation relative to two different sets of basis vectors. And this is one reason why the latter topic was discussed in Section 4.3.5. *However, the distinguishing feature of an eigenvector basis is that the nature of the transformation assumes a particularly simple geometric form, such as the stretch noted above.*

[4] As shown in Fig. 5.3, $\mathbf{x}^\circ = \left[\begin{smallmatrix}1\\2\end{smallmatrix}\right]$ is stretched along the $\mathbf{f}_1$ axis in the ratio 1.37 : 1 but compressed along the $\mathbf{f}_2$ axis in the ratio 0.67 : 2 (or 0.33 : 1).

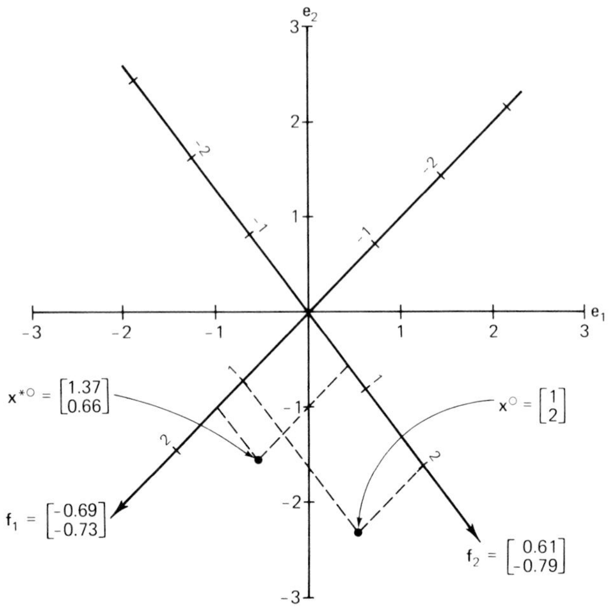

Fig. 5.3 Basis vector transformation involving eigenvectors of **T**.

5.2.5 Recapitulation

In summary, obtaining the eigenstructure of a (square) matrix entails solving the characteristic equation

$$|\mathbf{A} - \lambda_i \mathbf{I}| = 0$$

If **A** is of order $n \times n$, then we shall obtain n roots of the equation; these roots are called eigenvalues. Each eigenvalue can then be substituted in

$$(\mathbf{A} - \lambda_i \mathbf{I})\mathbf{x}_i = \mathbf{0}$$

to obtain its associated eigenvector.

But what about the eigenvalues (and eigenvectors) of some arbitrary matrix **A**? All we have said up to this point is that if **A** is $n \times n$, then n eigenvalues and eigenvectors are obtained. However, we shall find it is possible that

1. some, or all, of the eigenvalues are complex, rather than real valued (even though **A** is real valued);
2. some, or all, of the eigenvectors have complex elements;
3. even if all eigenvalues (and their eigenvectors) are real, some eigenvalues may be zero;
4. even if all eigenvalues are real and nonzero, some may be repeated.

Moreover, so far we have not said very much about the new basis of column eigenvectors in **U**, other than to indicate that it must be nonsingular in order for the relationship

$$\mathbf{D} = \mathbf{U}^{-1}\mathbf{A}\mathbf{U}$$

to hold. Furthermore, by the following algebraic operations:

$$\mathbf{U}\mathbf{D} = \mathbf{U}\mathbf{U}^{-1}\mathbf{A}\mathbf{U} = \mathbf{A}\mathbf{U}$$

and

$$\mathbf{U}\mathbf{D}\mathbf{U}^{-1} = \mathbf{A}\mathbf{U}\mathbf{U}^{-1}$$

we can express **A** as

$$\boxed{\mathbf{A} = \mathbf{U}\mathbf{D}\mathbf{U}^{-1}}$$

Other than the conditions that **U** is nonsingular and **D** is diagonal, however, no further requirements have been imposed. We might well wonder if situations exist in which **U** turns out to be orthogonal as well as nonsingular. Also, we recall that the illustrative matrices, whose eigenstructures were found above, are not symmetric. Do special properties exist in the case of symmetric transformation matrices?

We shall want to discuss these questions, and related ones as well, as we proceed to a consideration of eigenstructures in the context of multivariate analysis.

Here a complementary approach to the study of eigenstructures is adopted. Emphasis is now placed on *symmetric* matrices and the role that eigenstructures play in reorienting an original data space with correlated dimensions to uncorrelated axes, along which the objects are maximally separated, that is, display the highest variance. While it may seem that we are starting on a brand-new topic, it turns out that we are still interested in basis vector changes in order to achieve certain simplifications in the transformation. Hence we shall return to the present topic in due course, but now in the context of *symmetric* transformation matrices. As it turns out, the eigenstructure of a symmetric matrix displays properties that can be described more simply than those associated with the nonsymmetric case. Accordingly, we cover this simpler case first and then proceed to the situation involving nonsymmetric matrices.

5.3 TRANSFORMATIONS OF COVARIANCE MATRICES

At the end of Chapter 1, a small sample problem with hypothetical data was introduced in order to illustrate some of the more commonly used multivariate tools. The basic data, shown in Table 1.2, have already been employed as a running example to show

1. how matrix notation can be used to summarize various statistical operations in a concise manner (Chapter 2);
2. the geometric aspects of such statistical measures as standard deviations, correlations, and the generalized dispersion of a covariance matrix (Chapter 3);
3. how the pivotal method can be used to find matrix inverses and solutions to sets of simultaneous equations (Chapter 4).

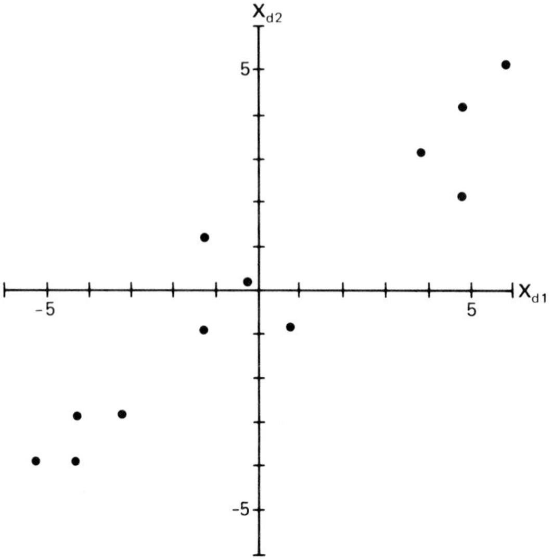

Fig. 5.4 Scatter plot of mean-corrected predictors from sample problem of Chapter 1.

In this chapter we continue to use this small data bank for expository purposes.

Suppose we were to start with a plot of the mean-corrected scores of the two predictors X_{d2} (years employed by the company) versus X_{d1} (attitude toward the company). Figure 5.4 shows the scatter plot obtained from the mean-corrected data of Table 1.2.

We note that X_{d2} and X_{d1} are positively associated (their correlation is 0.95). In line with the discussion of Chapter 1, suppose we wished to replace X_{d1} and X_{d2} by a single linear composite

$$\boxed{z_i = t_1 X_{di1} + t_2 X_{di2}}$$

for $i = 1, 2, \ldots, 12$, the total number of employees, so that the variance of this linear composite

$$\boxed{\operatorname{Var}(z_i) = \sum_{i=1}^{12} (z_i - \bar{z})^2/12}$$

subject to $\mathbf{t}'\mathbf{t} = 1$, is maximized. By this is meant that the twelve employees are maximally separated along the linear composite.

The motivation for doing this sort of thing is often based on parsimony, that is, the desire to replace two or more correlated variables with a single linear composite that, in a sense, attempts to account for as much as possible of the variation shared by the contributory variables. The vector of weights **t** is constrained to be of unit length so that $\mathrm{Var}(z_i)$ cannot be made indefinitely large by making the entries of **t** arbitrarily large.

If our desire is to maximize the variance of the linear composite, how should the weights t_1 and t_2 be chosen so as to bring this about?

To answer this question we need, of course, some kind of criterion. For example, one approach might be to bring the external variable Y into the problem and choose t_1 and t_2 so as to result in a linear composite whose values maximally correlate with Y. As was shown in our discussion of multiple regression in Chapter 4, this involves finding a set of predicted values $\hat{Y}_i$ whose sum of squared deviations from Y_i is minimized.

However, in the present case, let us assume that we choose some "internal" criterion that ignores the external variable Y. The approach suggested earlier is to find a vector that maximizes the variance (or a quantity proportional to this, such as the sum of squares) of the twelve points if they are projected onto this vector. This is also equivalent to minimizing the sum of the squared distances between all pairs of points in which one member of each pair is the to-be-found projection and the other member is the original point.

It is relevant to point out that we are really concerned with two types of vectors. The vector $\mathbf{t} = \left[\begin{smallmatrix} t_1 \\ t_2 \end{smallmatrix}\right]$ is a vector of direction cosines or direction numbers in terms of the original basis. The vector **z** comprises the particular point projections whose variance we are trying to maximize through the particular choice of **t**.

We have, of course, several possible candidates for measuring the original association between X_{d1} and X_{d2}, such as **S**, the SSCP matrix, **C**, the covariance matrix (which is proportional to **S**), and **R**, the correlation matrix.

Illustratively, let us develop the argument in terms of the covariance matrix which, in the sample problem, is

$$\mathbf{C} = \begin{array}{c} \\ X_1 \\ X_2 \end{array} \begin{array}{c} \begin{array}{cc} X_1 & X_2 \end{array} \\ \begin{bmatrix} 14.19 & 10.69 \\ 10.69 & 8.91 \end{bmatrix} \end{array}$$

Since, as it turns out, we shall be finding two linear composites, we shall refer to these new variables as $\mathbf{z}_1$ and $\mathbf{z}_2$, respectively.

Finding the first of these linear composites $\mathbf{z}_1$ represents the motivation for introducing a new use for computing the eigenstructure of a matrix. That is, we shall wish to find a vector in the space shown in Fig. 5.4 with the property of maximizing the variance of the twelve points if they are projected onto this vector. We might then wish to find a second vector in the same space that obeys certain other properties. If so, what we shall be doing is changing the original basis vectors to a *new* set of basis vectors. (These new basis vectors are often called principal axes.) And, in the course of doing this, it will turn out that we are also decomposing the covariance matrix into the product of simpler matrices from a geometric standpoint in just the same spirit as described in Section 5.2.

5.4 EIGENSTRUCTURE OF A SYMMETRIC MATRIX

Let us now focus on the covariance matrix of the two predictors

$$\mathbf{C} = \begin{matrix} & X_1 & X_2 \\ X_1 & 14.19 & 10.69 \\ X_2 & 10.69 & 8.91 \end{matrix}$$

The first thing to be noticed about **C** is that it is not only square but also symmetric. Many derived matrices in multivariate analysis exhibit these characteristics. We now wish to consider various linear composites of X_{d1} and X_{d2} that have some chance of maximally separating individuals.

Suppose, arbitrarily, we consider the following, overly simple, linear combinations of $\mathbf{x}_{d1}$ and $\mathbf{x}_{d2}$:

$$\mathbf{z}_1 = 0.707\mathbf{x}_{d1} + 0.707\mathbf{x}_{d2} = 0.707(\mathbf{x}_{d1} + \mathbf{x}_{d2})$$

$$\mathbf{z}_2 = 0.707\mathbf{x}_{d2} - 0.707\mathbf{x}_{d1} = 0.707(\mathbf{x}_{d2} - \mathbf{x}_{d1})$$

In the case of $\mathbf{z}_1$ we are giving equal weight to $\mathbf{x}_{d1}$ and $\mathbf{x}_{d2}$, while in the case of $\mathbf{z}_2$ we are concerned with their difference, that is, the "increment" (component by component) of $\mathbf{x}_{d2}$ over $\mathbf{x}_{d1}$. Notice, further, that (a) the transformation matrix representing these linear combinations, which we shall call **T**, is orthogonal and (b) we shall be postmultiplying each point represented as a row vector in $\mathbf{X}_d$ by the matrix

$$\mathbf{T} = \begin{bmatrix} 0.707 & -0.707 \\ 0.707 & 0.707 \end{bmatrix}$$

As surmised, 0.707 is chosen so that $(0.707)^2 + (0.707)^2 = 1$, and we have a set of direction cosines. Since **T** is orthogonal, the following relationships hold:

$$\mathbf{T}'\mathbf{T} = \mathbf{T}\mathbf{T}' = \mathbf{I}$$

Finally, we also note that the determinant $|\mathbf{T}| = 1$. Hence, a proper rotation, as a matter of fact, a 45° rotation, is entailed since $\cos 45^\circ = 0.707$. We first ask: Suppose one were to consider the mean-corrected matrix $\mathbf{X}_d$ of the two predictors. What is the relationship of the derived matrix **Z**, found by the rotation $\mathbf{X}_d\mathbf{T}$, to the matrix $\mathbf{X}_d$?

Panel I of Table 5.1 shows the linear composites for $\mathbf{z}_1$ and $\mathbf{z}_2$, respectively, as obtained from the 45° rotation matrix. Since we shall be considering a second transformation subsequently, we use the notation $\mathbf{Z}_a$ and $\mathbf{T}_a$ to denote the particular rotation (involving a 45° angle) that is now being applied.

The solid-line vectors, $\mathbf{z}_{1a}$ and $\mathbf{z}_{2a}$, of Fig. 5.5 show the results of rotating the original basis vectors 45° counterclockwise. If we project the twelve points onto the new basis vectors $\mathbf{z}_{1a}$ and $\mathbf{z}_{2a}$, we find the projections shown in the matrix $\mathbf{Z}_a$ in Panel I of Table 5.1. Note further that, within rounding error, the mean of each column of $\mathbf{Z}_a$ is zero. (In general, if a set of vectors is transformed by a linear function, we will find that the means of the transformed vectors are given by that linear function applied to the means of the original variables.) A more interesting aspect of the transformation is: What is the relationship of the *new* covariance matrix, derived from the transformed matrix $\mathbf{Z}_a$, to that derived from the original matrix $\mathbf{X}_d$?

TABLE 5.1

Linear Composites Based on 45° and 38° Rotations of Original Basis Vectors

I 45° rotation						II 38° rotation					
$\mathbf{Z}_a$		$\mathbf{X}_d$		$\mathbf{T}_a$		$\mathbf{Z}_b$		$\mathbf{X}_d$		$\mathbf{T}_b$	
−6.48	0.94	−5.25	−3.92	0.707	−0.707	−6.55	0.15	−5.25	−3.92	0.787	−0.617
−5.78	0.23	−4.25	−3.92	0.707	0.707	−5.76	−0.46	−4.25	−3.92	0.617	0.787
−5.07	0.94	−4.25	−2.92			−5.15	0.32	−4.25	−2.92		
−4.36	0.23	−3.25	−2.92			−4.36	−0.29	−3.25	−2.92		
−1.53	0.23	−1.25	−0.92			−1.55	0.05	−1.25	−0.92		
−0.12	1.65	−1.25	1.08			−0.32	1.62	−1.25	1.08		
−0.12	0.23	−0.25	0.08			−0.15	0.09	−0.25	0.08		
−0.12	−1.19	0.75	−0.92			0.02	−1.19	0.75	−0.92		
4.83	−0.47	3.75	3.08			4.85	0.11	3.75	3.08		
4.83	−1.89	4.75	2.08			5.02	−1.29	4.75	2.08		
6.24	−0.47	4.75	4.08			6.26	0.28	4.75	4.08		
7.66	−0.47	5.75	5.08			7.66	0.45	5.75	5.08		
Variances						Variances					
22.23	0.87					22.56	0.54				

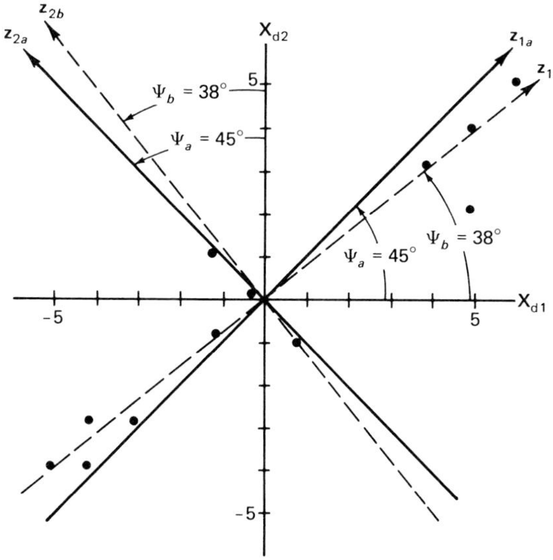

Fig. 5.5 Applying 45° and 38° counterclockwise rotations to axes of Fig. 5.4.

We recall from Chapter 2 that the covariance matrix can be computed from the raw-data matrix $\mathbf{X}$ by the expression

$$\mathbf{C} = 1/m[\mathbf{X}'\mathbf{X} - (1/m)(\mathbf{X}'\mathbf{1})(\mathbf{1}'\mathbf{X})]$$

where $\mathbf{X}'\mathbf{X}$ is called the minor product moment of the raw-data matrix $\mathbf{X}$, and the second term in the brackets is the correction for means.

In the present case by using $\mathbf{X}_d$ the mean of each column is already zero and similarly so for the columns of the transformed matrix $\mathbf{Z}_a$. Hence, the second term on the right of the above equation consists of a null matrix. The original covariance matrix can then be restated as

$$\mathbf{C}(X) = \begin{bmatrix} 14.19 & 10.69 \\ 10.69 & 8.91 \end{bmatrix}$$

5.4.1 The Behavior of the Covariance Matrix under Linear Transformation

If $\mathbf{C}(X)$ is computed as shown above, we could find $\mathbf{C}(Z_a)$ by the same procedure, namely,

$$\mathbf{C}(Z_a) = 1/m\, \mathbf{Z}_a{}'\mathbf{Z}_a$$

since, as before, the mean of each column of $\mathbf{Z}_a$ (within rounding error) is also zero and, hence $1/m(\mathbf{Z}_a{}'\mathbf{1})(\mathbf{1}'\mathbf{Z}_a) = \phi$.

Actually, however, a much more direct way to find $\mathbf{C}(Z_a)$ is available. Since $\mathbf{Z}_a = \mathbf{X}_d \mathbf{T}_a$ and $\mathbf{C}(Z_a) = 1/m\, \mathbf{Z}_a{}'\mathbf{Z}_a$, we have

$$\mathbf{C}(Z_a) = [(\mathbf{X}_d\mathbf{T}_a)'(\mathbf{X}_d\mathbf{T}_a)]/m = [\mathbf{T}_a{}'\mathbf{X}_d{}'\mathbf{X}_d\mathbf{T}_a]/m = \mathbf{T}_a{}'\,[\mathbf{C}(X)]\,\mathbf{T}_a$$

That is, we can find $\mathbf{C}(Z_a)$ *through knowledge of the transformation matrix* $\mathbf{T}_a$ *and* $\mathbf{C}(X)$, *the covariance matrix prior to transformation.*

To find $\mathbf{C}(Z_a)$ we compute

$$\mathbf{C}(Z_a) = \underset{\mathbf{T}_a{}'}{\begin{bmatrix} 0.707 & 0.707 \\ -0.707 & 0.707 \end{bmatrix}} \underset{\mathbf{C}(X)}{\begin{bmatrix} 14.19 & 10.69 \\ 10.69 & 8.91 \end{bmatrix}} \underset{\mathbf{T}_a}{\begin{bmatrix} 0.707 & -0.707 \\ 0.707 & 0.707 \end{bmatrix}}$$

$$= \begin{bmatrix} 22.23 & -2.64 \\ -2.64 & 0.87 \end{bmatrix}$$

The first thing to be noticed about $\mathbf{C}(Z_a)$ is that the sum of the main diagonal entries (22.23 + 0.87) is, within rounding, equal to the sum of the main diagonal entries of $\mathbf{C}(X)$. The second thing of interest is that $\mathbf{C}(Z_a)$ continues to remain symmetric, but now *the off-diagonal entries are much smaller in absolute value* (2.64) *than their counterpart entries* (10.69) *in* $\mathbf{C}(X)$.

What has happened here is that the rotation of axes, effected by $\mathbf{T}$, has resulted in a new set of basis vectors in which projections on the first of the new axes display a *considerably larger variance* than either contributing dimension $\mathbf{x}_{d1}$ or $\mathbf{x}_{d2}$.

We might next inquire if we can do still better in variance maximization. Does some *other* rotation result in a still higher sum of squares for the *first* transformed dimension? What we could do, of course, is to proceed by brute force. Based on what we have seen so far, we could try other orthogonal transformations in the vicinity of a 45° angle and duly note how the parceling out of variance between $\mathbf{z}_1$ and $\mathbf{z}_2$ behaves under each transformation. Fortunately, however, an analytical approach is available—one that again utilizes the concept of the eigenstructure of a matrix.

5.4.2 The Characteristic Equation

In preceding chapters we have talked about changing the basis of a vector space. We have also discussed transformations that involve a rotation (proper and improper) and a stretch. Finally, we know that a linear combination such as $z_1 = 0.707X_{d1} + 0.707X_{d2}$ can be so expressed that the sum of the squared weights equals unity. That is, we can—and have done so in the preceding illustration—normalize the weights of the linear combination so that they appear as direction cosines.

Suppose we take just one column vector of some new rotation matrix $\mathbf{T}$ (for the moment we drop the subscript for ease of exposition) and wish to maximize the expression

$$\mathbf{t}_1'[\mathbf{C}(X)]\mathbf{t}_1$$

subject to the normalization constraint that $\mathbf{t}_1'\mathbf{t}_1 = 1$. This restriction on the length of $\mathbf{t}_1$ will ensure that our transformation meets the unit length condition for a rotation. And, incidentally, this restriction will also ensure that the resultant scalar $\mathbf{t}_1'[\mathbf{C}(X)]\mathbf{t}_1$ cannot be made arbitrarily large by finding entries of $\mathbf{t}_1$ with arbitrarily large values.

The above problem is a more or less standard one in the calculus, namely, optimizing

$$F = \mathbf{t}_1'[\mathbf{C}(X)]\mathbf{t}_1 - \lambda(\mathbf{t}_1'\mathbf{t}_1 - 1)$$

This normalizes — where did this come from?

where λ is an additional unknown in the problem, called a *Lagrange multiplier.*

While we shall not go into details (see Appendix A for these), we can briefly sketch out their nature by differentiating F with respect to the elements of $\mathbf{t}_1$ and setting this partial derivative equal to the zero vector. The appropriate (symbolic) derivative is

$$\frac{\partial F}{\partial \mathbf{t}_1} = 2[\mathbf{C}(X)\mathbf{t}_1 - \lambda \mathbf{t}_1]$$

Setting this expression equal to the zero vector, dividing both sides by 2, and factoring out $\mathbf{t}_1$ leads to

$$[\mathbf{C}(X) - \lambda\mathbf{I}]\mathbf{t}_1 = \mathbf{0}$$

where $\mathbf{I}$ is the identity matrix. In terms of our specific problem, we have

$$\underset{\mathbf{C}(X)-\lambda\mathbf{I}}{\begin{bmatrix} 14.19-\lambda & 10.69 \\ 10.69 & 8.91-\lambda \end{bmatrix}} \underset{\mathbf{t}_1}{\begin{bmatrix} t_{11} \\ t_{21} \end{bmatrix}} = \underset{\mathbf{0}}{\begin{bmatrix} 0 \\ 0 \end{bmatrix}}$$

We may also recall from the calculus that satisfaction of the above equation is a necessary condition for a maximum (or minimum).[5]

Immediately we are struck by the resemblance of the above expression to the matrix equation of Section 5.2:

$$\boxed{(\mathbf{A} - \lambda\mathbf{I})\mathbf{x} = \mathbf{0}}$$

[5] Although not shown here, sufficiency conditions are also met.

that was used in finding the eigenstructure of **A**. Indeed, the only basic differences here are that $\mathbf{C}(X)$ is symmetric, and the eigenvectors are to be normalized to unit length.

If the matrix $\mathbf{C}(X) - \lambda\mathbf{I}$, for a fixed value of λ, were nonsingular (i.e., possessed an inverse), it would always be the case that the only possible solution to the equation involves setting $\mathbf{t}_1$ equal to the zero vector. Hence, in line with the discussion of Section 5.2, we want to accomplish *just the opposite*, namely, to find a λ that will make $\mathbf{C}(X) - \lambda\mathbf{I}$ *singular.* But we recall that singular matrices have determinants of zero. Hence we want to find a value for λ that satisfies the characteristic equation:

$$\boxed{|\mathbf{C}(X) - \lambda\mathbf{I}| = 0}$$

Another way of putting things is that we wish to find $\mathbf{t}_1$ such that

$$\boxed{\mathbf{C}(X)\mathbf{t}_1 = \lambda\mathbf{t}_1}$$

where $\mathbf{t}_1$ is a vector, which if premultiplied by $\mathbf{C}(X)$, results in a vector $\lambda\mathbf{t}_1$ whose components are proportional to those of $\mathbf{t}_1$.

This, of course, is the same line of reasoning that we discussed earlier in the chapter in the context of finding a new basis in which the matrix of the transformation in terms of that new basis could be denoted by a stretch or, possibly, by a stretch followed by a reflection.

As recalled, however, for an $n \times n$ matrix one obtains n roots in solving the characteristic equation. Since we wish to maximize F, we shall be on the lookout for the largest λ_i obtained from solving the characteristic equation. *That is, we shall order the roots from large to small and choose that eigenvector* $\mathbf{t}_i$ *corresponding to the largest* λ_i.

Either the approach described in Section 5.2 or the current approach leads to the same type of result so long as we remember to *order* the roots of $|\mathbf{C}(X) - \lambda| = 0$ from large to small.[6] As observed, $|\mathbf{C}(X) - \lambda| = 0$ is simply the characteristic equation of the covariance matrix $\mathbf{C}(X)$.

Now, while we initially framed the problem in terms of a single eigenvector $\mathbf{t}_1$ and a single eigenvalue λ_1, the characteristic equation, as formulated above, will enable us to solve for two eigenvalues λ_1 and λ_2, and two eigenvectors $\mathbf{t}_1$ and $\mathbf{t}_2$. As already noted in Section 5.2, if $\mathbf{C}(X)$ is of order $n \times n$, we shall obtain n eigenvalues and n associated eigenvectors.

5.4.3 Finding the Eigenvalues and Eigenvectors of C(X)

It is rather interesting that following either (a) the (present) variance-maximizing path or (b) the basis vector transformation path that seeks a new basis in which vectors are mapped onto scalar multiples of themselves leads to the same result—the characteristic equation. However, let us now concentrate our attention on the variance-maximizing path as the one that appears more appropriate from an intuitive standpoint in the context of the current problem.

[6] It should be mentioned, however, that $\mathbf{C}(X)$, the covariance matrix, exhibits special properties in that it is symmetric and represents the minor product moment of another matrix (in this case $\mathbf{X}_d/\sqrt{m}$). As such, all of its eigenvalues will be real and nonnegative and $\mathbf{T}$ will be orthogonal. We discuss these special properties later on.

The problem now is to solve for the eigenstructure of $\mathbf{C}(X)$. First, we shall want to find the eigenvalues of the characteristic equation

$$|\mathbf{C}(X)-\lambda_i\mathbf{I}| = \begin{vmatrix} 14.19-\lambda_i & 10.69 \\ 10.69 & 8.91-\lambda_i \end{vmatrix} = 0$$

Expansion of the second-order determinant is quite easy. In terms of the above problem we can express the characteristic equation as

$$(14.19-\lambda_i)(8.91-\lambda_i)-(10.69)^2 = 0$$

$$\lambda_i^2-23.1\lambda_i + 126.433-114.276 = 0$$

$$\lambda_i^2-23.1\lambda_i + 12.157 = 0$$

The simplest way of solving the above equation is to use the quadratic formula of the general form $y = ax^2 + bx + c$. First, we note that the coefficients in the preceding expression are

$$a = 1; \qquad b = -23.1; \qquad c = 12.157$$

Next, let us substitute these in the general quadratic formula:

$$\lambda_i = \frac{-b \pm \sqrt{b^2-4ac}}{2a} = \frac{23.1 \pm \sqrt{(-23.1)^2-4(12.157)}}{2} = \frac{23.1 \pm \sqrt{484.982}}{2}$$

$$\lambda_1 = 22.56; \qquad \lambda_2 = 0.54$$

As could be inferred from the sign of the discriminant $(b^2 - 4ac)$ of the general quadratic, we have a case in which the roots λ_1 and λ_2 are both real and are unequal. If we go back to our original problem of trying to optimize $F[\mathbf{C}(X)]$ subject to $\mathbf{t}'\mathbf{t} = 1$, we see that $\lambda_1 = 22.56$ denotes the *maximum variance achievable* along one dimension by a linear composite of the original basis vectors.

In the 2 x 2 case, solving for the eigenvectors $\mathbf{t}_1$ and $\mathbf{t}_2$ is rather simple. Let us first substitute the value of $\lambda_1 = 22.56$ in the expression $\mathbf{C}(X) - \lambda\mathbf{I}$:

$$\begin{bmatrix} 14.19-22.56 & 10.69 \\ 10.69 & 8.91-22.56 \end{bmatrix} = \overset{\mathbf{C}(X)-\lambda_1\mathbf{I}}{\begin{bmatrix} -8.37 & 10.69 \\ 10.69 & -13.65 \end{bmatrix}}$$

The next step is to set up the simultaneous equations needed to solve for $\mathbf{t}_1$, the first eigenvector:

$$\overset{\mathbf{C}(X)-\lambda_1\mathbf{I}}{\begin{bmatrix} -8.37 & 10.69 \\ 10.69 & -13.65 \end{bmatrix}} \begin{bmatrix} t_{11} \\ t_{21} \end{bmatrix} = \begin{bmatrix} 0 \\ 0 \end{bmatrix}$$

If we set $t_{21} = 1$ in the first of the two equations:

$$-8.37t_{11} + 10.69t_{21} = 0$$

we obtain $t_{11} = 1.277$. (Note that these values also satisfy the second equation.) This gives us the first (nonnormalized) eigenvector:

$$\mathbf{t}_1 = \begin{bmatrix} 1.277k \\ 1k \end{bmatrix}$$

As we know, we can then divide the components of $\mathbf{t}_1$ by $\|\mathbf{t}_1\| = 1.622$, its length, to get the normalized version:

$$(\text{norm})\mathbf{t}_1 = \begin{bmatrix} 0.787 \\ 0.617 \end{bmatrix}$$

In exactly the same way, we substitute $\lambda_2 = 0.54$ and perform the following calculations:

$$\begin{bmatrix} 14.19-0.54 & 10.69 \\ 10.69 & 8.91-0.54 \end{bmatrix} = \overset{\mathbf{C}(X)-\lambda_2\mathbf{I}}{\begin{bmatrix} 13.65 & 10.69 \\ 10.69 & 8.37 \end{bmatrix}}$$

$$\overset{\mathbf{C}(X)-\lambda_2\mathbf{I}}{\begin{bmatrix} 13.65 & 10.69 \\ 10.69 & 8.37 \end{bmatrix}} \begin{bmatrix} t_{12} \\ t_{22} \end{bmatrix} = \begin{bmatrix} 0 \\ 0 \end{bmatrix}$$

Next, we set $t_{22} = 1$ in the first of the two equations:

$$13.65t_{12} + 10.69t_{22} = 0$$

to obtain $t_{12} = -0.783$ and note, further, that these values also satisfy the second equation. We then obtain

$$\mathbf{t}_2 = \begin{bmatrix} -0.783k \\ 1k \end{bmatrix}; \qquad (\text{norm})\mathbf{t}_2 = \begin{bmatrix} -0.617 \\ 0.787 \end{bmatrix}$$

Let us now assemble the two normalized eigenvectors into the matrix, which we shall denote by $\mathbf{T}_b$:

$$\mathbf{T}_b = \begin{bmatrix} 0.787 & -0.617 \\ 0.617 & 0.787 \end{bmatrix}$$

Then, as before, we can collect the various terms of the decomposition into the following triple product:

$$\mathbf{Z}_b = \mathbf{D}_b = \begin{bmatrix} 22.56 & 0 \\ 0 & 0.54 \end{bmatrix} = \mathbf{T}_b^{-1}\mathbf{C}(X)\mathbf{T}_b$$

While we have just found $\mathbf{Z}_b$ and $\mathbf{T}_b$, and we know $\mathbf{C}(X)$ to begin with, we must still solve for $\mathbf{T}_b^{-1}$. We obtain $\mathbf{T}_b^{-1}$ from

$$\mathbf{T}_b = \begin{bmatrix} 0.787 & -0.617 \\ 0.617 & 0.787 \end{bmatrix}$$

by the now-familiar adjoint matrix method, first described in Chapter 4.

$$\mathbf{T}_b^{-1} = \frac{1}{|\mathbf{T}_b|} \operatorname{adj}(\mathbf{T}_b) = \frac{1}{1}\begin{bmatrix} 0.787 & 0.617 \\ -0.617 & 0.787 \end{bmatrix} = \begin{bmatrix} 0.787 & 0.617 \\ -0.617 & 0.787 \end{bmatrix}$$

Here we see the somewhat surprising result that $\mathbf{T}_b^{-1} = \mathbf{T}_b'$.

However, as recalled from Chapter 4, one of the properties of an orthogonal matrix is that its inverse equals its transpose:

$$\boxed{\mathbf{T}^{-1} = \mathbf{T}'}$$

and we see that such is the case here. Moreover, it is easily seen that $\mathbf{T}_b$ exhibits the orthogonality conditions of $\mathbf{t}_{1b}'\mathbf{t}_{2b} = 0$ and $\mathbf{t}_{1b}'\mathbf{t}_{1b} = \mathbf{t}_{2b}'\mathbf{t}_{2b} = 1$.

Thus, in the case of a *symmetric* matrix, illustrated by $\mathbf{C}(X)$, the general diagonal representation

$$\boxed{\mathbf{D} = \mathbf{U}^{-1}\mathbf{A}\mathbf{U}}$$

of Section 5.2 now can be written as

$$\boxed{\mathbf{D} = \mathbf{T}'\mathbf{C}(X)\mathbf{T}}$$

where $\mathbf{T}$ is orthogonal and $\mathbf{D}$ continues to be diagonal.

By the same token we can write

$$\boxed{\mathbf{A} = \mathbf{U}\mathbf{D}\mathbf{U}^{-1}}$$

in the case of a symmetric matrix as

$$\boxed{\mathbf{C}(X) = \mathbf{T}\mathbf{D}\mathbf{T}'}$$

Thus, the rather interesting outcome of all of this is that starting with a *symmetric* matrix $\mathbf{C}(X)$, we obtain an eigenstructure in which the matrix of eigenvectors is orthogonal. That is, not only is the transformation represented by a stretch (as also found in Section 5.2), but the basis vectors themselves, in the symmetric matrix case, are orthonormal. We shall return to this finding in due course. For the moment, however, let us pull together the results so far, particularly as they relate to the problem of computing some "best" linear composite for the sample problem.

5.4.4 Comparing the Results

When we rather arbitrarily tried a 45° rotation of the original axes in order to obtain the linear composites $\mathbf{Z}_a$ shown in Panel I of Table 5.1, we noted that the covariance matrix derived from this transformation was

$$\mathbf{C}(Z_a) = \mathbf{T}_a'\mathbf{C}(X)\mathbf{T}_a = \begin{bmatrix} 22.23 & -2.64 \\ -2.64 & 0.87 \end{bmatrix}$$

Let us now compare this result with the maximum variance attainable from $\mathbf{T}_b$. From the immediately preceding discussion, we know that the comparable results are

$$\mathbf{C}(Z_b) = \overset{\mathbf{T}_b'}{\begin{bmatrix} 0.787 & 0.617 \\ -0.617 & 0.787 \end{bmatrix}} \overset{\mathbf{C}(X)}{\begin{bmatrix} 14.19 & 10.69 \\ 10.69 & 8.91 \end{bmatrix}} \overset{\mathbf{T}_b}{\begin{bmatrix} 0.787 & -0.617 \\ 0.617 & 0.787 \end{bmatrix}}$$

$$= \begin{bmatrix} 22.56 & 0 \\ 0 & 0.54 \end{bmatrix}$$

The matrix $\mathbf{T}_b$ in the present case involves a counterclockwise rotation of 38°.

Panel II of Table 5.1 shows the computed $\mathbf{Z}_b$ values. Also, Fig. 5.5 shows this optimal rotation as a dashed line for comparison with the 45° rotation tried earlier. Clearly, the two rotations are quite close to each other. Moreover, when we compare the first (selected to be the largest) eigenvalues above:

$$\lambda_1 = 22.23 \qquad \text{(for 45° rotation)}$$

$$\lambda_1 = 22.56 \qquad \text{(for 38° optimal rotation)}$$

we again see how close the 45° rotation, which was just guessed at for expository purposes, is to the optimal.

Another point to note is that the off-diagonal elements of $\mathbf{C}(\mathbf{Z}_b)$, after the optimal rotation, are zero, and hence, $\mathbf{z}_{1b}$ and $\mathbf{z}_{2b}$ the two linear composites obtained from the optimal rotation, are uncorrelated. This is a bonus provided by the fact that $\mathbf{C}(X)$ is symmetric and, hence, $\mathbf{T}_b$ is orthogonal.

To recapitulate, we have demonstrated how a set of vectors represented by the matrix $\mathbf{X}_d$ and a transformation of those vectors $\mathbf{C}(X) = (1/m)\mathbf{X}_d'\mathbf{X}_d$ can be rotated such that the projections of the points $\mathbf{X}_d\mathbf{T}_b$ (with $\mathbf{T}_b$ orthogonal) onto the first axis $\mathbf{z}_{1b}$ have the property of maximum variance. Moreover, at the same time it turns out that the new covariance matrix

$$\mathbf{C}(Z_b) = 1/m[\mathbf{T}_b'\mathbf{X}_d'\mathbf{X}_d\mathbf{T}_b] = \begin{bmatrix} 22.56 & 0 \\ 0 & 0.54 \end{bmatrix}$$

is *diagonal.* That is, all vectors referenced according to this new basis will be mapped onto scalar multiples of themselves by a stretch. This means that all cross products in the $\mathbf{C}(Z_b)$ matrix vanish, as noted above.

Hence, we have diagonalized the original transformation $\mathbf{C}(X)$ by finding a rotation $\mathbf{T}_b$ of the $\mathbf{X}_d$ space that has the effect of making $\mathbf{C}(Z_b)$ a diagonal matrix. The second axis $\mathbf{z}_{2b}$ will be orthogonal to the first or variance-maximizing axis $\mathbf{z}_{1b}$.

Reflecting a bit on the above example and observing the configuration of points in Fig. 5.5, it would appear that the point pattern roughly resembles an ellipse. Furthermore, the new axes, $\mathbf{z}_{1b}$ and $\mathbf{z}_{2b}$, correspond, respectively, to the major and minor axes of that ellipse, called principal axes in multivariate analysis.

If the distribution of points is multivariate normal, it turns out that the loci of equal probability are represented by a family of concentric ellipses (in two dimensions) or ellipsoids or hyperellipsoids (in higher dimensions). The "ellipse" in Fig. 5.5 could be construed as an estimate of one of these concentration ellipses.

It also turns out that by solving for the eigenstructure of $\mathbf{C}(X)$, we also obtain the axes of the "ellipse." This reorientation of the plane along the axes of the implied ellipse in Fig. 5.5 (via the 38° rotation of basis vectors) will also be relevant to quadratic forms, a topic that is discussed later in the chapter.

5.5 PROPERTIES OF MATRIX EIGENSTRUCTURES

At this point we have discussed eigenstructures from two different, and complementary, points of view:

1. finding a new basis of some linear transformation so that the transformation relative to that new basis assumes a particularly simple form, such as a stretch or a stretch followed by reflection;
2. finding a new basis—by means of a rotation—so that the variance of a set of points is maximized if they are projected onto the first axis of the new basis; the second axis maximizes residual variance for that dimensionality, and so on.

In the first approach no mention was made of any need for the basis transformation to be symmetric. In the second case the presence of a symmetric matrix possessed the advantage of producing an orthonormal basis of eigenvectors (a rotation).

In the recapitulation of Section 5.2.5, we alerted the reader to a number of problems concerning the eigenstructure of nonsymmetric matrices. In general, even though we assume throughout that $\mathbf{A}$ has all real-valued entries, if $\mathbf{A}$ is nonsymmetric,

1. we may not be able to diagonalize it via $\mathbf{U}^{-1}\mathbf{A}\mathbf{U}$;[7]
2. even if it can be diagonalized, the eigenvalues and eigenvectors of $\mathbf{A}$ need not all be real;
3. even if the eigenvalues (and eigenvectors) of $\mathbf{A}$ are all real valued, they need not be all nonzero;[8]
4. even if they are all nonzero, they need not be all distinct.[9]

[7] However, any matrix *can* be made similar to an upper triangular matrix (a square matrix with all zeros below the main diagonal). We do not pursue this more general topic here.

[8] If $\mathbf{A}$ has at least one zero eigenvalue, it is singular, a point that will be discussed in more detail in Section 5.6.

[9] Points 3 and 4 pertain to symmetric matrices as well.

Delving into the properties of eigenstructures involving complex eigenvalues and eigenvectors would take us too far afield in this book.

Fortunately for the reader all nonsymmetric matrices of interest to us in multivariate analysis will have *real* eigenvalues and *real* eigenvectors. However, if **A** is nonsymmetric, then **U**, the new basis of eigenvectors, is not orthogonal. Moreover, the problem of dealing with zero (or nonzero but nondistinct) eigenvalues must be contended with in any case, and will be discussed in the context of symmetric matrices.

5.5.1 Properties of Symmetric Matrices

As could be inferred from earlier discussion, in order to satisfy the expression

$$\boxed{\mathbf{D} = \mathbf{U}^{-1}\mathbf{A}\mathbf{U}}$$

$\mathbf{U}^{-1}$ must, of course, exist. However, it turns out that in order for some (not necessarily symmetric) matrix $\mathbf{A}_{n \times n}$ to be made diagonal, it is *necessary and sufficient that* **U** *consist of linearly independent vectors* and, hence, forms a basis (and possesses an inverse). If this condition is not met, then **A** is not diagonalizable. However, even if a matrix is diagonalizable, it may not necessarily be orthogonally so. And, even if a matrix can be made diagonal, it need not consist of eigenvalues and eigenvectors that are all real valued.

Symmetric matrices take care of these problems. If **A** is symmetric, it is not only *always* diagonalizable but, in addition, it is *orthogonally* diagonalizable where we have the relation

$$\boxed{\mathbf{U}^{-1} = \mathbf{U}'}$$

This is a very important condition since it states that for any pair of *distinct* eigenvectors $\mathbf{u}_i$, $\mathbf{u}_j$ their scalar product $\mathbf{u}_i'\mathbf{u}_j = 0$.

Notice that this was the situation in Section 5.4.3 in which we had the result

$$\mathbf{C}(Z_b) = \overset{\mathbf{T}_b'}{\begin{bmatrix} 0.787 & 0.617 \\ -0.617 & 0.787 \end{bmatrix}} \overset{\mathbf{C}(X)}{\begin{bmatrix} 14.19 & 10.69 \\ 10.69 & 8.91 \end{bmatrix}} \overset{\mathbf{T}_b}{\begin{bmatrix} 0.787 & -0.617 \\ 0.617 & 0.787 \end{bmatrix}}$$

$$= \begin{bmatrix} 22.56 & 0 \\ 0 & 0.54 \end{bmatrix}$$

Not only is $\mathbf{C}(Z_b)$ diagonal, but $\mathbf{T}_b$ is orthogonal. Since orthonormal basis vectors are highly convenient to work with in multivariate analysis, orthogonally diagonalizable matrices are useful to have.

A further differentiating property for symmetric matrices versus their nonsymmetric counterparts is also useful in multivariate applications. *If a symmetric matrix* **A** *consists of all real-valued entries, then all of its eigenvalues and associated eigenvectors will be real valued.*

In practice, however, even the nonsymmetric matrices that we encounter in multivariate analysis–such as those that arise in multiple discriminant analysis and canonical correlation–will have real eigenvalues. Hence, in the kinds of applications of relevance to multivariate analysis, the researcher does not need to worry very much about cases involving complex eigenvalues and eigenvectors. Still it is nice to know that the problem of complex eigenvalues and eigenvectors does not arise if **A** is symmetric.

Now, let us next examine the case of equal eigenvalues in symmetric matrices. Suppose we have tied λ_i's of multiplicities l_k for blocks $k = 1, 2, \ldots, s$ where

$$\sum_{k=1}^{s} l_k = n$$

First, it is comforting to know that the orthogonality property is maintained *across* subsets of eigenvectors associated with tied eigenvalues. That is, if $\mathbf{t}_i$ and $\mathbf{t}_j$ are drawn from *different* sets, then $\mathbf{t}_i'\mathbf{t}_j = 0$. The problem, then, is to obtain a set of orthogonal eigenvectors *within* each tied set of eigenvalues. Since this can usually be done in an infinity of ways, the solution will not be unique.

To illustrate, suppose we have eigenvalues $\lambda_1 = 1$, $\lambda_2 = 2$, $\lambda_3 = 2$ with associated eigenvectors $\mathbf{t}_1$, $\mathbf{t}_2$, and $\mathbf{t}_3$. We note that λ_2 and λ_3 are tied. In the present case there is, in a sense, too much freedom with regard to the eigenvectors associated with λ_2 and λ_3. What can be done, however, is

1. Find the eigenvector $\mathbf{t}_1$ associated with the distinct eigenvalue λ_1; this is done routinely in the course of substituting λ_1 in the equation $(\mathbf{A}-\lambda_1\mathbf{I})\mathbf{t}_1 = \mathbf{0}$.
2. Choose eigenvectors $\mathbf{t}_2$ and $\mathbf{t}_3$ (e.g., via Gram–Schmidt orthonormalization) so that they form an orthonormal set *within* themselves. Each, of course, will already be orthogonal to $\mathbf{t}_1$.

While the above orientation is arbitrary, in view of the equality of λ_2 and λ_3, it does represent a way to deal with the problem of subsets of eigenvalues that are equal to each other.

If it turned out that *all* eigenvalues were equal, that is, $\lambda_1 = \lambda_2 = \lambda_3$, then we have a case in which the transformation of **A** is a scalar transformation with coefficient λ (i.e., just the scalar λ times an identity transformation). As such, all (nonzero) vectors in the original space can serve as eigenvectors. Thus, *any* set of mutually orthonormal vectors–including the original orthonormal basis that could lead to this condition–can serve as a new basis.

In general, if we have k eigenvalues, all with the same value λ, then we must first find k linearly independent eigenvectors, all having eigenvalue λ. Then we orthonormalize them via some process like Gram–Schmidt. If we have two or more subsets of tied eigenvalues, the orthonormalizing process is done separately within set. As noted earlier, all eigenvectors in different sets, where the eigenvalues differ, will already be orthogonal.

However, tied eigenvalues arise only rarely in data-based product-moment matrices, such as $\mathbf{C}(X)$ and $\mathbf{R}(X)$. However, if they do, the analyst should be aware that the representation of **A** in terms of its eigenstructure is not unique, even though **A** may be nonsingular.

In summary, if $\mathbf{A}_{n\times n}$ is symmetric, we can say the following:

1. An orthogonal transformation **T** can be found such that **A** can be made diagonal by

$$\boxed{\mathbf{D} = \mathbf{T}'\mathbf{A}\mathbf{T}}$$

where **D** is diagonal, and the columns of **T** are eigenvectors of **A**.

2. All eigenvalues and eigenvectors are real.
3. If $\lambda_i \neq \lambda_j$, then $\mathbf{t}_i'\mathbf{t}_j = 0$.
4. If tied eigenvalues occur, of multiplicity l_k for some block k, then **A** has l_k but not more than l_k mutually orthogonal eigenvectors corresponding to the kth block of tied eigenvalues. In such cases, the eigenstructure of **A** will not be unique, but **T**, the $n \times n$ matrix of eigenvectors, will still be nonsingular.

The topic of *zero eigenvalues*—and their relationship to matrix rank—is so important that a special section in the chapter is reserved for it. For the moment, however, we turn to some additional properties of eigenstructures, appropriate (unless stated otherwise) for both the symmetric and nonsymmetric cases.

5.5.2 Additional Properties of Eigenstructures

Two quite general properties of eigenstructures that apply to either the nonsymmetric or symmetric cases are:

1. The sum of the eigenvalues of the eigenstructure of a matrix equals the sum of the main diagonal elements (called the *trace*) of the matrix. That is, for some matrix $\mathbf{W}_{n\times n}$,

$$\boxed{\sum_{i=1}^{n} \lambda_i = \sum_{i=1}^{n} w_{ii}}$$

2. The product of the eigenvalues of **W** equals the determinant of **W**:

$$\boxed{\prod_{i=1}^{n} \lambda_i = |\mathbf{W}|}$$

Notice here that if **W** is singular, at least one of its eigenvalues must be zero in order that $|\mathbf{W}| = 0$, the condition that must be met in order for **W** to be singular. However, even though **W** may be singular, **T** in the expression

$$\mathbf{D} = \mathbf{T}'\mathbf{W}\mathbf{T}$$

is still orthogonal (and, hence, nonsingular).

In addition to the above, a number of other properties related to sums, products, powers, and roots should also be mentioned. (The last two properties that are listed pertain only to symmetric matrices with nonnegative eigenvalues.)

3. If we have the matrix $\mathbf{B} = \mathbf{A} + k\mathbf{I}$, where k is a scalar, then the eigenvectors of **B** are the same as those of **A**, and the ith eigenvalue of **B** is

$$\boxed{\lambda_i + k}$$

where λ_i is the ith eigenvalue of **A**.

4. If we have the matrix $\mathbf{C} = k\mathbf{A}$, where k is a scalar, then **C** has the same eigenvectors as **A** and

$$\boxed{k\lambda_i}$$

is the ith eigenvalue of **C**, where λ_i is the ith eigenvalue of **A**.

5. If we have the matrix $\mathbf{A}^p$ (where p is a positive integer), then $\mathbf{A}^p$ has the same eigenvectors as **A** and

$$\boxed{\lambda_i{}^p}$$

is the ith eigenvalue of $\mathbf{A}^p$, where λ_i is the ith eigenvalue of **A**.

6. If $\mathbf{A}^{-1}$ exists, then $\mathbf{A}^{-p}$ has the same eigenvectors as **A** and

$$\boxed{\lambda_i{}^{-p}}$$

is the eigenvalue of $\mathbf{A}^{-p}$ corresponding to the ith eigenvalue of **A**. In particular, $1/\lambda_i$ is the eigenvalue of $\mathbf{A}^{-1}$ corresponding to λ_i, the ith eigenvalue of **A**.

7. If a symmetric matrix **A** can be written as the product

$$\mathbf{A} = \mathbf{TDT}'$$

where **D** is diagonal with all entries nonnegative and **T** is an orthogonal matrix of eigenvectors, then

$$\mathbf{A}^{1/2} = \mathbf{TD}^{1/2}\mathbf{T}'$$

and it is the case that $\mathbf{A}^{1/2}\mathbf{A}^{1/2} = \mathbf{A}$.[10]

8. If a symmetric matrix $\mathbf{A}^{-1}$ can be written as the product

$$\mathbf{A}^{-1} = \mathbf{TD}^{-1}\mathbf{T}'$$

where $\mathbf{D}^{-1}$ is diagonal with all entries nonnegative and **T** is an orthogonal matrix of eigenvectors, then

$$\mathbf{A}^{-1/2} = \mathbf{TD}^{-1/2}\mathbf{T}'$$

and it is the case that $\mathbf{A}^{-1/2}\mathbf{A}^{-1/2} = \mathbf{A}^{-1}$.

[10] In Chapter 2 the square root of a diagonal matrix was defined as $\mathbf{D}^{1/2}\mathbf{D}^{1/2} = \mathbf{D}$ where the diagonal elements of $\mathbf{D}^{1/2}$ are $\sqrt{d_{ii}}$ and it was assumed that all d_{ii} in **D** are nonnegative to begin with. In the present case $\mathbf{A}^{1/2}\mathbf{A}^{1/2} = \mathbf{A}$, where **A** need not be diagonal, but the conditions stated in point 7 must be met. In the present context we see that $(\mathbf{TD}^{1/2}\mathbf{T}')(\mathbf{TD}^{1/2}\mathbf{T}') = \mathbf{TD}^{1/2}\mathbf{D}^{1/2}\mathbf{T}' = \mathbf{TDT}' = \mathbf{A}$.

We can illustrate these properties of eigenstructures by means of a simple example:

$$\mathbf{A} = \begin{bmatrix} 2 & 1 \\ 1 & 2 \end{bmatrix}$$

The eigenvalues of $\mathbf{A}$ are $\lambda_1 = 3$ and $\lambda_2 = 1$. The associated (and normalized) eigenvectors are

$$\mathbf{t}_1 = \begin{bmatrix} 0.707 \\ 0.707 \end{bmatrix}; \qquad \mathbf{t}_2 = \begin{bmatrix} 0.707 \\ -0.707 \end{bmatrix}$$

Since $\mathbf{A}$ is symmetric, we can write the decomposition as

$$\mathbf{D} = \mathbf{T}'\mathbf{A}\mathbf{T} = \begin{bmatrix} 0.707 & 0.707 \\ 0.707 & -0.707 \end{bmatrix} \begin{bmatrix} 2 & 1 \\ 1 & 2 \end{bmatrix} \begin{bmatrix} 0.707 & 0.707 \\ 0.707 & -0.707 \end{bmatrix} = \begin{bmatrix} 3 & 0 \\ 0 & 1 \end{bmatrix}$$

The various properties, discussed above, are now illustrated:

Trace of $\mathbf{A}$

$$\operatorname{tr}(\mathbf{A}) = a_{11} + a_{22} = \lambda_1 + \lambda_2 = 3 + 1 = 4$$

Determinant of $\mathbf{A}$

$$|\mathbf{A}| = \lambda_1\lambda_2 = 3(1) = 3$$

Eigenvalues of $\mathbf{A} + 2\mathbf{I}$

$$\mathbf{A} + 2\mathbf{I} = \begin{bmatrix} 4 & 1 \\ 1 & 4 \end{bmatrix}$$

$$\lambda^2 - 8\lambda + 15 = 0$$

$$\lambda_1 = 5; \quad \lambda_2 = 3$$

Eigenvalues of $2\mathbf{A}$

$$2\mathbf{A} = \begin{bmatrix} 4 & 2 \\ 2 & 4 \end{bmatrix}$$

$$\lambda^2 - 8\lambda + 12 = 0$$

$$\lambda_1 = 6; \quad \lambda_2 = 2$$

Eigenvalues of $\mathbf{A}^2$

$$\mathbf{A}^2 = \begin{bmatrix} 5 & 4 \\ 4 & 5 \end{bmatrix}$$

$$\lambda^2 - 10\lambda + 9 = 0$$

$$\lambda_1 = 9, \quad \lambda_2 = 1$$

Eigenvalues of $\mathbf{A}^{-2}$

$$\mathbf{A}^{-1} = \begin{bmatrix} 2/3 & -1/3 \\ -1/3 & 2/3 \end{bmatrix}; \qquad \mathbf{A}^{-2} = \begin{bmatrix} 5/9 & -4/9 \\ -4/9 & 5/9 \end{bmatrix}$$

$$\lambda^2 - 4/3\lambda + 1/3 = 0 \qquad \lambda^2 - 10/9\lambda + 1/9 = 0$$

$$\lambda_1 = 1/3, \quad \lambda_2 = 1 \qquad \lambda_1 = 1/9, \quad \lambda_2 = 1$$

The Square Root of $\mathbf{A}$

$$\mathbf{A}^{1/2} = \mathbf{T}\mathbf{D}^{1/2}\mathbf{T}'$$

$$= \begin{bmatrix} 0.707 & 0.707 \\ 0.707 & -0.707 \end{bmatrix} \begin{bmatrix} \sqrt{3} & 0 \\ 0 & \sqrt{1} \end{bmatrix} \begin{bmatrix} 0.707 & 0.707 \\ 0.707 & -0.707 \end{bmatrix}$$

$$= \begin{bmatrix} 1.366 & 0.366 \\ 0.366 & 1.366 \end{bmatrix}; \qquad \mathbf{A}^{1/2}\mathbf{A}^{1/2} = \begin{bmatrix} 2 & 1 \\ 1 & 2 \end{bmatrix}$$

The Square Root of $\mathbf{A}^{-1}$

$$\mathbf{A}^{-1/2} = \mathbf{T}\mathbf{D}^{-1/2}\mathbf{T}'$$

$$= \begin{bmatrix} 0.707 & 0.707 \\ 0.707 & -0.707 \end{bmatrix} \begin{bmatrix} 1/\sqrt{3} & 0 \\ 0 & 1/\sqrt{1} \end{bmatrix} \begin{bmatrix} 0.707 & 0.707 \\ 0.707 & -0.707 \end{bmatrix}$$

$$= \begin{bmatrix} 0.788 & -0.211 \\ -0.211 & 0.788 \end{bmatrix}; \qquad \mathbf{A}^{-1/2}\mathbf{A}^{-1/2} = \begin{bmatrix} 2/3 & -1/3 \\ -1/3 & 2/3 \end{bmatrix}$$

$$\underset{\mathbf{A}^{1/2}}{\begin{bmatrix} 1.366 & 0.366 \\ 0.366 & 1.366 \end{bmatrix}} \underset{\mathbf{A}^{-1/2}}{\begin{bmatrix} 0.788 & -0.211 \\ -0.211 & 0.788 \end{bmatrix}} = \underset{\mathbf{I}}{\begin{bmatrix} 1 & 0 \\ 0 & 1 \end{bmatrix}}$$

The preceding properties are quite useful in various aspects of multivariate analysis, and we shall return to a discussion of some of them later in the chapter.

5.6 EIGENSTRUCTURES AND MATRIX RANK

In Chapter 4 we described two procedures for finding the rank of a matrix, square or rectangular, as the case may be:

1. The examination of various square submatrices in order to find that one with the largest order for which the determinant is nonzero.
2. The echelon matrix approach followed by a count of the number of rows with at least one nonzero entry.

Eigenstructures are computed only for square matrices. However, by some procedures to be described in this section, we shall see how eigenstructures also provide a way to determine the rank of *any* matrix, even if the matrix is rectangular.

In addition, it is now time to discuss the topic of zero eigenvalues in solving for the eigenstructure of a matrix.[11] As noted in Section 5.5, the presence of one or more zero eigenvalues is sufficient evidence that the matrix **A** is singular.

5.6.1 Square Matrices

First, we recall that if $\mathbf{A}_{n \times n}$ is symmetric, then all of its eigenvalues are real. It is possible, of course, that some may be positive, others negative, and some even zero. Also, from the previous section we know that

$$|\mathbf{A}| = \prod_{i=1}^{n} \lambda_i$$

Hence, if any λ_i is zero, **A** is singular. But what about the rank of **A**? Or, if **A** is rectangular, how can its rank be found by means of eigenstructures?

In the case where $\mathbf{A}_{n \times n}$ is symmetric, finding the rank of **A** is simple. We first find the eigenstructure of **A** and then count the number—including multiple values, if any are present—of *nonzero* (either positive or negative) eigenvalues. The number of nonzero eigenvalues of **A** is equal to its rank. Since we assume here that any square matrix (symmetric or nonsymmetric) of interest to us in multivariate analysis will have real eigenvalues—this, of course, must be the case if **A** is symmetric—we can use this same procedure for finding the rank of any **A** as long as **A** is square.

[11] Our remarks will pertain to nonsymmetric as well as symmetric matrices.

5.6.2 Rectangular Matrices

Finding the rank of $\mathbf{A}_{m \times n}$, where $m \neq n$, by means of eigenstructures rests on an important fact about the minor and major product moments of a matrix:

$\mathbf{A}'\mathbf{A}$; minor product moment of $\mathbf{A}$

$\mathbf{A}\mathbf{A}'$; major product moment of $\mathbf{A}$

and that fact is

$$\boxed{r(\mathbf{A}'\mathbf{A}) = r(\mathbf{A}\mathbf{A}') = r(\mathbf{A})}$$

The rank of either a minor or major product moment is the same as the rank of the matrix itself.

Since $r(\mathbf{A}'\mathbf{A}) = r(\mathbf{A}\mathbf{A}')$, we should, of course, find the eigenstructure of the *smaller* of these two orders, so as to reduce computational time and effort. And, if the researcher finds it easier to work with the eigenstructures of symmetric matrices, if $\mathbf{A}_{n \times n}$ is square but nonsymmetric, one can also compute its product moment, either minor or major, and find the eigenstructure of the symmetrized matrix.

Another virtue attaches to finding the eigenstructure of a product moment matrix, $\mathbf{A}'\mathbf{A}$ or $\mathbf{A}\mathbf{A}'$, and that is that *all eigenvalues will be either positive or zero*. In the process of finding the product moment, any negative eigenvalues of $\mathbf{A}$ will become positive in the case of either $\mathbf{A}'\mathbf{A}$ or $\mathbf{A}\mathbf{A}'$, as we shall note later on.

For the moment, however, let us set down the procedure for rank determination in a step-by-step way. First, if $\mathbf{A}$ is originally nonsymmetric or rectangular, we can always find the minor product moment $(\mathbf{A}'\mathbf{A})$ or the major product moment $(\mathbf{A}\mathbf{A}')$ of $\mathbf{A}$, whichever is of smaller order. The product-moment matrix will be square and symmetric. Furthermore, the eigenstructure of the product-moment matrix will exhibit either positive or zero eigenvalues, and a general procedure for rank determination can be followed. This general procedure, applicable to finding the rank of square but nonsymmetric and rectangular matrices alike, is as follows:

1. Find $\mathbf{A}'\mathbf{A}$ or $\mathbf{A}\mathbf{A}'$, whichever is of smaller order in the case of rectangular matrices. Their product will be symmetric, and *all* eigenvalues will be nonnegative. The number of positive eigenvalues of $\mathbf{A}'\mathbf{A}$ (or $\mathbf{A}\mathbf{A}'$) equals the rank of $\mathbf{A}$.
2. If $r(\mathbf{A}) = n$ and $\mathbf{A}$ is $n \times n$, then the vectors, either row or column vectors in $\mathbf{A}$, are linearly independent.
3. If $r(\mathbf{A}) = n$ and $n < m$ (where $\mathbf{A}$ is of order $m \times n$), then the row vectors are linearly dependent. If $r(\mathbf{A}) = m < n$, then the column vectors are linearly dependent.
4. If $r(\mathbf{A}) < n \leqslant m$, then the set of either row or column vectors are linearly dependent and $r(\mathbf{A}) = k$ is the largest number of linearly independent vectors in $\mathbf{A}$. (The number k is the number of positive eigenvalues in $\mathbf{A}'\mathbf{A}$ or $\mathbf{A}\mathbf{A}'$.)

Next, suppose that the symmetric matrix being examined is still of the form $\mathbf{A}'\mathbf{A}$ or $\mathbf{A}\mathbf{A}'$, where we have adopted this form because $\mathbf{A}$ is either rectangular or nonsymmetric. However, even if $\mathbf{A}'\mathbf{A}$ (or $\mathbf{A}\mathbf{A}'$) is nonsingular, some of the (positive) λ_i may be equal to each other. If so, their multiplicities are still counted up in finding the rank of $\mathbf{A}$. If $\mathbf{A}$ (or $\mathbf{A}'\mathbf{A}$ or $\mathbf{A}\mathbf{A}'$) is singular with a subset of l_k nondistinct eigenvalues, we can still find a mutually orthonormal set of eigenvectors of rank l_k by some process, such as the Gram–Schmidt orthonormalization process, for the tied block k.[12]

In summary, finding the eigenstructure of a matrix—either symmetric to begin with or else a derived product-moment matrix—is a highly useful procedure for determining matrix rank. If the $n \times n$ original symmetric matrix has rank $r(\mathbf{A}) = k$, then $k \leqslant n$ nonzero eigenvalues will be found. If the *derived* matrix

$$\mathbf{A}'\mathbf{A} \quad \text{or} \quad \mathbf{A}\mathbf{A}' \quad \text{(whichever is of smaller order)}$$

has $r(\mathbf{A}'\mathbf{A})$ or $r(\mathbf{A}\mathbf{A}')$ equal to k, then $k \leqslant \min(m, n)$ positive eigenvalues will be found. In short, one can *always* find the rank of a matrix via the eigenstructure approach.

5.6.3 Special Characteristics of Product-Moment Matrices

Product-moment matrices, like the SSCP, covariance, and correlation matrices, play a unique role in multivariate analysis. For example, let us return to the covariance matrix used in the eigenstructure problem of Section 5.4:

$$\mathbf{C}(X) = \begin{array}{c} \\ X_1 \\ X_2 \end{array}\begin{array}{c} \begin{array}{cc} X_1 & \quad X_2 \end{array} \\ \begin{bmatrix} 14.19 & 10.69 \\ 10.69 & 8.91 \end{bmatrix} \end{array}$$

We recall that this represents the minor product moment found from

$$\mathbf{C}(X) = (1/m)\mathbf{X}_\mathrm{d}'\mathbf{X}_\mathrm{d}$$

where $\mathbf{X}_\mathrm{d}$ is the matrix of deviations about column means.

By way of summarizing some aspects of matrix rank and their relationship to eigenstructures, let us set down a few properties of product-moment matrices, such as $\mathbf{C}(X)$. We can illustrate the properties in terms of the covariance matrix:

1. If $\mathbf{C}(X)$, the covariance matrix, has all distinct eigenvalues, it can be written uniquely as the triple product

$$\boxed{\mathbf{C}(X) = \mathbf{TDT}'}$$

where $\mathbf{D}$ is diagonal, and $\mathbf{T}$ is an orthogonal matrix of associated eigenvectors.

2. Since $\mathbf{C}(X)$ is of product-moment form, all of its eigenvalues are nonnegative, and we can always order the eigenvalues of $\mathbf{C}(X)$ from large to small.

3. Whether nonsingular or not, the rank of $\mathbf{C}(X)$ equals the number of positive eigenvalues in its eigenstructure (since all eigenvalues in this case are nonnegative).

4. It is generally the case that if *any* square matrix $\mathbf{A}$ is symmetric with nonnegative eigenvalues, then $\mathbf{A} = \mathbf{B}'\mathbf{B}$. One way of writing the relationship involves defining $\mathbf{B}'$ as

$$\boxed{\mathbf{B}' = \mathbf{TD}^{1/2}}$$

[12] In a sense, then, the problem of dealing with tied (but nonzero) eigenvalues is independent of the problem of determining the rank of a matrix.

where $\mathbf{D}^{1/2}$ is a diagonal matrix of the square roots of the eigenvalues of $\mathbf{A}$, and $\mathbf{T}$ is the orthogonal matrix whose columns are the associated eigenvectors of $\mathbf{A}$.

5. Since all of the eigenvalues of $\mathbf{C}(X)$ are either positive or zero, there exists a matrix $\mathbf{B}$ of order $k \times n$ such that

$$\boxed{\mathbf{C}(X) = \mathbf{B}'\mathbf{B}}$$

6. The preceding definition of $\mathbf{B}' = \mathbf{T}\mathbf{D}^{1/2}$ is, however, not unique. If $\mathbf{B}_1 = \mathbf{V}\mathbf{B}$ is the product of an orthogonal matrix $\mathbf{V}$ and $\mathbf{B}$, then $\mathbf{A}$ can also be written, equally appropriately, as

$$\mathbf{A} = \mathbf{B}_1{}'\mathbf{B}_1 = (\mathbf{V}\mathbf{B})'(\mathbf{V}\mathbf{B}) = \mathbf{B}'\mathbf{V}'\mathbf{V}\mathbf{B}$$

The last three points can be illustrated by returning to the eigenstructure of $\mathbf{C}(X)$ in the sample problem. First, we can write $\mathbf{C}(X)$ as

$$\mathbf{C}(X) = \overset{\mathbf{T}_b}{\begin{bmatrix} 0.787 & -0.617 \\ 0.617 & 0.787 \end{bmatrix}} \overset{\mathbf{D}_b}{\begin{bmatrix} 22.56 & 0 \\ 0 & 0.54 \end{bmatrix}} \overset{\mathbf{T}_b{}'}{\begin{bmatrix} 0.787 & 0.617 \\ -0.617 & 0.787 \end{bmatrix}}$$

Next, we define $\mathbf{B}' = \mathbf{T}_b\,\mathbf{D}_b^{1/2}$ and $\mathbf{B} = \mathbf{D}_b^{1/2}\,\mathbf{T}_b'$ and, furthermore, restate $\mathbf{C}(X)$:

$$\mathbf{B}' = \begin{bmatrix} 3.74 & -0.45 \\ 2.93 & 0.58 \end{bmatrix}; \quad \mathbf{B} = \begin{bmatrix} 3.74 & 2.93 \\ -0.45 & 0.58 \end{bmatrix}; \quad \mathbf{B}'\mathbf{B} = \begin{bmatrix} 14.19 & 10.69 \\ 10.69 & 8.81 \end{bmatrix}$$

We can then check to see that $\mathbf{B}'\mathbf{B} = \mathbf{C}(X)$.

Next, however, let us take some orthogonal matrix $\mathbf{V}$ and write

$$\mathbf{B}_1 = \overset{\mathbf{V}}{\begin{bmatrix} 0.707 & 0.707 \\ -0.707 & 0.707 \end{bmatrix}} \overset{\mathbf{B}}{\begin{bmatrix} 3.74 & 2.93 \\ -0.45 & 0.58 \end{bmatrix}} = \begin{bmatrix} 2.33 & 2.48 \\ -2.96 & -1.66 \end{bmatrix}$$

Then, we can write

$$\mathbf{C}(X) = \overset{\mathbf{B}_1{}'}{\begin{bmatrix} 2.33 & -2.96 \\ 2.48 & -1.66 \end{bmatrix}} \overset{\mathbf{B}_1}{\begin{bmatrix} 2.33 & 2.48 \\ -2.96 & -1.66 \end{bmatrix}} = \begin{bmatrix} 14.19 & 10.69 \\ 10.69 & 8.91 \end{bmatrix}$$

and see that $\mathbf{C}(X)$ can be reproduced in this way just as well as the original way. *This latter property:*

$$\boxed{\mathbf{C}(X) = \mathbf{B}'\mathbf{B} = \mathbf{B}'\mathbf{V}'\mathbf{V}\mathbf{B}}$$

where $\mathbf{V}$ *is orthogonal* $(\mathbf{V}'\mathbf{V} = \mathbf{V}\mathbf{V}' = \mathbf{I})$ *is of particular relevance to factor analysis and has to do with the so-called rotation problem.* That is, suppose we were first to find the eigenstructure of $\mathbf{C}(X)$ and then write $\mathbf{C}(X)$ in the context of the sample problem as

$$\mathbf{C}(X) = \mathbf{B}'\mathbf{B} = [\mathbf{T}_b\mathbf{D}_b^{1/2}\mathbf{D}_b^{1/2}\mathbf{T}_b{}']$$

By means of the preceding argument, $\mathbf{B}' = \mathbf{T}_b\mathbf{D}_b^{1/2}$ is *not* unique, and we are free to rotate **B** as we please. This, of course, introduces a certain ambiguity into the question of what factor dimensions are "correct." As recalled, the principal components axes found from the *unique* representation:

$$\mathbf{C}(X) = \mathbf{T}_b\mathbf{D}_b\mathbf{T}_b'$$

are unambiguous in the sense of *maximizing variance* in the derived covariance matrix $\mathbf{C}(Z_b)$.

In summary, data-based product-moment matrices exhibit a number of virtues, such as real-valued eigenstructures and orthogonal transformations for rotating the matrix to diagonal form. As we saw in the sample problem, one can order the eigenvalues from largest to smallest in the process of transforming a data matrix into a set of linear composites that are mutually orthogonal. Finally, the rank of product-moment matrices is easily discerned by simply counting up the number of positive eigenvalues. In most data-based problems the rank of $\mathbf{C}(X)$, and other types of derived product-moment matrices, will equal the order of the (minor) product-moment matrix.

5.6.4 Recapitulation

At this point we have covered quite a bit of ground regarding eigenstructures and matrix rank. In the case of nonsymmetric matrices in general, we noted that even if a (square) matrix **A** could be diagonalized via

$$\boxed{\mathbf{D} = \mathbf{U}^{-1}\mathbf{A}\mathbf{U}}$$

the eigenvalues and eigenvectors need not be all real valued. (Fortunately, in the types of matrices encountered in multivariate analysis, we shall always be dealing with real-valued eigenvalues and eigenvectors.)

In the case of symmetric matrices, any such matrix **A** is diagonalizable, and orthogonally so, via

$$\boxed{\mathbf{D} = \mathbf{T}'\mathbf{A}\mathbf{T}}$$

where $\mathbf{T}' = \mathbf{T}^{-1}$ since **T** is orthogonal.[13] Moreover, all eigenvalues and eigenvectors are necessarily real. If the eigenvalues are not all distinct, an orthogonal basis—albeit not a unique one—can still be constructed. Furthermore, eigenvectors associated with distinct eigenvalues are already orthogonal to begin with.

The rank of any matrix **A**, square or rectangular, can be found from its eigenstructure. If $\mathbf{A}_{n \times n}$ is symmetric, we merely count up the number of nonzero eigenvalues k and note that $r(\mathbf{A}) = k \leqslant n$. If **A** is nonsymmetric or rectangular, we can find its minor (or major) product moment and then compute the eigenstructure. In this case, all eigenvalues are real and nonnegative. To find the rank of **A**, we simply count up the number of positive eigenvalues k and observe that $r(\mathbf{A}) = k \leqslant \min(m, n)$ if **A** is rectangular or $r(\mathbf{A}) = k \leqslant n$ if **A** is square.

[13] Note also that $\mathbf{A} = \mathbf{T}\mathbf{D}^{1/2}\mathbf{D}^{1/2}\mathbf{T}' = \mathbf{T}\mathbf{D}\mathbf{T}'$, as desired.

Finally, if **A** is of product-moment form to begin with, or if **A** is symmetric with nonnegative eigenvalues, then it can be written—although not uniquely so—as $\mathbf{A} = \mathbf{B}'\mathbf{B}$. The lack of uniqueness of **B** was illustrated in the context of axis rotation in factor analysis.

5.7 THE BASIC STRUCTURE OF A MATRIX

With some oversimplification, we can summarize the intent of the chapter so far by saying that, given a transformation matrix **A**, we would like to find a representation of it in some way that simplifies its geometric nature.

In the case of a square, but nonsymmetric, matrix **A**, we found that under fairly general circumstances (in which the vectors of **U** are linearly independent), **A** could be written as

$$\boxed{\mathbf{A} = \mathbf{U}\mathbf{D}\mathbf{U}^{-1}}$$

where **U** is nonsingular and **D** is diagonal. The eigenvalues and eigenvectors of **A** need not all be real valued, however.

In the case of a symmetric matrix **A**, *orthogonal* diagonalization can be achieved. In this case the above equation is satisfied and, in addition, we have

$$\boxed{\mathbf{A} = \mathbf{T}\mathbf{D}\mathbf{T}'}$$

since, given an orthogonal matrix **T**, we know that $\mathbf{T}' = \mathbf{T}^{-1}$. Moreover, all real-valued symmetric matrices are orthogonally diagonalizable with real-valued eigenvalues and eigenvectors.

Not all matrix transformations are symmetric, however, and not all matrices are square. Thus, with the exception of the preceding discussion of matrix rank, we have said relatively little on the topics of (a) the eigenstructure of square, nonsymmetric matrices and (b) the role of eigenstructure analysis in dealing with rectangular matrices which, by definition, do not have eigenstructures. It is now time to bring up these aspects, particularly the latter one.

This section of the chapter deals with *basic structure*, the most general decomposition of a transformation matrix that is discussed in the book. As we shall see, *any* matrix can be decomposed into its basic structure (although not necessarily uniquely so). As such, basic structure represents an extremely powerful concept in multivariate analysis and unifies much of our earlier discussion of matrix decomposition.[14] Again, we shall tie in the current material with some comments made on related matters in Chapter 4.

In Section 4.5.5 we demonstrated that an arbitrary nonsingular matrix

$$\mathbf{V} = \begin{bmatrix} 1 & 2 \\ 3 & 4 \end{bmatrix}$$

[14] While eigenstructure analysis plays a central role in finding the basic structure of a matrix, it should be stressed that the concepts are distinct.

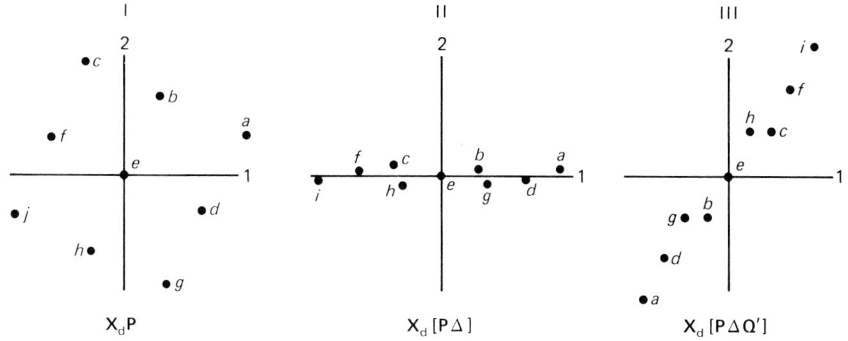

Fig. 5.6 Decomposition of general linear transformation $\mathbf{V} = \mathbf{P}\Delta\mathbf{Q}'$ (reproduced from Fig. 4.13).

could be uniquely decomposed into the triple product

$$\mathbf{V} = \mathbf{P}\Delta\mathbf{Q}' = \overset{\mathbf{P}}{\begin{bmatrix} -0.41 & -0.91 \\ -0.91 & 0.41 \end{bmatrix}} \overset{\Delta}{\begin{bmatrix} 5.47 & 0 \\ 0 & 0.37 \end{bmatrix}} \overset{\mathbf{Q}'}{\begin{bmatrix} -0.58 & -0.82 \\ 0.82 & -0.58 \end{bmatrix}}$$

where $\mathbf{P}$ is orthogonal (specifically, an improper rotation), Δ is diagonal (a stretch), and $\mathbf{Q}'$ is orthogonal (a proper rotation).

For ease of reference, Fig. 5.6 reproduces Fig. 4.13 in which a square lattice of points (shown in Panel II of Fig. 4.10) was transformed, in three stages, by the matrices making up the specific decomposition of $\mathbf{V}$. Each stage is shown in Fig. 5.6. Moreover, at that point we indicated that *any* nonsingular matrix could be uniquely decomposed into the product of (a) a rotation–stretch–rotation or (b) a rotation–reflection–stretch–rotation. It turns out, however, that this type of decomposition is a very general type of decomposition. *It is so general, in fact, that any matrix, square or rectangular, nonsingular or singular, symmetric or nonsymmetric, can be so decomposed, albeit not necessarily uniquely so.*

In the case of symmetric matrices, we know, of course, that the geometric relationship does hold:

$$\boxed{\mathbf{A} = \mathbf{T}\mathbf{D}\mathbf{T}'}$$

since, in this case, $\mathbf{T}$ and $\mathbf{T}'$ are orthogonal (rotations), and $\mathbf{D}$ is diagonal (a stretch).[15] Thus, if $\mathbf{A}$ is symmetric, the above decomposition holds as a special case.

However, as already observed in describing the eigenstructure of nonsymmetric matrices, there is no requirement that $\mathbf{T}$ be orthogonal. Furthermore, no such decomposition—orthogonal or otherwise—has been discussed for rectangular matrices.

We now consider the cases in which $\mathbf{A}$ is either square but nonsymmetric or $\mathbf{A}$ is rectangular. As it turns out, both cases can be handled by the same procedure, and we shall illustrate the approach by assuming that $\mathbf{A}$ is rectangular, of order $m \times n$. For convenience, assume that $\mathbf{A}$ is "vertical" with $m > n$, although this is not essential. We shall also assume for the moment that the rank of $\mathbf{A}$ is k $(k < n < m)$.

[15] Or, possibly, a stretch preceded by a reflection.

We first examine some of the algebra of this type of decomposition. Then we apply the decomposition to an illustrative problem. Let us set down our objective at the outset. And that is: Given an arbitrary rectangular transformation matrix **A**, we wish to find a way to express **A** in terms of the product of three, relatively simple, matrices:

1. an $m \times k$ matrix **P** which is orthonormal by columns and, hence, satisfies the relation $\mathbf{P}'\mathbf{P} = \mathbf{I}$;
2. a $k \times k$ matrix Δ which is diagonal and consists of k positive diagonal entries ordered from large to small (with ties allowed);
3. an $n \times k$ matrix **Q** which is orthonormal by columns and, hence, satisfies the relation $\mathbf{Q}'\mathbf{Q} = \mathbf{I}$.

The representation of the matrix **A** *as the triple product* $\mathbf{P}\Delta\mathbf{Q}'$ *is called its basic structure* (Horst, 1963). It is also called "singular value decomposition" by numerical analysts.

5.7.1 The Algebra of Basic Structure

The mathematical aspects of basic structure become rather complex, and so we shall settle for a less technical discussion. Given an arbitrary rectangular matrix $\mathbf{A}_{m \times n}$, where $m > n$ and $r(\mathbf{A}) = k < n < m$, its basic structure involves the triple product

$$\mathbf{A}_{m \times n} = \mathbf{P}_{m \times k}\Delta_{k \times k}\mathbf{Q}'_{k \times n}$$

where **P** and **Q** are each orthonormal by columns ($\mathbf{P}'\mathbf{P} = \mathbf{I}_{k \times k}$; $\mathbf{Q}'\mathbf{Q} = \mathbf{I}_{k \times k}$), and $\Delta_{k \times k}$ is diagonal with ordered positive entries.

Note, in particular, that **Q** cannot be orthogonal (where $\mathbf{Q}\mathbf{Q}' = \mathbf{I}_{n \times n}$) unless $k = n$. Moreover, **P** cannot be orthogonal (where $\mathbf{P}\mathbf{P}' = \mathbf{I}_{m \times m}$) unless $k = n = m$. As such, $\mathbf{P}_{m \times k}$ and $\mathbf{Q}_{n \times k}$ might be called *orthonormal sections* in which all columns are of unit length and mutually orthogonal.

Next, let us comment on $\Delta_{k \times k}$, which is called the *basic diagonal.* First, as we shall see, all elements of $\Delta_{k \times k}$ can be

1. taken to be positive;
2. ordered from large to small (with ties allowed).

Moreover, it will turn out that there is one and only one basic diagonal for any given matrix; that is, the basic diagonal part of the decomposition is always unique, and this will be true regardless of whether **A** is of full rank, square, or rectangular. It may happen, however, that some entries in $\Delta_{k \times k}$ are tied. If such is the case, only those portions of **P** and **Q** that relate to distinct entries in $\Delta_{k \times k}$ will be unique, a point to which we return later. *Finally, the rank of* **A** *is given quite simply by the number of positive entries in* Δ, *the basic diagonal.*

For the moment, let us examine the relationship of $\mathbf{A} = \mathbf{P}\Delta\mathbf{Q}'$ to its major and minor product moments, $\mathbf{A}\mathbf{A}'$ and $\mathbf{A}'\mathbf{A}$, respectively:

$$\mathbf{A}\mathbf{A}' = (\mathbf{P}\Delta\mathbf{Q}')(\mathbf{P}\Delta\mathbf{Q}')' = \mathbf{P}\Delta\mathbf{Q}'\mathbf{Q}\Delta\mathbf{P}'$$

but since $\mathbf{Q}'\mathbf{Q} = \mathbf{I}$ and letting $\Delta^2 = \mathbf{D}$, we see that **D** is still diagonal. Thus, we get

$$\mathbf{A}\mathbf{A}' = \mathbf{P}\mathbf{D}\mathbf{P}'$$

Furthermore,

$$\mathbf{A}'\mathbf{A} = (\mathbf{P\Delta Q}')'(\mathbf{P\Delta Q}') = \mathbf{Q\Delta P'P\Delta Q}'$$

but since $\mathbf{P'P} = \mathbf{I}$ and $\Delta^2 = \mathbf{D}$, we get

$$\mathbf{A}'\mathbf{A} = \mathbf{QDQ}'$$

Note that in both cases we have the eigenstructure formulation shown earlier for the case of symmetric matrices, namely, the triple product of an orthogonal, diagonal, and transposed orthogonal matrix. This is not surprising since both $\mathbf{AA}'$ and $\mathbf{A}'\mathbf{A}$ are symmetric. However, the diagonal matrix $\mathbf{D}$ of each triple product *is the same* for both product moments $\mathbf{AA}'$ and $\mathbf{A}'\mathbf{A}$. Furthermore,

1. all entries of $\mathbf{D}$ are real;
2. all entries of $\mathbf{D}$ are nonnegative;
3. all positive entries of $\mathbf{D}$ can be ordered from large to small (with possible ties, of course).

We take advantage of these facts in describing Δ, the $k \times k$ portion of $\mathbf{D}$ that has positive (as opposed to zero) entries, in terms of the following definition:

$$\boxed{\Delta_{k \times k} = \mathbf{D}_{k \times k}^{1/2}}$$

At this point, then, we are starting to get some hints about how to find $\mathbf{P}_{m \times k}$, $\Delta_{k \times k}$, and $\mathbf{Q}'_{k \times n}$ As observed above, $\mathbf{P}_{m \times k}$ represents the first k columns of the orthogonal matrix $\mathbf{P}_{m \times m}$ obtained from the eigenstructure of $\mathbf{AA}'$, while $\mathbf{Q}_{n \times k}$ represents the first k columns of the orthogonal matrix $\mathbf{Q}_{n \times n}$ obtained from the eigenstructure of $\mathbf{A}'\mathbf{A}$. Their common diagonal matrix $\mathbf{D}$ has all nonnegative entries. Furthermore, we can order (with ties allowed) the positive entries of $\mathbf{D}$ from large to small, until we get k of them. The remaining entries on the main diagonal will all be zero. The columns of $\mathbf{P}$ and $\mathbf{Q}$ can, of course, be made to correspond to the ordered diagonal elements in D and, hence, to their square roots in Δ.

Next, let us take the argument one step further. If we let Δ be the first k rows and k columns embedded in a (larger) $m \times n$ *rectangular* matrix, with $m - k$ rows and $n - k$ columns of zeros elsewhere, both $\mathbf{P}$ and $\mathbf{Q}'$ could be made fully orthogonal and, in this sense, properly constitute rotation matrices of order $m \times m$ and $n \times n$, respectively. This generalization can be called the *full* basic structure of a matrix.

The preceding generalization is a significant one. It tells us that *any* matrix—not just square, nonsingular ones—with real-valued entries can be represented as the product of

1. a rotation (possibly followed by a reflection), followed by
2. a stretch, followed by
3. a rotation.

Note further that if, indeed, $\mathbf{A}$ is *symmetric* to begin with, we have the special case

$$\mathbf{A} = \mathbf{TDT}'$$

since $\mathbf{AA}' = \mathbf{A}'\mathbf{A}$, and therefore $\mathbf{P}' = \mathbf{Q}' = \mathbf{T}'$. Hence, this same approach to matrix decomposition can be applied even to the more familiar case of a symmetric matrix.

However, the diagonal **D** is to be interpreted as Δ in the current context since we refer to **A** rather than to its product moments **AA**′ or **A**′**A**.

In summary, *any* matrix of real entries has a basic structure and can be written as

$$\mathbf{A} = \mathbf{P}\Delta\mathbf{Q}'$$

where Δ is diagonal, and **P** and **Q** are orthonormal by columns. The concept of *full* basic structure embeds the $k \times k$ diagonal matrix of positive entries in an $m \times n$ matrix in which $m - k$ rows and $n - k$ columns are zeros. Alternatively, we can require Δ to be $k \times k$ with all positive entries and, hence, **P** and **Q** will not, in general, be rotation matrices, although they will still be orthonormal by columns and constitute orthonormal sections.

Finally, a special case of the above involves the case in which **A** is symmetric to begin with. If so, it can be written as

$$\mathbf{A} = \mathbf{T}\mathbf{D}\mathbf{T}'$$

where **T**, of course, is a rotation matrix and **D** = Δ in the present discussion.

Figure 5.7 shows schematically the two cases that we have been considering. Panel I shows the case in which **P** is orthonormal by columns ($\mathbf{P}'\mathbf{P} = \mathbf{I}$) but is not orthogonal. Similarly, **Q** is orthonormal by columns ($\mathbf{Q}'\mathbf{Q} = \mathbf{I}$) but is not orthogonal. The diagonal matrix Δ has k positive eigenvalues, where $k < n < m$.

Panel II shows the full basic structure in the sense that **P** and **Q** can be made orthogonal by embedding Δ in a larger matrix, consisting of $m - k$ rows and $n - k$ columns of zeros (in addition, of course, to the zeros in the off-diagonal entries of the $k \times k$ portion). Not surprisingly, the $m - k$ columns of **P** and the $n - k$ rows of **Q**—while they could be made orthogonal—are mainly of theoretical interest since those dimensions would be annihilated by the $m - k$ rows of zeros and the $n - k$ columns of zeros in the $m \times n$ matrix in which Δ is embedded.

5.7.2 Conditions for Basic Matrices

The foregoing discussion applies to any matrix of interest since any matrix possesses a basic structure, written as the triple product $\mathbf{P}\Delta\mathbf{Q}'$. While any matrix can be decomposed into its basic structure, not all matrices are basic. This distinction needs to be made in discussing the rank of a matrix.

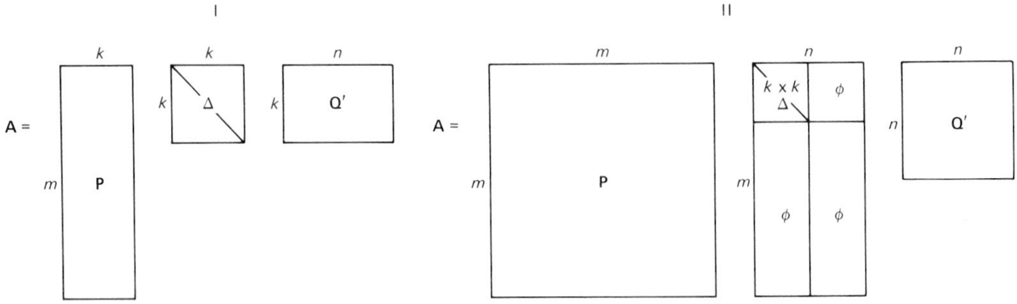

Fig. 5.7 Alternative formulations of the basic structure of $\mathbf{A}_{m \times n}$

A *basic* matrix **A** is one whose rank equals its smaller dimension. If **A** is $m \times n$ (and $m > n$), and if **A** is basic, then $r(\mathbf{A}) = k = n$.

If **A** is square and basic, then $r(\mathbf{A})$ equals its (common) order, and we have called this kind of matrix *nonsingular.* If **A** is rectangular and basic, then the rank of the major product moment $\mathbf{AA}'$ or the minor product moment $\mathbf{A}'\mathbf{A}$ is equal to the *smaller order* of **A**. If **A** is nonbasic, then its rank is $k (k < n \leqslant m)$.

This can all be summarized by saying that any matrix **A** can be decomposed into the basic structure

$$\boxed{\mathbf{A}_{m \times n} = \mathbf{P}_{m \times k} \mathbf{\Delta}_{k \times k} \mathbf{Q}'_{k \times n}}$$

where $k \leqslant \min(m, n)$. *Then* **A** *is basic if and only if* $k = \min(m, n)$.

5.7.3 Finding the Basic Structure

It is one thing to define the basic structure of a matrix **A** and quite another to solve for its representation as $\mathbf{A} = \mathbf{P\Delta Q}'$. Finding the basic structure of a matrix **A** makes use of concepts already discussed under the topic of eigenstructure. First, as previously discussed, if $\mathbf{A} = \mathbf{P\Delta Q}'$, then

$$\mathbf{A}'\mathbf{A} = \mathbf{Q\Delta P}'\mathbf{P\Delta Q}'$$

and, since **P** is orthonormal by columns, we have

$$\mathbf{A}'\mathbf{A} = \mathbf{Q\Delta}^2\mathbf{Q}'$$

Then, after we solve for $\mathbf{\Delta}^2$ and **Q** by finding the eigenstructure of the symmetric matrix $\mathbf{A}'\mathbf{A}$, we can find $\mathbf{\Delta}$ and then find **P** from

$$\boxed{\mathbf{P} = \mathbf{AQ\Delta}^{-1}}$$

Since $\mathbf{\Delta}$ is diagonal, $\mathbf{\Delta}^{-1}$ consists simply of the reciprocals of the diagonal entries of $\mathbf{\Delta}$. At this point, then, we have a procedure for solving for the triple product

$$\mathbf{A} = \mathbf{P\Delta Q}'$$

The procedure involves the following steps, assuming first that **A** is of order $m \times n$ with $n \leqslant m$:

1. Compute the minor product moment $\mathbf{A}'\mathbf{A}$ which results in a square symmetric matrix of order $n \times n$.
2. Find the eigenstructure of $\mathbf{A}'\mathbf{A}$, thus yielding the matrix of eigenvalues $\mathbf{\Delta}^2$ of rank k ($k \leqslant n$) and the matrix **Q** where **Q** is the matrix of associated eigenvectors that are orthonormal by columns.
3. Find the square roots of the diagonal entries of $\mathbf{\Delta}^2$.
4. Find $\mathbf{\Delta}^{-1}$, the reciprocals of the diagonal entries of $\mathbf{\Delta}$.
5. Find $\mathbf{P} = \mathbf{AQ\Delta}^{-1}$.

It turns out that if $k = n$, then **A** is basic. If $k < n$, then **A** is nonbasic and of rank k.

On the other hand, if $n > m$ we apply the same type of procedure to $\mathbf{A}'$ and transpose the result. That is, let

$$\mathbf{A}' = \mathbf{P}_1\boldsymbol{\Delta}_1\mathbf{Q}_1{}'$$

Then its transpose is

$$\mathbf{A} = (\mathbf{P}_1\boldsymbol{\Delta}_1\mathbf{Q}_1{}')' = \mathbf{Q}_1\boldsymbol{\Delta}_1\mathbf{P}_1{}'$$

Then, if we define

$$\mathbf{P} \equiv \mathbf{Q}_1; \qquad \boldsymbol{\Delta} \equiv \boldsymbol{\Delta}_1; \qquad \mathbf{Q} \equiv \mathbf{P}_1$$

it turns out that we have the desired decomposition of $\mathbf{A}$ into the basic structure

$$\mathbf{A} = \mathbf{P}\boldsymbol{\Delta}\mathbf{Q}'$$

5.7.4 Illustrating the Basic Structure Decomposition Procedure

To illustrate the procedure described above, let us take a particularly simple case involving a 3 x 2 matrix:

$$\mathbf{A} = \begin{bmatrix} 1 & 2 \\ 0 & 2 \\ -1 & 1 \end{bmatrix}$$

Since $m > n$, we first find the (smaller) minor product-moment matrix:

$$\mathbf{A}'\mathbf{A} = \begin{bmatrix} 2 & 1 \\ 1 & 9 \end{bmatrix} = \begin{bmatrix} 1 & 0 & -1 \\ 2 & 2 & 1 \end{bmatrix} \begin{bmatrix} 1 & 2 \\ 0 & 2 \\ -1 & 1 \end{bmatrix}$$

Table 5.2 shows a summary of the computations involved in finding the eigenstructure of $\mathbf{A}'\mathbf{A}$. Note that this part of the problem is a standard one in finding the eigenstructure of a symmetric matrix.

After finding Δ^2 and $\mathbf{Q}$, by means of solving the characteristic equation, we find Δ and then Δ^{-1}. The last step is to solve for $\mathbf{P}$ in the equation

$$\boxed{\mathbf{P} = \mathbf{A}\mathbf{Q}\boldsymbol{\Delta}^{-1}}$$

These results also appear in Table 5.2. Finally, we assemble the triple product

$$\mathbf{A} = \begin{bmatrix} 1 & 2 \\ 0 & 2 \\ -1 & 1 \end{bmatrix} = \overset{\mathbf{P}}{\begin{bmatrix} -0.70 & 0.52 \\ -0.65 & -0.20 \\ -0.28 & -0.83 \end{bmatrix}} \overset{\boldsymbol{\Delta}}{\begin{bmatrix} 3.02 & 0 \\ 0 & 1.36 \end{bmatrix}} \overset{\mathbf{Q}'}{\begin{bmatrix} -0.14 & -0.99 \\ 0.99 & -0.14 \end{bmatrix}}$$

TABLE 5.2

Finding the Eigenstructure of **A′A**

Minor product-moment matrix

$$\mathbf{A}'\mathbf{A} = \begin{bmatrix} 2 & 1 \\ 1 & 9 \end{bmatrix}$$

Characteristic equation

$$|(\mathbf{A}'\mathbf{A}) - \lambda\mathbf{I}| = \begin{vmatrix} 2-\lambda & 1 \\ 1 & 9-\lambda \end{vmatrix} = 0$$

$$\lambda^2 - 11\lambda + 17 = 0$$

Quadratic formula

$$y = ax^2 + bx + c$$

Substitution in general quadratic

$$\frac{-b \pm \sqrt{b^2 - 4ac}}{2a} = \frac{11 \pm \sqrt{(-11)^2 - 4(17)}}{2}$$

Eigenvalues of **A′ A**

$$\lambda_1 = 9.14$$

$$\lambda_2 = 1.86$$

Matrix of eigenvectors of **A′A**

$$\mathbf{Q} = \begin{bmatrix} -0.14 & 0.99 \\ -0.99 & -0.14 \end{bmatrix}$$

Basic diagonal

$$\mathbf{\Delta} = \begin{bmatrix} (9.14)^{1/2} & 0 \\ 0 & (1.86)^{1/2} \end{bmatrix} = \begin{bmatrix} 3.02 & 0 \\ 0 & 1.36 \end{bmatrix}$$

Solving for the matrix **P**

$$\mathbf{P} = \overset{\mathbf{A}}{\begin{bmatrix} 1 & 2 \\ 0 & 2 \\ -1 & 1 \end{bmatrix}} \overset{\mathbf{Q}}{\begin{bmatrix} -0.14 & 0.99 \\ -0.99 & -0.14 \end{bmatrix}} \overset{\mathbf{\Delta}^{-1}}{\begin{bmatrix} 1/3.02 & 0 \\ 0 & 1/1.36 \end{bmatrix}} = \begin{bmatrix} -0.70 & 0.52 \\ -0.65 & -0.20 \\ -0.28 & -0.83 \end{bmatrix}$$

As can be noted, after taking the transpose of **Q**, the matrix **A** has been decomposed into the product of an orthonormal (by columns) matrix times a diagonal times a (square) orthogonal matrix. In general, however, $\mathbf{Q}'$ will be orthonormal by rows if **A** is not basic.

Figure 5.8 shows the separate aspects of the decomposition. We first start with the matrix **P**. **P** is then differentially stretched in accordance with **Δ**. Finally, the points of **PΔ** are rotated in accordance with $\mathbf{Q}'$, leading to a reproduction of what we started out with, namely, the matrix **A** shown at the top of the figure.

5.7.5 The Basic Structure of the Sample Problem Matrix of Predictors

In Section 5.4 we found the eigenstructure of the covariance matrix of the two predictor variables in the sample problem. As recalled,

$$\mathbf{C}(X) = \mathbf{X}_d'\mathbf{X}_d/m = \mathbf{TDT}'$$

$$= \overset{\mathbf{T}}{\begin{bmatrix} 0.787 & 0.617 \\ 0.617 & -0.787 \end{bmatrix}} \overset{\mathbf{D}}{\begin{bmatrix} 22.56 & 0 \\ 0 & 0.54 \end{bmatrix}} \overset{\mathbf{T}'}{\begin{bmatrix} 0.787 & 0.617 \\ 0.617 & -0.787 \end{bmatrix}} = \begin{bmatrix} 14.19 & 10.69 \\ 10.69 & 8.91 \end{bmatrix}$$

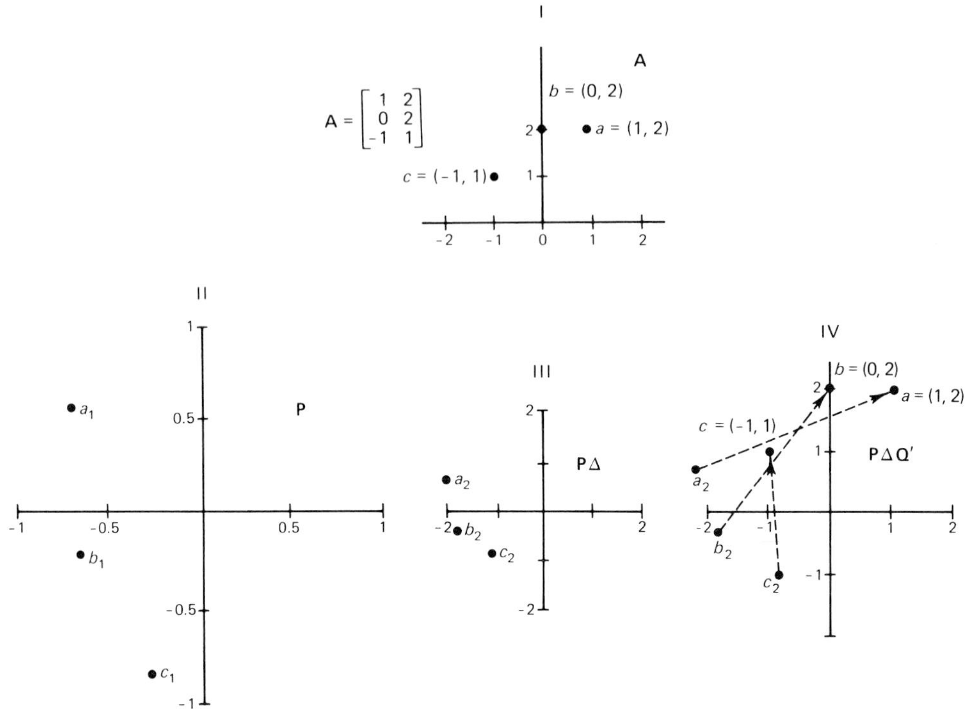

Fig. 5.8 Decomposition of $\mathbf{A}$ to basic structure $\mathbf{A} = \mathbf{P}\boldsymbol{\Delta}\mathbf{Q}'$. Key: I, "target" configuration as defined by $\mathbf{A} = \mathbf{P}\boldsymbol{\Delta}\mathbf{Q}'$; II, orthonormal-by-columns matrix $\mathbf{P}$; III, application of stretch (defined by $\boldsymbol{\Delta}$) to $\mathbf{P}$; IV, rotation of $\mathbf{P}\boldsymbol{\Delta}$ via matrix $\mathbf{Q}'$.

where $\mathbf{T}$ and $\mathbf{T}'$ are orthogonal and $\mathbf{D}$ is diagonal.

Suppose, now, that we wished to find the basic structure of $\mathbf{X}_d/\sqrt{m}$, the mean-corrected matrix of predictors, scaled by the square root of the sample size.

Based on what we have just covered, we know how to proceed. First, we find

$$\boldsymbol{\Delta}^{-1} = \mathbf{D}^{-1/2} = \begin{bmatrix} 1/\sqrt{22.56} & 0 \\ 0 & 1/\sqrt{0.54} \end{bmatrix} = \begin{bmatrix} 0.211 & 0 \\ 0 & 1.361 \end{bmatrix}$$

Next, analogous to solving for $\mathbf{P}$ in the expression $\mathbf{P} = \mathbf{A}\mathbf{Q}\boldsymbol{\Delta}^{-1}$, we now solve for $\mathbf{U}$ in the expression

$$\mathbf{U} = \mathbf{X}_d/\sqrt{m}\,\mathbf{T}\boldsymbol{\Delta}^{-1}$$

which leads to the basic structure of $\mathbf{X}_d/\sqrt{m}$ in the present notation as

$$\boxed{\mathbf{X}_d/\sqrt{m} = \mathbf{U}\boldsymbol{\Delta}\mathbf{T}'}$$

Since no new principles are involved, we do not go through the extensive computations to find $\mathbf{U}$, which is of order 12 x 2. What can be said, however, is that any data matrix, $\mathbf{X}$, $\mathbf{X}_d$, $\mathbf{X}_s$, $\mathbf{X}/\sqrt{m}$, $\mathbf{X}_s/\sqrt{m}$ can be expressed in terms of basic structure in just the way described above.

5.7.6 Recapitulation

The concept of basic structure represents the most general of decompositions that are considered in this chapter. We have only provided introductory material on the topic, but, having done so, it seems useful to recapitulate the main results and add a few more comments as well:

1. Any matrix **A** can be decomposed into the triple product:

$$\mathbf{A} = \mathbf{P}\boldsymbol{\Delta}\mathbf{Q}'$$

where **P** and **Q** are each orthonormal by columns, and Δ is diagonal with ordered positive elements.

2. The number of positive elements in Δ, the basic diagonal, is equal to the rank of **A**. Moreover, the basic diagonal is unique.

3. If all entries of Δ are distinct, then **P** and $\mathbf{Q}'$ are also unique (up to a possible reflection).

4. If some entries of Δ are tied, then those portions of **P** and $\mathbf{Q}'$ corresponding to tied blocks of entries are not unique. The portions of **P** and **Q** corresponding to the subset of distinct entries of Δ are unique (up to a reflection), however.

5. Mutually orthogonal vectors for tied blocks of entries in Δ can also be found by the Gram–Schmidt process, after first finding a set of r linearly independent vectors in the tied block. (These are unique up to orthogonal rotation within the r-dimensional subspace corresponding to the r tied eigenvalues.)

6. A basic matrix is one whose rank equals its smaller dimension.

7. A square basic matrix is one whose rank equals its order. More commonly, a square basic matrix is called nonsingular.

8. If **A** is nonsingular, then $\mathbf{P}'\mathbf{P} = \mathbf{P}\mathbf{P}' = \mathbf{I}$; $\mathbf{Q}'\mathbf{Q} = \mathbf{Q}\mathbf{Q}' = \mathbf{I}$, and we have the case of rotation–stretch–rotation or rotation–reflection–stretch–rotation.

9. If **A** is rectangular or square but singular, the concept of full basic structure, in which $\Delta_{k\times k}$ is embedded in a larger $m \times n$ matrix (see Fig. 5.7), still involves the sequence of transformations shown immediately above. Some dimensions of **P** and/or **Q** are annihilated, however.

10. The orthogonal diagonalization of a symmetric matrix

$$\mathbf{A} = \mathbf{T}\mathbf{D}\mathbf{T}'$$

was shown to be a special case of basic structure.

11. The concepts of matrix nonsingularity and decomposition uniqueness should be kept separate. A square matrix $\mathbf{A}_{n\times n}$ can be nonsingular but still nonunique in terms of its basic structure if it contains tied (positive) entries in $\Delta_{n\times n}$.

12. A square matrix $\mathbf{A}_{n\times n}$ can be singular but still unique in terms of its basic structure if it contains all distinct entries in $\Delta_{n\times n}$, thus implying that only one entry is zero.

13. A square matrix $\mathbf{A}_{n\times n}$ can, of course, be both nonsingular and unique in terms of basic structure if all entries in $\Delta_{n\times n}$ are positive and distinct.

As we know at this point, if a matrix is of rank k, then the basic structure procedure will reproduce it in terms of the triple product $\mathbf{P}\boldsymbol{\Delta}\mathbf{Q}'$, where the basic diagonal of Δ is $k \times k$.

What has not been covered is the case in which we would like to approximate **A** with a triple product whose diagonal is of order *less* than $k \times k$. This type of problem crops up in principal components analysis, among other things, when we wish to reduce the original space to fewer dimensions with the least loss of information.

Fortunately, basic structure decomposition provides a reduced-rank approximation to **A** whose sum of squared deviations between **A** and $\mathbf{P\Delta Q}'$ is minimal for the order of the diagonal being retained. While it would take us too far afield to explore the topic of matrix approximation via basic structure, this turns out to be another valuable aspect of the technique. Not surprisingly, the fact that the entries of Δ are ordered from large to small figures prominently in this type of approximation.

5.8 QUADRATIC FORMS

In multivariate analysis one often encounters situations in which the mapping of some vector entails a quadratic, rather than linear, function.[16] At first blush it may seem surprising that matrix algebra is relevant for this situation. After all, thus far we have emphasized the applicability of matrices to linear transformations. It is now time to expand the topic and consider quadratic functions and, in particular, quadratic forms.

5.8.1 Linear Forms

We have already encountered linear forms in our discussion of simultaneous equations in Chapter 4. If we have a set of variables x_i and a set of coefficients a_i, a linear form can be written in scalar notation as

$$g(x) = a_1x_1 + a_2x_2 + \cdots + a_nx_n = \sum_{i=1}^{n} a_ix_i$$

in which all x_i, as noted, are of the first degree. In vector notation we have

$$g(\mathbf{x}) = \mathbf{a}'\mathbf{x} = (a_1, a_2, \ldots, a_n)\begin{bmatrix} x_1 \\ x_2 \\ \vdots \\ x_n \end{bmatrix}$$

which, of course, equals some scalar, once we assign numerical values to **a** and **x**.

[16] Clearly, the idea of the *variance* of some variable entails a quadratic function, and variances represent a central concept in statistical analysis.

Next, suppose we consider a set of several linear forms, with the matrix of coefficients given by **A** and the vector of constants given by **c**. Then we have

$$\mathbf{Ax} = \mathbf{c} = \overset{\mathbf{A}}{\begin{bmatrix} a_{11} & a_{12} & \cdots & a_{1n} \\ a_{21} & a_{22} & \cdots & a_{2n} \\ \vdots & \vdots & & \vdots \\ a_{n1} & a_{n2} & \cdots & a_{nn} \end{bmatrix}} \overset{\mathbf{x}}{\begin{bmatrix} x_1 \\ x_2 \\ \vdots \\ x_n \end{bmatrix}} = \overset{\mathbf{c}}{\begin{bmatrix} c_1 \\ c_2 \\ \vdots \\ c_n \end{bmatrix}}$$

This, of course, represents a set of simultaneous linear equations. Hence, a linear form is simply a linear function in a set of variables x_i

5.8.2 Bilinear Forms

Bilinear forms involve only a slight extension of the above. Here we have two sets of variables x_i and y_j, each of the first degree, as illustrated specifically by

$$f(x, y) = x_1y_1 + 6x_2y_1 - 4x_3y_1 + 2x_1y_2 + 3x_2y_2 + 2x_3y_2$$

in which exactly one x_i and one y_j (each of the first degree) appears in each term. More generally, expressions of this type can be written in scalar notation as

$$f(x, y) = \sum_{i=1}^{m} \sum_{j=1}^{n} a_{ij} x_i y_j$$

and are called bilinear forms in x_i and y_j. If we write the vectors $\mathbf{x}' = (x_1, x_2, \ldots, x_m)$ and $\mathbf{y}' = (y_1, y_2, \ldots, y_n)$, a bilinear form involves terms in which every possible combination of vector components is formed. In matrix notation we can write a bilinear form as

$$f(\mathbf{x}, \mathbf{y}) = \mathbf{x}'\mathbf{Ay}$$

In the numerical example above, we have

$$a_{11} = 1; \quad a_{12} = 2; \quad a_{21} = 6; \quad a_{22} = 3; \quad a_{31} = -4; \quad a_{32} = 2$$

and the function can be expressed as

$$f(\mathbf{x}, \mathbf{y}) = \overset{\mathbf{x}'}{(x_1, x_2, x_3)} \overset{\mathbf{A}}{\begin{bmatrix} 1 & 2 \\ 6 & 3 \\ -4 & 2 \end{bmatrix}} \overset{\mathbf{y}}{\begin{bmatrix} y_1 \\ \\ y_2 \end{bmatrix}}$$

$$= (x_1 + 6x_2 - 4x_3, 2x_1 + 3x_2 + 2x_3) \begin{bmatrix} y_1 \\ y_2 \end{bmatrix}$$

$$= x_1y_1 + 6x_2y_1 - 4x_3y_1 + 2x_1y_2 + 3x_2y_2 + 2x_3y_2$$

The matrix **A** is called the matrix of the bilinear form, and it determines the form completely. Note that, in general, **A** need not be square.

By assigning different values to **x** and **y** one obtains different values of the bilinear form, each of which is a scalar. The set of all such scalars, for a given domain of **x** and **y**, is the range of the bilinear form.

5.8.3 Quadratic Forms

Next, let us specialize the bilinear form to the specific case in which $\mathbf{x} = \mathbf{y}$. In this case we assume that **y** can be replaced by **x** and, given their same dimensionality, the matrix of coefficients **A** will be square rather than rectangular. For example,

$$f(x_1, x_2) = 2x_1^2 + 5x_1x_2 + 3x_1x_2 + 6x_2^2$$

can now be written in matrix form as

$$f(\mathbf{x}) = (x_1, x_2)\begin{bmatrix} 2 & 3 \\ 5 & 6 \end{bmatrix}\begin{bmatrix} x_1 \\ x_2 \end{bmatrix} = (2x_1 + 5x_2, 3x_1 + 6x_2)\begin{bmatrix} x_1 \\ x_2 \end{bmatrix}$$

$$= 2x_1^2 + 5x_1x_2 + 3x_1x_2 + 6x_2^2$$

and the result, again, is a scalar, once numerical values are assigned to x_1 and x_2. Also, by assigning different values to **x** over its domain, we can obtain the range of $f(\mathbf{x})$, the quadratic form.

By way of formal definition, *a quadratic form is a polynomial function of* x_1, $x_2, \ldots, x_n$ that is homogeneous and of second degree. For example, in the case of two variables, we have

$$f(x_1, x_2) = x_1^2 + 6x_1x_2 + 9x_2^2$$

However, we can also write this as

$$f(\mathbf{x}) = x_1^2 + 6x_1x_2 + 9x_2^2$$

in which the vector $\mathbf{x}' = (x_1, x_2)$ is mapped from a two-dimensional space into a one-dimensional space. Similarly, an example of a quadratic form in three variables is

$$f(\mathbf{x}) = x_1^2 + x_2^2 + x_3^2$$

where the vector $\mathbf{x}' = (x_1, x_2, x_3)$ is mapped from three dimensions to one dimension.

In general, a quadratic form in n dimensions can be written in scalar notation as

$$q(\mathbf{u}) = \sum_{i,j}^{n} a_{ij}u_iu_j$$

where $\mathbf{u}' = (u_1, u_2, \ldots, u_n)$, the a_{ij} are real-valued coefficients and the $u_i\,u_j$ are the preimages of the mapping. If $i = j$, we obtain the squared term $a_{ii}u_i^2$, and if $i \neq j$, we obtain the cross-product term $a_{ij}u_iu_j$.

By "homogeneous" we mean that all terms are of the above form and, in particular, there are no linear terms in the u_i's nor is there a constant term. While the function

$$v = x_1^2 + 2x_2^2 + x_1x_2 + x_1 + 3x_2$$

is a second-degree polynomial, it is not a quadratic form since the last two terms are not of the general form $a_{ij}u_iu_j$.

Quadratic forms are of particular interest to multivariate data analysis inasmuch as we are often concerned with what happens to variances and covariances under various linear functions of a set of multivariate data.

While we did not bring up the topic of quadratic forms at that time, our diagonalization of the sample problem covariance matrix in Sections 5.3 and 5.4 involved a quadratic form, with matrix $\mathbf{C}(X)$. Indeed, *all of the cross-product matrices employed in multivariate analysis,* such as the raw cross product, SSCP, covariance, and correlation matrices, are illustrations of quadratic forms. In these cases the diagonal entries are some measure of single-variable dispersion, and the off-diagonal entries are some measure of covariation between a pair of variables.

In working with quadratic forms, our motivation is similar to diagonalizing transformation matrices. That is, we shall wish to find a linear function of the original data that has the effect of leading to a cross-products matrix in which two things are desired: (a) an arrangement of the linear composites so that the main diagonal entries in the cross-product matrix decrease in size and (b) off-diagonal entries of the cross-products matrix being zero, indicating uncorrelatedness of all pairs of composites. This, of course, is the same motivation underlying principal components analysis, as illustrated in Section 5.4.

5.8.4 An Illustrative Problem

Suppose we have the quadratic form $q(\mathbf{x}) = 66x_1^2 + 24x_1x_2 + 59x_2^2$. This can be expressed in matrix product form as

$$\mathbf{q}'\mathbf{A}\mathbf{q} = (x_1, x_2)\begin{bmatrix} 66 & 12 \\ 12 & 59 \end{bmatrix}\begin{bmatrix} x_1 \\ x_2 \end{bmatrix}$$

$$= (66x_1 + 12x_2, 12x_1 + 59x_2)\begin{bmatrix} x_1 \\ x_2 \end{bmatrix}$$

$$= 66x_1^2 + 12x_1x_2 + 12x_1x_2 + 59x_2^2$$

$$\mathbf{q}'\mathbf{A}\mathbf{q} = 66x_1^2 + 24x_1x_2 + 59x_2^2$$

Notice that we set the off-diagonal entries of **A** to half the coefficient of x_1x_2 which is $24/2 = 12$.

Notice further that a quadratic form involves a transformation into one dimension of an n-component vector in which the transformation is characterized by an $n \times n$ symmetric matrix.[17] On the other hand, a linear mapping of a vector in n dimensions into one dimension entails a single vector (either a $1 \times n$ or an $n \times 1$ matrix), whose entries are usually expressed as direction cosines.

[17] The matrix of a quadratic form does not have to be represented by a symmetric matrix. However, the original matrix can always be symmetrized by setting each off-diagonal entry equal to half the sum of the original off-diagonal entries; that is, $(a_{ij} + a_{ji})/2$.

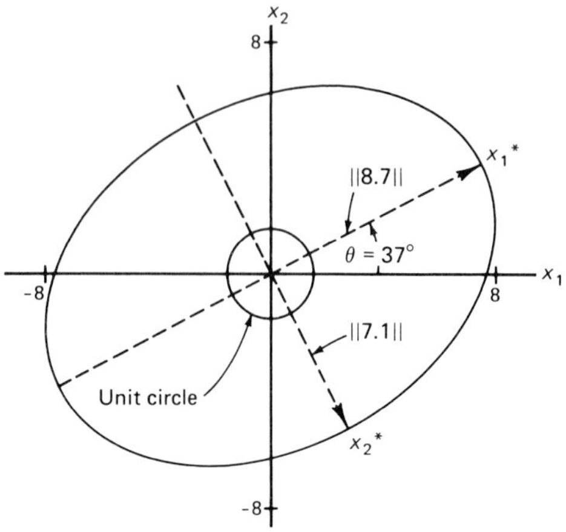

Fig. 5.9 Change of basis vectors of matrix representing quadratic form.

Now let us see what happens when we take various values of x_1 and x_2 and substitute them in $\mathbf{q}'\mathbf{A}\mathbf{q}$. The way this can be done graphically, as shown in Fig. 5.9, is to take various pairs of x_1, x_2 values on the unit circle in which we have the condition

$$(x_1, x_2)\begin{bmatrix} x_1 \\ x_2 \end{bmatrix} = 1$$

For example let

$x_1 = 1;$	$x_2 = 0$	$\mathbf{q}'\mathbf{A}\mathbf{q} = 66$	$(\mathbf{q}'\mathbf{A}\mathbf{q})^{1/2} = 8.1$
$x_1 = 0;$	$x_2 = 1$	$= 59$	$= 7.7$
$x_1 = -0.707;$	$x_2 = -0.707$	$= 74.5$	$= 8.7$
$x_1 = 0.707;$	$x_2 = 0.707$	$= 74.5$	$= 8.7$
$x_1 = -0.707;$	$x_2 = 0.707$	$= 50.5$	$= 7.1$
$x_1 = 0.707;$	$x_2 = -0.707$	$= 50.5$	$= 7.1$

We can select still other vectors of points on the unit circle and multiply the length or distance from the origin of each by $(\mathbf{q}'\mathbf{A}\mathbf{q})^{1/2}$. This "stretching" of $\mathbf{q}$ on the unit circle into the point $[(\mathbf{q}'\mathbf{A}\mathbf{q})^{1/2}]\mathbf{q}$ results in the ellipse shown in Fig. 5.9. This ellipse may be viewed as a geometric representation of the quadratic form.

Now, however, suppose we consider another quadratic form:

$$\mathbf{u}'\mathbf{B}\mathbf{u} = 75x_1^{*2} + 50x_2^{*2}$$

that can, in turn, be represented as

$$\mathbf{u}'\mathbf{B}\mathbf{u} = (x_1^*, x_2^*)\begin{bmatrix} 75 & 0 \\ 0 & 50 \end{bmatrix}\begin{bmatrix} x_1^* \\ x_2^* \end{bmatrix}$$

We see from Fig. 5.9 that if we rotate the coordinate system 37° to the axes x_1*and x_2*, then *this function also lies on the previously obtained ellipse.* Vector lengths of the major and minor semiaxes of the ellipse are $\sqrt{75} = 8.7$ and $\sqrt{50} = 7.1$, respectively. That is, by a change in orientation of the axes, we obtain a *new* representation of the quadratic form in which the $x_1 x_2$ cross product vanishes. Moreover, one axis of this form coincides with the major axis of the ellipse, while the other corresponds to the minor axis of the ellipse. These axes are usually referred to as principal axes. By eliminating the cross-product term the second matrix is seen to be a simpler representation of the quadratic form than the first. Moreover, the entries of the diagonal matrix **B** are in decreasing order.

5.8.5 Finding the New Basis Vectors

As the reader has probably surmised already, the new representation of the quadratic form $\mathbf{u}'\mathbf{B}\mathbf{u}$ is obtained by solving for the eigenstructure of **A**. Primarily in the nature of review we set up the characteristic equation

$$|\mathbf{A}-\lambda\mathbf{I}| = \begin{vmatrix} 66-\lambda & 12 \\ 12 & 59-\lambda \end{vmatrix} = 0$$

and solve for its eigenvalues by finding the second-order determinant and setting it equal to zero:

$$\lambda^2 - 125\lambda + 3750 = 0$$

$$(\lambda-75)(\lambda-50) = 0$$

$$\lambda_1 = 75; \qquad \lambda_2 = 50$$

After substitution of λ_1 and λ_2, we find the normalized eigenvectors $\left[\begin{smallmatrix}-0.8\\-0.6\end{smallmatrix}\right]$ and $\left[\begin{smallmatrix}0.6\\-0.8\end{smallmatrix}\right]$, which can be arranged in the matrix **Q**.

$$\mathbf{Q} = \begin{bmatrix} -0.8 & 0.6 \\ -0.6 & -0.8 \end{bmatrix}$$

Notice that $|\mathbf{Q}| = 1$ and $\mathbf{Q}\mathbf{Q}' = \mathbf{Q}'\mathbf{Q} = \mathbf{I}$. That is, **Q** is orthogonal and represents a proper rotation.

We then have the relationship

$$\mathbf{Q}'\mathbf{A}\mathbf{Q} = \mathbf{D}$$

$$\begin{bmatrix} -0.8 & -0.6 \\ 0.6 & -0.8 \end{bmatrix} \begin{bmatrix} 66 & 12 \\ 12 & 59 \end{bmatrix} \begin{bmatrix} -0.8 & 0.6 \\ -0.6 & -0.8 \end{bmatrix} = \begin{bmatrix} 75 & 0 \\ 0 & 50 \end{bmatrix}$$

Finally, we see that the last matrix on the right is equal to **B**, the diagonal matrix of the new quadratic form that reorients the axes. Furthermore, if we reflect the first column of **Q**, we note that cos 37° = 0.8, indicating that $\mathbf{x}_1$* makes an angle of 37° with the horizontal axis, while $\mathbf{x}_2$*makes an angle of −53° with the horizontal axis.

In brief, no new principles are involved in the present diagonalization process. As noted, **A** is symmetric to begin with, so all of our previous discussion about diagonalizing symmetric matrices is relevant here. We note that the present formula

$$\mathbf{D} = \mathbf{Q}'\mathbf{A}\mathbf{Q}$$

is the same as that found in Section 5.2:

$$\mathbf{D} = \mathbf{T}'\mathbf{A}\mathbf{T}$$

where **D** is diagonal. The matrix **Q**, an orthogonal matrix, is the same as **T** in the context of Section 5.2.

5.8.6 Types of Quadratic Forms

Quadratic forms can be classified according to the nature of the eigenvalues of the matrix of the quadratic form:

1. If all λ_i are positive, the form is said to be *positive definite.*
2. If all λ_i are negative, the form is said to be *negative definite.*
3. If all λ_i are nonnegative (positive or zero), the form is said to be *positive semidefinite.*
4. If all λ_i are nonpositive (zero or negative), the form is said to be *negative semidefinite*.
5. If the λ_i represent a mixture of positive, zero, and negative values, the form is said to be *indefinite*

In multivariate analysis we are generally interested in forms that are either positive definite or positive semidefinite. For example, if a symmetric matrix is of product-moment form (either $\mathbf{A}'\mathbf{A}$ or $\mathbf{A}\mathbf{A}'$), then it is *either positive definite or positive semidefinite.* Since various types of cross-products matrices are of this form, the cases of positive definite or positive semidefinite are of most interest to us in multivariate analysis.

5.8.7 Relating Quadratic Forms to Matrix Transformations

As might be surmised from our earlier discussion of matrix eigenstructure and basic structure, quadratic forms are intimately connected with much of the preceding material. For example, suppose we have the point transformation

$$\mathbf{u} = \mathbf{X}\mathbf{v}$$

where **X**, whose rows are sets of direction cosines, maps **v**, considered as a column vector, onto **u** in some space of interest.

To illustrate, we let

$$\mathbf{X} = \begin{bmatrix} 0.8 & 0.6 \\ 0.71 & 0.71 \end{bmatrix}$$

Hence, if $\mathbf{v} = \begin{bmatrix}1\\0\end{bmatrix}$, we have

$$\mathbf{u} = \mathbf{X}\mathbf{v} = \begin{bmatrix} 0.8 & 0.6 \\ 0.71 & 0.71 \end{bmatrix} \begin{bmatrix} 1 \\ 0 \end{bmatrix} = \begin{bmatrix} 0.8 \\ 0.71 \end{bmatrix}$$

Now suppose we want to find the squared length of $\mathbf{u}$.

The squared length of $\mathbf{u}$ is defined to be $\mathbf{u}'\mathbf{u}$. Given $\mathbf{v}$ and the linear transformation $\mathbf{X}$, we set up the expression

$$\mathbf{u}'\mathbf{u} = (\mathbf{X}\mathbf{v})'(\mathbf{X}\mathbf{v}) = \mathbf{v}'\mathbf{X}'\mathbf{X}\mathbf{v}$$

But now we see that $\mathbf{X}'\mathbf{X}$ is just the minor product moment of $\mathbf{X}$ which we have already discussed. We can denote this as $\mathbf{A}$. Thus, we have

$$\mathbf{u}'\mathbf{u} = \mathbf{v}'\mathbf{A}\mathbf{v} = (1, 0)\overset{\mathbf{A}}{\begin{bmatrix} 1.14 & 0.98 \\ 0.98 & 0.86 \end{bmatrix}} \begin{bmatrix} 1 \\ 0 \end{bmatrix}$$

$$\mathbf{u}'\mathbf{u} = 1.14$$

Hence, product-moment matrices, which were discussed earlier in the context of eigenstructure and basic structure, also appear in the present context as matrices defining quadratic forms. That is, $\mathbf{A} = \mathbf{X}'\mathbf{X} = \mathbf{S}(X)$ is *the matrix of the quadratic form that finds the squared length of* $\mathbf{v}$ *under the linear transformation* $\mathbf{X}$.

Up to this point we have said relatively little about the process of finding eigenstructures of *nonsymmetric* matrices. We did indicate, however, that for the matrices of interest to us in multivariate analysis their eigenstructures will involve real-valued eigenvalues and eigenvectors. Be that as it may, it is now time to discuss their eigenstructure computation.

5.9 EIGENSTRUCTURES OF NONSYMMETRIC MATRICES IN MULTIVARIATE ANALYSIS

In multivariate analysis it is not infrequently the case that we encounter various types of nonsymmetric matrices for which we desire to find an eigenstructure. Canonical correlation, multiple discriminant analysis, and multivariate analysis of variance are illustrative of techniques where this may occur.

As a case in point, let us examine the third sample problem presented in Section 1.6.4. As recalled, the twelve employees were divided into three groups on the basis of level of absenteeism. While an underlying variable, degree of absenteeism, is present in this example, let us assume that the three groups represent only an unordered polytomy.

The two predictor variables were X_1 (attitude toward the firm) and X_2 (number of years employed by the firm). Figure 5.10 reproduces the scatter plot of the mean-corrected values of X_{d2} versus X_{d1}, as first shown in Fig. 1.5. The three groups have been appropriately coded by dots, circles, and small x's. We note from the figure that the individuals in the three groups show some tendency to cluster.

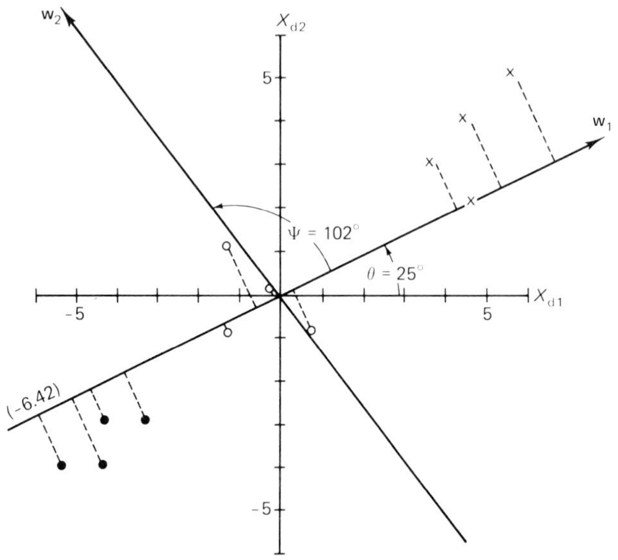

Fig. 5.10 Mean-corrected predictor variables (from Fig. 1.5). Key: • Group 1; ○ Group 2; x Group 3.

However, we wonder if a linear composite of X_{d1} and X_{d2} could be found that would have the property of maximally separating the three groups in the sense of maximizing their among-group variation relative to their within-group variation on this composite. Somewhat more formally, we seek a linear composite with values

$$w_{i(1)} = v_1 X_{di1} + v_2 X_{di2}$$

with the intent of maximizing the ratio

$$\boxed{\lambda_1 = \frac{SS_A(\mathbf{w}_1)}{SS_W(\mathbf{w}_1)}}$$

where SS_A and SS_W denote the among-group and within-group sums of squares of the linear composite $\mathbf{w}_1$.

We can rewrite the preceding expression in terms of quadratic forms by means of

$$\boxed{\lambda_1 = \frac{\mathbf{v}_1'\mathbf{A}\mathbf{v}_1}{\mathbf{v}_1'\mathbf{W}\mathbf{v}_1}}$$

where **A** and **W** denote among-group and (pooled) within-group SSCP matrices, respectively. Thus, we wish to find a new axis in Fig. 5.10, that can be denoted $\mathbf{w}_1$, with the property of maximizing the among- to within-group variation of the twelve points, when they are projected onto it.

The reader will note the similarity of this problem to the motivation underlying principal components analysis. Again we wish to maximize a quantity λ_1, with respect to $\mathbf{v}_1$. However, λ_1 is now considered as a *ratio of two different quadratic forms.* As such, this problem differs from principal components analysis in several significant ways.

As shown in Appendix A, the following matrix equation

$$\boxed{(\mathbf{A}-\lambda_1\mathbf{W})\mathbf{v}=\mathbf{0}}$$

is involved in the present maximization task. However, if $\mathbf{W}$ is nonsingular and hence $\mathbf{W}^{-1}$ exists, we can multiply both sides by $\mathbf{W}^{-1}$:

$$(\mathbf{W}^{-1}\mathbf{A}-\lambda_1\mathbf{I})\mathbf{v}=\mathbf{0}$$

with the resulting characteristic equation

$$\boxed{|\mathbf{W}^{-1}\mathbf{A}-\lambda_1\mathbf{I}|=0}$$

and the problem now is to solve for the eigenstructure of $\mathbf{W}^{-1}\mathbf{A}$.

So far, nothing new except for the important fact that $\mathbf{W}^{-1}\mathbf{A}$ is nonsymmetric, even though both $\mathbf{W}^{-1}$ and $\mathbf{A}$ are symmetric. Up to this point relatively little has been said about finding the eigenstructure of a nonsymmetric matrix. We can, however, proceed in two different, but related, ways.[18] First, we solve directly for the eigenstructure of the matrix involved in the current sample problem. This approach is a straightforward extension of earlier discussion involving the eigenstructure of symmetric matrices (as well as material covered in Section 5.3).

Second, we can show geometrically and algebraically an equivalent approach that involves the simultaneous diagonalization of two different quadratic forms. This presentation ties in some of the material here with previous comments on principal components analysis.

5.9.1 The Eigenstructure of $\mathbf{W}^{-1}\mathbf{A}$

Probably the most popular approach to solving for the eignestructure of $\mathbf{W}^{-1}\mathbf{A}$ is to find the eigenvalues and eigenvectors directly, in the same general way as discussed earlier for symmetric matrices. In this case, however, $\mathbf{W}^{-1}\mathbf{A}$ is nonsymmetric; hence $\mathbf{V}$, the matrix of eigenvectors, will *not* be orthogonal.

Table 5.3 shows the preliminary calculations needed for finding the (pooled) within-group SSCP matrix $\mathbf{W}$ and the among-group SSCP matrix $\mathbf{A}$.

Table 5.4 shows the various quantities needed to solve for the eigenstructure of $\mathbf{W}^{-1}\mathbf{A}$ in terms of the characteristic equation

$$|\mathbf{W}^{-1}\mathbf{A}-\lambda_i\mathbf{I}|=0$$

As noted in Table 5.4, we first compute the (pooled) within-group SSCP matrix $\mathbf{W}$ and the among-group SSCP matrix $\mathbf{A}$.

[18] As a matter of fact, still other ways are available to solve this problem. The interested reader can see McDonald (1968).

TABLE 5.3

Preliminary Calculations for Multiple Discriminant Analysis

Employee		X_{d1}	X_{d2}	$\mathbf{X}_k - \bar{\mathbf{X}}_k$ Within-group deviations		$\bar{\mathbf{X}}_k - \bar{\bar{\mathbf{X}}}$ Among-group deviations	
1	*a*	−5.25	−3.92	−1	−0.5	−4.25	−3.42
	b	−4.25	−3.92	0	−0.5	−4.25	−3.42
	c	−4.25	−2.92	0	0.5	−4.25	−3.42
	d	−3.25	−2.92	1	0.5	−4.25	−3.42
Mean		−4.25	−3.42				
2	*e*	−1.25	−0.92	−0.75	−0.75	−0.5	−0.17
	f	−1.25	1.08	−0.75	1.25	−0.5	−0.17
	g	−0.25	0.08	0.25	0.25	−0.5	−0.17
	h	0.75	−0.92	1.25	−0.75	−0.5	−0.17
Mean		−0.50	−0.17				
3	*i*	3.75	3.08	−1	−0.50	4.75	3.58
	j	4.75	2.08	0	−1.50	4.75	3.58
	k	4.75	4.08	0	0.50	4.75	3.58
	l	5.75	5.08	1	1.50	4.75	3.58
Mean		4.75	3.58				

Within-group SSCP matrix

$$\mathbf{W} = (\mathbf{X}_k - \bar{\mathbf{X}}_k)'(\mathbf{X}_k - \bar{\mathbf{X}}_k)$$

Among-group SSCP matrix

$$\mathbf{A} = (\bar{\mathbf{X}}_k - \bar{\bar{\mathbf{X}}})'(\bar{\mathbf{X}}_k - \bar{\bar{\mathbf{X}}})$$

TABLE 5.4

Finding the Eigenstructure of $\mathbf{W}^{-1}\mathbf{A}$

SSCP matrices of sample problem

Within-group SSCP matrix; Among-group SSCP matrix

$$\mathbf{W} = \begin{bmatrix} 6.75 & 1.75 \\ 1.75 & 8.75 \end{bmatrix} \qquad \mathbf{A} = \begin{bmatrix} 163.50 & 126.50 \\ 126.50 & 98.17 \end{bmatrix}$$

Solving for the eigenstructure of $\mathbf{W}^{-1}\mathbf{A}$

$$\mathbf{W}^{-1} = \begin{bmatrix} 0.156 & -0.031 \\ -0.031 & 0.121 \end{bmatrix}; \qquad \mathbf{W}^{-1}\mathbf{A} = \begin{bmatrix} 21.594 & 16.698 \\ 10.138 & 7.880 \end{bmatrix}$$

Eigenvalues of $\mathbf{W}^{-1}\mathbf{A}$; Eigenvectors of $\mathbf{W}^{-1}\mathbf{A}$

$$\Lambda = \begin{bmatrix} 29.444 & 0 \\ 0 & 0.0295 \end{bmatrix}; \qquad \mathbf{V} = \begin{bmatrix} 0.905 & -0.612 \\ 0.425 & 0.791 \end{bmatrix}$$

One then finds $\mathbf{W}^{-1}$ and the matrix product $\mathbf{W}^{-1}\mathbf{A}$. From here on, the same general procedure applies for finding the eigenvalues. These turn out to be

$$\lambda_1 = 29.444; \qquad \lambda_2 = 0.0295$$

which appear in Table 5.4 along with the matrix $\mathbf{V}$ whose columns are eigenvectors of $\mathbf{W}^{-1}\mathbf{A}$. And, as indicated earlier, $\mathbf{V}$ is, in general, not orthogonal.

Returning to Fig. 5.10, we note that the first column of $\mathbf{V}$ entails direction cosines related to a 25° angle with the horizontal axis. The resulting linear composite $\mathbf{w}_1$ has scores that maximize among- to within-group variation. The second discriminant axis $\mathbf{w}_2$ (with an associated eigenvalue of only 0.0295) produces very little separation and, in cases of practical interest, would no doubt be discarded.

Other parallels with the principal components analysis of Sections 5.3 and 5.4 are found here. For example, discriminant scores–analogous to component scores–are found by projecting the points onto the discriminant axes. The discriminant score of the first observation on $\mathbf{w}_1$ is

$$w_{1(1)} = 0.905(-5.25) + 0.425(-3.92) = -6.42$$

as shown in Fig. 5.10.

However, unlike principal components analysis, we can observe from the figure that $\mathbf{v}_1$ and $\mathbf{v}_2$ are *not* orthogonal, even though the scores on $\mathbf{w}_1$ versus $\mathbf{w}_2$ are uncorrelated. From the $\mathbf{V}$ matrix in Table 5.4 we can compute the cosine between $\mathbf{v}_1$ and $\mathbf{v}_2$ as follows:

$$\cos \Psi = (0.905 \quad 0.425)\begin{bmatrix} -0.612 \\ 0.791 \end{bmatrix} = -0.21$$

The angle Ψ separating $\mathbf{v}_1$ and $\mathbf{v}_2$ is 90° + 12° = 102°, as shown in Fig. 5.10.

In summary, finding the eigenstructure of the nonsymmetric matrix $\mathbf{W}^{-1}\mathbf{A}$ proceeds in an analogous fashion to the procedure followed in the case of symmetric matrices. Note, however, that the matrix of eigenvectors $\mathbf{V}$ is not orthogonal even though the discriminant scores on $\mathbf{w}_1$ and $\mathbf{w}_2$ are uncorrelated.

5.9.2 Diagonalizing Two Different Quadratic Forms

The preceding solution, while straightforward and efficient, does not provide much in the way of an intuitive guide to what goes on in the simultaneous diagonalization of two different quadratic forms:

$$\mathbf{v}_1'\mathbf{W}\mathbf{v}_1; \qquad \mathbf{v}_1'\mathbf{A}\mathbf{v}_1$$

However, we can sketch out briefly a complementary geometric and algebraic approach that relates this diagonalization problem to the earlier discussion of principal components analysis.

As recalled from Chapter 2, variances and covariances can be represented as vector lengths and angles in person space. Moreover, as shown in Fig. 5.9, quadratic forms can be pictured geometrically as ellipses in two variables, or ellipsoids in three variables, or hyperellipsoids in more than three variables.[19] The thinner the ellipse, the greater the correlation between the two variables. The tilt of the ellipse and the relative lengths of its

[19] Of course, if more than three dimensions are involved, a literal "picture" is not possible.

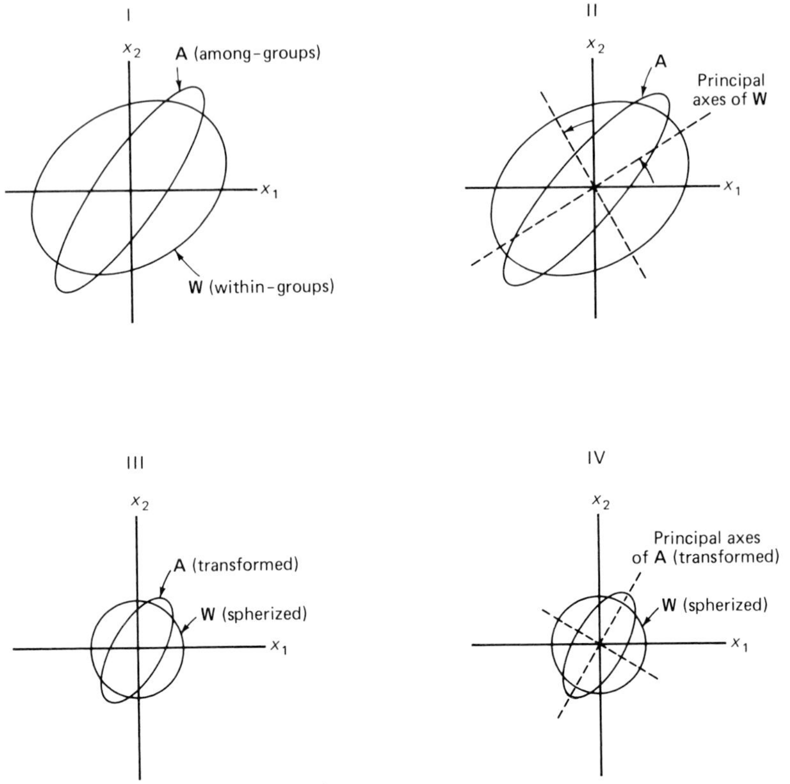

Fig. 5.11 Simultaneous diagonalization of two different quadratic forms.

axes are functions of the covariances and variances of the two variables. As we know, larger variances lead to longer (squared) lengths and also tilt the ellipse in the direction of the variable with the larger variance.

For what follows we shall use the matrix $\mathbf{Q}_1$ to refer to the matrix of eigenvectors of $\mathbf{W}^{-1/2}$ and the matrix $\mathbf{Q}_2$ to refer to the matrix of eigenvectors of the transformed matrix $\mathbf{W}^{-1/2}\mathbf{A}\mathbf{W}^{-1/2}$ (as will be explained).

As motivation for this discussion, suppose we wished to find a single change of basis vectors in Panel I of Fig. 5.11 that diagonalizes both quadratic forms.[20] One quadratic form could involve a pooled within-group SSCP matrix **W**. Similarly, the second form could be represented by an among-group SSCP matrix **A**. We assume that the quadratic form denoting the within-group variation is positive definite. (One of the two forms *must* be positive definite for what follows.)

Geometrically, what is involved is first to rotate the within-group ellipse in Panel I to principal axes, as shown in Panel II. We then change scale, deforming the reoriented ellipse **W** to a circle in Panel III. This, in general, is called "spherizing." After this is done, *any* direction can serve as a principal axis. Hence, we can rotate the new axes to line them

[20] By diagonalization we mean a transformation in which off-diagonal elements vanish in the matrix of the quadratic form.

up with the principal axes of the *second* ellipse **A**, representing the among-group SSCP matrix in Panel IV. And that, basically, is what simultaneous diagonalization is all about.

Let us see what these geometric operations mean algebraically. First, any observation $X_{\text{d}ij}$ in Table 5.3 can be represented as the sum of

$$X_{\text{d}ij} = (\mathbf{X}_k - \bar{\mathbf{X}}_k) + (\bar{\mathbf{X}}_k - \bar{\bar{\mathbf{X}}}_k)$$

$$\mathbf{X}_\text{d} = \mathbf{J} + \mathbf{G}$$

where **J** denotes the matrix of within-group deviations and **G** the matrix of among-group deviations. For example, the first observation on variable X_1 in Table 5.3 is

$$-5.25 = -1 + (-4.25)$$

where -1 indicates that it is one unit less than its group mean, and -4.25 indicates that its group mean is 4.25 units less than the grand mean. If we can find a transformation of $\mathbf{X}_\text{d}$ that spherizes the **J** portion (the within-group variation), we could then find the eigenstructure of the *adjusted* cross-products matrix.

The **J** portion can be readily spherized by the transformation:

$$\boxed{\mathbf{X}_\text{d}\mathbf{W}^{-1/2}}$$

where $\mathbf{W}^{-1/2}$, in turn, can be written as $\mathbf{Q}_1\boldsymbol{\Lambda}^{-1/2}\mathbf{Q}_1'$. In this case $\boldsymbol{\Lambda}^{-1/2}$ is a diagonal matrix of the reciprocals of the square roots of the eigenvalues of **W**, and $\mathbf{Q}_1$ is an orthogonal matrix of associated eigenvectors (since **W** is symmetric).[21]

Note, then, that what is being done here is to find the "square root" of $\mathbf{W}^{-1}$, the inverse of the (pooled) within-group SSCP matrix. To do so we recall that if **W** is symmetric and possesses an inverse $\mathbf{W}^{-1}$, we can write

$$\mathbf{W}^{-1/2}\mathbf{W}^{-1/2} = \mathbf{W}^{-1}$$

where

$$\mathbf{W}^{-1/2} = \mathbf{Q}_1\boldsymbol{\Lambda}^{-1/2}\mathbf{Q}_1'$$

Geometrically, the multiplication of $\mathbf{X}_\text{d}$ by $\mathbf{W}^{-1/2}$ has the effect of normalizing the within-group portion of the vectors in $\mathbf{X}_\text{d}$ to unit length, *after* rotation to the principal axes of **W** by means of the direction cosines represented by $\mathbf{Q}_1$. Subsequent rotation by $\mathbf{Q}_1'$ has no effect on what happens next, since the spherizing has already occurred.

Next we set up the equation

$$\boxed{[\mathbf{W}^{-1/2}\mathbf{A}\mathbf{W}^{-1/2}]\mathbf{Q}_2 = \mathbf{Q}_2\boldsymbol{\Lambda}}$$

where $\mathbf{Q}_2$ is the matrix of eigenvectors, and $\boldsymbol{\Lambda}$ the matrix of eigenvalues of $[\mathbf{W}^{-1/2}\mathbf{A}\mathbf{W}^{-1/2}]$. This, in turn, follows from

$$(\mathbf{X}_\text{d}\mathbf{W}^{-1/2})'(\mathbf{X}_\text{d}\mathbf{W}^{-1/2}) = \mathbf{W}^{-1/2}\mathbf{X}_\text{d}'\mathbf{X}_\text{d}\mathbf{W}^{-1/2} = \mathbf{W}^{-1/2}\mathbf{J}'\mathbf{J}\mathbf{W}^{-1/2} + \mathbf{W}^{-1/2}\mathbf{G}'\mathbf{G}\mathbf{W}^{-1/2}$$

$$= \mathbf{I} + \mathbf{W}^{-1/2}\mathbf{G}'\mathbf{G}\mathbf{W}^{-1/2} = \mathbf{I} + \mathbf{W}^{-1/2}\mathbf{A}\mathbf{W}^{-1/2}$$

[21] The square root of a symmetric matrix was discussed in Section 5.5.2.

where the within-group portion has been transformed to an identity matrix **I**, as desired.[22]

We then find the eigenstructure of $[\mathbf{W}^{-1/2}\mathbf{A}\mathbf{W}^{-1/2}]$ which, given the preliminary spherization, is tantamount to a rotation to principal axes orientation. A nice feature of this procedure is that $[\mathbf{W}^{-1/2}\mathbf{A}\mathbf{W}^{-1/2}]$ is *also symmetric.* The final transformation to be applied to the *original matrix* of mean-corrected scores $\mathbf{X}_\mathrm{d}$ involves

$$\boxed{\mathbf{Y} = \mathbf{X}_\mathrm{d}\mathbf{W}^{-1/2}\mathbf{Q}_2}$$

which effects the desired simultaneous diagonalization of **W** and **A**. Note, however, that $\mathbf{W}^{-1/2}\mathbf{Q}_2$ is *not* a rotation since the data are rescaled so that the **J** portion is spherized. In summary, then, a principal components analysis of data that are first spherized in terms of pooled within-group variation provides a counterpart approach to the direct attack on finding the eigenstructure of $\mathbf{W}^{-1}\mathbf{A}$.

5.9.3 Recapitulation

While two methods have been discussed for solving

$$(\mathbf{A} - \lambda\mathbf{W})\,\mathbf{v} = \mathbf{0}$$

the first method, utilizing a direct approach to computing the eigenstructure of a nonsymmetric matrix, is probably the better known. The second procedure appears useful in its own right, however, as well as serving as an alternative method to the more usual decomposition.

If we return to the ratio of quadratic forms, stated earlier:

$$\lambda_1 = \frac{\mathbf{v}_1'\mathbf{A}\mathbf{v}_1}{\mathbf{v}_1'\mathbf{W}\mathbf{v}_1}$$

the problem of multiple discriminant analysis can be stated as one of finding extreme values of the above function where **V**, the matrix of discriminant weights, exhibits the properties:

$$\boxed{\begin{aligned} \mathbf{V}'\mathbf{A}\mathbf{V} &= \mathbf{\Lambda} \\ \mathbf{V}'\mathbf{W}\mathbf{V} &= \mathbf{I} \end{aligned}}$$

Thus, **V** diagonalizes **A** (since $\mathbf{\Lambda}$ is diagonal) and converts **W** to an identity matrix. Notice, then, that the correlation of group means on any of the linear composite(s) is zero, since $\mathbf{\Lambda}$ is diagonal. Similarly, the *average* within-group correlation of individuals on each discriminant function (linear composite) is also zero, since **I** is an identity.

However, it should be remembered, as shown in Fig. 5.10, that **V** is *not* orthogonal since $\mathbf{W}^{-1}\mathbf{A}$, the matrix to be diagonalized, is not symmetric. Moreover, a preliminary transformation such as that applied in the second method described above, *still* ends up with a **V** that is not orthogonal.

[22] Since $\mathbf{W}^{-1/2}$ is symmetric, $\mathbf{W}^{-1/2} = (\mathbf{W}^{-1/2})'$. Since $\mathbf{J}'\mathbf{J} = \mathbf{W}$, $\mathbf{W}^{-1/2}\,\mathbf{J}'\mathbf{J}\mathbf{W}^{-1/2} = \mathbf{I}$.

5.10 SUMMARY

This chapter has primarily been concerned with various types of matrix decompositions–eigenstructures, basic structures, and quadratic forms. The common motivation has been to search for special kinds of basis vector transformations that can be expressed in simple ways, for example, as the product of a set of matrices that individually permit straightforward geometric interpretations. In addition, such decomposition provides new perspectives on the concepts of matrix singularity and rank.

The topic was introduced by first reviewing the nature of point and basis vector changes. This introduction led to a discussion of the role of eigenstructures in rendering a given matrix (not necessarily symmetric) diagonal via nonsingular transformations. The geometric aspects of eigenstructures were stressed at this point.

We next discussed eigenstructures from a complementary view, one involving the development of linear composites with the property of maximizing the variance of point projections, subject to all composites being orthogonal with previously found composites. This time we discussed the eigenstructure of symmetric matrices with real-valued entries. In such cases all eigenvalues and eigenvectors of the matrix are real.

Since eigenstructures are not defined for rectangular matrices and their computation can present problems in the case of square, nonsymmetric matrices, we discussed these cases next in the context of basic structure. Matrix decomposition in this case involves finding a triple matrix product by which *any* matrix can be expressed as

$$\boxed{\mathbf{A} = \mathbf{P}\boldsymbol{\Delta}\mathbf{Q}'}$$

where $\mathbf{P}$ and $\mathbf{Q}$ are orthonormal by columns and $\boldsymbol{\Delta}$ is diagonal. This form of matrix decomposition represents a powerful organizing concept in matrix algebra, since it can be applied to any matrix of full, or less than full, rank. Furthermore, it shows that any matrix transformation can be considered as the product of a rotation–stretch–rotation or rotation–reflection–stretch–rotation under a suitable change in basis vectors.

By using product-moment matrices–$\mathbf{A}'\mathbf{A}$ or $\mathbf{A}\mathbf{A}'$, whichever has the smaller order–we were able to relate the basic structure of a matrix to earlier ideas involving symmetric matrices. One can solve for the eigenstructure of $\mathbf{A}'\mathbf{A}$ (or $\mathbf{A}\mathbf{A}'$) in order to find $\mathbf{Q}'$ and $\boldsymbol{\Delta}$ and then solve finally for $\mathbf{P}$. The net result is the determination of matrix rank as well as the specific geometric character of the transformation. And, if $\mathbf{A}$ is symmetric to begin with, the general procedure leads to the special case of $\mathbf{Q}'\mathbf{A}\mathbf{Q} = \mathbf{D} = \boldsymbol{\Delta}^2$.

Related ideas were presented in our discussion of quadratic forms, a function that maps n-dimensional vectors into one dimension. Again, the motivation is to find a new set of basis vectors, via rotation, in which the function assumes a particularly simple form, namely, one in which cross-product terms vanish.

The last main section of the chapter dealt with ways of finding the eigenstructure of nonsymmetric matrices as they may arise in the simultaneous diagonalization of two different quadratic forms. The geometric character of this type of transformation was described and illustrated graphically.

The material of this chapter represents a major part of the more basic mathematical aspects of multivariate procedures. Typically, in multivariate analysis we are trying to find linear combinations of the original variables that optimize some quantity of interest to the researcher. As Appendix A shows, function optimization subject to certain constraints, such as Lagrange multipliers, is used time and time again in many of the statistical techniques that appear in multivariate data analysis.

REVIEW QUESTIONS

1. Form the characteristic equations of the following matrices and determine their eigenvalues and eigenvectors:

a. $$\mathbf{A} = \begin{bmatrix} 5 & 0 \\ 7 & 8 \end{bmatrix}$$

b. $$\mathbf{A} = \begin{bmatrix} -8 & 8 \\ -1 & -2 \end{bmatrix}$$

c. $$\mathbf{A} = \begin{bmatrix} -3 & 5 \\ 4 & -2 \end{bmatrix}$$

d. $$\mathbf{A} = \begin{bmatrix} 2 & -1 \\ 6 & 7 \end{bmatrix}$$

e. $$\mathbf{A} = \begin{bmatrix} 2 & 2 & -4 \\ -2 & 4 & 2 \\ 0 & 2 & -2 \end{bmatrix}$$

f. $$\mathbf{A} = \begin{bmatrix} 0 & 3 & 0 \\ 0 & 0 & 3 \\ 3 & -9 & 9 \end{bmatrix}$$

2. Calculate the trace and determinant of each of the first four of the matrices above and verify that

$$\operatorname{tr}(\mathbf{A}) = \sum_{i=1}^{n} \lambda_i; \qquad |\mathbf{A}| = \prod_{i=1}^{n} \lambda_i$$

3. Find the invariant vectors (i.e., eigenvectors) under the following transformations:

a. A shear

$$\begin{bmatrix} x_1^* \\ x_2^* \end{bmatrix} = \begin{bmatrix} 1 & 0 \\ 3 & 1 \end{bmatrix} \begin{bmatrix} x_1 \\ x_2 \end{bmatrix}$$

b. A stretch

$$\begin{bmatrix} x_1^* \\ x_2^* \end{bmatrix} = \begin{bmatrix} 3 & 0 \\ 0 & 2 \end{bmatrix} \begin{bmatrix} x_1 \\ x_2 \end{bmatrix}$$

c. A central dilation

$$\begin{bmatrix} x_1^* \\ x_2^* \end{bmatrix} = \begin{bmatrix} 3 & 0 \\ 0 & 3 \end{bmatrix} \begin{bmatrix} x_1 \\ x_2 \end{bmatrix}$$

d. A rotation

$$\begin{bmatrix} x_1^* \\ x_2^* \end{bmatrix} = \begin{bmatrix} 0.707 & 0.707 \\ -0.707 & 0.707 \end{bmatrix} \begin{bmatrix} x_1 \\ x_2 \end{bmatrix}$$

4. Starting with the matrix

$$\mathbf{A} = \begin{bmatrix} 5 & 1 \\ 1 & 3 \end{bmatrix}$$

find the eigenstructures of the following matrices:

a. $3\mathbf{A}$ b. $\mathbf{A} + 2\mathbf{I}$ c. $\mathbf{A} - 3\mathbf{I}$

d. $\mathbf{A}^3$ e. $\mathbf{A}^{-1}$ f. $\mathbf{A}^{1/2}$ g. $\mathbf{A}^{-1/2}$

5. Given the set of linearly independent vectors

$$\mathbf{a}_1 = \begin{bmatrix} 1 \\ 1 \\ 1 \end{bmatrix}; \qquad \mathbf{a}_2 = \begin{bmatrix} 2 \\ -1 \\ 0 \end{bmatrix}; \qquad \mathbf{a}_3 = \begin{bmatrix} 1 \\ 0 \\ -1 \end{bmatrix}$$

and the matrix

$$\mathbf{B} = \begin{bmatrix} 2 & 2 & 1 \\ 1 & 3 & 1 \\ 1 & 2 & 2 \end{bmatrix}$$

show that **B** can be made diagonal via the 3 x 3 matrix **A** (made up from the linearly independent vectors) and its inverse.

6. Find an orthogonal matrix **U** such that $\mathbf{U}^{-1}\mathbf{AU}$ is diagonal, where

$$\mathbf{A} = \begin{bmatrix} 7 & -2 & 1 \\ -2 & 10 & -2 \\ 1 & -2 & 7 \end{bmatrix}$$

7. Using the minor product-moment procedure, and subsequent calculation of eigenstructure, what is the rank of the following matrices:

a. $$\mathbf{A} = \begin{bmatrix} 1 & 2 & 2 \\ 2 & 1 & 1 \\ 3 & 4 & 1 \end{bmatrix}$$

b. $$\mathbf{A} = \begin{bmatrix} 0 & 2 & 3 \\ 1 & 0 & 1 \\ 0 & 0 & 0 \\ 0 & 1 & -1 \end{bmatrix}$$

c. $$\mathbf{A} = \begin{bmatrix} 1 & 2 \\ 2 & 2 \\ 3 & 0 \\ 4 & 0 \\ 5 & 3 \end{bmatrix}$$

d. $$\mathbf{A} = \begin{bmatrix} 1 & 2 \\ 2 & 4 \\ 3 & 6 \\ 5 & 10 \end{bmatrix}$$

8. Find the basic structure and rank of each of the following matrices:

a. $$\mathbf{A} = \begin{bmatrix} 0 & 1 \\ 1 & 0 \end{bmatrix}$$

b. $$\mathbf{A} = \begin{bmatrix} 2 & 2 \\ -1 & 1 \\ 3 & 0 \end{bmatrix}$$

c. $$\mathbf{A} = \begin{bmatrix} 0.707 & 0.707 \\ -0.707 & 0.707 \end{bmatrix}$$

d. $$\mathbf{A} = \begin{bmatrix} 1 & 2 \\ 3 & 0 \\ 4 & 2 \\ 6 & -1 \\ 7 & 1 \end{bmatrix}$$

9. In the first two examples of Question 8, what is the basic structure of

a. $[\mathbf{A}'\mathbf{A}]^2$ b. $[\mathbf{A}'\mathbf{A}]^{1/2}$

10. Compute $\mathbf{A}^{1/2}$ and $\mathbf{A}^{-1/2}$, by finding eigenstructures for the matrices:

a. $\mathbf{A} = \begin{bmatrix} 4 & 3 \\ 3 & 4 \end{bmatrix}$ b. $\mathbf{A} = \begin{bmatrix} 7 & 1 \\ 1 & 7 \end{bmatrix}$ c. $\mathbf{A} = \begin{bmatrix} 8 & 2 \\ 2 & 5 \end{bmatrix}$ d. $\mathbf{A} = \begin{bmatrix} 2 & 0 \\ 0 & 2 \end{bmatrix}$

11. Represent each of the following quadratic forms by a real symmetric matrix and determine its rank

a. $x_1^2 + 2x_1x_2 + x_2^2$ b. $x_1^2 + 2x_2^2 - 4x_1x_2$

c. $9x_1^2 - 6x_1x_2 + x_2^2$ d. $2x_1^2 - 3x_1x_2 + 3x_2^2$

12. Diagonalize the matrix of each quadratic form in Question 11 and describe its geometric character.

13. In the sample problem, whose mean-corrected data appear in Table 1.2:

a. Find the covariance matrix of the full set of three variables.

b. Compute the principal components of the three-variable covariance matrix and the matrix of component scores.

c. Compare these results with those found in the present chapter.

14. Simplify the following quadratic forms and indicate the type of definiteness of each form:

a. $\mathbf{x}'\begin{bmatrix} 9 & -3 \\ -3 & 1 \end{bmatrix}\mathbf{x}$ b. $\mathbf{x}'\begin{bmatrix} -4 & 2 & -1 \\ 2 & -4 & 2 \\ -1 & 2 & -4 \end{bmatrix}\mathbf{x}$

c. $\mathbf{x}'\begin{bmatrix} 2 & -1.5 \\ -1.5 & 3 \end{bmatrix}\mathbf{x}$ d. $\mathbf{x}'\begin{bmatrix} 4 & 2 & -2 \\ 2 & 4 & 2 \\ -2 & 2 & 4 \end{bmatrix}\mathbf{x}$

15. Returning to the multiple discriminant function problem considered in Section 5.9:

a. Spherize the (pooled) within-group SSCP matrix and compute the eigenstructure in accordance with the procedure outlined in Section 5.9.2.

b. Compare these results with those found from the procedure used in Section 5.9.1.

CHAPTER 6

Applying the Tools to Multivariate Data

6.1 INTRODUCTION

In this chapter we come around full circle to the three substantive problems first introduced in Chapter 1. As recalled, each problem was based on a "toy" data bank and formulated in terms of three commonly used techniques in multivariate analysis: (a) multiple regression, (b) principal components analysis, and (c) multiple discriminant analysis.

We discuss the multiple regression problem first. The problem is structured so as to require the solution of a set of linear equations, called "normal" equations from least-squares theory. These equations are first set up in terms of the original data, and the parameters are found by matrix inversion. We then show how the same problem can be formulated in terms of either a covariance or a correlation matrix.

R^2, a measure of overall goodness of fit, and other regression statistics such as partial correlation coefficients, are also described. The results are interpreted in terms of the substantive problem of interest, and comments are made on the geometric aspects of multiple regression.

We then discuss variations on the general linear model of multiple regression: analysis of variance and covariance, two-group discriminant analysis, and binary-valued regression (in which all variables, criterion and predictors, are expressed as zero–one dummies). This discussion is presented as another way of showing the essential unity among single-criterion, multiple-predictor models.

Discussion then turns to the second substantive problem, formulated as a principal components model. Here the solution is seen to entail finding the eigenstructure of a covariance matrix. Component loadings and component scores are also defined and computed in terms of the sample problem.

After solving this sample problem, some general comments are made about other aspects of factor analysis, such as the factoring of other kinds of cross-product matrices, rotation of component solutions, and dimension reduction methods other than the principal components procedure.

The three-group multiple discriminant problem of Chapter 1 is taken up next. This problem is formulated in terms of finding the eigenstructure of a nonsymmetric matrix which, in turn, represents the product of two symmetric matrices. The discriminant functions are computed, and significance tests are conducted. The results are interpreted in the context of the third sample problem.

We then turn to other aspects of multiple discriminant analysis (MDA), including classification matrices, alternative ways to scale the discriminant functions, and the relationship of MDA to principal components analysis. Finally, some summary-type comments are made about other techniques for dealing with multiple-criterion, multiple-predictor association.

The last major section of the chapter is, in some respects, a prologue to textbooks that deal with multivariate analysis per se. In particular, the concepts of transformational geometry, as introduced in earlier chapters, are now brought together as another type of descriptor by which multivariate techniques can be classified. Under this view multivariate methods are treated as procedures for matching one set of numbers with some other set or sets of numbers. Techniques can be distinguished by the nature of the transformation(s) used to effect the matching and the characteristics of the transformed numbers.

This organizing principle is described in some detail and suggests a framework that can be useful for later study of multivariate procedures as well as suggestive of new models in this field.

6.2 THE MULTIPLE REGRESSION PROBLEM

We are now ready to work through the details of the sample problem in Chapter 1 dealing with the relationship of employee absenteeism Y, to attitude toward the firm X_1 and number of years employed by the firm X_2. To simplify our discussion, the basic data, first shown in Table 1.2, are reproduced in Table 6.1.

As recalled from the discussion in Chapter 1, here we are interested in

1. finding a regression equation for estimating values of the criterion variable Y from a linear function of the predictor variables X_1 and X_2;
2. determining the strength of the overall relationship;
3. testing the significance of the overall relationship;
4. determining the relative importance of the two predictors X_1 and X_2 in accounting for variation in Y.

6.2.1 The Estimating Equation

As again recalled from Chapter 1, the multiple regression equation

$$\boxed{\hat{Y}_i = b_0 + b_1 X_{i1} + b_2 X_{i2}}$$

is a linear equation for predicting values of Y that minimize the sum of the squared errors

$$\sum_{i=1}^{12} e_i^2 = \sum_{i=1}^{12} (Y_i - \hat{Y}_i)^2$$

TABLE 6.1

Basic Data of Sample Problem (from Table 1.2)

	Number of days absent			Attitude rating			Years with company		
Employee	Y	Y_d	Y_s	X_1	X_{d1}	X_{s1}	X_2	X_{d2}	X_{s2}
a	1	−5.25	−0.97	1	−5.25	−1.39	1	−3.92	−1.31
b	0	−6.25	−1.15	2	−4.25	−1.13	1	−3.92	−1.31
c	1	−5.25	−0.97	2	−4.25	−1.13	2	−2.92	−0.98
d	4	−2.25	−0.41	3	−3.25	−0.86	2	−2.92	−0.98
e	3	−3.25	−0.60	5	−1.25	−0.33	4	−0.92	−0.31
f	2	−4.25	−0.78	5	−1.25	−0.33	6	1.08	0.36
g	5	−1.25	−0.23	6	−0.25	−0.07	5	0.08	0.03
h	6	−0.25	−0.05	7	0.75	0.20	4	−0.92	−0.31
i	9	2.75	0.51	10	3.75	0.99	8	3.08	1.03
j	13	6.75	1.24	11	4.75	1.26	7	2.08	0.70
k	15	8.75	1.61	11	4.75	1.26	9	4.08	1.37
l	16	9.75	1.80	12	5.75	1.53	10	5.08	1.71
Mean	6.25			6.25			4.92		
Standard deviation	5.43			3.77			2.98		

Appendix A shows how the set of normal equations, used to find b_0, b_1, and b_2, are derived. In terms of the specific problem here, we have in matrix notation:[1]

$$\mathbf{y} = \begin{bmatrix} 1 \\ 0 \\ 1 \\ \vdots \\ 16 \end{bmatrix}; \quad \mathbf{X} = \overset{\begin{matrix} C & X_1 & X_2 \end{matrix}}{\begin{bmatrix} 1 & 1 & 1 \\ 1 & 2 & 1 \\ 1 & 2 & 2 \\ \vdots & \vdots & \vdots \\ 1 & 12 & 10 \end{bmatrix}}; \quad \mathbf{b} = \begin{bmatrix} b_0 \\ b_1 \\ b_2 \end{bmatrix}; \quad \mathbf{e} = \begin{bmatrix} e_1 \\ e_2 \\ e_3 \\ \vdots \\ e_{12} \end{bmatrix}$$

The model being fitted by least squares is

$$\boxed{\mathbf{y} = \mathbf{X}\mathbf{b} + \mathbf{e}}$$

Notice that the model, in matrix form, starts off with the observed vector y and the observed matrix **X**. As will be shown later, the device of including a column of ones as the first column of **X** (called C) is employed for estimating the intercept b_0.

We wish to solve for **b**, the vector of parameters, so that $\sum_{i=1}^{12} e_i^2 = \mathbf{e}'\mathbf{e}$ is minimized. As can be checked in Appendix A, the problem is a standard one in the calculus and leads to the so-called normal equations which, expressed in matrix form, are

$$\boxed{\mathbf{b} = (\mathbf{X}'\mathbf{X})^{-1}\mathbf{X}'\mathbf{y}}$$

[1] The sample entries in y and **X** are taken from Table 6.1.

That is, we first need to find the minor product moment of **X**, which is **X**′**X**. Next, we find the inverse of **X**′**X** and postmultiply this inverse by **X**′**y**.

In terms of the specific problem of Table 6.1, we have

$$\mathbf{b} = \begin{bmatrix} b_0 \\ b_1 \\ b_2 \end{bmatrix} = \left\{ \overset{\mathbf{X}'}{\begin{bmatrix} 1 & 1 & 1 & \cdots & 1 \\ 1 & 2 & 2 & \cdots & 12 \\ 1 & 1 & 2 & \cdots & 10 \end{bmatrix}} \overset{\mathbf{X}}{\begin{bmatrix} 1 & 1 & 1 \\ 1 & 2 & 1 \\ 1 & 2 & 2 \\ \vdots & \vdots & \vdots \\ 1 & 12 & 10 \end{bmatrix}} \right\}^{-1} \overset{\mathbf{X}'}{\begin{bmatrix} 1 & 1 & 1 & \cdots & 1 \\ 1 & 2 & 2 & \cdots & 12 \\ 1 & 1 & 2 & \cdots & 10 \end{bmatrix}} \overset{\mathbf{y}}{\begin{bmatrix} 1 \\ 0 \\ 1 \\ \vdots \\ 16 \end{bmatrix}}$$

$$\mathbf{b} = \begin{bmatrix} -2.263 \\ 1.550 \\ -0.239 \end{bmatrix}$$

Hence, in terms of the original data of Table 6.1, we have the estimating equation

$$\hat{Y}_i = -2.263 + 1.550X_{i1} - 0.239X_{i2}$$

The 12 values of $\hat{Y}_i$ appear in the lower portion of Table 6.2, along with the residuals e_i.

If one adds up the squared residuals, one obtains (within rounding error) the residual term shown in the analysis of variance table of Table 6.2:

$$\text{residual} = 34.099$$

The total sum of squares is obtained from

$$\sum_{i=1}^{12} (Y - \bar{Y})^2 = 354.25$$

and the difference

$$\text{due to regression} = 320.15$$

6.2.2 Strength of Overall Relationship and Statistical Significance

The squared multiple correlation coefficient is R^2, and this measures the portion of variance in Y (as measured about its mean) that is accounted for by variation in X_1 and X_2. As mentioned in Chapter 1, the formula is

$$\boxed{R^2 = 1 - \frac{\sum_{i=1}^{12} e_i^2}{\sum_{i=1}^{12} (Y_i - \bar{Y})^2}}$$

$$R^2 = 1 - \frac{34.099}{354.25} = 0.904$$

TABLE 6.2

Selected Output from Multiple Regression

$R^2 = 0.904$; $R = 0.951$; variance of estimate 3.789

Analysis of Variance for Multiple Regression

Source	df	Sums of squares	Mean squares	F ratio
Due to regression	2	320.151	160.075	42.25
Residual	9	34.099	3.789	
Total	11	354.250		

Variable	Regression coefficients	Standard errors	t values	Partial correlations	Proportion of cumulative variance
X_1	1.550	0.481	3.225	0.732	0.902
X_2	−0.239	0.606	−0.393	−0.130	0.002

Y intercept −2.263

Table of Residuals

Employee	Y	$\hat{Y}$	e	Employee	Y	$\hat{Y}$	e
a	1	−0.95	1.95	g	5	5.85	−0.84
b	0	0.60	−0.60	h	6	7.63	−1.63
c	1	0.36	0.64	i	9	11.33	−2.33
d	4	1.91	2.09	j	13	13.11	−0.11
e	3	4.53	−1.53	k	15	12.64	2.36
f	2	4.05	−2.05	l	16	13.95	2.05

The statistical significance of R, the positive square root of R^2, is tested via the analysis of variance subtable of Table 6.2 by means of the F ratio:

$$F = 42.25$$

which, with 2 and 9 degrees of freedom, is highly significant at the $\alpha = 0.01$ level. Thus, as described in Chapter 1, the equivalent null hypotheses

$$\boxed{\begin{gathered} R_p = 0 \\ \beta_1 = \beta_2 = 0 \end{gathered}}$$

are rejected at the 0.01 level, and we conclude that the multiple correlation is significant.

Up to this point, then, we have established the estimating equation and measured, via R^2, the strength of the overall relationship between Y versus X_1 and X_2.

If we look at the equation again

$$\hat{Y}_i = -2.263 + 1.550X_{i1} - 0.239X_{i2}$$

we see that the intercept is negative. In terms of the current problem, a negative 2.263 days of absenteeism is impossible, illustrating, of course, the possible meaninglessness of extrapolation beyond the range of the predictor variables used in developing the parameter values.

The partial regression coefficient for X_1 seems reasonable; it says that predicted absenteeism increases 1.55 days per unit increase in attitude rating. This is in accord with the scatter plot (Fig. 1.2) that shows the association of Y with X_1 alone.

The partial regression coefficient for X_2, while small in absolute value, is negative, even though the scatter plot of Y on X_2 alone (Fig. 1.2) shows a positive relationship. The key to this seeming contradiction lies in the strong positive relationship between the predictors X_1 and X_2 (also noted in the scatter plot of Fig. 1.2). Indeed, the correlation between X_1 and X_2 is 0.95. The upshot of all of this is that once X_1 is in the equation, X_2 is so redundant with X_1 that its inclusion leads to a negative partial regression coefficient that effectively is zero (given its large standard error).

6.2.3 Other Statistics

The redundancy of X_2, once X_1 is in the equation, is brought out in Table 6.2 under the column

	Proportion of cumulative variance
X_1	0.902
X_2	0.002

That is, of the total $R^2 = 0.904$, the contribution of X_1 alone represents 0.902. The increment due to X_2 (0.002) is virtually zero, again reflecting its high redundancy with X_1.

This same type of finding is reinforced by examining the t values and the partial correlations in Table 6.2. These are

	t Values	Partial correlations
X_1	3.225	0.732
X_2	−0.393	−0.130

The Student t value is the ratio of a predictor variable's partial regression coefficient to its standard error. The standard error, in turn, is a measure of how well the predictor variable itself can be predicted from a linear combination of the other predictors. The higher the standard error, the more redundant (better predicted) that predictor variable is with the others. Hence, the less contribution it makes to Y on its own and the lower its t value.

We see that the ratio of b_1 to its own standard error is

$$t(b_1) = \frac{1.550}{0.481} = 3.225$$

which is significant at the 0.01 level. The t value for X_2 of -0.393 is not significant, however. Without delving into formulas, the t test is a test of the separate significance of each predictor variable X_j, when included in a regression model, versus the same regression model with all predictors included except it. We note here that only X_1 is needed in the equation.

The partial correlations also suggest the importance of X_1 rather than X_2 in accounting for variation in Y. The partial correlation of Y with some predictor X_j is their simple correlation when both variables are expressed on a residual basis, that is, net of the linear association of each with all of the other predictors. In the present problem, the partial correlation of Y with X_1 is considerably higher than Y with X_2, supporting the earlier conclusions.

But what if X_2 is entered first in the regression? What happens in this case to the various statistics reported in Table 6.2? As it turns out, the only statistic that changes if X_2 is credited with as much variance as it can account for before X_1 is allowed to contribute to criterion variance is the last column, proportion of cumulative variance. If X_2 is entered first, it is credited with 0.79, while X_1 is credited with only 0.11 of the 0.90 total. The rest of the output does not change, and X_2 is still eliminated from the regression on the basis of the t test results.

What this example points out is that in the usual case of correlated predictors, the question of "relative importance" of predictors is ambiguous. Many researchers interpret relative importance in terms of the change in R^2 occurring when the predictor in question is the last to enter. Other importance measures are also available, as pointed out by Darlington (1968). However, in the case of correlated predictors, no measure is entirely satisfactory.

6.2.4 Other Formulations of the Problem

In the sample problem of Table 6.1, the normal equations were formulated in terms of the original data. Alternatively, suppose we decided to work with the mean-corrected scores Y_d, X_{d1}, X_{d2}. In this case we would compute the covariance matrix

$$\mathbf{C} = \mathbf{X}_d'\mathbf{X}_d/m$$

and the vector of partial regression parameters would be found from

$$\boxed{\mathbf{b} = \mathbf{C}^{-1}\mathbf{a}(y)}$$

where $\mathbf{a}(y)$ is the vector of covariances between the criterion and each predictor in turn with elements

$$\boxed{\begin{aligned} a_1 &= \mathbf{y}_d'\mathbf{x}_{d1}/m \\ a_2 &= \mathbf{y}_d'\mathbf{x}_{d2}/m \end{aligned}}$$

in the sample problem.

The preceding formula for computing $\mathbf{b}$ would find only b_1 and b_2 since all data would be previously mean centered. To work back to original data, we can find the intercept of the equation by the simple formula:

$$\boxed{b_0 = \bar{Y} - b_1\bar{X}_1 - b_2\bar{X}_2}$$

If we decided to work with the standardized data Y_s, X_{s1}, and X_{s2}, the appropriate minor product moment is the correlation matrix

$$R = \mathbf{X}_s{}'\mathbf{X}_s/m$$

and the vector of parameters $\mathbf{b}^*$ (often called beta weights) would be found from

$$\boxed{\mathbf{b}^* = \mathbf{R}^{-1}\mathbf{r}(y)}$$

where $\mathbf{r}(y)$ is the vector of product-moment correlations between the criterion and each predictor in turn, with elements

$$\boxed{\begin{aligned} r_1 &= \mathbf{y}_s{}'\mathbf{x}_{s1}/m \\ r_2 &= \mathbf{y}_s{}'\mathbf{x}_{s2}/m \end{aligned}}$$

in the sample problem.

The vector $\mathbf{b}^*$ measures the change in Y per unit change in each of the predictors, when all variables are expressed in standardized units. To find the elements of $\mathbf{b}$, we use the conversion equations

$$\boxed{\begin{aligned} b_1 &= b_1{}^* \frac{s_y}{s_{x_1}} \\ b_2 &= b_2{}^* \frac{s_y}{s_{x_2}} \end{aligned}}$$

These simple transformations, involving ratios of standard deviations, enable us to express changes in Y per unit change in X_1 and X_2 in terms of the original Y units. Having done this, we can then solve for the intercept term in exactly the same way:

$$b_0 = \bar{Y} - b_1\bar{X}_1 - b_2\bar{X}_2$$

as shown in the covariance matrix case. Many computer routines for performing multiple regression operate on the correlation matrix. As seen here, any of the cross-product matrices—raw cross products, covariances, or correlations—can be used and, in the latter two cases, modified for expressing regression results in terms of original data.

6.2.5 Geometric Aspects—the Response Surface Model

Figure 1.2 showed two-dimensional scatter plots of Y versus X_1, Y versus X_2, and X_2 versus X_1 It is also a relatively simple matter to plot a three-dimensional diagram of Y versus X_1 and X_2. This is shown in Fig. 6.1.

We also show the fitted regression plane, as computed by least squares. This type of model, in which observations are represented by points and variables by dimensions, is often called the response surface or point model.

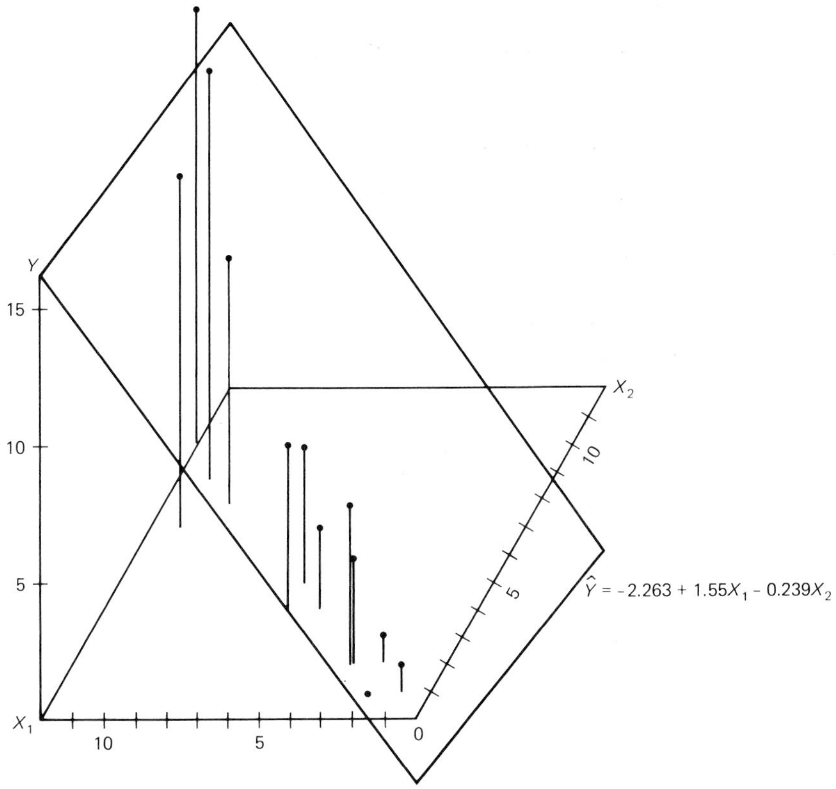

Fig. 6.1 Three-dimensional plot and fitted regression plane.

The intersection of the regression plane in Fig. 6.1 with the Y axis provides the estimate b_0, the intercept term. If we next imagine constructing a plane perpendicular to the X_1 axis, we, in effect, hold X_1 constant; hence b_2 represents the estimated contribution of a unit change in X_2 to a change in Y. Similar remarks pertain to the interpretation of b_1.

The regression plane itself is oriented so as to minimize the sum of squared deviations between each Y_i and its counterpart value on the fitted plane, where these deviations are taken along directions parallel to the Y axis. Similarly, we can find the sum of squared deviations about the mean of the Y_i's by imagining a plane perpendicular to the Y axis passing through the value $\bar{Y}$. Total variation in Y is thus partitioned into two parts. As indicated earlier, these separate parts are found by

1. subtracting *unaccounted-for variation,* involving squared deviations $(Y_i - \hat{Y}_i)^2$ about the fitted regression plane, from
2. *total* variation involving squared deviations $(Y_i - \bar{Y})^2$ from the plane imagined to be passing through $\bar{Y}$.

The quantity $\sum_{i=1}^{12}(Y_i - \hat{Y}_i)^2$ represents the unaccounted-for sum of squares, and the quantity $[\sum_{i=1}^{12}(Y_i - \bar{Y})^2 - \sum_{i=1}^{12}(Y_i - \hat{Y}_i)^2]$ represents the accounted-for sum of squares. If no variation is accounted for, then we note that using $\bar{Y}$ is just as good at predicting Y as introducing the variation in X_1 and X_2.

6.2.6 An Alternative Representation

The foregoing representation of the 12 responses in variable space considers the 12 observations as points in three dimensions, where each variable, Y, X_1, or X_2, denotes a dimension. Alternatively, we can imagine that each of the 12 employees represents a dimension, and each of the variables constitutes a vector in this 12-dimensional person space. As we know from the discussion of matrix rank in Chapter 5, the three vectors will not span the whole 12-dimensional space but, rather, will lie in (at most) a three-dimensional subspace that is embedded in the 12-dimensional person space.

We also remember that if the vectors are translated to a mean-centered origin and are assumed to be of unit length, the (product-moment) correlation between each pair of vectors is given by the cosine of their angle. In this case we have three two-variable correlations: r_{yx_1}, r_{yx_2}, and $r_{x_1x_2}$.

This concept is pictured, in general terms, in Fig. 6.2. In the left panel of the figure are two unit length vectors $\mathbf{x}_1$ and $\mathbf{x}_2$ emanating from the origin. Each is a 12-component vector of unit length, embedded in the "person" space. The cosine of the angle Ψ separating $\mathbf{x}_1$ and $\mathbf{x}_2$ is the simple correlation $r_{x_1x_2}$.

Since the criterion vector $\mathbf{y}$ is not perfectly correlated with $\mathbf{x}_1$ and $\mathbf{x}_2$, it must extend into a third dimension. The cosines of its angular separation between $\mathbf{x}_1$ and $\mathbf{x}_2$ are each measured, respectively, by its simple correlations r_{yx_1} and r_{yx_2}. However, one can project $\mathbf{y}$ onto the plane formed by $\mathbf{x}_1$ and $\mathbf{x}_2$. The projection of $\mathbf{y}$ onto this plane is denoted by $\hat{\mathbf{y}}$.

In terms of this viewpoint, the idea behind multiple regression is to find the particular vector in the $\mathbf{x}_1$, $\mathbf{x}_2$ plane that minimizes the angle θ with $\mathbf{y}$. This vector will be the projection $\hat{\mathbf{y}}$ onto the plane formed by $\mathbf{x}_1$, $\mathbf{x}_2$. Since any vector in the $\mathbf{x}_1$, $\mathbf{x}_2$ plane is a linear combination of $\mathbf{x}_1$ and $\mathbf{x}_2$, it follows that we want the vector $\hat{\mathbf{y}} = b_1^*\mathbf{x}_1 + b_2^*\mathbf{x}_2$, where the b_j^*'s are beta weights, that minimizes the angle, or maximizes the cosine of the angle with $\mathbf{y}$. The cosine of this angle θ (see Fig. 6.2) is R, the multiple correlation. The problem then is to find a set of b_j^*'s that define a linear combination of the vectors $\mathbf{x}_1$

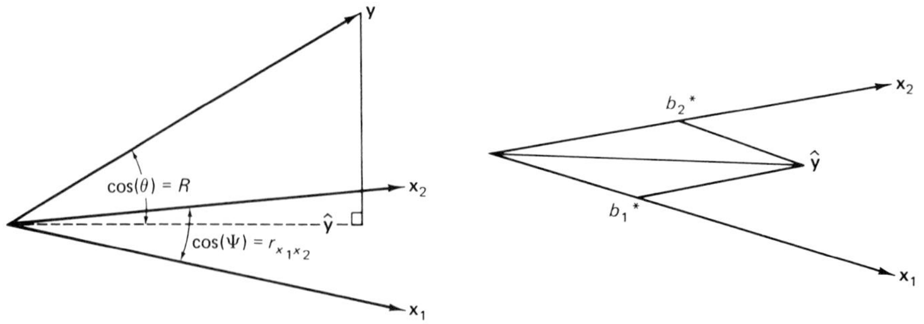

Fig. 6.2 Geometric relationship of $\hat{\mathbf{y}}$ to $\mathbf{y}$ in vector space. The graph on the right shows geometric interpretation of partial regression weights in vector space.

and $\mathbf{x}_2$ maximizing the cosine R of θ, the angle separating y and $\hat{\mathbf{y}}$. However, this is equivalent to minimizing the square of the distance from the terminus of y to its projection $\hat{\mathbf{y}}$. This criterion

$$\text{minimize}\left[\sum_{i=1}^{12} (y_{si} - \sum_{j=1}^{n} b_j^* x_{sij})^2\right]$$

again leads to the least-squares equations. (Since all variables are assumed to be measured in standardized form, the intercept b_0 is zero.)

In general, the $\mathbf{x}_1$, $\mathbf{x}_2$ axes will be oblique, as noted in Fig. 6.2. The right panel shows that a linear combination of $\mathbf{x}_1$ and $\mathbf{x}_2$, which results in the predicted vector $\hat{\mathbf{y}}$, involves combining oblique axes via $b_1^*$and b_2^*. In this case, $b_1^*$and $b_2^*$are direction cosines.

Figure 6.3 shows some conditions of interest. In Panel I we see that y is uncorrelated with $\mathbf{x}_1$ and $\mathbf{x}_2$. This lack of correlation is indicated by the 90° angle between y and the $\mathbf{x}_1$, $\mathbf{x}_2$ plane. Panel II shows the opposite situation where y is perfectly correlated with $\mathbf{x}_1$ and $\mathbf{x}_2$ and, hence, can be predicted without error by a linear combination of $\mathbf{x}_1$ and $\mathbf{x}_2$.

Panel III shows the case where $\mathbf{x}_1$ and $\mathbf{x}_2$ are uncorrelated and y evinces some correlation with $\mathbf{x}_1$ and none with $\mathbf{x}_2$. Panel IV shows the case in which $\mathbf{x}_1$ and $\mathbf{x}_2$ are uncorrelated, but the projection of y lies entirely along $\mathbf{x}_2$.

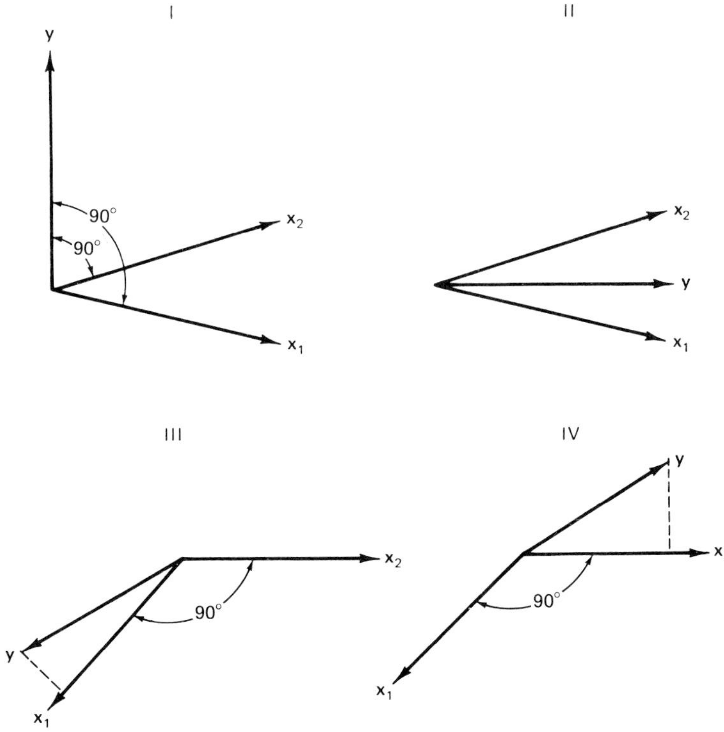

Fig. 6.3 Some illustrative cases involving multiple correlation. Key: I, no correlation; II, perfect correlation; III, y correlated with $\mathbf{x}_1$ only; IV, y correlated with $\mathbf{x}_2$ only.

In summary, the multiple correlation coefficient R is the cosine of the angle θ made by $\mathbf{y}$ and $\hat{\mathbf{y}}$. The b_j^*'s are normalized beta weights and represent coordinates of $\hat{\mathbf{y}}$ in the oblique space of the predictor variables. If more than two predictors are involved, the same geometric reasoning applies, although in this case the predictors involve higher-dimensional hyperplanes.

Partial correlations between $\mathbf{y}$ and $\mathbf{x}_1$ and $\mathbf{x}_2$, respectively, can also be interpreted. For example, if we consider a plane perpendicular to $\mathbf{x}_2$ and project $\mathbf{y}$ and $\mathbf{x}_1$ onto this plane, $r_{yx_1 \cdot x_2}$, the partial correlation of $\mathbf{y}$ with $\mathbf{x}_1$ (with $\mathbf{x}_2$ partialed out) is represented by the cosine of the angle separating them on this plane. Similar remarks pertain to the partial correlation of $\mathbf{y}$ with $\mathbf{x}_2$ and would involve a projection onto a space that is orthogonal to $\mathbf{x}_1$. The same general idea holds for larger numbers of predictors.

6.3 OTHER FORMS OF THE GENERAL LINEAR MODEL

The typical multiple regression model considers each variable as intervally scaled. This representation is overly restrictive. Indeed, by employing the dummy-variable device, as introduced in Chapter 1, we can extend the linear regression model to a more general model that subsumes the techniques of

1. analysis of variance
2. analysis of covariance
3. two-group discriminant analysis
4. binary-valued regression

All of these cases are developed from two basic concepts: (a) the least-squares criterion for matching one set of data with some transformation of another set and (b) the dummy variable.

Figure 6.4 shows, in a somewhat abstract sense, various special cases in terms of the response surface or point model involving m observations in three dimensions.

Panel I of Fig. 6.4 shows each of the three columns of a data matrix as a dimension and each row of the matrix as a point. If we were then to append to the $m \times 2$ matrix of predictors a unit vector, we have the familiar matrix expression for fitting a plane or, more generally, a response surface, in the three-dimensional space shown in Panel I. Predicted values $\hat{\mathbf{y}}$ of the criterion variable $\mathbf{y}$ are given by

$$\boxed{\hat{\mathbf{y}} = \mathbf{Xb}}$$

where $\mathbf{b}$ is a 3×1 column vector with entries b_0, b_1, b_2 denoting, respectively, the intercept, partial regression coefficient for $\mathbf{x}_1$, and partial regression coefficient for $\mathbf{x}_2$.

However—and this is the key point—nothing in the least-squares procedure precludes $\mathbf{y}$ (or $\mathbf{x}_1$ or $\mathbf{x}_2$ for that matter) from taking on values that are just zero or one. Panel II shows the case where $\mathbf{y}$ assumes only binary values, but $\mathbf{x}_1$ and $\mathbf{x}_2$ are allowed to be continuous. Panel III shows the opposite situation. Panel IV shows a "mixed" case where $\mathbf{y}$ and $\mathbf{x}_1$ are continuous and $\mathbf{x}_2$ is binary valued. Panel V shows a case where all three variables are binary valued.

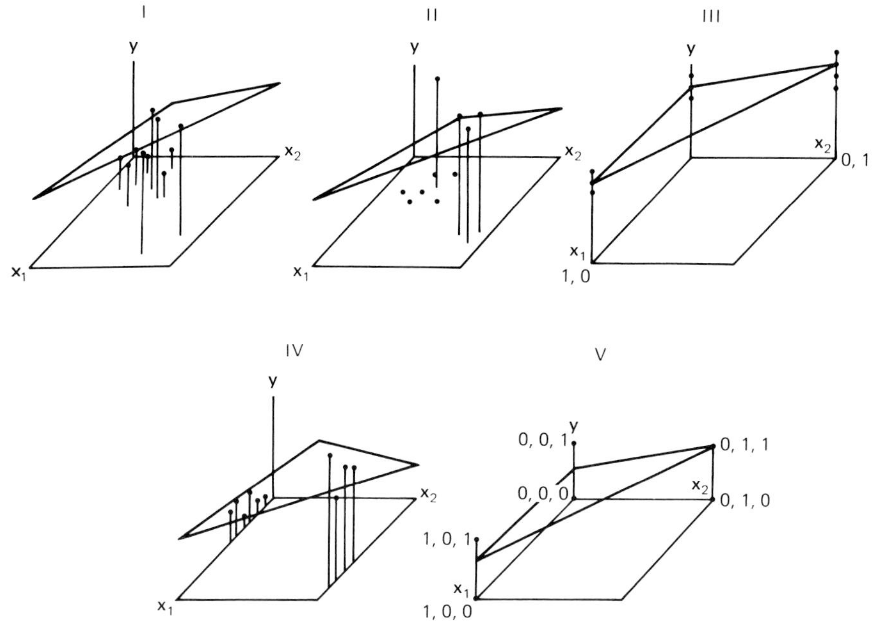

Fig. 6.4 Variations of the response surface model.

Not surprisingly, from Chapter 1 we recognize Panel I as a traditional multiple regression formulation. Panel II appears as a two-group discriminant function. Panel III appears as a one-way analysis of variance design with one treatment variable at three levels. Panel IV represents a simple analysis of covariance design with a single two-level treatment variable ($\mathbf{x}_2$) and one continuous covariate ($\mathbf{x}_1$). Panel V seems less familiar, but could be viewed as a type of binary-valued regression where the criterion and the predictors are each dichotomies (e.g., predicting high versus low attitude toward the firm as a function of sex and marital status).

As observed from Fig. 6.4, we can now conclude that all of these models are variations on a common theme—namely, one in which we are attempting to find some type of linear transformation that results in a set of scores that best match, in the sense of minimum sum of squared deviations, a set of criterion scores. In each case we are fitting a plane in the three-dimensional space of $\mathbf{x}_1$, $\mathbf{x}_2$, and $\mathbf{y}$ and then finding estimates $\hat{\mathbf{y}}$ of $\mathbf{y}$ that result in a minimum sum of squared deviations.

All of the cases depicted in Fig. 6.4 are characterized by the fact that a single-criterion variable, either 0–1 coded or intervally scaled, is involved. Our interest is in finding some linear combination of predictors, where $\mathbf{b}$ denotes the set of combining weights, that leads to a set of predicted scores $\hat{\mathbf{y}}$ that are most congruent with the original scores $\mathbf{y}$.

Extension of the multiple regression model to handle binary-valued predictors is described in various texts (e.g., Neter and Wasserman, 1974) in terms of a general linear model.

If a further extension is made in order to allow for a binary-valued criterion, least squares can still be used to estimate parameter values, although the usual statistical tests are not strictly appropriate since the normality and constant variance assumptions are missing. Still, as a descriptive device least squares can be used to find estimating equations for all of the cases depicted in Fig. 6.4.

Discussion of the multiple regression problem has thus resulted in a much wider scope of application than might first have been imagined. Through the dummy-variable coding device, one can subsume all cases of interest—analysis of variance and covariance, two-group discrimination, binary-valued regression—that involve a single-criterion variable and multiple predictors. Moreover, although detailed discussion of the geometric aspects of the models was more or less confined to multiple regression, all of these methods can be represented by either

1. the response surface or point model in variable space, or
2. the vector model in person or object space.

From the standpoint of matrix algebra, all of the preceding models entail solutions based on a set of linear (the normal) equations from least-squares theory. As such, the operation of matrix inversion becomes germane, as does the concept of matrix rank and related ideas such as determinants. In brief, the algebraic underpinnings of single-criterion, multiple-predictor association are concepts of matrix rank and inversion. Thus, it is no accident that much of the discussion in earlier chapters was devoted to these topics.

6.4 THE FACTOR ANALYSIS PROBLEM

If matrix inversion and rank are the hallmarks of single-criterion, multiple-predictor association, then eigenstructures are the key concepts in dimension-reducing methods like factor analysis. Eigenstructures are also essential in multiple-criterion, multiple-predictor association, as we shall see later in the chapter.

In Chapter 1 we introduced a small-scale problem in principal components analysis in the context of developing a reduced space for the two predictors: (a) X_1, attitude score and (b) X_2, number of years with company. Using the X_{d1} and X_{d2} data of Table 6.1 we wish to know if a change of basis vectors can be made that will produce an axis whose variance of point projections is maximal.

This is a standard problem in finding the eigenstructure of a symmetric matrix. Here we employ the covariance matrix, although in some cases one might want to use some other type of cross-products matrix. Table 6.3 details the steps involved in finding the eigenstructure of **C**, the simple 2×2 covariance matrix of the sample problem of predictor variables in Table 6.1. (Supporting calculations appear in Chapter 5.)

As observed from Table 6.3, the first eigenvalue $\lambda_1 = 22.56$ accounts for nearly all, actually 98 percent, of the variance of **C**, the covariance matrix. The linear composite $\mathbf{z}_1$, developed from $\mathbf{t}_1$, makes an angle of approximately 38° with the horizontal axis, as noted in Fig. 6.5. Thus, if we wished to combine the vectors of scores X_{di1} and X_{di2} into a single linear composite, we would have, in scalar notation,

$$\boxed{z_{i(1)} = 0.787X_{di1} + 0.617X_{di2}}$$

Note also that the second linear composite $\mathbf{z}_2$ is at a right angle to $\mathbf{z}_1$.

TABLE 6.3

Finding the Eigenstructure of the Covariance Matrix (Predictor Variables in Table 6.1)

Covariance matrix

$$\mathbf{C} = \begin{array}{c} X_1 \\ X_1 \end{array} \overset{\begin{array}{cc} X_1 & X_2 \end{array}}{\begin{bmatrix} 14.19 & 10.69 \\ 10.69 & 8.91 \end{bmatrix}}$$

Matrix equation

$$(\mathbf{C} - \lambda_i \mathbf{I})\mathbf{t}_i = \mathbf{0}$$

Characteristic equation

$$|\mathbf{C} - \lambda_i \mathbf{I}| = \begin{vmatrix} 14.19 - \lambda_i & 10.69 \\ 10.69 & 8.91 - \lambda_i \end{vmatrix} = 0$$

Expansion of determinant

$$\lambda_i^2 - 23.1\lambda_i + 126.433 - 114.276 = 0$$

Eigenvalues

$\lambda_1 = 22.56$; $\lambda_2 = 0.54$

Eigenvectors

$$\mathbf{t}_1 = \begin{bmatrix} 0.787 \\ 0.617 \end{bmatrix}; \quad \mathbf{t}_2 = \begin{bmatrix} 0.617 \\ -0.787 \end{bmatrix}$$

6.4.1 Component Scores

Component scores are the projections of the twelve points on each new axis, $\mathbf{z}_1$ and $\mathbf{z}_2$, in turn. For example, the component score of the first point on $\mathbf{z}_1$ is

$$z_{1(1)} = 0.787(-5.25) + 0.617(-3.92) = -6.55$$

as shown in Fig. 6.5. The full set of component scores appears in Table 5.1.

The variance of each column of $\mathbf{Z}$ will equal its respective eigenvalue. If one wishes to find a matrix of component scores with unit variance, this is done quite simply by a transformation involving the matrix of eigenvectors $\mathbf{T}$ and the reciprocals of the square roots of the eigenvalues:[2]

$$\boxed{\mathbf{Z}_s = \mathbf{X}_d \mathbf{T} \mathbf{\Lambda}^{-1/2}}$$

In the sample problem, the product of $\mathbf{T}$ and $\mathbf{\Lambda}^{-1/2}$ is given by

$$\mathbf{S} = \overset{\mathbf{T}}{\begin{bmatrix} 0.787 & 0.617 \\ 0.617 & -0.787 \end{bmatrix}} \overset{\mathbf{\Lambda}^{-1/2}}{\begin{bmatrix} 0.211 & 0 \\ 0 & 1.361 \end{bmatrix}}; \quad \mathbf{S} = \begin{bmatrix} 0.166 & 0.840 \\ 0.130 & -1.071 \end{bmatrix}$$

In the sample problem, $\mathbf{Z}_s$ denotes the 12×2 matrix of unit-variance component scores; $\mathbf{X}_d$ is the 12×2 matrix of mean-centered predictor variables; $\mathbf{T}$ is the matrix of

[2] In this illustration we use $\mathbf{\Lambda}$ to denote the diagonal matrix of eigenvalues of $\mathbf{C}$, the covariance matrix. Accordingly, $\mathbf{\Lambda}^{-1/2}$ is a diagonal matrix whose main diagonal elements are the reciprocals of the square roots of the main diagonal elements of $\mathbf{\Lambda}$.

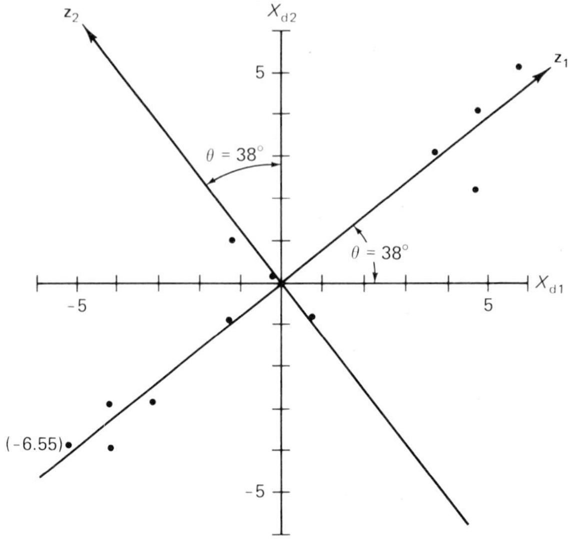

Fig. 6.5 Principal components rotation of mean-corrected predictor variables.

eigenvectors from Table 6.3; and $\Lambda^{-1/2}$ is a diagonal matrix of the reciprocals of the square roots of the eigenvalues. By application of the transformation matrix **S** (instead of **T**), we would obtain unit-variance component scores. That is, in this case,

$$\boxed{\mathbf{Z}_s'\mathbf{Z}_s/m = \mathbf{I}}$$

Geometrically, then, postmultiplication of $\mathbf{X}_d$ by **S** has the effect of transforming the ellipsoidal-like swarm of points in Fig. 6.5 into a circle, along the axes of the ellipse.

6.4.2 Component Loadings

Component loadings are simply product-moment correlations of each original variable X_{d1} and X_{d2} with each set of component scores.

To illustrate, the (unit-variance) component scores $\mathbf{z}_{si(1)}$ on the first principal component are

a	−1.38	b	−1.21	c	−1.08	d	−0.92	e	−0.33	f	−0.07
g	−0.03	h	0.01	i	1.02	j	1.06	k	1.32	l	1.62

These represent the first column of $\mathbf{Z}_s$. For example,

$$z_{si(1)} = 0.166X_{di1} + 0.130X_{di2} = 0.166(-5.25) + 0.130(-3.92) = -1.38$$

The product-moment correlation of $\mathbf{z}_{s(1)}$ with $\mathbf{x}_{d1}$ is 0.99, and the product-moment correlation of $\mathbf{z}_{s(1)}$ with $\mathbf{x}_{d2}$ is 0.98. Not surprisingly, given the high variance accounted for by the first component, both loadings are high.

A more general definition of a component loading considers a loading as a weight, obtained for each variable, whose square measures the contribution that the variable makes to variation in the component. However, usually in applied work it is the correlation matrix that is factored rather than the covariance, or some other type of cross products, matrix. Hence, the simpler definition of loading, namely, as the correlation of a variable with a component, is most prevalent.

In the present problem, the principal components analysis of a 2 x 2 correlation matrix would necessarily effect a 45° rotation, rather than the 38° rotation shown in Fig. 6.5. Hence the loadings of X_1 and X_2 on each component would necessarily be equal. However, this will not, in general, be the case with correlations based on three or more variables being analyzed by principal components.

The matrix of component "loadings" for the covariance matrix in the present problem is found quite simply from the relationship

$$\boxed{\mathbf{F} = \mathbf{T}\Lambda^{1/2}}$$

$$\mathbf{F} = \overset{\mathbf{T}}{\begin{bmatrix} 0.787 & 0.617 \\ 0.617 & -0.787 \end{bmatrix}} \overset{\Lambda^{1/2}}{\begin{bmatrix} 4.75 & 0 \\ 0 & 0.73 \end{bmatrix}}$$

$$\mathbf{F} = \begin{matrix} & X_1 & X_2 & \\ X_1 & 3.74 & 0.45 & 14.19 \\ X_2 & 2.93 & 0.57 & 8.91 \\ \lambda_i & 22.56 & 0.54 & 23.10 \end{matrix}$$

An interesting property of **F** is that the sum of the squared "loadings" of each component (column) equals its respective eigenvalue. For example,

$$\lambda_1 = (3.74)^2 + (2.93)^2 = 22.56$$

the variance of the first component, within rounding error.

Similarly, the sum of the squared entries of each variable (row) equals its respective variance. For example,

$$(3.74)^2 + (0.45)^2 = 14.19$$

the variance of X_1.

Finally, we see that both components together exhaust the total variance in the covariance matrix **C**. Furthermore, the first component itself accounts for 22.56/23.10 = 0.98 of the total variance.[3] Clearly, little is gained by inclusion of the second component insofar as the sample problem is concerned.

6.4.3 The Basic Structure of $\mathbf{X}_d$

Another way of looking at the principal components problem is in terms of the basic structure of a matrix, as described in Chapter 5. In line with our earlier discussion, suppose we wished to find the basic structure of

$$\boxed{\mathbf{X}_d/\sqrt{m} = \mathbf{U}\Delta\mathbf{T}'}$$

[3] Of additional interest is the fact that X_1 accounts for 14.19/23.10 or 0.61 of the total variance.

where, as we know, the minor product of the left-hand side of the equation represents the covariance matrix

$$\mathbf{C} = \mathbf{X}_d{}'\mathbf{X}_d/m$$

As shown in Chapter 5, **C** is symmetric and, hence, orthogonally decomposable into the triple product

$$\mathbf{C} = \mathbf{T\Lambda T}'$$

$$\begin{matrix} \mathbf{T} & & \mathbf{\Lambda} & & \mathbf{T}' \end{matrix}$$

$$= \begin{bmatrix} 0.787 & 0.617 \\ 0.617 & -0.787 \end{bmatrix} \begin{bmatrix} 22.56 & 0 \\ 0 & 0.54 \end{bmatrix} \begin{bmatrix} 0.787 & 0.617 \\ 0.617 & -0.787 \end{bmatrix} = \begin{bmatrix} 14.19 & 10.69 \\ 10.69 & 8.91 \end{bmatrix}$$

where **T** is orthogonal, and **Λ** is diagonal. Note that **Λ** is the matrix of eigenvalues of $\mathbf{X}_d{}'\mathbf{X}_d/m$, and **T** is the matrix of eigenvectors, as shown in Table 6.3. As shown in Chapter 5, we can next solve for the orthonormal-by-columns matrix **U** by the equation

$$\mathbf{U} = \mathbf{X}_d/\sqrt{m}\,\mathbf{T\Delta}^{-1}$$

where $\mathbf{\Delta}^{-1} = \mathbf{\Lambda}^{-1/2}$. This, in turn, leads to the basic structure of $\mathbf{X}_d/\sqrt{m}$:

$$\mathbf{X}_d/\sqrt{m} = \mathbf{U\Delta T}'$$

As recalled, **U** is orthonormal by columns; **Δ** is diagonal (a stretch transformation); and **T′** is orthogonal (a rotation). Moreover, as also pointed out in Chapter 5, if the eigenstructure of the major product moment $\mathbf{X}_d\mathbf{X}_d{}'/m$ is found instead, the matrix of its eigenvalues **Λ** will still be the same, and the representation is now

$$\boxed{\mathbf{X}_d\mathbf{X}_d{}'/m = \mathbf{U\Lambda U}'}$$

where **U** is the same matrix found above. One then goes on to solve for **T′** in a manner analogous to that shown above.[4]

Finally, by similar procedures we could find the basic structure of any of the following matrices of interest in Table 6.1:

$$\mathbf{X}; \quad \mathbf{X}_d; \quad \mathbf{X}_s; \quad \mathbf{X}/\sqrt{m}; \quad \text{or} \quad \mathbf{X}_s/\sqrt{m}$$

by procedures identical to those shown above. As we know, division of **X**, $\mathbf{X}_d$, or $\mathbf{X}_s$ by the scalar $\sqrt{m}$ has no effect on the eigenvectors of either the minor or major product moments of **X**, $\mathbf{X}_d$, or $\mathbf{X}_s$. Corresponding eigenvalues of the product-moment matrix are changed by multiplication by $1/m$, which, in this case, represents the sample size.

6.4.4 Other Aspects of Principal Components Analysis

The example of Table 6.3 represents only one type of principal components analysis, namely, a components analysis of the covariance matrix **C**. As indicated above, one can component-analyze the averaged raw cross-products matrix $\mathbf{X}'\mathbf{X}/m$ or the correlation

[4] Alternatively, we could find **U** and **T′** simply by finding the eigenvectors of $\mathbf{X}_d\mathbf{X}_d{}'/m$ and $\mathbf{X}_d{}'\mathbf{X}_d/m$ separately.

matrix $\mathbf{X}_s'\mathbf{X}_s/m$. In general, the eigenstructures of these three matrices will differ. That is, unlike some factoring methods, such as canonical factor analysis (Van de Geer, 1971), principal components analysis is *not* invariant over changes in scale or origin.

Principal components analysis does exhibit the orthogonality of axes property in which the axes display sequentially maximal variance. That is, the first axis displays the largest variance, the second (orthogonal) axis, the next largest variance, and so on. In problems of practical interest a principal components analysis might involve a set of 30 or more variables, rather than the two variables X_1 and X_2, used here for illustrative purposes. Accordingly, the opportunity to replace a large number of highly correlated variables with a relatively small number of uncorrelated variables, with little loss of information, represents an attractive prospect. It is little wonder that principal components analysis has received much attention by researchers working in the behavioral and administrative sciences.

It is also not surprising that a large variety of other kinds of factoring methods have been developed to aid the researcher in reducing the dimensionality of his data space. Still, principal components represents one of the most common procedures for factoring matrices and, if anything, its popularity is on the rise.

However, from a substantive viewpoint, the orientation obtained from principal components may not be the most interpretable. Accordingly, applications researchers often rotate the component axes that they desire to retain to a more meaningful orientation from a content point of view. A number of procedures (Harman, 1967) are available to accomplish this task. Generally, the applied researcher likes to rotate component axes with a view to having each variable project highly on only one rotated dimension and nearly zero on others.

Another problem in any type of factoring procedure concerns the number of components (or factors, generally) to retain. Most data matrices will be full rank; hence, assuming that the number of objects exceeds the number of variables, one will obtain as many components as there are variables. Often the "lesser" components (those with lesser variance) are discarded; one often keeps only the first r $(<n)$ components that account for some appreciable proportion (e.g., 80 to 90 percent) of the total variance in the data. Other rules for deciding how many factors to retain are also in use, including various statistical and graphical criteria. Still, the decision is largely a judgmental one, and factor analysis remains something of an ad hoc set of procedures.

Factor analysis—either principal components or other type of factoring procedure—represents only one class of methods for effecting dimensional reduction of one's data. More recently, new classes of techniques, such as multidimensional scaling (Green and Wind, 1973), have been used to develop reduced spaces. Many of these newer methods require only rank order input data. For example, the elements of a covariancelike matrix need only be ranked in order for these "nonmetric" procedures to be used.

However, insofar as the metric procedure of principal components analysis is concerned, we see that the major mathematical tool involves the eigenstructure of symmetric matrices. Related concepts such as the singular value decomposition of a matrix into its basic structure, quadratic forms, and matrix rank are also of interest.

From a geometric viewpoint we seek a rotation of the original basis of the space that coincides with the axes of the hyperellipsoid of points, assumed to represent the objects, in the original n-dimensional space. The eigenvalues correspond to the variances of these new axes, and the normalized eigenvectors of the particular cross-product matrix employed are the direction cosines that define the rotation.

6.5 THE MULTIPLE DISCRIMINANT ANALYSIS PROBLEM

The third problem described in Chapter 1 concerns the development of linear composites of X_{d1} and X_{d2} with the property of maximally separating the three groups (shown in Fig. 1.5). That is, in this multivariate application, the 12 employees, based on the data of Table 6.1, were split into three groups with regard to degree of absenteeism:

Group 1–low	(employees a, b, c, d)
Group 2–intermediate	(employees e, f, g, h)
Group 3–high	(employees i, j, k, l)

While it happens to be the case here that the three groups are ordered with respect to extent of absenteeism, this is not a requirement of multiple discriminant analysis (MDA). Any polytomy consisting of a set of mutually exclusive and collectively exhaustive groups is sufficient for application of MDA.

In the sample problem application of MDA, we wish to find a linear composite of X_{d1} and X_{d2} with the property of maximizing among-group variation relative to (pooled) within-group variation. Like principal components analysis, this involves finding the eigenstructure of a matrix. However, in this case the matrix is nonsymmetric, although it, in turn, represents the product of two symmetric matrices.

6.5.1 Finding the Eigenstructure

The quantity to be maximized in MDA consists of the ratio

$$\boxed{\lambda_1 = \frac{\mathbf{v}_1'\mathbf{A}\mathbf{v}_1}{\mathbf{v}_1'\mathbf{W}\mathbf{v}_1} = \frac{SS_{\mathrm{A}}(\mathbf{w}_1)}{SS_{\mathrm{W}}(\mathbf{w}_1)}}$$

where, as it turns out, λ_1 is an eigenvalue, and $\mathbf{A}$ and $\mathbf{W}$ denote among-group and pooled within-group SSCP matrices, respectively. The vector $\mathbf{v}_1$ denotes the set of weights used to develop the linear composite (denoted as $\mathbf{w}_1$) of the original mean-corrected score matrix $\mathbf{X}_{\mathrm{d}}$, while SS_{A} and SS_{W} denote the among-group and within-group sums of squares of the linear composite. In scalar notation,

$$\boxed{w_{i(1)} = v_1 X_{\mathrm{d}i1} + v_2 X_{\mathrm{d}i2}}$$

Let us develop these concepts, step by step.

Figure 6.6 shows the first linear composite $\mathbf{w}_1$ that we seek. We see that $\mathbf{w}_1$ makes an angle of 25° with the horizontal axis. We can project the 12 points onto $\mathbf{w}_1$ and find their discriminant scores $w_{i(1)}$. (The grand mean of these scores will be zero.) Also, we can find the three group means on $\mathbf{w}_1$ and the associated among-group and pooled

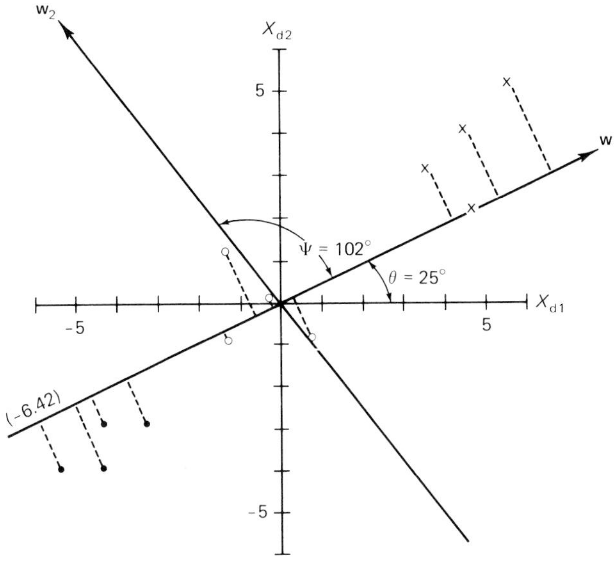

Fig. 6.6 Discriminant transformation of the mean-corrected predictor variables from Table 6.1. Key: ● Group 1; ○ Group 2; x Group 3.

within-group sums of squares. According to the preceding criterion, we have found a set of scores $w_{i(1)}$ with the property that

$$\lambda_1 = \frac{SS_A(\mathbf{w}_1)}{SS_W(\mathbf{w}_1)}$$

is maximal. That is, if we find (a) the sum of squares of the three group means from the grand mean, which is zero in the case of mean-corrected data and (b) the pooled within-group sum of squares of each of the scores about their respective group means on $\mathbf{w}_1$, then (c) the ratio λ_1 of these two sums of squares is greater than that found by any other suitably normalized[5] axis in the space of Fig. 6.6.

Table 6.4 shows the preliminary calculations of interest. First, we compute each group mean on X_{d1} and X_{d2}, respectively; with equal-size groups these means, of course, sum to zero, within rounding error. Then we find the matrix of within-group deviations and the matrix of among-group deviations from group and grand mean, respectively.

From here, we compute the minor product moment of

$\mathbf{X}_k - \bar{\mathbf{X}}_k$, the matrix of within-group deviations and, similarly, the minor product moment of

$\bar{\mathbf{X}}_k - \bar{\bar{\mathbf{X}}}_k$, the matrix of among-group deviations (for $k = 1,2, \ldots, K = 3$ groups)

[5] That is, we seek a linear composite in which the coefficients v_1 and v_2 are direction cosines. Also, it should be remembered in the computation of $SS_A(\mathbf{w}_1)$ that each group mean is based on four observations.

TABLE 6.4

Preliminary Calculations for Multiple Discriminant Analysis

Employee		X_{d1}	X_{d2}	$\mathbf{X}_k - \bar{\mathbf{X}}_k$ Within-group deviations		$\bar{\mathbf{X}}_k - \bar{\bar{\mathbf{X}}}$ Between-group deviations	
1	a	−5.25	−3.92	−1	−0.5	−4.25	−3.42
	b	−5.25	−3.92	0	−0.5	−4.25	−3.42
	c	−4.25	−2.92	0	0.5	−4.25	−3.42
	d	−3.25	−2.92	1	0.5	−4.25	−3.42
Mean		−4.25	−3.42				
2	e	−1.25	−0.92	−0.75	−0.75	−0.5	−0.17
	f	−1.25	1.08	−0.75	1.25	−0.5	−0.17
	g	−0.25	0.08	0.25	0.25	−0.5	−0.17
	h	0.75	−0.92	1.25	−0.75	−0.5	−0.17
Mean		−0.50	−0.17				
3	i	3.75	3.08	−1	−0.50	4.75	3.58
	j	4.75	2.08	0	−1.50	4.75	3.58
	k	4.75	4.08	0	0.50	4.75	3.58
	l	5.75	5.08	1	2.50	4.75	3.58
Mean		4.75	3.58				

$$\mathbf{W} = (\mathbf{X}_k - \bar{\mathbf{X}}_k)'(\mathbf{X}_k - \bar{\mathbf{X}}_k); \qquad \mathbf{A} = (\bar{\mathbf{X}}_k - \bar{\bar{\mathbf{X}}})'(\bar{\mathbf{X}}_k - \bar{\bar{\mathbf{X}}})$$

$$\mathbf{W} = \begin{bmatrix} 6.75 & 1.75 \\ 1.75 & 8.75 \end{bmatrix}; \qquad \mathbf{A} = \begin{bmatrix} 163.50 & 126.50 \\ 126.50 & 98.17 \end{bmatrix}$$

$$\mathbf{T} = \mathbf{W} + \mathbf{A} = \begin{bmatrix} 170.25 & 128.25 \\ 128.25 & 106.92 \end{bmatrix}$$

to find **W**, the pooled within-group SSCP matrix, and **A**, the among-group SSCP matrix, as shown in Table 6.4. Their sum equals the total sample SSCP matrix **T**, which is also shown in Table 6.4.

The problem, as shown in Appendix A, is to maximize λ_1 with respect to $\mathbf{v}_1$. The resulting matrix equation is

$$\boxed{(\mathbf{A} - \lambda_1 \mathbf{W})\mathbf{v}_1 = \mathbf{0}}$$

Assuming that **W** is nonsingular and, hence, that $\mathbf{W}^{-1}$ exists, we can premultiply both sides of the above equation to get

$$(\mathbf{W}^{-1}\mathbf{A} - \lambda_1 \mathbf{I})\mathbf{v}_1 = \mathbf{0}$$

with characteristic equation

$$|\mathbf{W}^{-1}\mathbf{A} - \lambda_1 \mathbf{I}| = 0$$

Note, then, that we have another eigenstructure problem, one now involving the nonsymmetric matrix $\mathbf{W}^{-1}\mathbf{A}$.

TABLE 6.5

Finding the Eigenstructure of the $\mathbf{W}^{-1}\mathbf{A}$ *Matrix*

$$\mathbf{W}^{-1} = \begin{bmatrix} 0.156 & -0.031 \\ -0.031 & 0.121 \end{bmatrix}; \qquad \mathbf{W}^{-1}\mathbf{A} = \begin{bmatrix} 21.594 & 16.698 \\ 10.138 & 7.880 \end{bmatrix}$$

Eigenvalues of $\mathbf{W}^{-1}\mathbf{A}$ — Eigenvectors of $\mathbf{W}^{-1}\mathbf{A}$

$$\Lambda = \begin{bmatrix} 29.444 & 0 \\ 0 & 0.0295 \end{bmatrix}; \qquad \mathbf{V} = \begin{bmatrix} 0.905 & -0.612 \\ 0.425 & 0.791 \end{bmatrix}$$

As is the case with principal components analysis, generally the characteristic equation will have more than a single root. In fact, in this problem we shall be able to find two eigenvalues, λ_1 and λ_2, and their associated eigenvectors.

Table 6.5 shows the eigenvalues and eigenvectors obtained for this sample application. Note the parallel between this problem and the principal components problem. In each case we are finding the eigenstructure of a matrix, but here the matrix is nonsymmetric.

From Table 6.5 we see that the first discriminant function displays a relatively large eigenvalue of

$$\lambda_1 = 29.444$$

with associated, and normalized, eigenvector

$$\mathbf{v}_1 = \begin{bmatrix} 0.905 \\ 0.425 \end{bmatrix}$$

representing an angle of 25° from the horizontal axis.

Also, similar to principal components analysis, we can obtain a second discriminant function $\mathbf{w}_2$ with scores that are uncorrelated with those of the first function. The eigenvalue associated with $\mathbf{w}_2$ is

$$\lambda_2 = 0.0295$$

with normalized eigenvector

$$\mathbf{v}_2 = \begin{bmatrix} -0.612 \\ 0.791 \end{bmatrix}$$

Note that λ_2 is much smaller than λ_1; for all practical purposes it appears that a single discriminant function might account for these data.

In general, with K groups and n predictors one obtains

$$\boxed{\min(K-1, n)}$$

different discriminant functions; here, of course $K - 1 = n = 2$, and we note that two functions are obtained. Usually, however, the number of predictors will greatly exceed the number of groups, and a great deal of parsimony can often be achieved by the use of discriminant scores.

As before, discriminant scores are found by projecting the points onto the discriminant axes. For example, the discriminant score of the first observation on $\mathbf{w}_1$ is

$$w_{1(1)} = 0.905(-5.25) + 0.425(-3.92) = -6.42$$

as shown in Fig. 6.6.

We also observe in Fig. 6.6 that $\mathbf{v}_1$ and $\mathbf{v}_2$ are not orthogonal, even though the scores on $\mathbf{w}_1$ versus those on $\mathbf{w}_2$ are uncorrelated. From the matrix of eigenvectors in Table 6.5 we can compute the cosine between $\mathbf{v}_1$ and $\mathbf{v}_2$ as follows:

$$\cos \Psi = (0.905 \quad 0.424)\begin{bmatrix} -0.612 \\ 0.791 \end{bmatrix} = -0.21, \qquad \text{so that} \quad \Psi = 102^\circ$$

Thus, the angle Ψ separating $\mathbf{v}_1$ and $\mathbf{v}_2$ is $90^\circ + 12^\circ = 102^\circ$, as shown in Fig. 6.6.

6.5.2 Statistical Significance and Classification

It is one thing, of course, to find linear composites with the properties described above; it is quite another to test their statistical significance and to use them for classifying observations. Accordingly, each of these problems is taken up, in turn. At this point we have found two eigenvalues:

$$\lambda_1 = 29.444; \qquad \lambda_2 = 0.0295$$

Bartlett (1947) has proposed a statistic that can be used to test the significance of the discriminant functions (actually, their eigenvalues).

Bartlett's statistic starts out by testing the null hypothesis that group centroids are all equal in the *full* discriminant space, in this case involving both the $\mathbf{w}_1$ and $\mathbf{w}_2$ axes.

Bartlett's statistic is expressed as follows:

$$\boxed{V = 2.3026[m-1-(n+K)/2] \sum_{i=1}^{r} \log(1+\lambda_i)}$$

where m denotes sample size, n denotes number of predictor variables, K denotes number of groups, and λ_i denotes the ith eigenvalue ($i = 1, 2, \ldots, r$). In terms of the sample problem,

$$\begin{aligned} V &= 2.3026[12-1-(2+3)/2](\log 30.444 + \log 1.0295) \\ &= 2.3026(8.5)(1.48350 + 0.01263) \\ &= 29.035 + 0.247 \\ &= 29.282 \end{aligned}$$

Bartlett's V statistic is approximately distributed as chi square with $n(K-1) = 4$ degrees of freedom. In the sample problem, V is significant beyond the 0.01 alpha level.

However, one wonders whether the second discriminant function, whose eigenvalue is almost zero, adds anything beyond the first. Fortunately, V can be decomposed into the separate parts

$$V_1 = 29.035; \qquad V_2 = 0.247$$

The first portion V_1 has already been tested in the context of V. However, if V_1 is "partialed out," is V_2 statistically significant? As it turns out, V_2 can be tested in the same way that V was tested: $V_2 = 0.247$ is also approximately distributed as chi square with

$$n(K-1)-(n+K-2) = (n-1)(K-2) = 1$$

degree of freedom. This approximate chi square ($V_2 = 0.247$) is clearly nonsignificant. Not surprisingly, we conclude that only the first discriminant function need be retained.

Bartlett's procedure can be used for more than two discriminant functions in a similar manner. Had a third discriminant function been involved, its associated degrees of freedom would be $(n-2)(K-3)$; those associated with a fourth discriminant function would be $(n-3)(K-4)$, and so on. However, the reader interested in applying this test in substantive research should be aware of its assumptions (Harris, 1975; pp. 109-113).

In the sample problem it is not hard to see why only the first discriminant function is significant. The following ratio:

$$\frac{\lambda_1}{\lambda_1 + \lambda_2} = \frac{29.444}{29.444 + 0.0295} = 0.999$$

shows that $\mathbf{w}_1$ exhausts virtually all of the variation in the discriminant space.

Classifying the twelve observations by means of $\mathbf{w}_1$, the retained and significant discriminant function, is quite straightforward. All that is entailed is to compute a discriminant score for each observation, according to

$$w_{(1)} = v_1 X_{d1} + v_2 X_{d2}$$

One also computes the discriminant scores for the three group means

$$\bar{w}_1(\text{Group 1}) = 0.905(-4.25) + 0.425(-3.42) = -5.30$$

$$\bar{w}_1(\text{Group 2}) = 0.905(-0.50) + 0.425(-0.17) = -0.52$$

$$\bar{w}_1(\text{Group 3}) = 0.905(4.75) + 0.425(3.58) = 5.82$$

One then assigns each observation to that group whose mean score on $\mathbf{w}_1$ is closest to the individual score, $w_{i(1)}$.

When this procedure is implemented for the sample problem, it turns out that all twelve cases are correctly assigned to their respective groups. Had $\mathbf{w}_2$ also been statistically significant and retained for classification purposes, the classification procedure would have been modified to involve the computation of Euclidean distances between each individual observation and each group centroid in discriminant function space.[6] Each observation would then be assigned to the group whose centroid, in discriminant function space, was nearest.

It should be mentioned, however, that the use of Bartlett's statistic and the classification rules enumerated above only scratch the surface of the topics of statistical

[6] It should be noted that in discriminant function space the pooled within-group SSCP matrix would first be spherized by means of the procedure described in Section 5.9.2; it is *this* space in which ordinary Euclidean distance is appropriate (given equal prior probabilities and equal costs of misclassification) for assigning objects to groups.

significance and assignment. For example, Bartlett's statistic can be modified by Schatzoff's tables (Schatzoff, 1966) to produce an exact test. Also, other tests (Rao, 1952) are available for testing the null hypothesis of group centroid equality in the full discriminant space.

The classification rules described above also need modification in cases where the prior probabilities of inclusion differ across the groups or where the costs of misclassification differ. Modern approaches to the problem (e.g., Eisenbeis and Avery, 1972) formulate the classification task in terms of statistical decision theory. As such, both prior probabilities of an observation belonging to each of the groups and costs of misclassification can be explicitly introduced into the assignment procedure.

6.5.3 Other Aspects of Multiple Discriminant Analysis

One of the questions posed in Chapter 1 concerned the relative importance of the two predictors X_{d1} and X_{d2} in effecting group discrimination. In the case of correlated predictors, this represents an ambiguous question and shares, along with multiple regression and other multivariate techniques, the difficulties of parceling out variance among nonorthogonal predictors. While we do not go into this question in detail, a few procedures that have been suggested for ascribing relative importance to X_{d1} versus X_{d2} can be mentioned.

First, the entries in the normalized eigenvector $\mathbf{v}_1$ are 0.905 and 0.425 for X_{d1} and X_{d2}, respectively. These are analogous to partial regression coefficients in multiple regression. To convert them into standardized (beta-type) numbers, each is multiplied by that predictor variable's pooled within-group standard deviation:[7]

$$\text{Standardized weight } (X_{d1}); \qquad 0.905 \times \sqrt{0.75} = 0.783$$

$$\text{Standardized weight } (X_{d2}); \qquad 0.425 \times \sqrt{0.972} = 0.419$$

As can be noted, on either a standardized or nonstandardized basis, X_{d1} receives the larger weight.

Cooley and Lohnes (1971) recommend what they call structure correlations to ascertain predictor importance. These are merely the product-moment correlations between scores on each original variable and the discriminant scores. In this example they turn out to be

$$\text{Structure correlation } (X_{d1}) = 0.998$$

$$\text{Structure correlation } (X_{d2}) = 0.976$$

In this case both predictors correlate highly with the retained discriminant function $\mathbf{w}_1$, although the correlation for X_{d1} is slightly greater.

Still other procedures, such as Bock and Haggard's (1968) step-down F ratios, can be employed to measure the relative importance of various predictors. However, we do not delve into these more esoteric methods, other than to say that the question of ascribing "relative importance" remains ambiguous in the case of correlated predictor variables no matter what procedure is used.

[7] Other standardization procedures, based on multiplication of each discriminant coefficient by the total-sample (as opposed to pooled within-group) standard deviation of the variables of interest, are also in use.

Another topic of interest concerns the relationship of MDA to other multivariate techniques. For example, an intimate connection exists between MDA and principal components analysis. Without delving deeply into the technical details, it turns out that a preliminary "spherizing" of the data matrix via

$$\boxed{X_d W^{-1/2}}$$

results in a new set of coordinates with spherical (pooled) within-group variation.

One can then find the eigenstructure of the matrix

$$\boxed{W^{-1/2} A W^{-1/2}}$$

so as to satisfy the equation

$$\boxed{[W^{-1/2} A W^{-1/2}] Q = Q\Lambda}$$

where Q is orthogonal and Λ is diagonal.[8] The final transformation is then

$$\boxed{X_d W^{-1/2} Q}$$

which, of course, is a spherizing transformation followed by a rotation of the spherized within-group variation to principal axes, on the basis of among-group variation. This idea was described in the discussion in Chapter 5 of the simultaneous diagonalization of two different quadratic forms.[9]

Probably the most important point to mention, however, is that MDA is one member of the same general family that includes

1. canonical correlation,
2. multivariate analysis of variance and covariance,
3. categorical canonical correlation.

The linkage among these multiple-criterion techniques is provided by a generalized canonical correlation model that allows for dummy variables on one or both sides of the equation. For example, one could have developed a multiple discriminant function for the sample problem by means of a canonical correlation in which the criterion variables were represented by the dummies

1	0	(Group 1)
0	1	(Group 2)
0	0	(Group 3)

By a similar judicious choice of dummy and continuous variables, one can find linear composites of both the criterion and predictor batteries that relate to any of the specific multivariate techniques described above, and in Chapter 1 as well.

[8] In this illustration Λ denotes the diagonal matrix of eigenvalues of $W^{-1/2} A W^{-1/2}$, while Q denotes the associated matrix of eigenvectors.

[9] Still other procedures are available for finding the eigenstructure of $W^{-1}A$ (see Overall and Klett, 1972).

Full discussion of the interrelationships among techniques would take us far beyond the scope of the book. As we have illustrated in Chapter 5, however, the eigenstructure of nonsymmetric matrices and the simultaneous diagonalization of two different quadratic forms figure prominently in the computation of discriminant functions for three or more groups. These concepts are also central in canonical correlation, multivariate analysis of variance and covariance, and categorical canonical correlation; in the last case, all variables are expressed as dummies.

6.6 A PARTING LOOK AT MULTIVARIATE TECHNIQUE CLASSIFICATION

In Chapter 1 a number of characteristics were enumerated that provided guidance for classifying the large, and still growing, variety of multivariate techniques. In particular, the following descriptors represented the main bases of classification:

1. whether the data matrix is kept intact versus partitioned into criterion and predictor subsets;
2. the number of variables in each subset (if partitioning is undertaken);
3. the types of scales by which the variables are measured.

At this point, however, the various types of linear transformations described in Chapters 4 and 5 are behind us. And, even in the introductory material of Chapter 1, it was suggested that multivariate analysis is largely concerned with transformations for matching one set of numbers, such as a data vector, a data matrix, or a linear composite, with some other set of numbers.

The degree of matching is usually assessed by a residual sum of squared deviations or some other measure that can be related to this. This idea was illustrated at the beginning of the present chapter in the context of multiple regression. Here we desired to minimize the quantity

$$\boxed{\sum_{i=1}^{m} e_i^2 = (Y_i - \hat{Y}_i)^2}$$

where Y_i denotes a datum, and $\hat{Y}_i$ denotes a predicted value of Y_i. *As a further aid to technique classification, we now take the view that multivariate techniques may differ according to the nature of the allowable transformations and the properties that the transformed matrices exhibit in the matching process.*

Partly by way of review of Chapters 4 and 5 and partly by way of prologue, let us list the major classes of transformations that vectors or matrices can undergo in the course of achieving various types of matching. For illustration, let us assume a general data matrix, denoted by **X**, of m rows and n columns ($m \geqslant n$).

Our objective here will be to recapitulate various types of transformations described in earlier chapters as a way to make explicit the present descriptor, the nature of the linear transformation, for characterizing multivariate techniques.

6.6.1 Types of Transformations

By way of an overview, Fig. 6.7 shows a directed graph of the transformations that are considered. This list of transformations is not meant to be exhaustive. However, those shown in Fig. 6.7 appear to be the most frequently encountered ones in multivariate analysis. We consider the more general classes first, followed by the more restricted transformations.

We shall let **T** denote an arbitrary matrix. The matrices **U** and **V**′ denote either orthogonal matrices or orthonormal sections, while Δ denotes a diagonal matrix. From Section 5.7 we know that **T** can always be decomposed into the triple product

$$\boxed{\mathbf{T} = \mathbf{U}\boldsymbol{\Delta}\mathbf{V}'}$$

We shall take advantage of this type of singular value (Eckart–Young, 1936) decomposition in describing various special cases of a general linear (or affine) transformation.

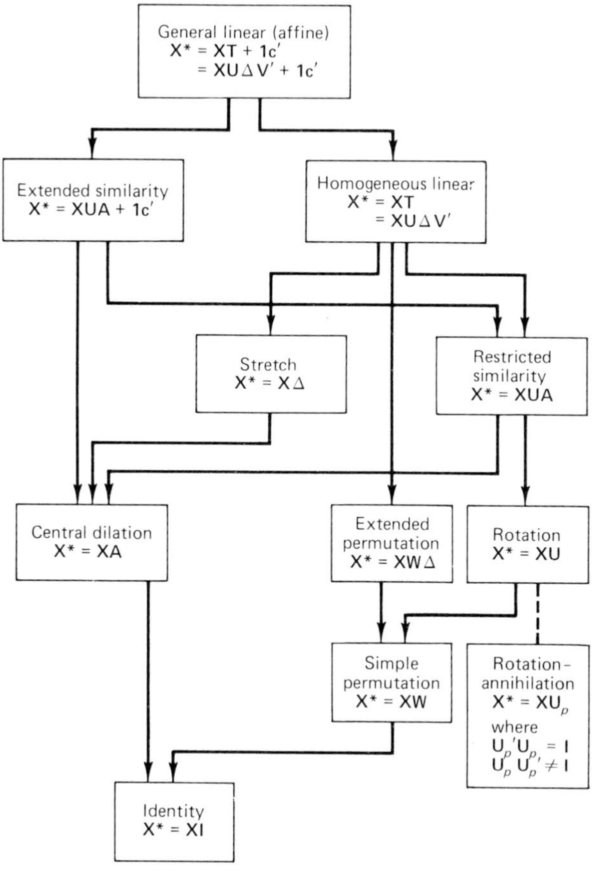

Fig. 6.7 A directed graph of various types of linear transformations.

6.6.1.1 General Linear (Affine) Transformation The most general transformation of $\mathbf{X}$ to be considered is an affine transformation, defined as

$$\boxed{\mathbf{X}^* = \mathbf{XT} + \mathbf{1c}'}$$

where $\mathbf{T}(=\mathbf{U\Delta V}')$ denotes an arbitrary linear transformation. The matrix product of the $m \times 1$ unit column vector $\mathbf{1}$ and $\mathbf{c}'$ a $1 \times n$ row vector of constants defines the permissible shift of origin (as illustrated in Chapter 4 in the case of a centroid-centered orientation).[10]

6.6.1.2 Homogeneous Linear Transformation A homogenous linear transformation can be defined as

$$\boxed{\mathbf{X}^* = \mathbf{XT}}$$

with no shift in origin, but $\mathbf{T}$, defined as before, is otherwise not restricted.

6.6.1.3 Similarity Transformations An extended similarity transformation involves a rotation, achieved by the orthogonal matrix $\mathbf{U}$, a central dilation, effected by the scalar matrix $\mathbf{A}$, and a shift in origin:

$$\boxed{\mathbf{X}^* = \mathbf{XUA} + \mathbf{1c}'}$$

where $\mathbf{1}$ and $\mathbf{c}'$ are defined as before.

A restricted similarity transformation is a special case of this in which no shift in origin is permitted:

$$\boxed{\mathbf{X}^* = \mathbf{XUA}}$$

where $\mathbf{U}$ and $\mathbf{A}$ are defined as before.

6.6.1.4 Rotation As illustrated in earlier chapters, a rotation is a transformation that is carried out by an orthogonal matrix. This type of matrix is denoted by $\mathbf{U}$, where $\mathbf{U}'\mathbf{U} = \mathbf{UU}' = \mathbf{I}$. The transformation is written as

$$\boxed{\mathbf{X}^* = \mathbf{XU}}$$

If the determinant $|\mathbf{U}| = 1$, then a proper rotation of $\mathbf{X}$ is entailed. If $|\mathbf{U}| = -1$, then a rotation of $\mathbf{X}$ followed by a reflection is entailed (i.e., an improper rotation).

6.6.1.5 Rotation–Annihilation One type of transformation stipulates that only the condition $\mathbf{U}_p{}'\mathbf{U}_p = \mathbf{I}$ be met; that is, $\mathbf{U}_p$ can be an orthonormal section (rectangular rather than square) whose columns are mutually orthogonal and of unit length. This amounts to a rotation followed by annihilation of some dimensions.

In Fig. 6.7 we show this transformation with a dotted rather than solid line. This is because $\mathbf{U}_p$ is not a special case of $\mathbf{U}$ for the reason that $\mathbf{U}_p \mathbf{U}_p{}' \neq \mathbf{I}$. As such, $\mathbf{U}_p$ is rather tangentially related to the overall schema of Fig. 6.7.

[10] Note that an affine transformation is nonhomogeneous in the sense that there are no fixed points (e.g., the $\mathbf{0}$ or origin vector) under this type of mapping. However, Section 4.4.1 shows how it can be carried out via matrix multiplication.

6.6.1.6 Permutation An extended permutation permits both a reordering of dimensions and a stretch or rescaling of the configuration. This is written as

$$\boxed{\mathbf{X}^* = \mathbf{X}\mathbf{W}\mathbf{\Delta}}$$

where $\mathbf{W}$ is a permutation matrix and $\mathbf{\Delta}$ is diagonal. As recalled, a permutation matrix is an orthogonal matrix, all of whose entries are either 0 or 1, that changes the order of dimensions.

A simple permutation is written as

$$\boxed{\mathbf{X}^* = \mathbf{X}\mathbf{W}}$$

with $\mathbf{W}$ defined as before.

6.6.1.7 Stretch A stretch transformation involves a simple rescaling of the configuration by a diagonal matrix $\mathbf{\Delta}$. That is,

$$\boxed{\mathbf{X}^* = \mathbf{X}\mathbf{\Delta}}$$

6.6.1.8 Central Dilation A special case of a stretch transformation involves a central dilation, given by the scalar matrix $\mathbf{A}$. That is,

$$\boxed{\mathbf{X}^* = \mathbf{X}\mathbf{A}(=\mathbf{A}\mathbf{X})}$$

6.6.1.9 Identity Transformation A special case of the central dilation transformation is the identity transformation

$$\boxed{\mathbf{X}^* = \mathbf{I}\mathbf{X} = \mathbf{X}\mathbf{I} = \mathbf{X}}$$

where $\mathbf{I}$, of course, is the identity matrice.

While still other combinations are possible, the preceding ones cover most cases of practical interest.

6.6.2 Constructing the Classification

With the various geometric illustrations presented in the preceding section, it is now appropriate to discuss the nature of multivariate techniques from the standpoint of configuration matching. We consider the following classes:

1. vector-matrix matching,
2. matching of two matrices,
3. matching of three or more matrices,
4. matching of a data-based and an internally derived matrix.

Within each of these classes, two additional aspects are discussed:

1. types of scores—continuous or binary (dummy variable),
2. type of permissible transformation applicable to each matrix or vector.

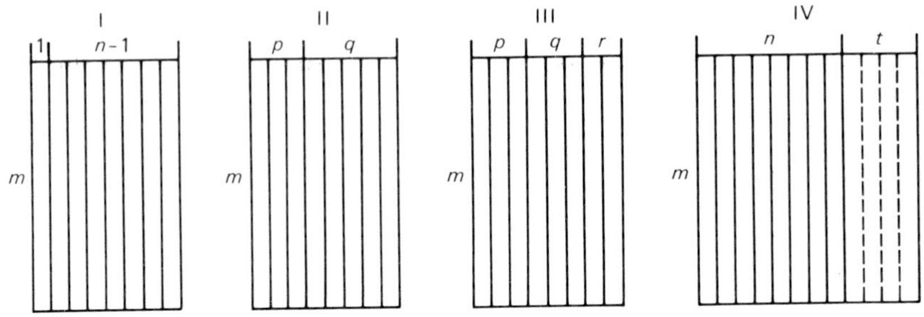

Fig. 6.8 Illustrative partitionings of data matrix.

Each of the above major classes is examined in turn. To assist us in this regard, we reproduce in Fig. 6.8 the schema that appeared as in Fig. 1.1. However, now we emphasize the nature of the linear transformation.

6.6.2.1 Vector-Matrix Matching Panel I of Fig. 6.8 is the prototype of the family of multivariate techniques illustrated by the matrix equation

$$\boxed{\hat{\mathbf{y}} = \mathbf{X}\mathbf{b}}$$

where $\mathbf{b}$ is a vector of combining weights, and $\hat{\mathbf{y}}$ is a set of predicted values for $\mathbf{y}$, the criterion variable. As illustrated earlier, by allowing $\mathbf{y}$ or $\mathbf{X}$ to be mixtures of continuous or binary-coded variables, this family is broad enough to include

1. multiple (and simple) regression;
2. two-group discriminant analysis, where $\mathbf{y}$ is binary valued;
3. analysis of variance and covariance, where some columns of $\mathbf{X}$ are binary valued;
4. binary-valued regression, where both $\mathbf{y}$ and $\mathbf{X}$ are binary valued.

In least-squares theory the scalars R^2 or η^2 (eta squared) are usually the quantities being maximized.[11] Both R^2, in the context of regression, and η^2, in the context of analysis of variance, are invariant over linear transformations of $\mathbf{y}$ or $\mathbf{X}$. Moreover, both R^2 and η^2 can be simply related to the criterion of minimizing the sum of the squares of $\mathbf{y} - \hat{\mathbf{y}}$, as described earlier.

6.6.2.2 Matching Two Matrices In Panel II of Fig. 6.8 we have the case of two matrices, $\mathbf{Y}_{m \times p}$ and $\mathbf{X}_{m \times q}$, and are interested in the association between these two batteries of variables. If we assume that both sets of variables represent continuous values, the canonical correlation problem can be represented by separate affine transformations of $\mathbf{Y}$ and $\mathbf{X}$ such that each pair of linear composites is most congruent with each other, subject to being uncorrelated with previously "extracted" composites. This uncorrelatedness condition is an illustration of the kinds of restrictions that may be placed on the transformed values.

[11] The scalar η^2 (eta squared) is computed in just the same way as R^2 except for the fact that all predictor values are dummy variables. This is equivalent to

$$\eta^2 = \frac{SS_A}{SS_T}$$

while SS_A denotes the among-group sum of squares and SS_T denotes the total-group sum of squares.

Other possibilities come to mind, however. For example, one could allow only a separate homogenous linear transformation of each matrix with no shift in origin permitted. Or, one could permit a shift in origin but require each transformation to be an extend similarity transformation which, as shown earlier, is less general than an affine transformation.

In other kinds of applications we may desire **Y** to remain fixed (i.e., transformed by an identity matrix) but permit **X** to be transformed by an affine transformation, extended similarity, or a similarity transformation. Some "procrustes" solutions, as used in matching factor score solutions from different studies, are of this general type (Rummel, 1970).

Still other restrictions are possible. Schönemann and Carroll (1970) describe a matching procedure in which one matrix undergoes an extended similarity transformation, while the other undergoes either (a) an extended similarity, (b) a similarity, (c) a rotation, or (d) an identity transformation. Cliff's procedure (Cliff, 1966) allows a similarity transformation on one side and a similarity, rotation, or identity transformation on the other.

If one matrix consists of two or more binary-valued variables, we have an instance of either multiple discriminant analysis or multivariate analysis of variance, depending upon how one frames the problem. From the standpoint of permissible transformations, however, the techniques are similar. That is, one can formulate either a multiple discriminant problem or a multivariate analysis of variance problem in terms of the canonical correlation model with one of the two matrices represented by binary-valued dummies. Generally, however, we are interested in special kinds of output that are related to the particular procedure employed. Therefore, while one *could* use a canonical correlation program to find discriminant weights, ordinarily we would not do so since we would be interested in various ancillary outputs as well.

If both data sets consist of dummy variables, we may have a case of categorical canonical correlation or categorical conjoint measurement (Carroll, 1973). Insofar as the solution to the problem is concerned, these techniques are special cases of canonical correlation in which *both* matrices consist of dummy variables.

Variations can be developed, however. For example, Horst (1956) describes a type of multiple discriminant analysis in which the dummy-variable criterion matrix, defining group membership, remains fixed. The predictor matrix is transformed linearly to best match it, subject to the predicted values maintaining the same column means and variances as the columns of the criterion-variable matrix.

6.6.2.3 Three or More Matrices Heretofore, we have described multivariate analysis of covariance in terms of a matrix of criterion variables and a matrix of predictor variables. The latter matrix consists of a mixture of dummy variables, the design variables, and covariates, whose effect on the criterion variables we desire to remove. Alternatively, we can partition the data matrix into three matrices: criterion, design dummies, and covariates, as illustrated in Panel III of Fig. 6.8.

Problems involving three, or more, matrices fall into two major types:

1. a multivariate analysis of covariance situation, or multiple, partial correlation (Cooley and Lohnes, 1971), in which one of the data matrices consists of a set of

covariates, moderators, or contingency variables whose effect is to be removed before considering the association between the remaining matrices;

2. a generalized canonical correlation situation where the status of all three, or more, matrices is considered to be the same (in this case, we extend two-group canonical correlation to cover three or more matrices).

Multivariate analysis of covariance problems occurs frequently in the behavioral and administrative sciences. For example, one may set up various experiments in which several response measures are sought from the subjects and, furthermore, certain covariates like task familiarity, education, and IQ level are also included in the analysis.

In multivariate analysis of covariance one matrix, the response matrix, is typically made up of continuous scores, while the design matrix is typically made up of dummy variables. The matrix of covariates is usually made up of continuous scores. However, this is not necessary. In principle, any (or all three) of the matrices could consist of continuous or binary-valued scores, or, indeed, as mixtures. In this class of problems one generally allows affine transformations to be applied to any of the three matrices, in the spirit of two-set canonical correlation.

Generalized canonical correlation, employing three or more data-based matrices of equal status, is concerned primarily with configuration matching. Horst (1961), Carroll (1968), and Kettenring (1972) have all proposed models for this type of problem. For example, in the Carroll and Chang approach, an $r + 1$st space is defined such that the r original spaces, each consisting of the same m observations on r sets of variables, are transformed to match it as well as possible. This procedure allows an affine transformation of each "contributing" configuration.

While generalized canonical correlation has usually been considered in the context of all scores being continuous, this, again, is not necessary provided that the researcher's interest is centered on data description and summarization, rather than on statistical inference. Binary-valued scores, or mixtures of continuous and binary valued, can be dealt with just as readily. Again, affine transformations would generally be permitted.

6.6.2.4 Matching Based on an Internal Criterion Multivariate techniques can also cover the possibility of deriving a matrix (e.g., a "latent" matrix) that best reproduces the scores of a data-based matrix, or some matrix derived from it, subject to meeting certain internal criteria. For example, in our earlier discussion of principal components analysis employing the covariance matrix as input, we found a rotation of the space whose successive dimensions accounted for the greatest amount of residual variance. This can also be viewed as defining successively higher-dimensional subspaces that maximize variance for that dimensionality.

As pointed out earlier, most factor analytic techniques (e.g., principal components analysis) are not independent of scale. That is, different results are obtained depending upon whether the averaged raw sums of squares and cross products, covariance, or correlation matrix is the one being factored. A major exception to this is canonical factor analysis (McDonald, 1968). This technique produces results that are comparable across various types of data scaling. That is, the solution obtained from one type of scaling can be readily transformed to a solution obtained from a different scaling of the original data matrix. Maximum likelihood factor analysis (Van de Geer, 1971) also yields results that are independent of scale.

As pointed out earlier, some types of factor rotation (e.g., Varimax) are based on achieving internal criteria of "simple" structure (Horst, 1966). Simple structure entails the idea of a hypothetical zero–one matrix in which each variable is, ideally, supposed to load with unity on one factor only (i.e., with zeros appearing elsewhere). In this sense a type of "matching" of one matrix to another is also involved.

In brief, a useful descriptor in characterizing multivariate methods is the type of linear transformation involved in the matching process and the restrictions placed on the nature of the transformed data. As we have illustrated, if only briefly, the various possibilities are extensive. Combined with the descriptors of Chapter 1, the type of linear transformation descriptor provides a rather comprehensive system for characterizing all current multivariate techniques. Moreover, it can be suggestive of still other combinations to be invented.

6.7 SUMMARY

In this chapter we have tried to show how the mathematical tools developed in the foregoing chapters and the appendixes underlie the formulation and solution of various multivariate techniques. In particular, multiple regression, principal components analysis, and multiple discriminant analysis were presented as prototypical techniques.

In multiple regression, the concepts of matrix inversion, determinants, and matrix rank figured prominently in the solution. We also showed how the multiple regression problem could be described geometrically, both from the standpoint of a response surface or point model and from the standpoint of a vector model. Finally, the notion of generalized regression, as a least-squares model that encompasses analysis of variance and covariance, two-group discrimination, and binary-valued regression, was illustrated graphically.

Principal components, the technique described next, entailed the rotation of a set of basis vectors to a new orthogonal basis with projections whose variance was sequentially maximal. The concepts of matrix eigenstructure of a symmetric matrix, matrix rank, and quadratic forms were most important here.

Multiple discriminant analysis then provided us with a procedure for extending our discussion to cover the eigenstructure of a nonsymmetric matrix. The simultaneous diagonalization of two different quadratic forms represented the central concept from matrix algebra. Geometrically, this entailed a rotation to align the configuration with the principal axes of the within-group SSCP matrix, a spherizing along these axes and then a further rotation to principal axes of the transformed among-group SSCP matrix.

The various matrix transformations of Chapter 4 were then recapitulated and organized into a framework within which various multivariate techniques could be described. In conjunction with the descriptors of Chapter 1, the specific nature of the linear transformation provided a useful way to characterize various multivariate procedures.

This chapter (and the entire book) has served as something of a prologue for textbooks dealing with multivariate methods per se. A large number of such texts are listed in the references, although no attempt has been made to be exhaustive. We do hope, however, that this book will make the going a bit easier as the reader delves more deeply into the subject matter of multivariate analysis.

REVIEW QUESTIONS

1. Using the data of Table 6.1,

a. compute the parameter values of Y regressed on X_1 and X_2 by means of the covariance matrix;

b. repeat the process, now employing the correlation matrix;

c. regress Y on X_1 and X_1^2 (in place of X_2) by means of a raw cross-products matrix. How does the R^2 of this compare with the simple squared correlation found from the regression of Y on X_1 alone?

2. Again using the data of Table 6.1,

a. find the principal components of the correlation matrix **R**, obtained from X_1 and X_2. How do the eigenvectors compare with those obtained from **C**, the covariance matrix?

b. find the principal components of the averaged raw cross-products matrix $\mathbf{X}'\mathbf{X}/m$, obtained from X_1 and X_2;

c. returning to Section 6.4.1, find the multiple regression of Y on the two columns of component scores computed from the covariance matrix **C**. How does the value of this R^2 compare to that obtained by regressing Y on X_1 and X_2 originally? What happens to the squared correlation if only the scores on the first component are used?

d. find the eigenstructure of **C**, the covariance matrix based on all three variables, Y, X_1, and X_2. Compare the eigenvectors obtained here with those appearing in Fig. 6.5.

3. Again using the data of Table 6.1,

a. perform a three-group discriminant analysis on the standardized columns $\mathbf{x}_{s1}$ and $\mathbf{x}_{s2}$ using the same group designation as before. How do these discriminant weights compare with those found earlier?

b. using the procedure of Chapter 5, perform a simultaneous diagonalization of the **W** and **A** matrices in Table 6.4 and compare your results with those of Table 6.5.

c. split the mean-corrected columns, X_{d1} and X_{d2} in Table 6.1, into two groups (viz., the first six versus the second six employees) and compute a two-group discriminant function. What simplifications in the computations are noted in this case?

4. Regress Y on X_1 and X_2, where the latter predictor is now dichotomized with $X_2 < 5$ receiving the code value 0 and $X_2 \geqslant 5$ receiving the code value 1.

a. How does the regression equation compare to the original shown in Table 6.2?

b. What is the effect on R^2 and the proportion of cumulative variance column, as illustrated in Table 6.2?

APPENDIX A

Symbolic Differentiation and Optimization of Multivariable Functions

A.1 INTRODUCTION

In our earlier discussions of multiple regression, principal components analysis, and multiple discriminant analysis, matrix equations were employed to solve for the various parameter values of interest. However, relatively little has been said so far about the characteristics of the functions being optimized and the process by which the matrix equations are derived.

In each of the three preceding cases, it is the calculus that provides the rationale and specific techniques for optimization. Accordingly, this appendix provides a selective review of those topics from the calculus that bear on problems of optimizing functions of multivariable arguments. No exhaustive treatment is attempted; rather, we confine our discussion to specific aspects of optimization involving only the case of differentiable variables where all appropriate partial derivatives can be assumed to exist.

We first provide a rapid review of formulas from the calculus that involve functions of one argument. This is followed by a similar discussion that covers functions of two variables. At this point, optimization subject to side conditions is introduced, and the topic of Lagrange multipliers is described and illustrated numerically.

The next main section of the appendix deals with symbolic differentiation of multivariable functions, with respect to vectors and matrices. Constrained optimization in this most general of cases is also discussed.

We then turn to each of the three major techniques described in the book:

1. multiple regression,
2. principal components analysis,
3. multiple discriminant analysis,

and show how their respective matrix equations are obtained from application of optimization procedures drawn from the calculus.

A.2 DIFFERENTIATION OF FUNCTIONS OF ONE ARGUMENT

In Section 6.2.1, a matrix equation (involving the so-called normal equations of multiple regression) was described:

$$\boxed{\mathbf{b} = (\mathbf{X}'\mathbf{X})^{-1}\mathbf{X}'\mathbf{y}}$$

where **b** is the to-be-solved-for vector of regression coefficients, **X** is the data matrix of predictor variables (augmented by a column of unities), and y is the data vector representing the criterion variable. The parameter vector **b** is chosen so as to minimize the sum of squared deviations between the original criterion vector and the fitted values obtained from the regression equation.

In Section 6.4, the matrix equation

$$\boxed{(\mathbf{C} - \lambda_1 \mathbf{I})\mathbf{t}_1 = \mathbf{0}}$$

was set up to find the eigenvalue λ_1 and its associated eigenvector $\mathbf{t}_1$ that maximized the variance of point projections of the deviation-from-mean data. The axis itself was obtained by considering the entries of $\mathbf{t}_1$ as direction cosines defining the first principal component. **C** denotes the covariance matrix.

In Section 6.5 we sought a vector $\mathbf{v}_1$ that maximized the ratio

$$\boxed{\lambda_1 = \frac{\mathbf{v}_1'\mathbf{A}\mathbf{v}_1}{\mathbf{v}_1'\mathbf{W}\mathbf{v}_1}}$$

where λ_1 is a scalar (actually an eigenvalue), **A** denotes the among-group SSCP matrix, and **W** denotes the pooled within-group SSCP matrix. One solves for λ_1 via the matrix equation

$$\boxed{(\mathbf{A} - \lambda_1 \mathbf{W})\mathbf{v}_1 = \mathbf{0}}$$

The discriminant analysis problem involves finding the eigenstructure of $\mathbf{W}^{-1}\mathbf{A}$, a matrix that is nonsymmetric.[1]

Note that in all three cases we are trying to optimize some function that involves multiple arguments. Also recall that in the case of principal components and multiple discriminant analysis, certain side conditions, such as $\mathbf{t}_1'\mathbf{t}_1$ or $\mathbf{v}_1'\mathbf{v}_1 = 1$, are imposed. Appendix A is motivated by the desire to provide a rationale for the preceding matrix equations. As such, we shall need to draw upon various tools from the calculus, starting with the simplest case of functions involving one argument and then working up to more complex problems involving several variables.

[1] In the cases of principal components and (multiple) discriminant analysis, we shall generally find successive λ_i's, subject to meeting stated side conditions with regard to their associated eigenvectors.

A.2.1 Derivatives of Functions of One Argument

By way of introduction, assume that we have some function of one argument, such as the quadratic

$$y = f(x) = 2x^2$$

We can find the value of $y = f(x)$ for each value x of interest. For a given value of x, let us next imagine taking a somewhat larger value, such as $x_1 = x_0 + \Delta x$. If so, the function y will change, as well, from y to $y + \Delta y$. That is,

$$y + \Delta y = f(x_1) = f(x_0 + \Delta x)$$

If we plot y versus x, the ratio $\Delta y/\Delta x$ can be viewed as the tangent of the angle between the x axis and the chord joining the point (x_0, y) to the point $(x_1, y + \Delta y)$. Furthermore, if Δx is made smaller and smaller, the angle that the chord makes with the x axis will approximate the angle between the x axis and the tangent line of the point (x_0, y). This appears as a dotted line in Fig. A.1.

If we let dy/dx denote the limit of the ratio $\Delta y/\Delta x$ as x_1 approaches x_0, then we can call this limit the (first) derivative of $f(x)$, denoted variously as dy/dx, y', or $f'(x)$:

$$\boxed{\frac{dy}{dx} = y' = f'(x) = \lim_{x_1 \Rightarrow x_0} \frac{f(x_1) - f(x_0)}{x_1 - x_0}}$$

Note that as $x_1 \Rightarrow x_0$, $\Delta x \Rightarrow 0$.

To illustrate this notion numerically, let us go through the computations for the preceding example:

$$y = f(x) = 2x^2$$
$$y + \Delta y = 2(x + \Delta x)^2$$
$$= 2x^2 + 4x\,\Delta x + 2(\Delta x)^2$$

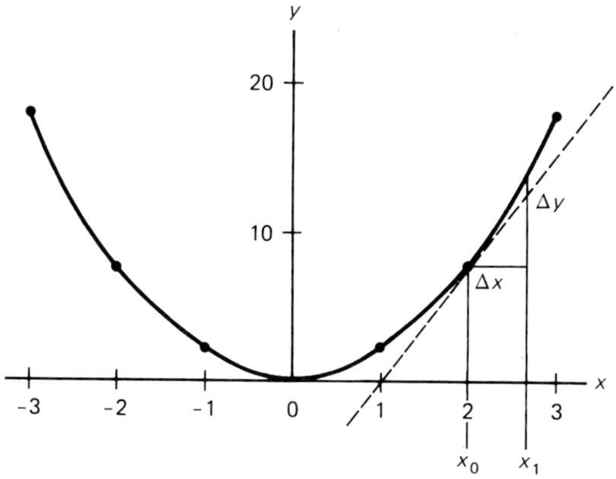

Fig. A.1 Graph of function $y = 2x^2$ $(-3 \leqslant x \leqslant 3)$. Dashed line is tangent line.

TABLE A.1

Derivatives of Elementary Functions

Constant

The derivative of a constant is 0. For example, if

$$f(x) = 12$$

then

$$f'(x) = 0$$

Algebraic

The derivative of ax^n is nax^{n-1} (for $n \neq 0$). For example, if

$$f(x) = 3x^4$$

then

$$f'(x) = 12x^3$$

Exponential

The derivative of the exponential e^x is e^x. The derivative of $e^{f(x)}$ is $f'(x)e^{f(x)}$. The derivative of a^x is $(\ln a)(a^x)$. As one example, if

$$f(x) = e^{-2x}$$

then

$$f'(x) = -2e^{-2x}$$

Logarithmic

The derivative of the natural logarithm $\ln x$ is $1/x$. The derivative of $\ln v(x)$ is $1/v \cdot dv/dx$. For example, if

$$v = x^2$$

then

$$\frac{d}{dx}(\ln v) = \left[\frac{1}{v}\right]\left[\frac{dv}{dx}\right] = \left[\frac{1}{x^2}\right](2x) = \frac{2}{x}$$

Since $y = 2x^2$, we can simplify the above expression to

$$\Delta y = 4x\,\Delta x + 2(\Delta x)^2$$

Dividing both sides by Δx gives us

$$\Delta y/\Delta x = 4x + 2\,\Delta x$$

However, as Δx becomes smaller and smaller, the second term on the right can be neglected, and we get the (first) derivative

$$\frac{dy}{dx} = f'(x) = 4x$$

TABLE A.2

Basic Rules of Differentiation Involving Simple Functions

Sum or Difference of Two Functions

The derivative of a sum (difference) of two functions $u(x)$ and $v(x)$ equals the sum (difference) of their derivatives. If u and v are functions of x, then

$$\frac{d}{dx}(u \pm v) = \frac{du}{dx} \pm \frac{dv}{dx}$$

Let

$$u = 2x + 2; \qquad v = 3x - 4$$

$$\frac{du}{dx} = 2; \qquad \frac{dv}{dx} = 3$$

$$\frac{d}{dx}(u + v) = \frac{d}{dx}(5x - 2) = 5 = 2 + 3$$

Product of Two Functions

The derivative of the product of two functions $u(x)$ and $v(x)$ is given by

$$\frac{d}{dx}(uv) = v\frac{du}{dx} + u\frac{dv}{dx}$$

Let

$$u = 2x; \qquad v = 3x - 4$$

$$\frac{du}{dx} = 2; \qquad \frac{dv}{dx} = 3$$

$$\frac{d}{dx}(uv) = \frac{d}{dx}(6x^2 - 8x) = 12x - 8 = 2(3x - 4) + 3(2x)$$

Quotient of Two Functions

The derivative of the quotient of two functions $u(x)$ and $v(x)$ is given by

$$\frac{d}{dx}\left[\frac{u}{v}\right] = \frac{v(du/dx) - u(dv/dx)}{v^2}$$

$$u = 2x^4; \qquad v = 3x^3$$

$$\frac{du}{dx} = 8x^3; \qquad \frac{dv}{dx} = 9x^2$$

$$\frac{d}{dx}\left[\frac{u}{v}\right] = \frac{d}{dx}\left[\frac{2x^4}{3x^3}\right] = 2/3 = \frac{3x^3(8x^3) - 2x^4(9x^2)}{(3x^3)^2}$$

We could, in turn, take the derivative of the function $f'(x) = 4x$ and obtain the second derivative $f''(x) = 4$. (Higher-order derivatives are all zero since 4 is a constant.) That is, second-order derivatives are simply derivatives of first-order derivatives; third-order derivatives are derivatives of second-order derivatives, and so on.

The above procedure can be generalized to the case of finding the derivative of any algebraic expression of the form $y = ax^n$. In general, if $y = ax^n$, we have[2]

$$\frac{dy}{dx} = nax^{n-1}$$

By way of review, Table A.1 lists this formula and others that involve various elementary functions of interest to the applied researcher.[3]

Not only are derivatives of the more common elementary functions of use, but we may also be interested in the derivatives of simple functions of these. Accordingly, Table A.2 lists the basic rules that are applicable to simple functions of two elementary functions, such as their sum or product.

A.2.2 The Chain Rule

Another concept of the elementary calculus that should be reviewed is the chain rule. The chain rule applies to functions of functions. Suppose we have a function $f(g)$ where g, in turn, is $g(x)$, a function of x. If such is the case, the chain rule states that

$$\frac{df}{dx} = \left[\frac{df}{dg}\right]\left[\frac{dg}{dx}\right]$$

To illustrate application of the chain rule, consider the function $f(x) = 2(x^2 + 3x)^2 - 1$, which in turn, can be written as

$$f(g) = 2g^2 - 1$$

$$g(x) = x^2 + 3x$$

The chain rule states that

$$\frac{df}{dx} = \left[\frac{df}{dg}\right]\left[\frac{dg}{dx}\right] = 4g(2x + 3)$$

Next, we substitute the expression for $g(x)$ to get

$$\frac{df}{dx} = 4(x^2 + 3x)(2x + 3) = 8x^3 + 36x^2 + 36x$$

We can verify this result directly by making the substitution of $g(x)$ in $f(g)$ to get

$$f(x) = 2(x^2 + 3x)^2 - 1 = 2x^4 + 12x^3 + 18x^2 - 1$$

and

$$\frac{df}{dx} = 8x^3 + 36x^2 + 36x$$

as desired.

[2] We assume that n is a real number not equal to 0 and that $f(x)$ is defined and differentiable.

[3] Although not shown in Table A.1, the derivative of $\sin x$ is $\cos x$, the derivative of $\cos x$ is $-\sin x$, and the derivative of $\tan x$ is $1/\cos^2 x$.

As a second example, consider the function $f(x) = \ln(x^2 + 3)$. This, in turn, can be expressed as

$$f(g) = \ln g; \qquad g(x) = x^2 + 3$$

Then, by application of the chain rule, we have

$$\frac{df}{dx} = \left[\frac{df}{dg}\right]\left[\frac{dg}{dx}\right] = \frac{1}{g}(2x)$$

and, substituting $g(x) = x^2 + 3$ for g, we have

$$\frac{df}{dx} = \frac{2x}{x^2 + 3}$$

The chain rule can be easily extended to three or more functions in terms of the following:

$$\frac{df}{dx} = \left[\frac{df}{dg}\right]\left[\frac{dg}{dh}\right]\left[\frac{dh}{dx}\right]$$

and so on, for additional functions.

As an example of the case involving three functions, consider the expression

$$f(x) = [\ln(x + 1)]^2$$

This, in turn, can be expressed as

$$f(g) = g^2; \qquad g(h) = \ln h; \qquad h(x) = x + 1$$

Applying the chain rule leads to

$$\frac{df}{dx} = \left[\frac{df}{dg}\right]\left[\frac{dg}{dh}\right]\left[\frac{dh}{dx}\right] = 2g\left(\frac{1}{h}\right)(1) = \frac{2\ln(x + 1)}{x + 1}$$

The chain rule, augmented by the formulas shown in Tables A.1 and A.2, provides a flexible procedure for differentiating the more common functions encountered in applied research.

A.2.3 Optimization of Functions of One Argument

As recalled from the elementary calculus, a function of one argument has a local maximum at some point x_0 if the values of the function on either side of x_0 are less than $f(x_0)$. On the other hand, if the values of the function on either side of x_0 are greater than $f(x_0)$, the function has a local minimum. Maxima and minima are called extreme values, and the values of x for which $f(x)$ takes on an extreme value are called *extreme points.*

Suppose that $f(x)$ has a continuously varying first derivative in an interval that includes x_0. If $f(x_0)$ is a maximum, the first derivative must then change from positive to negative. Conversely, if $f(x_0)$ is a minimum, the first derivative must change from negative to positive. These facts relate, of course, to the basic definition of $f'(x)$ as the slope of the curve $y = f(x)$ at the point x_0. Under the preceding conditions then, $f'(x)$ is zero at the point x_0 where the curvature of $f(x)$ changes direction.

More generally, $f'(x) = 0$ is the necessary condition for a *stationary point* (that includes the instances of maxima and minima as special cases). At a stationary point, the function may have either a maximum, a minimum, or neither. For example, each of the following functions displays a stationary point at $x = 0$, but we note that for

$f(x) = x^2$, the stationary point is a minimum

$f(x) = -(x^2)$, the stationary point is a maximum

$f(x) = x^3$, the stationary point is neither a maximum nor minimum[4]

Finding local extreme points for differentiable functions of one argument involves the following steps:

1. Since local extreme points can only occur at stationary points, where $f'(x) = 0$, first find all solutions to the equation

$$f'(x) = 0$$

2. For each of the stationary points obtained from the above solutions, compute higher-order derivatives $f''(x)$, $f'''(x)$, etc., as needed, so as to find the value of the *lowest*-order derivative that is not zero at the stationary point in question.
3. Examine the lowest-order derivative that is nonzero to determine if

   a. its order is even.
      (i) If the value of the derivative of this order is positive, the function exhibits a local minimum at the stationary point under evaluation.
      (ii) If the value of the derivative of this order is negative, the function exhibits a local maximum at the stationary point under evaluation.
   b. its order is odd; if so, the stationary point is an inflection point.[5]

As a simple numerical illustration of the above procedure, consider the function

$$f(x) = x^3 + 2x^2 + x$$

Its first derivative is

$$f'(x) = 3x^2 + 4x + 1$$

Next, we solve for the stationary points by setting $f'(x)$ equal to zero:

$$f'(x) = 3x^2 + 4x + 1 = 0$$

and find, as solutions,

$$x_1 = -1; \qquad x_2 = -\tfrac{1}{3}$$

Next, let us find the second derivative of $f(x)$. This is, of course, the derivative of $f'(x)$. That is,

$$f''(x) = 6x + 4$$

[4] In this case the stationary point is an inflection point where the first derivative $f'(x_0)$ is zero and, furthermore, the second derivative $f''(x)$ changes sign as the function goes through x_0.

[5] As noted earlier, the necessary condition for x_0 to be a stationary point is that $f'(x_0) = 0$. The sufficient conditions appearing above can be obtained by examining successive terms of the Taylor series expansion (Lang, 1964).

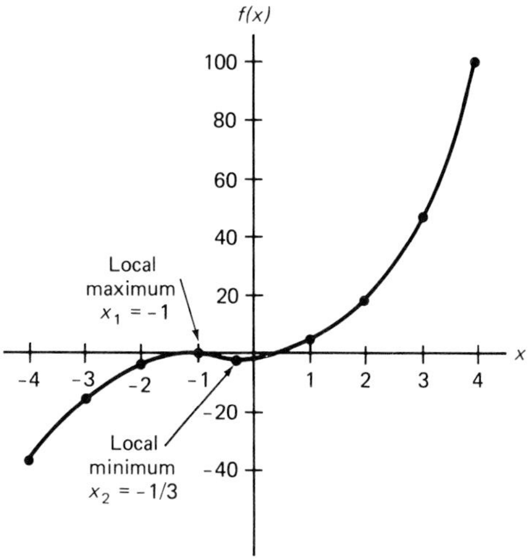

Fig. A.2 Graph of function $f(x) = x^3 + 2x^2 + x$ $(-4 \leqslant x \leqslant 4)$.

We then evaluate $f''(x)$ at the first stationary point x_1 to see if the value of $f''(x)$ is nonzero:

$$f''(-1) = -6 + 4 = -2$$

Since $f''(-1) = -2$ is nonzero, the order of the first nonzero derivative is even; moreover, we have a local maximum at this point since $f''(-1)$ is negative.

Next, the same thing is done for the second stationary point x_2:

$$f''(-\tfrac{1}{3}) = -2 + 4 = 2$$

Since $f''(-1/3) = 2$ is also nonzero, the stationary point is an extreme point; since $f''(-1/3)$ is positive, we have a local minimum at this point.

Figure A.2 shows a plot of the function

$$f(x) = x^3 + 2x^2 + x$$

over the (illustrative) domain $-4 \leqslant x \leqslant 4$. At the point $x_1 = -1, f(x) = 0$, which is a local maximum. At the point $x_2 = -\frac{1}{3}, f(x) = -4/27$, which is a local minimum.

It should be kept in mind that what is being found are *local* stationary points in which interest centers on the behavior of the function in a relatively small interval. As indicated in Fig. A.2, neither x_1 nor x_2 represents a global extremum over the domain $(-4 \leqslant x \leqslant 4)$ of interest.

Figure A.3 further illustrates the distinction between global and local extrema. If we consider local extrema *within* the interval of $x_0 < x < x_6$, a local maximum is found at x_2 and a local minimum at x_4. Also, other stationary points (viz., inflection points) are found at x_1, x_3, and x_5. However, when the end points x_0 and x_6 are also considered, the global maximum turns out to be at x_6, while the global minimum is still at x_4. The

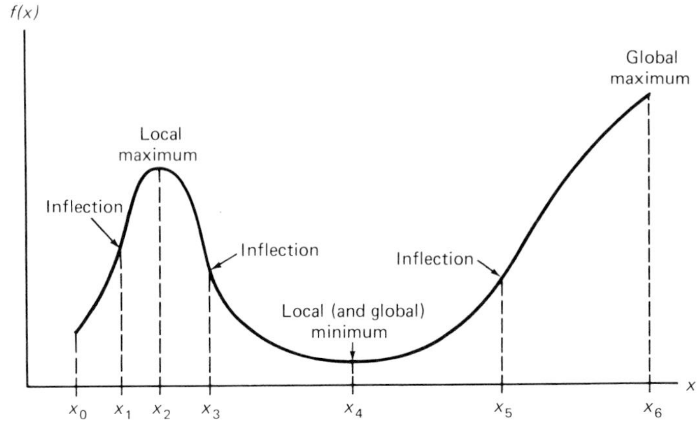

Fig. A.3 Global versus local extrema.

search for global extrema can be quite tedious, particularly if various discontinuities appear in the function.[6] However, further discussion of this specialized process exceeds our intended coverage.

A.3 DIFFERENTIATION OF FUNCTIONS OF TWO ARGUMENTS

The next step in this review discussion of the calculus involves functions of two arguments:

$$\boxed{z = f(x, y)}$$

In the case of functions of one argument we were able to represent $f(x)$ versus x by a curve in two dimensions. Analogously, in the present case $f(x, y)$ versus x and y is represented by a surface embedded in three dimensions. Graphical devices, such as contour lines and projective drawings, are useful in portraying certain three-dimensional relationships in two-dimensional space.

In this part of the appendix we discuss the concepts of level curve, partial differentiation, unconstrained optimization, and optimization subject to equality constraints.

A.3.1 Level Curves and Partial Differentiation

The notion of level curves is employed in a variety of applications. For example, in map making one may use a series of contour lines to represent altitudes, as schematized in Panel I of Fig. A.4. We note that the level of 300 feet consists of all of those points on the hill that involve

$$f(x, y) = 300$$

[6] If discontinuities appear, the function must be examined at each of these points (as well as at end points and stationary points).

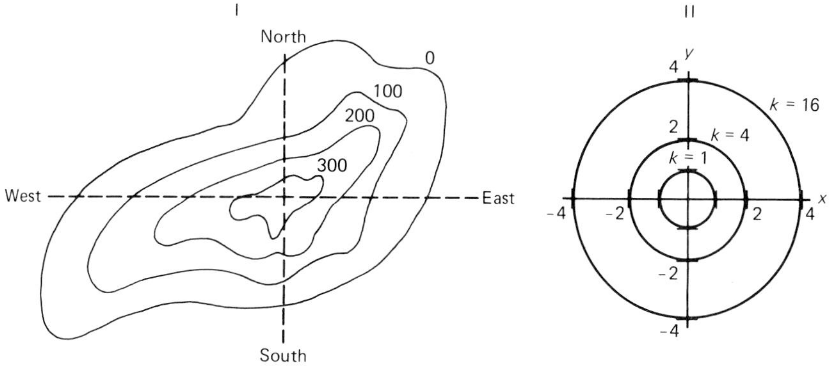

Fig. A.4 Illustrations of level curves. Key: I, altitude of land; II, $z = x^2 + y^2$.

If one were to hike around the hill on that level curve, one would remain at a constant height of 300 feet. Similarly, other contour lines show other altitudes, such as 200 feet, 100 feet, and sea level.

Level curves appear in many fields, such as pictorial displays of air pressure (as isobars), temperature (as isotherms), and consumer utility (as indifference curves). As inferred from Panel I, we could "build up" the hill if we imagined that each level curve were made of cardboard and we stacked one piece on top of another in building up the surface.

Panel II shows level curves for the function

$$z = f(x, y) = x^2 + y^2$$

In this case we can find any level curve of interest by choosing a fixed number k and then finding the set of points x_i and y_j for which

$$k = f(x, y) = x^2 + y^2$$

For example, if we set $k = 9$, we have

$$x^2 + y^2 = 9; \qquad x = \pm\sqrt{9 - y^2}; \quad y = \pm\sqrt{9 - x^2}$$

If we let $y = 0$, then $x = \pm 3$; conversely if $x = 0$, then $y = \pm 3$. As can be inferred from examining the various level curves of Panel II in Fig. A.4, the surface of the function looks like that of a bowl with its center point at the origin of the x, y plane. The level curves are, of course, circles of varying radius.

As a third example of level curves, consider the surface, depicted in Panel I of Fig. A.5, representing the function

$$f(x, y) = 3xy$$

where we assume that $x, y \geqslant 0$. Panel I shows the surface itself as a conelike figure that is cut in half.[7] Panel II shows selected level curves for

$$k_1 = 0; \qquad k_2 = 1; \qquad k_3 = 3; \qquad k_4 = 6; \qquad k_5 = 9; \qquad k_6 = 12$$

We return to this function after a brief review of partial differentiation.

[7] The plane ABC, depicted as a slice through the surface in Panel I, and dotted line AC in Panel II are discussed later (in Section A.3.3) in the context of Lagrange multipliers.

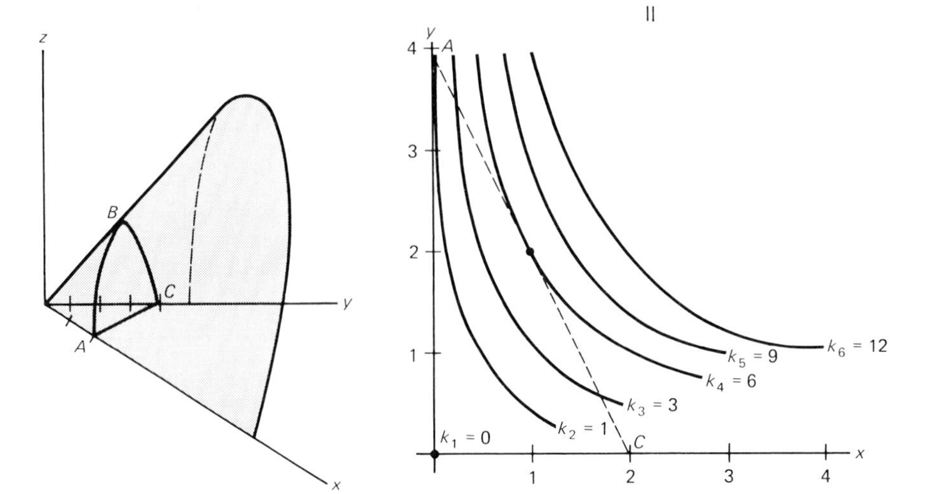

Fig. A.5 Surface and level curve plots of $f(x, y) = 3xy$ $(x, y \geqslant 0)$. Key: I, response surface, $f(x, y) = 3xy$ $(x, y \geqslant 0)$; II, level curves, $k = 3xy$.

Partial differentiation is a straightforward generalization of simple differentiation. Partial derivatives are found by differentiating the function $f(x, y)$ with respect to each variable separately. The variable not involved in the differentiation is treated as a constant. For example, for the function portrayed in Fig. A.5, we have

$$f(x, y) = 3xy; \qquad \frac{\partial f}{\partial x} = 3y; \quad \frac{\partial f}{\partial y} = 3x$$

Partial derivatives are usually denoted by $\partial f/\partial x$ or f_x.

Second-order partial derivatives involve differentiation of first-order partial derivatives, since partial derivatives are, themselves, functions. That is,

$$\frac{\partial}{\partial x}\left[\frac{\partial f}{\partial x}(x, y)\right] \qquad \text{and} \qquad \frac{\partial}{\partial y}\left[\frac{\partial f}{\partial y}(x, y)\right]$$

are second-order derivatives, denoted by the symbols

$$\frac{\partial^2 f}{\partial x^2}(x, y) \quad \text{or} \quad f_{xx}(x, y); \quad \frac{\partial^2 f}{\partial y^2}(x, y) \quad \text{or} \quad f_{yy}(x, y)$$

Moreover, the function $(\partial f/\partial x)(x, y)$ may also be differentiated with respect to y, and $(\partial f/\partial y)(x, y)$ may be differentiated with respect to x. These are called *mixed* partial derivatives and are usually denoted by

$$\frac{\partial^2 f}{\partial x\, \partial y} \quad (\text{or } f_{xy}) \qquad \text{and} \qquad \frac{\partial^2 f}{\partial y\, \partial x} \quad (\text{or } f_{yx})$$

respectively. If the mixed partial derivatives of $f(x, y)$ are continuous, then

$$\frac{\partial^2 f}{\partial x\, \partial y} = \frac{\partial^2 f}{\partial y\, \partial x}$$

and the order in which differentiation proceeds is irrelevant. Continuing with the example

$$f(x, y) = 3xy$$

we have

$$\frac{\partial f}{\partial x} = 3y; \qquad \frac{\partial^2 f}{\partial x^2} = 0; \qquad \frac{\partial^2 f}{\partial x\,\partial y} = 3$$

$$\frac{\partial f}{\partial y} = 3x; \qquad \frac{\partial^2 f}{\partial y^2} = 0; \qquad \frac{\partial^2 f}{\partial y\,\partial x} = 3$$

As an additional example, consider the more elaborate polynomial

$$f(x, y) = 2x^2 + y^2 + 3xy + x - y + 3$$

The first- and second-order partial derivatives are

$$\frac{\partial f}{\partial x} = 4x + 3y + 1; \qquad \frac{\partial^2 f}{\partial x^2} = 4; \qquad \frac{\partial^2 f}{\partial x\,\partial y} = 3$$

$$\frac{\partial f}{\partial y} = 3x + 2y - 1; \qquad \frac{\partial^2 f}{\partial y^2} = 2; \qquad \frac{\partial^2 f}{\partial y\,\partial x} = 3$$

Since all second-order partial derivatives are constants, all third- (and higher) order partial derivatives are zero.

A.3.2 Unconstrained Optimization of Functions of Two Arguments

Analogous to the case involving functions of a single argument, conditions for local extrema can be listed for the two-argument case. To be specific, let us continue to consider the case of the function

$$f(x, y) = 2x^2 + y^2 + 3xy + x - y + 3$$

A necessary condition that (x_0, y_0) be a local stationary point is that the following equations are satisfied:

$$\frac{\partial f}{\partial x}(x_0, y_0) = 0; \qquad \frac{\partial f}{\partial y}(x_0, y_0) = 0$$

In the above example we have

$$\frac{\partial f}{\partial x}(x, y) = 4x + 3y + 1 = 0; \qquad \frac{\partial f}{\partial y}(x, y) = 3x + 2y - 1 = 0$$

On solving these equations simultaneously we get

$$x = 5; \qquad y = -7$$

Therefore, $(5, -7)$ is a stationary point.

Sufficiency conditions for local extrema involving functions of two arguments are somewhat more complex than the counterpart case for one variable. To summarize what these are, one first sets up the determinant of second-order partial derivatives as follows:

$$\delta_2 = \begin{vmatrix} \dfrac{\partial^2 f}{\partial x^2} & \dfrac{\partial^2 f}{\partial x\, \partial y} \\ \dfrac{\partial^2 f}{\partial y\, \partial x} & \dfrac{\partial^2 f}{\partial y^2} \end{vmatrix}$$

which, in the illustrative problem, is

$$|\mathbf{A}| = \begin{vmatrix} 4 & 3 \\ 3 & 2 \end{vmatrix} = (4 \times 2) - (3 \times 3) = -1$$

Next, we examine $\delta_1 = \partial^2 f/\partial x^2 = 4$ and note that its sign is positive. Sufficiency rules, for the general case, can now be stated in terms of δ_1 and δ_2:

1. A local extreme point exists if $\delta_2 > 0$.
   a. The local extreme point is a minimum if $\delta_1 > 0$.
   b. The local extreme point is a maximum if $\delta_1 < 0$.
2. A local extreme point does not exist if $\delta_2 < 0$.
3. A local extreme point may or may not exist if $\delta_2 = 0$. In general, additional examination of the function at the stationary point values is needed to see if an extreme point exists and, if so, whether it is a minimum or a maximum.[8]

In the illustrative problem we note that $\delta_2 = -1$ (<0), and, hence, no local extreme point exists for this function.

Since quadratic functions of the general form

$$f(x, y) = ax^2 + bxy + cy^2 + dx + ey + f$$

appear so frequently in multivariate statistical work, it is useful to examine their properties more generally. First, we shall find, as noted earlier, that all second-order partial derivatives are constants. Stationary points are found by solving the equations $\partial f/\partial x = 0$, $\partial f/\partial y = 0$, simultaneously:

$$\frac{\partial f}{\partial x} = 2ax + by + d = 0; \qquad \frac{\partial^2 f}{\partial x^2} = 2a$$

$$\frac{\partial f}{\partial y} = 2cy + bx + e = 0; \qquad \frac{\partial^2 f}{\partial y^2} = 2c$$

which can be expressed as

$$2ax + by = -d; \qquad bx + 2cy = -e$$

[8] In particular, if both $\partial^2 f/\partial x^2$ and $\partial^2 f/\partial y^2$ are equal to zero, the stationary point is not a local extreme point. If $\partial^2 f/\partial x\, \partial y = \partial^2 f/\partial y\, \partial x = 0$ and if $\partial^2 f/\partial x^2$ and $\partial^2 f/\partial y^2$ have the same sign, the stationary point is a local extreme point. Still, specific examination of the function is generally needed to see whether the point is a minimum or a maximum.

to obtain the solutions

$$x_0 = \frac{2cd - eb}{b^2 - 4ac}; \qquad y_0 = \frac{2ae - bd}{b^2 - 4ac}$$

Moreover, since the second-order partial derivatives are

$$\frac{\partial^2 f}{\partial x^2} = 2a; \qquad \frac{\partial^2 f}{\partial y^2} = 2c; \qquad \frac{\partial^2 f}{\partial x\, \partial y} = \frac{\partial^2 f}{\partial y\, \partial x} = b$$

we have, as the expression for δ_2,

$$\delta_2 = \begin{vmatrix} \dfrac{\partial^2 f}{\partial x^2} & \dfrac{\partial^2 f}{\partial x\, \partial y} \\ \dfrac{\partial^2 f}{\partial y\, \partial x} & \dfrac{\partial^2 f}{\partial y^2} \end{vmatrix} = 4ac - b^2$$

If $4ac - b^2 > 0$, an extreme point is indicated. That is,

1. if $\delta_2 = 4ac - b^2 > 0$, the stationary point is an extreme point.
   a. If $\delta_1 = \partial^2 f/\partial x^2 = 2a > 0$, then the extreme point is a local minimum.
   b. If $\delta_1 = 2a < 0$, then the extreme point is a local maximum.
2. if $\delta_2 < 0$, no extreme point exists.
3. if $\delta_2 = 0$, additional examination of the function is needed to see if an extreme point exists and, if so, whether it is a minimum or a maximum.

In the preceding numerical case, $x_0 = 5$ and $y_0 = -7$. Substituting these values leads to

$$4ac - b^2 = 4(2)(1) - (3)^2 = -1$$

and the stationary point $(5, -7)$ is neither a minimum nor a maximum, as was indicated earlier.

A.3.3 Constrained Optimization and Lagrange Multipliers

In the previous section our concern was with finding local extrema without constraining the domain over which x and y, the two arguments of $f(x, y)$, might vary. However, cases frequently arise where we are interested in setting up certain side conditions that must be satisfied in the course of optimizing some function $f(x, y)$.

Optimization of functions subject to constraints is a vast topic which goes well beyond our coverage. Here we concern ourselves with only one technique, the method of Lagrange multipliers for optimizing functions subject to equality constraints.

The basic idea behind Lagrange multipliers involves setting up a more general function *that includes the constraint* and optimizing this more general function. Suppose we have a function $f(x, y)$ and a side condition, expressed as $g(x, y) = 0$. If so, we can define a new function, composed of *three* variables:

$$u(x, y, \lambda) = f(x, y) - \lambda g(x, y)$$

where λ is the Lagrange multiplier. (The variable λ is an artificial variable that is employed to provide as many unknowns as there are equations.) Having set up the general function $u(x, y, \lambda)$, we can find its partial derivatives with respect to x, y, and λ and set each of them equal to zero:

$$\frac{\partial u}{\partial x} = 0; \qquad \frac{\partial u}{\partial y} = 0; \qquad \frac{\partial u}{\partial \lambda} = 0$$

Any point that satisfies the above necessary conditions is a stationary point. We continue to assume, of course, that both $f(x, y)$ and $g(x, y)$, the latter being the constraint equation, are differentiable in the neighborhood of the stationary point.

Let us illustrate the Lagrange multiplier technique by returning to the function

$$f(x, y) = 3xy$$

as first depicted in Fig. A.5. However, now let us impose the constraint equation that

$$g(x, y) = 2x + y = 4$$

or, equivalently,

$$g(x, y) = 2x + y - 4 = 0$$

As shown in Fig. A.5, application of this constraint results in a plane (labelled as ABC) intersecting the surface in Panel I. Furthermore, AC in Panel II represents the same linear constraint as a dotted straight line; this line is the locus of all points on the xy plane that satisfy the constraint equation $2x + y = 4$.

We next set up the function $u(x, y, \lambda)$ that incorporates the Lagrange multiplier

$$u(x, y, \lambda) = 3xy - \lambda(2x + y - 4)$$

Then $u(x, y, \lambda)$ is differentiated with respect to each argument, in turn, and each derivative is set equal to zero:

$$\frac{\partial u}{\partial x} = 3y - 2\lambda = 0; \qquad \frac{\partial u}{\partial y} = 3x - \lambda = 0; \qquad -\frac{\partial u}{\partial \lambda} = 2x + y - 4 = 0$$

Notice that the partial derivative with respect to λ is just the constraint equation itself. The next step is to find the stationary point by solving the three preceding equations for x, y, and λ. This is easily done by first expressing both y and x in terms of λ:

$$y = 2\lambda/3; \qquad x = \lambda/3$$

and solving for λ in the third equation:

$$2(\lambda/3) + 2\lambda/3 - 4 = 0; \qquad \lambda = 3$$

We then find x and y to be

$$x = 1; \quad y = 2$$

Also we note that in terms of the original function $f(x, y) = 3xy$, the value of the function at the stationary point $(1, 2)$ is

$$f(1, 2) = 3(1)(2) = 6$$

Panel II of Fig. A.5 provides a graphical representation of what is going on. First, we examine the constraint equation, represented by the dotted line $AC = g(x, y) = 2x + y = 4$. The one level curve for which AC represents the tangent line is

$$k_4 = f(x, y) = 6$$

Thus, when the stationary point $(x_0, y_0) = (1, 2)$ is found with $f(1, 2) = 6$, we see that its tangent line is represented by the constraint equation $g(x, y) = 2x + y = 4$. The value of the function $f(1, 2) = 6$ coincides with the highest level curve that can be reached via AC. Furthermore, in looking at the plane ABC, slicing through the surface in Panel I, we also see that the local extreme value is a maximum point B on the arch traced out by ABC.

One additional point of interest concerns the value of λ itself; in this case $\lambda = 3$. If we examine the original function

$$f(x, y) = 3xy$$

and assume that the (negative) constraint equation $2x + y = 4$ were "relaxed" to $2x + y = 5$, we would have an extra unit of (say) y to work with. If so, we could compute the value of $f(x, y)$ before and after allowing for one unit increase in y. Letting $x_0 = 1$, the x coordinate at the original stationary point, we have

$$f(x, y) = 3(1)y = 3y$$

$$f(x, y + 1) = 3(1)(y + 1) = 3y + 3$$

Hence, an increase in $f(x, y)$ of 3 units could be effected if one more unit of y were available. This is equal to λ, the value of the Lagrange multiplier, as found in the earlier computations.

In summary, Lagrange multipliers are useful in handling one (or possibly more) equality constraints in cases where it would be difficult to solve the problem via direct substitution of the constraint equation(s) into $f(x, y)$.[9] Since a necessary condition for a stationary point is that each first-order partial derivative equals zero, introduction of the Lagrange multiplier λ adds a needed artificial variable to balance out the number of equations with the number of unknowns. In terms of the numerical example

$$u(x, y, \lambda) = 3xy - \lambda(2x + y - 4)$$

we see that partial differentiation with respect to x, y, and λ separately leads to three equations in three unknowns.

Furthermore, if the constraint equation is always met, then the term $(2x + y - 4)$ in the general function $u(x, y, \lambda)$ will always equal zero so that $u(x, y, \lambda)$ will behave in the same way as $f(x, y)$, the original function.[10]

[9] With only two original variables, one would not generally deal with more than a single constraint equation since two constraint equations in two variables would normally have only a single point in common. With a large number of original variables, however, two (or more) equality constraints might be employed.

[10] Sufficiency conditions for local extrema in the context of Lagrange multipliers are found in Hancock (1960).

A.4 SYMBOLIC DIFFERENTIATION

The most complex cases in partial differentiation and function optimization involve functions of several arguments. Often, this situation is portrayed by vector and matrix notation. The term "symbolic differentiation" has been coined to refer to partial differentiation of vector or matrix functions whose results are also described in the same format. For example, symbolic differentiation of a function with respect to a vector involves finding the partial derivative of the function with respect to each entry of the vector; the partial derivatives themselves are then arranged in vector form.

To illustrate, if $y = f(\mathbf{x})$, where $\mathbf{x}$ is a column vector with elements $x_1, x_2, \ldots, x_n$, then one can express its symbolic derivative by means of the column vector

$$\frac{\partial f}{\partial \mathbf{x}} = \begin{bmatrix} \partial f/\partial x_1 \\ \partial f/\partial x_2 \\ \vdots \\ \partial f/\partial x_n \end{bmatrix}$$

Notice that each entry of $\partial f/\partial \mathbf{x}$ is a partial derivative of $f(\mathbf{x})$ with respect to a specific variable. By the same token, one can find a row vector of partial derivatives:[11]

$$\frac{\partial f}{\partial \mathbf{x}'} = (\partial f/\partial x_1, \partial f/\partial x_2, \ldots, \partial f/\partial x_n)$$

In applied multivariate analysis, we frequently have occasion to find the symbolic derivatives of functions that are bilinear or quadratic forms, such as

$u = \mathbf{x}'\mathbf{A}\mathbf{y}$ (where $\mathbf{A}$ may be rectangular)

$v = \mathbf{x}'\mathbf{A}\mathbf{x}$ (where $\mathbf{A}$ is square, nonsymmetric)

$w = \mathbf{x}'\mathbf{A}\mathbf{x}$ (where $\mathbf{A}$ is symmetric)

$t = \mathbf{x}'\mathbf{I}\mathbf{x}$ (where $\mathbf{I}$ is the identity matrix)

We first consider symbolic differentiation with respect to vectors. We can then turn to problems of optimization of functions involving multivariable arguments.

A.4.1 Symbolic Differentiation with Respect to Vectors

If we take one of the simplest cases first, namely, the linear combination $y = \mathbf{a}'\mathbf{x}$, where $\mathbf{a}'$ is a row vector of coefficients, the symbolic derivative of y with respect to $\mathbf{x}$ is simply the row vector

$$\frac{\partial y}{\partial \mathbf{x}'} = \mathbf{a}'$$

[11] Note that $\partial \mathbf{x}'$ appears as a row vector in the denominator since $\partial f/\partial \mathbf{x}'$ is being expressed explicitly in this form. That is, we shall adopt the notation of $\partial \mathbf{x}$ or $\partial \mathbf{x}'$ on the basis of how the final vector of derivatives is displayed—in column or row form, respectively.

This follows from the fact that we can write y explicitly as

$$y = a_1 x_1 + a_2 x_2 + \cdots + a_n x_n$$

Then, each partial derivative of y is found, in turn: $\partial y/\partial x_1 = a_1$, $\partial y/\partial x_2 = a_2, \ldots, \partial y/\partial x_n = a_n$. These elements can then be arranged in the row vector $\mathbf{a}'$.

Next, suppose we have the bilinear form

$$u = \mathbf{x}'\mathbf{A}\mathbf{y}$$

which can be written out explicitly (for $\mathbf{A}$ of order $m \times n$) as

$$\begin{aligned} u = {} & x_1 a_{11} y_1 + x_1 a_{12} y_2 + \cdots + x_1 a_{1n} y_n \\ & + x_2 a_{21} y_1 + x_2 a_{22} y_2 + \ldots + x_2 a_{2n} y_n \\ & + \cdots \\ & + x_m a_{m1} y_1 + x_m a_{m2} y_2 + \cdots + x_m a_{mn} y_n \end{aligned}$$

The partial derivative $\partial u/\partial x_1$ of $u = \mathbf{x}'\mathbf{A}\mathbf{y}$ with respect to the first element x_1 is

$$\frac{\partial u}{\partial x_1} = a_{11} y_1 + a_{12} y_2 + \cdots + a_{1n} y_n$$

which can be written as the scalar product

$$\frac{\partial u}{\partial x_1} = \mathbf{a}_1{}'\mathbf{y}$$

By the same procedure the other partial derivatives are obtained:

$$\frac{\partial u}{\partial x_2} = \mathbf{a}_2{}'\mathbf{y}$$

$$\vdots$$

$$\frac{\partial u}{\partial x_m} = \mathbf{a}_m{}'\mathbf{y}$$

which can all be arranged in the $m \times 1$ column vector

$$\frac{\partial u}{\partial \mathbf{x}} = \mathbf{A}\mathbf{y}$$

By a similar rationale we can obtain $\partial u/\partial \mathbf{y}$ as the $1 \times n$ row vector

$$\frac{\partial u}{\partial \mathbf{y}'} = \mathbf{x}'\mathbf{A}$$

By taking appropriate transposes of the two preceding equations, we could also write

$$\frac{\partial u}{\partial \mathbf{x}'} = \mathbf{y}'\mathbf{A}'; \qquad \frac{\partial u}{\partial \mathbf{y}} = \mathbf{A}'\mathbf{x}$$

as a row and column vector, respectively.

By similar reasoning we can find that the symbolic derivative of

$$v = \mathbf{x}'\mathbf{A}\mathbf{x} \qquad \text{(for square, nonsymmetric } \mathbf{A}\text{)}$$

with respect to **x** is the column vector

$$\frac{\partial v}{\partial \mathbf{x}} = (\mathbf{A} + \mathbf{A}')\mathbf{x}$$

Furthermore, the symbolic derivative of

$$w = \mathbf{x}'\mathbf{A}\mathbf{x} \qquad \text{(for symmetric } \mathbf{A}\text{, so that } \mathbf{A} = \mathbf{A}'\text{)}$$

is the column vector

$$\frac{\partial w}{\partial \mathbf{x}} = 2\mathbf{A}\mathbf{x}$$

In particular, if $\mathbf{A} = \mathbf{I}$, we have the special case of the sum of squares

$$t = \mathbf{x}'\mathbf{I}\mathbf{x} = \mathbf{x}'\mathbf{x}; \qquad \frac{\partial t}{\partial \mathbf{x}} = 2\mathbf{x}$$

As a numerical illustration of the case involving a bilinear form, consider the function

$$u = (x_1, x_2)\begin{bmatrix} 1 & 3 & 1 \\ 2 & 4 & 3 \end{bmatrix}\begin{bmatrix} y_1 \\ y_2 \\ y_3 \end{bmatrix}$$

$$= x_1y_1 + 2x_2y_1 + 3x_1y_2 + 4x_2y_2 + x_1y_3 + 3x_2y_3$$

If we differentiate u with respect to **x**, we have the column vector

$$\frac{\partial u}{\partial \mathbf{x}} = \mathbf{A}\mathbf{y} = \begin{bmatrix} y_1 + 3y_2 + y_3 \\ 2y_1 + 4y_2 + 3y_3 \end{bmatrix}$$

In a similar way, we could find symbolic derivatives of other functions with respect to **x** or **y**. As would be surmised, however, no new principles are involved.[12]

A.4.2 Some Aspects of Optimization in Matrix Notation

Extreme values can be found for functions of vectors in much the same way as described earlier for functions of scalars. In particular, suppose we had the function

$$y = 2x_1^2 + 3x_2^2$$

subject to the constraint equation

$$g(x) = x_1 - x_2 - 1 = 0$$

[12] Bilinear and quadratic forms can also be differentiated with respect to **A**, the matrix of the form. For example, for nonsymmetric **A**, $(\partial/\partial\mathbf{A})(\mathbf{x}'\mathbf{A}\mathbf{x})$ is $\mathbf{x}\mathbf{x}'$. However, this more advanced topic exceeds our scope. The interested reader is referred to Tatsuoka (1971).

We could, of course, apply the same Lagrange multiplier procedure described in Section A.3.3 to find the extreme values (if any) of this function. However, let us express the equations in vector or matrix form and work through their solution in this format:

$$y = \mathbf{x}'\mathbf{A}\mathbf{x} = (x_1, x_2)\begin{bmatrix} 2 & 0 \\ 0 & 3 \end{bmatrix}\begin{bmatrix} x_1 \\ x_2 \end{bmatrix}$$

subject to

$$g = \mathbf{c}'\mathbf{x} - 1 = 0 = \left\{ (1, -1)\begin{bmatrix} x_1 \\ x_2 \end{bmatrix} \right\} - 1 = 0$$

In matrix equation form, the general function is, analogously,

$$u = \mathbf{x}'\mathbf{A}\mathbf{x} - \lambda(\mathbf{c}'\mathbf{x} - 1)$$

We set its partial derivatives equal to the zero column vector:

$$\frac{\partial u}{\partial \mathbf{x}} = 2\mathbf{A}\mathbf{x} - \lambda\mathbf{c} = \mathbf{0}$$

and find the vector solution

$$\mathbf{x} = \lambda\mathbf{A}^{-1}\mathbf{c}/2$$

Moreover, since $\mathbf{c}'\mathbf{x} - 1 = 0$, then $\mathbf{c}'\mathbf{x} = 1$, so that after multiplying both sides of the preceding equation by $\mathbf{c}'$, we have

$$1 = \mathbf{c}'\mathbf{x} = \lambda\mathbf{c}'\mathbf{A}^{-1}\mathbf{c}/2$$

and λ can then be found from

$$\lambda = 2(\mathbf{c}'\mathbf{A}^{-1}\mathbf{c})^{-1}$$

Substituting this expression for λ in $\mathbf{x} = \lambda\mathbf{A}^{-1}\mathbf{c}/2$ gives us

$$\mathbf{x} = (\mathbf{c}'\mathbf{A}^{-1}\mathbf{c})^{-1}\mathbf{A}^{-1}\mathbf{c}$$

In the simple numerical illustration shown above, we have

$$\lambda = 2\left\{ \underset{\mathbf{c}'}{(1, -1)} \underset{\mathbf{A}^{-1}}{\begin{bmatrix} 1/2 & 0 \\ 0 & 1/3 \end{bmatrix}} \underset{\mathbf{c}}{\begin{bmatrix} 1 \\ -1 \end{bmatrix}} \right\}^{-1} = 12/5$$

$$\mathbf{x} = \frac{12}{5} \cdot \frac{1}{2} \begin{bmatrix} 1/2 & 0 \\ 0 & 1/3 \end{bmatrix}\begin{bmatrix} 1 \\ -1 \end{bmatrix} = \begin{bmatrix} 3/5 \\ -2/5 \end{bmatrix}$$

At the stationary point $\mathbf{x}' = (3/5, -2/5)$, the value of the function is

$$y = 2(3/5)^2 + 3(-2/5)^2 = 0$$

and it is noted that the constraint equation

$$g(\mathbf{x}) = 3/5 - (-2/5) - 1 = 0$$

is also satisfied.[13]

In summary, use of matrix notation provides a compact way to set down procedures for function optimization, in this case optimization under a constraint equation.

A.4.3 Conditions for the Optimization of Functions Involving Multivariable Arguments

In preceding sections of the appendix, necessary and sufficient conditions for identifying local extreme points have been listed for the single-argument and (unconstrained) two-argument cases. Things become considerably more complicated when we consider functions of multivariable arguments. Accordingly, we do not delve into the topic in much detail; in particular, all proofs are omitted. The reader interested in a more detailed discussion is referred to books by Beveridge and Schechter (1970) and Wilde and Beightler (1967).

In the case of a multivariable function, the necessary condition for a stationary point continues to be the vanishing of all first-order derivatives. That is, at a staionary point, we have the condition

$$\frac{\partial f}{\partial x_1} = 0; \quad \frac{\partial f}{\partial x_2} = 0; \ldots; \quad \frac{\partial f}{\partial x_n} = 0$$

This condition holds for either unconstrained or constrained functions (in the context of Lagrange multipliers).

However, as was observed in the cases of one or two arguments, a multivariable function does not necessarily have a local extremum at the stationary point of interest. To examine sufficiency conditions for a local extreme point, use is again made of the determinant of second-order partial derivatives:

$$\delta_n = \begin{vmatrix} \dfrac{\partial^2 f}{\partial x_1^2} & \dfrac{\partial^2 f}{\partial x_1\, \partial x_2} & \cdots & \dfrac{\partial^2 f}{\partial x_1\, \partial x_n} \\ \dfrac{\partial^2 f}{\partial x_2\, \partial x_1} & \dfrac{\partial^2 f}{\partial x_2^2} & \cdots & \dfrac{\partial^2 f}{\partial x_2\, \partial x_n} \\ \vdots & \vdots & & \vdots \\ \dfrac{\partial^2 f}{\partial x_n\, \partial x_1} & \dfrac{\partial^2 f}{\partial x_n\, \partial x_2} & \cdots & \dfrac{\partial^2 f}{\partial x_n^2} \end{vmatrix}$$

[13] Although we do not delve into details, the stationary point (3/5, −2/5) is a minimum.

As was the case with two arguments, we set up the principal minors of δ_n as follows:[14]

$$\delta_1 = \frac{\partial^2 f}{\partial x_1^2}; \quad \delta_2 = \begin{vmatrix} \dfrac{\partial^2 f}{\partial x_1^2} & \dfrac{\partial^2 f}{\partial x_1\, \partial x_2} \\ \dfrac{\partial^2 f}{\partial x_2\, \partial x_1} & \dfrac{\partial^2 f}{\partial x_2^2} \end{vmatrix}; \quad \delta_3 = \begin{vmatrix} \dfrac{\partial^2 f}{\partial x_1^2} & \dfrac{\partial^2 f}{\partial x_1\, \partial x_2} & \dfrac{\partial^2 f}{\partial x_1\, \partial x_3} \\ \dfrac{\partial^2 f}{\partial x_2\, \partial x_1} & \dfrac{\partial^2 f}{\partial x_2^2} & \dfrac{\partial^2 f}{\partial x_2\, \partial x_3} \\ \dfrac{\partial^2 f}{\partial x_3\, \partial x_1} & \dfrac{\partial^2 f}{\partial x_3\, \partial x_2} & \dfrac{\partial^2 f}{\partial x_3^2} \end{vmatrix}$$

..., up to, and including, δ_n.

Having done this, we evaluate each of the n determinants. *In order for a stationary point to be an extreme point, all of the n determinants must be nonzero.* A local minimum is distinguished from a local maximum in terms of the pattern of signs of the (evaluated) determinants:

1. If $\delta_j > 0$ for *all* $j = 1, 2, \ldots, n$, then the stationary point is a local minimum.
2. If $\delta_1 < 0$, $\delta_2 > 0$, $\delta_3 < 0$, $\delta_4 > 0$, ..., then the stationary point is a local maximum.
3. If neither situation occurs, one must examine the specific nature of the stationary point by computing values of the function in the neighborhood of the point.

Notice, then, that these conditions generalize what was discussed earlier for functions of one and two arguments.[15]

In case the function is subject to an equality constraint of the type illustrated in the context of Lagrange multipliers, the necessary condition for an extreme point involves finding a vector $\mathbf{x}_0$ that satisfies the $n + 1$ equations

$$\frac{\partial f(\mathbf{x})}{\partial \mathbf{x}} - \frac{\lambda\, \partial g(\mathbf{x})}{\partial \mathbf{x}} = \mathbf{0}; \qquad g(\mathbf{x}) = 0$$

that are obtained by setting the derivatives of

$$u(\mathbf{x}, \lambda) = f(\mathbf{x}) - \lambda g(\mathbf{x})$$

with respect to $\mathbf{x}$ and λ each equal to zero.

Sufficiency conditions for a local extremum in the case of Lagrange constraints become rather complex, particularly if more than one constraint equation is involved. Accordingly, the interested reader is referred to more specialized books on the subject, such as the book by Hancock (1960).

[14] By the term principal minors is meant successive determinants computed for submatrices of order 1, order 2, etc., formed along the main (principal) diagonal of the original $n \times n$ matrix.

[15] It should also be mentioned that a rather elegant approach to examining sufficiency conditions utilizes the matrix of partial derivatives as the matrix of a quadratic form. One then checks on whether the form is positive definite, negative definite, etc., and the type of definiteness is related to the type of extremum represented by the stationary point. This approach is fully compatible with the principal minor procedure, described above.

A.5 APPLICATION OF THE CALCULUS TO MULTIVARIATE ANALYSIS

At this point we have discussed (albeit selectively and rapidly) a number of concepts from the calculus that relate to the development of various matrix equations that arise in solving problems in multiple regression, principal components, and multiple discriminant analysis. It is now time to examine the specific nature of these central equations in multivariate analysis.

A.5.1 Multiple Regression Equations

The so-called normal equations of multiple regression theory represent a straightforward application of function minimization that utilizes the least-squares criterion. In multiple regression we have the case in which the matrix equation

$$\boxed{\mathbf{y} \cong \mathbf{Xb}}$$

has more equations (one for each case) than unknowns. As recalled from Chapters 1 and 6, $\mathbf{X}$ is the data matrix of predictors (augmented by a column vector of unities); $\mathbf{y}$ is the data vector representing the criterion variable; $\mathbf{b}$ is the to-be-solved-for vector of regression coefficients (including the intercept term); and $\cong$ denotes least-squares approximation.

The vector of prediction errors can be written as

$$\mathbf{e} = \mathbf{y} - \hat{\mathbf{y}}$$

where $\hat{\mathbf{y}}$ denotes the set of predicted values for $\mathbf{y}$. As we know, the least-squares criterion seeks a vector $\mathbf{b}$ that minimizes

$$f = (\mathbf{y} - \hat{\mathbf{y}})'(\mathbf{y} - \hat{\mathbf{y}})$$

Since $\hat{\mathbf{y}} = \mathbf{Xb}$, we have

$$\begin{aligned} f &= (\mathbf{y} - \mathbf{Xb})'(\mathbf{y} - \mathbf{Xb}) \\ &= \mathbf{y}'\mathbf{y} - \mathbf{b}'\mathbf{X}'\mathbf{y} - \mathbf{y}'\mathbf{Xb} + \mathbf{b}'\mathbf{X}'\mathbf{Xb} \\ &= \mathbf{b}'\mathbf{X}'\mathbf{Xb} - 2\mathbf{y}'\mathbf{Xb} + \mathbf{y}'\mathbf{y} \end{aligned}$$

where $\mathbf{y}'\mathbf{Xb} = \mathbf{b}'\mathbf{X}'\mathbf{y}$ since each term denotes the same scalar. Our objective is to find a vector of parameters $\mathbf{b}$ that minimizes f. This suggests finding the symbolic derivative and setting it equal to the $\mathbf{0}$ column vector:

$$\frac{\partial f}{\partial \mathbf{b}} = 2\mathbf{X}'\mathbf{Xb} - 2\mathbf{X}'\mathbf{y} = \mathbf{0}$$

We note that $\mathbf{X}'\mathbf{X}$ in $\mathbf{b}'\mathbf{X}'\mathbf{Xb}$ is symmetric, with derivative $2\mathbf{X}'\mathbf{Xb}$. Furthermore, we observe that the partial derivative with respect to the row vector $\mathbf{b}'$ is being found; hence, we take the transpose of $2\mathbf{y}'\mathbf{X}$ to obtain $2\mathbf{X}'\mathbf{y}$, the second term in the preceding equation. Dividing both sides by 2 and transposing leads to

$$\mathbf{X}'\mathbf{X}\mathbf{b} = \mathbf{X}'\mathbf{y}$$

and solving for $\mathbf{b}$, we get

$$\boxed{\mathbf{b} = (\mathbf{X}'\mathbf{X})^{-1}\mathbf{X}'\mathbf{y}}$$

We now recognize the matrix equation as that appearing in the discussion of multiple regression in Chapter 6. Although no check of sufficiency conditions has been made here, it turns out that $\mathbf{b}$ is the vector of parameters that does, indeed, minimize the function f.

A.5.2 Principal Components Analysis

In principal components analysis, we recall that interest centers on rotation of a deviation-from-mean data matrix $\mathbf{X}_d$ so as to maximize the quadratic form

$$f = \mathbf{t}'(\mathbf{X}_d'\mathbf{X}_d)\mathbf{t}$$

where we denote the SSCP matrix by $\mathbf{X}_d'\mathbf{X}_d$, the minor product moment of $\mathbf{X}_d$. (Alternatively, we could use the raw cross products, covariance, or correlation matrix.) Furthermore, we want to restrict the vector $\mathbf{t}$ to be a set of direction cosines that define the vector of linear composites:

$$\mathbf{y} = \mathbf{X}_d\mathbf{t}, \qquad \text{where} \quad \mathbf{t}'\mathbf{t} = 1$$

If we let $\mathbf{A} = \mathbf{X}_d'\mathbf{X}_d$, the principal components problem is to maximize

$$f = \mathbf{t}'\mathbf{A}\mathbf{t}$$

subject to the constraint that $\mathbf{t}'\mathbf{t} = 1$.

Based on our discussion of Lagrange multipliers, we can formalize the task by writing

$$u = \mathbf{t}'\mathbf{A}\mathbf{t} - \lambda(\mathbf{t}'\mathbf{t} - 1)$$

where $\mathbf{t}'\mathbf{t} - 1 = 0$ represents the constraint equation. As we know, the problem is to find the symbolic partial derivative of u with respect to $\mathbf{t}$ and set this equal to the $\mathbf{0}$ vector. Remembering that $\mathbf{A}$ is symmetric, we obtain

$$\frac{\partial u}{\partial \mathbf{t}} = 2\mathbf{A}\mathbf{t} - 2\lambda\mathbf{t} = \mathbf{0}$$

Next, dividing through by 2 and factoring out $\mathbf{t}$, we get

$$(\mathbf{A} - \lambda\mathbf{I})\mathbf{t} = \mathbf{0}$$

This represents the necessary condition to be satisfied by a stationary point $\mathbf{t}$ in which the constraint equation $\mathbf{t}'\mathbf{t} = 1$ is also satisfied. Again, we do not delve into the more complex topic of checking on sufficiency conditions, other than to say that the eigenvector $\mathbf{t}_1$ associated with the largest eigenvalue λ_1 of $\mathbf{A}$ is the vector of direction cosines that maximizes the function u.

A.5.3 Multiple Discriminant Analysis

As recalled from Chapter 6, in multiple discriminant analysis we seek a vector **v** with the property of maximizing the ratio

$$\lambda = \frac{\mathbf{v}'\mathbf{A}\mathbf{v}}{\mathbf{v}'\mathbf{W}\mathbf{v}}$$

where **A** is the among-group SSCP matrix and **W** is the pooled within-group SSCP matrix. (Again, we could place some restriction on the vector **v**, such as $\mathbf{v}'\mathbf{v} = 1$.) Note, however, that λ, the discriminant ratio in the present context, is simply the quotient of two functions (as illustrated in Table A.2). We can then find the symbolic derivative of λ with respect to **v**, by means of the quotient rule, and set it equal to the **0** vector:

$$\frac{\partial\lambda}{\partial\mathbf{v}} = \frac{2[(\mathbf{A}\mathbf{v})(\mathbf{v}'\mathbf{W}\mathbf{v}) - (\mathbf{v}'\mathbf{A}\mathbf{v})(\mathbf{W}\mathbf{v})]}{(\mathbf{v}'\mathbf{W}\mathbf{v})^2} = \mathbf{0}$$

This can be simplified by dividing numerator and denominator by $(\mathbf{v}'\mathbf{W}\mathbf{v})$ and making the substitution

$$\lambda = \frac{\mathbf{v}'\mathbf{A}\mathbf{v}}{\mathbf{v}'\mathbf{W}\mathbf{v}}$$

to obtain

$$\frac{2[\mathbf{A}\mathbf{v} - \lambda\mathbf{W}\mathbf{v}]}{\mathbf{v}'\mathbf{W}\mathbf{v}} = \mathbf{0}$$

Next, we divide both sides by the scalar 2 and further simplify to

$$(\mathbf{A} - \lambda\mathbf{W})\mathbf{v} = \mathbf{0}$$

Next, assuming that **W** is nonsingular, we have the familiar expression of Chapter 6:

$$(\mathbf{W}^{-1}\mathbf{A} - \lambda\mathbf{I})\mathbf{v} = \mathbf{0}$$

where, as we know, $\mathbf{W}^{-1}\mathbf{A}$ is nonsymmetric. Again, we omit discussion of the sufficiency conditions, indicating that a maximum has been found. Suffice it to say that all three procedures:

1. multiple regression,
2. principal components analysis, and
3. multiple discriminant analysis

involve aspects of the calculus that deal with the optimization of functions of multivariable arguments. The concept of symbolic differentiation is central to the topic as well as the techniques of function optimization, either unconstrained or constrained optimization, as the case may be.

A.6 SUMMARY

This appendix has dealt with those aspects of the calculus—particularly symbolic differentiation and optimization theory—related to the matrix equations that appear in various multivariate methods, such as multiple regression, principal components, and multiple discriminant analysis.

The review was brief and selective. We first discussed the differentiation of functions of one argument, including the statement of necessary and sufficient conditions for local extrema. This was followed by a similar discussion of the case involving functions of two arguments. Also, the technique of Lagrange multipliers was introduced at this point.

We next described the most general case of functions of multivariable arguments and the concept of symbolic differentiation. Symbolic derivatives of common matrix functions were illustrated, and necessary and sufficient conditions for local extrema of multivariable functions were also listed. We concluded the appendix with applications of the calculus to the derivation of matrix equations in multiple regression, principal components, and multiple discriminant analysis.

REVIEW QUESTIONS

1. By means of the chain rule, find the derivative of the following functions:

a. $\ln(2x - x^2)$ b. $\dfrac{1-2x}{x^2+4}$ c. $e^{x^{1/2}}$ d. $\left[\dfrac{x^3-1}{2x^3+1}\right]^4$

2. Find (and identify) extreme points for the function

$$f(x) = \frac{6x}{x^2+1}$$

over the domain $-2 \leqslant x \leqslant 2$.

3. Find the partial derivative of $f(x, y)$ at $f(1, 3)$ where

$$f(x, y) = x^2 + 2x + 4y + \ln(x^2 + y^2)$$

4. Find the minimum of $f(x, y) = 2x^2 + 4x + 8y + y^2$.
5. Find (and identify) a stationary point of the function

$$f(x, y) = y^2 + 4y + 2x - x^2$$

subject to the constraint

$$x + 2y = 2$$

6. Find the derivative with respect to **x** of the quadratic form

$$g(\mathbf{x}) = (x_1, x_2, x_3)\begin{bmatrix} 2 & 3 & 4 \\ 3 & 1 & 2 \\ 4 & 2 & 3 \end{bmatrix}\begin{bmatrix} x_1 \\ x_2 \\ x_3 \end{bmatrix}$$

Evaluate the derivative at $\mathbf{x}' = (3, 1, 2)$.

7. If $g(\mathbf{x}) = \mathbf{x}'\mathbf{A}\mathbf{x} + \mathbf{b}'\mathbf{x} + c$ where $\mathbf{A}$ is symmetric, then it can be shown that the derivative of this general quadratic function with respect to **x** is

$$\frac{\partial g}{\partial \mathbf{x}} = 2\mathbf{A}\mathbf{x} + \mathbf{b}$$

Furthermore, a stationary point is given by

$$2\mathbf{A}\mathbf{x} + \mathbf{b} = \mathbf{0}; \qquad \mathbf{x} = -\tfrac{1}{2}\mathbf{A}^{-1}\mathbf{b}$$

If

$$\underset{\mathbf{x}'}{(x_1, x_2)} \underset{\mathbf{A}}{\begin{bmatrix} 1 & -\frac{1}{2} \\ -\frac{1}{2} & 1 \end{bmatrix}} \underset{\mathbf{x}}{\begin{bmatrix} x_1 \\ x_2 \end{bmatrix}} + \underset{\mathbf{b}'}{(2, 2)} \underset{\mathbf{x}}{\begin{bmatrix} x_1 \\ x_2 \end{bmatrix}} \underset{c}{-4}$$

is the function $g(\mathbf{x})$, find (and identify) a stationary point of $g(\mathbf{x})$.

APPENDIX B

Linear Equations and Generalized Inverses

B.1 INTRODUCTION

In various sections of the book, and particularly in Section 4.6, we have discussed how one solves a set of simultaneous linear equations where the matrix of coefficients has a regular inverse. Moreover, the pivotal method has been described as an illustrative computational procedure for obtaining the desired inverse.

In this appendix interest centers on a matrix of coefficients $\mathbf{A}$ for which no regular inverse $\mathbf{A}^{-1}$ exists. That is, $\mathbf{A}$ may be rectangular or, even if square, it may be singular.[1] The *generalized inverse* is a concept that provides a way to solve a set of consistent linear equations in which a regular inverse does not exist. As we shall note later, several different types of generalized inverse have been defined, although we concentrate here on only two variations, the Moore–Penrose inverse (Penrose, 1955) and the g inverse (Rao, 1962).

Before discussing generalized inverses, we provide a review of the types of solutions: (a) none, (b) one, or (c) infinitely many, that one can obtain in attempting to solve a set of simultaneous linear equations. Aspects of homogeneous equations and nonhomogeneous equations are described and illustrated numerically. Finally, a general method is evolved for solving sets of equations.

We then introduce the topic of generalized inverse in terms of a set of properties that such inverses are designed to satisfy. Following this, the Moore–Penrose type of inverse is defined and related to the concept of basic structure (or singular value decomposition) discussed in Chapter 5.

The second type of inverse, called the g inverse, is then introduced. This inverse is required to satisfy only one of the Penrose properties and, in practice, is easier to compute. Illustrations of its computation are presented, and this type of generalized inverse is related to procedures for solving linear equations. In so doing, a general procedure for computing inverses—regular or generalized—is described and related to earlier material involving the solution of simultaneous equations.

[1] That is, its determinant is zero.

B.2 SIMULTANEOUS LINEAR EQUATIONS

In the examples considered in the book we often had occasion to solve the system of equations

$$\boxed{\mathbf{Ax} = \mathbf{b}}$$

by means of matrix inversion. Recall that **A** is the matrix of coefficients, **x** is the vector of unknowns, and **b** is the vector of constants. In this case **A** was $n \times n$ and $r(\mathbf{A}) = n$. That is, **A** was square and nonsingular. The pivotal method was employed as a general solution technique.

However, suppose **A** is either rectangular or square singular so that a regular inverse does not exist. What happens then? Before launching into this topic, let us review some of the basic results related to solving a system of simultaneous linear equations. First, let us consider the set of equations

$$3x_1 + x_2 = 5$$

$$5x_1 + 2x_2 = 9$$

This set of equations, as could be easily verified, has the solution $x_1 = 1$, $x_2 = 2$. Furthermore, the solution is unique—only that specific set of values satisfies the set of equations.

Let us next consider the simultaneous equations

$$3x_1 + x_2 = 5$$

$$6x_1 + 2x_2 = 11$$

If we try to eliminate x_1 by taking twice the first equation and subtracting it from the second, we get the result

$$0 = 1$$

and, of course, something is wrong. This is most easily observed by noting that insofar as the left-hand side of the equations is concerned, the second equation is twice that of the first, but this relationship is not true for the right-hand side. The equations in this case are said to be *inconsistent,* and no solution exists.

Next, let us take the three equations

$$x_1 + x_2 + 3x_3 = 8$$

$$x_1 + 2x_2 + 6x_3 = 14$$

$$x_2 + 3x_3 = 6$$

We first eliminate x_1 from the second equation by subtracting the first from the second to get the pair of equations

$$x_2 + 3x_3 = 6$$

$$x_2 + 3x_3 = 6$$

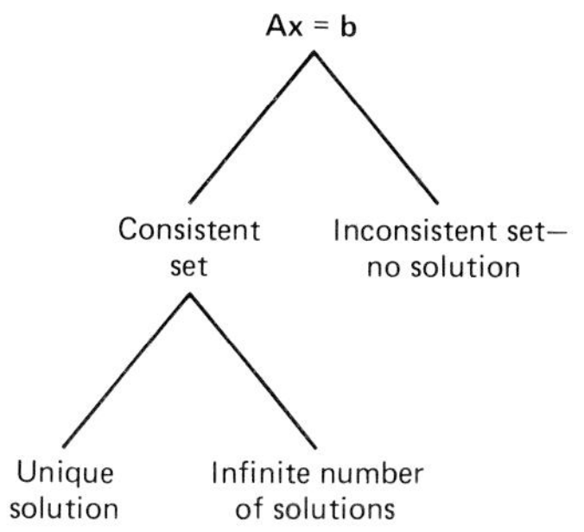

Fig. B.1 Tree diagram of types of solutions to a set of linear equations.

Note that these are identical. Thus, we have

$$x_2 = 6 - 3x_3$$

$$x_1 = 8 - (6 - 3x_3) - 3x_3 = 2$$

In this case, then, more than a single solution exists. For example, we have

$$x_1 = 2; \quad x_2 = 3; \quad x_3 = 1$$

$$x_1 = 2; \quad x_2 = 0; \quad x_3 = 2$$

$$x_1 = 2; \quad x_2 = -3; \quad x_3 = 3$$

and so on.

The tree diagram in Fig. B.1 shows the three cases of interest. We first want to examine whether the set of equations is consistent or not. If inconsistent, no solutions exist. If consistent, either a single (and unique) solution exists or an infinite number of solutions exist.

The three theorems[2] of interest in determining which condition prevails are:

1. A set of linear equations is consistent if and only if the rank of the augmented matrix (found by appending the **b** vector to the matrix of coefficients **A**) is equal to the rank of the original coefficients matrix.
2. A set of consistent linear equations has a unique solution if and only if the rank of the coefficients matrix **A** equals its order; that is, if and only if $r(\mathbf{A}) = n$, where **A** is of order $n \times n$ and n unknowns are present.
3. A consistent set of linear equations, where **A** is of rank k, can be solved for k unknowns in terms of the remaining $n - k$ unknowns if and only if the submatrix of coefficients (obtained from **A**) is of rank k.

With these theorems to guide us, let us return to the pair of equations

$$3x_1 + x_2 = 5$$

$$6x_1 + 2x_2 = 11$$

[2] Proofs can be bound in Graybill (1969).

and now form the matrix of coefficients **A** and the augmented matrix **M**:

$$\mathbf{A} = \begin{bmatrix} 3 & 1 \\ 6 & 2 \end{bmatrix}; \qquad \mathbf{M} = \begin{bmatrix} 3 & 1 & \vdots & 5 \\ 6 & 2 & \vdots & 11 \end{bmatrix}$$

By inspection we see that the second row of **A** is twice the first; hence $r(\mathbf{A}) = 1$. However, if we apply the reduction to echelon form procedure (from Section 4.7) to **M** we get the echelon matrix $\mathbf{H_M}$:

$$\mathbf{H_M} = \begin{bmatrix} 1 & 1/3 & \vdots & 5/3 \\ 0 & 0 & \vdots & 1 \end{bmatrix}$$

We note that both rows of $\mathbf{H_M}$ have at least one nonzero entry. Hence, $r(\mathbf{H_M}) = r(\mathbf{M}) = 2$ while $r(\mathbf{A}) = 1$; the set of equations is not consistent, and no solution exists.

Next, taking the equations

$$3x_1 + x_2 = 5$$

$$5x_1 + 2x_2 = 9$$

we have the matrices

$$\mathbf{A} = \begin{bmatrix} 3 & 1 \\ 5 & 2 \end{bmatrix}; \qquad \mathbf{M} = \begin{bmatrix} 3 & 1 & \vdots & 5 \\ 5 & 2 & \vdots & 9 \end{bmatrix}$$

After reduction to echelon form we obtain

$$\mathbf{H_A} = \begin{bmatrix} 1 & 1/3 \\ 0 & 1 \end{bmatrix}; \qquad \mathbf{H_M} = \begin{bmatrix} 1 & 1/3 & \vdots & 5/3 \\ 0 & 1 & \vdots & 2 \end{bmatrix}$$

and note, then, that $r(\mathbf{A}) = r(\mathbf{M}) = 2$, which is also equal to the order of **A**. In this case the equations are consistent, and a unique solution exists.

Finally, if we take the set of the three equations

$$x_1 + x_2 + 3x_3 = 8$$

$$x_1 + 2x_2 + 6x_3 = 14$$

$$x_2 + 3x_3 = 6$$

we have

$$\mathbf{A} = \begin{bmatrix} 1 & 1 & 3 \\ 1 & 2 & 6 \\ 0 & 1 & 3 \end{bmatrix}; \qquad \mathbf{M} = \begin{bmatrix} 1 & 1 & 3 & \vdots & 8 \\ 1 & 2 & 6 & \vdots & 14 \\ 0 & 1 & 3 & \vdots & 6 \end{bmatrix}$$

with associated echelon forms

$$\mathbf{H_A} = \begin{bmatrix} 1 & 1 & 3 \\ 0 & 1 & 3 \\ 0 & 0 & 0 \end{bmatrix}; \qquad \mathbf{H_M} = \begin{bmatrix} 1 & 1 & 3 & \vdots & 8 \\ 0 & 1 & 3 & \vdots & 6 \\ 0 & 0 & 0 & \vdots & 0 \end{bmatrix}$$

and, we note that the rank in each case is 2, while the order of **A** is 3. Thus, we can solve for $k = 2$ unknowns in terms of $n - k = 3 - 2 = 1$ remaining unknown. The results suggest a general approach to solving sets of simultaneous linear equations.

B.2.1 A General Procedure for Solving Linear Equations

As might be surmised at this point, in solving sets of linear equations we must determine whether a solution exists and, if so, whether the solution is unique or whether an infinity of solutions exists. The reduction of the matrix to echelon form via elementary row (or column) operations provides a practical way to find the rank of the coefficients matrix **A** and the rank of the augmented matrix **M**. As it turns out, however, reduction of a matrix to echelon form, followed by a few additional operations, provides us with a very general method for solving sets of simultaneous equations. As recalled, elementary row (column) operations permit

1. the interchange of two rows (columns);
2. the multiplication of each entry in a row (column) by any scalar $\lambda \neq 0$;
3. the addition, to each entry of some row (column), of λ times the corresponding element of some other row (column).

Each of these operations can be carried out on the rows of **A** by means of premultiplying **A** by a matrix that, in turn, can be obtained by performing the given elementary row operation on the *identity* matrix.[3] For example, let

$$\mathbf{A} = \begin{bmatrix} 1 & 4 \\ 2 & 5 \\ 3 & 6 \end{bmatrix}$$

and assume that we wish to

1. interchange rows 1 and 2;
2. multiply row 2 by the scalar 4;
3. add twice row 3 to row 1.

If these three operations are separately performed on the identity matrix **I**, of order 3 x 3, we have, respectively,

$$\begin{bmatrix} 0 & 1 & 0 \\ 1 & 0 & 0 \\ 0 & 0 & 1 \end{bmatrix}; \quad \begin{bmatrix} 1 & 0 & 0 \\ 0 & 4 & 0 \\ 0 & 0 & 1 \end{bmatrix}; \quad \begin{bmatrix} 1 & 0 & 2 \\ 0 & 1 & 0 \\ 0 & 0 & 1 \end{bmatrix}$$

The reader can convince himself that premultiplication of **A** by each of the three matrices above will effect the desired row operation. Similarly, elementary column operations can be carried out by performing the indicated operation on the columns of a 2 x 2 identity matrix and *postmultiplying* **A** by the appropriate matrix. Successive operations are represented, of course, by a set of matrices whose sequence is determined by the desired sequence in which the elementary operations are to be performed.

[3] In the case of elementary column operations, the matrix **A** is *postmultiplied* by the specified elementary column operation on the identity matrix.

To illustrate the notion of a sequence of elementary row operations, let us simultaneously transform **A** and **I** by the three operations noted above, in the order given:

Interchange rows 1 and 2:

$$\mathbf{A}_1 = \begin{bmatrix} 2 & 5 \\ 1 & 4 \\ 3 & 6 \end{bmatrix}; \qquad \mathbf{B}_1 = \begin{bmatrix} 0 & 1 & 0 \\ 1 & 0 & 0 \\ 0 & 0 & 1 \end{bmatrix}$$

Multiply row 2 by the scalar 4:

$$\mathbf{A}_2 = \begin{bmatrix} 2 & 5 \\ 4 & 16 \\ 3 & 6 \end{bmatrix}; \qquad \mathbf{B}_2 = \begin{bmatrix} 0 & 1 & 0 \\ 4 & 0 & 0 \\ 0 & 0 & 1 \end{bmatrix}$$

Add twice row 3 to row 1:

$$\mathbf{A}_3 = \begin{bmatrix} 8 & 17 \\ 4 & 16 \\ 3 & 6 \end{bmatrix}; \qquad \mathbf{B}_3 = \begin{bmatrix} 0 & 1 & 2 \\ 4 & 0 & 0 \\ 0 & 0 & 1 \end{bmatrix}$$

Finally, we note that $\mathbf{A}_3$ can be obtained from the combined row operations—in the *indicated order*—by

$$\mathbf{A}_3 = \overset{\mathbf{B}_3}{\begin{bmatrix} 0 & 1 & 2 \\ 4 & 0 & 0 \\ 0 & 0 & 1 \end{bmatrix}} \overset{\mathbf{A}}{\begin{bmatrix} 1 & 4 \\ 2 & 5 \\ 3 & 6 \end{bmatrix}} = \begin{bmatrix} 8 & 17 \\ 4 & 16 \\ 3 & 6 \end{bmatrix}$$

Moreover, we also recall from Chapter 4 that $\mathbf{B}_1, \mathbf{B}_2, \ldots,$ is each nonsingular and that the rank of **A** is unaffected by elementary row (column) operations.

With this review information out of the way, the formal method for solving sets of simultaneous equations can be stated. First, we start with the augmented matrix **M**. Then we carry out elementary operations to reduce **M** to echelon form $\mathbf{H_M}$. As a final step we carry out additional elementary row operations on $\mathbf{H_M}$ so as to obtain an identity matrix in the subset of columns corresponding to **A**, the matrix of coefficients. To illustrate, let us take the matrix **M**, as used earlier for the set of two simultaneous equations for which the unique solution was $x_1 = 1, x_2 = 2$. We then apply the echelon reduction procedure to get $\mathbf{H_M}$. That is,

$$\begin{aligned} 3x_1 + x_2 &= 5 \\ 5x_1 + 2x_2 &= 9 \end{aligned}; \qquad \mathbf{M} = \left[\begin{array}{cc:c} 3 & 1 & 5 \\ 5 & 2 & 9 \end{array}\right]; \qquad \mathbf{H_M} = \left[\begin{array}{cc:c} 1 & 1/3 & 5/3 \\ 0 & 1 & 2 \end{array}\right]$$

Next, subtract 1/3 of the second row of $\mathbf{H_M}$ from the first row to get

$$\mathbf{N} = \left[\begin{array}{cc:c} 1 & 0 & 1 \\ 0 & 1 & 2 \end{array}\right]$$

Now, let us consider $\mathbf{N}$ in its original context of two linear equations

$$1x_1 + 0x_2 = 1$$

$$0x_1 + 1x_2 = 2$$

with the desired solution $x_1 = 1, x_2 = 2$.

Next, let us take the case of the two inconsistent equations:

$$\begin{aligned} 3x_1 + x_2 &= 5 \\ 6x_1 + 2x_2 &= 11 \end{aligned}; \quad \mathbf{M} = \left[\begin{array}{cc:c} 3 & 1 & 5 \\ 6 & 2 & 11 \end{array}\right]; \quad \mathbf{H_M} = \left[\begin{array}{cc:c} 1 & 1/3 & 5/3 \\ 0 & 0 & 1 \end{array}\right]$$

In this case we need go no further since in the second row of $\mathbf{H_M}$ we note the inconsistency:

$$0x_1 + 0x_2 = 1$$

which, of course, shows that this set of equations has no solution. Further evidence for this is found by examining the left-hand submatrix of $\mathbf{H_M}$; it is of rank 1 while $\mathbf{H_M}$ itself is of rank 2.

Finally, let us take the third example:

$$\begin{aligned} x_1 + x_2 + 3x_3 &= 8 \\ x_1 + 2x_2 + 6x_3 &= 14; \\ x_2 + 3x_3 &= 6 \end{aligned} \quad \mathbf{M} = \left[\begin{array}{ccc:c} 1 & 1 & 3 & 8 \\ 1 & 2 & 6 & 14 \\ 0 & 1 & 3 & 6 \end{array}\right]; \quad \mathbf{H_M} = \left[\begin{array}{ccc:c} 1 & 1 & 3 & 8 \\ 0 & 1 & 3 & 6 \\ 0 & 0 & 0 & 0 \end{array}\right]$$

If we subtract the second row of $\mathbf{H_M}$ from the first, we get

$$\mathbf{N} = \left[\begin{array}{ccc:c} 1 & 0 & 0 & 2 \\ 0 & 1 & 3 & 6 \\ 0 & 0 & 0 & 0 \end{array}\right]; \quad \begin{aligned} 1x_1 + 0x_2 + 0x_3 &= 2 \\ 0x_1 + 1x_2 + 3x_3 &= 6 \\ 0x_1 + 0x_2 + 0x_3 &= 0 \end{aligned}$$

In this case the best we can do is obtain a 2 x 2 identity matrix for the *first two* rows and columns of $\mathbf{N}$. As illustrated earlier, we can then transfer x_3 to the right-hand side, giving us

$$x_1 = 2; \qquad x_2 = 6 - 3x_3$$

If we then treat x_3 as a parameter, by setting it equal to (say) γ_3, we have

$$x_1 = 2; \qquad x_2 = 6 - 3\gamma_3; \qquad x_3 = \gamma_3$$

or, in vector form,

$$\mathbf{x} = \begin{bmatrix} 2 \\ 6 - 3\gamma_3 \\ \gamma_3 \end{bmatrix}$$

and an infinity of solutions exists depending upon what value we choose for γ_3.

B.2.2 Other Cases

In the cases examined so far we dealt with square matrices **A** of order $n \times n$ and vectors **x** and **b**, each of order $n \times 1$.

In the more general case, **A** can be of order $m \times n$. First, let us assume that $r(\mathbf{A}) = r(\mathbf{M}) = k$. If so, then at least one solution must exist. Next, let us assume that $k = n$, the number of unknowns. Since k cannot exceed m, the number of rows of **A** (or **M**), and $k = n$, then either $m = n$ or $m > n$. If $m = n = k$, then we know that **A** is square and nonsingular, and the solution is unique. However, if $m > n$, the echelon matrix $\mathbf{H_M}$ will have $m - k$ rows of zeros, and we can say that $m - k$ equations are redundant. If so, we proceed as before and solve k equations in $n = k$ variables. The submatrix **N**, of order $k \times k$, is still nonsingular, and the solution is still unique.

Next, suppose that $k < n$. If this case exists, then either $k = m$ or $k < m$. (We know that k cannot exceed m.) If we assume that $k = m < n$, we shall have an infinite number of solutions, as illustrated earlier. That is, $n - k$ of the unknowns can be treated as parameters.

Finally, assume that $k < m$ (and $k < n$). In this case, we have not only unknowns to spare but redundant equations as well. Not surprisingly, by following the formal method outlined earlier, we shall end up with an infinity of solutions *and* redundant equations in the bargain. To illustrate,

$$\begin{aligned} x_1 + x_2 + 3x_3 &= 8 \\ x_1 + 2x_2 + 6x_3 &= 14 \\ x_2 + 3x_3 &= 6 \\ 2x_1 + 3x_2 + 9x_3 &= 22 \end{aligned} ; \quad \mathbf{M} = \begin{bmatrix} 1 & 1 & 3 & \vdots & 8 \\ 1 & 2 & 6 & \vdots & 14 \\ 0 & 1 & 3 & \vdots & 6 \\ 2 & 3 & 9 & \vdots & 22 \end{bmatrix}$$

If we then reduce **M** to echelon form, we get

$$\mathbf{H_M} = \begin{bmatrix} 1 & 1 & 3 & \vdots & 8 \\ 0 & 1 & 3 & \vdots & 6 \\ 0 & 0 & 0 & \vdots & 0 \\ 0 & 0 & 0 & \vdots & 0 \end{bmatrix}$$

As can be noted from $\mathbf{H_M}$, $r(\mathbf{A}) = r(\mathbf{M}) = 2$, and at least one solution exists. The next step is to find an identity submatrix by further elementary row operations on $\mathbf{H_M}$, giving us

$$\mathbf{N} = \begin{bmatrix} 1 & 0 & 0 & \vdots & 2 \\ 0 & 1 & 3 & \vdots & 6 \\ 0 & 0 & 0 & \vdots & 0 \\ 0 & 0 & 0 & \vdots & 0 \end{bmatrix}$$

with the solution

$$\mathbf{x} = \begin{bmatrix} 2 \\ 6 - 3\gamma_3 \\ \gamma_3 \end{bmatrix}$$

as found earlier. We see in this case that the fourth equation is redundant with the others. As a matter of fact, it is simply the sum of the first two equations (whereas the third equation represents their difference).

Thus, if we have m equations in n unknowns (of the form $\mathbf{Ax} = \mathbf{b}$) in which $r(\mathbf{A}) = r(\mathbf{M}) = k$, while $k < n$ and $k < m$, we have an infinite number of solutions in which $n - k$ variables can be treated as parameters and $m - k$ equations are redundant. In effect, then, the relationship between k and n deals with the question of a single versus infinite number of solutions, while the relationship between k and m concerns whether some of the equations (viz., $m - k$) are redundant.

B.2.3 Homogeneous Equations

Up to this point we have been discussing the case of $\mathbf{Ax} = \mathbf{b}$, involving nonhomogeneous equations. Sometimes the multivariate analyst will encounter sets of linear equations of the form

$$\boxed{\mathbf{Ax} = \mathbf{0}}$$

These are called homogeneous equations. First of all, we note that one possible solution is to let $\mathbf{x} = \mathbf{0}$. That is, if we assign 0 to each unknown, the equation is satisfied, since $\mathbf{A0} = \mathbf{0}$. This is called the *trivial* solution.

Viewed another way, if we append the zero vector to $\mathbf{A}$ to get the augmented matrix $\mathbf{M}$, then $r(\mathbf{A})$ will always equal $r(\mathbf{M})$, and the equations will always be consistent. Hence, we shall always have at least one solution, namely, the trivial solution.

The basic question, then, becomes one of determining the conditions under which solutions other than the trivial one exist. As with the case for nonhomogeneous equations, the answer depends on the relationship between $r(\mathbf{A}) = r(\mathbf{M}) = k$ and n, the number of unknowns. Since k cannot exceed n, we are left with the two cases: (a) $k = n$ and (b) $k < n$. The results in each case are contained in the following assertions:

1. Given a set of homogeneous linear equations $\mathbf{Ax} = \mathbf{0}$, involving m equations in n unknowns, a unique (and trivial) solution $\mathbf{x} = \mathbf{0}$ exists if $r(\mathbf{A}) = k = n$.
2. Given a set of homogeneous linear equations $\mathbf{Ax} = \mathbf{0}$, involving m equations in n unknowns, an infinite number of solutions exist if $r(\mathbf{A}) = k < n$.

We can illustrate these cases by the following set of $m = 3$ equations in $n = 2$ unknowns:

$$\begin{aligned} 2x_1 + x_2 &= 0 \\ 3x_1 + 2x_2 &= 0; \\ 7x_1 + 4x_2 &= 0 \end{aligned} \qquad \mathbf{A} = \begin{bmatrix} 2 & 1 \\ 3 & 2 \\ 7 & 4 \end{bmatrix}$$

With homogeneous equations, there is no point in obtaining **M**, the augmented matrix, since the appended column would be the zero vector. Rather, we can reduce **A** itself to echelon form, so as to get

$$\mathbf{H_A} = \begin{bmatrix} 1 & 1/2 \\ 0 & 1 \\ 0 & 0 \end{bmatrix}$$

From $\mathbf{H_A}$ we see that $r(\mathbf{A}) = k = n = 2$. Next, if we subtract $\frac{1}{2}$ times row 2 from row 1, we get

$$\mathbf{N} = \begin{bmatrix} 1 & 0 \\ 0 & 1 \\ 0 & 0 \end{bmatrix}; \qquad \begin{aligned} 1x_1 + 0x_2 &= 0 \\ 0x_1 + 1x_2 &= 0 \\ 0x_1 + 0x_2 &= 0 \end{aligned}$$

with the trivial solution $x_1 = 0, x_2 = 0$. Moreover, the third original equation is redundant and, as a matter of fact, equals twice the first equation plus the second.

An illustration of the second case, $r(\mathbf{A}) = k < n$, is the following set of $m = 2$ equations in $n = 3$ unknowns:

$$\begin{aligned} 2x_1 + x_2 + 3x_3 &= 0 \\ x_1 + x_2 - x_3 &= 0 \end{aligned}$$

As before we find the echelon form of **A** as

$$\mathbf{H_A} = \begin{bmatrix} 1 & 1/2 & 3/2 \\ 0 & 1 & -5 \end{bmatrix}$$

and note that $r(\mathbf{H_A}) = r(\mathbf{A}) = 2$ and, hence, $k<n$. Next, we find the identity submatrix for the first two rows and two columns by subtracting $\frac{1}{2}$ of row 2 from row 1:

$$\mathbf{N} = \begin{bmatrix} 1 & 0 & 4 \\ 0 & 1 & -5 \end{bmatrix}$$

The equations can now be written[4] as

$$\begin{aligned} 1x_1 + 0x_2 &= -4x_3 \\ 0x_1 + 1x_2 &= 5x_3 \end{aligned}$$

and if we let $x_3 = \gamma_3$, we have the general solution, in vector form, as

$$\mathbf{x} = \begin{bmatrix} -4\gamma_3 \\ 5\gamma_3 \\ \gamma_3 \end{bmatrix} = \gamma_3 \begin{bmatrix} -4 \\ 5 \\ 1 \end{bmatrix}$$

[4] Note here that the implied vector $\begin{bmatrix} 4x_3 \\ -5x_3 \end{bmatrix}$ in the third column of the preceding matrix has simply been transposed to the right-hand side of the equation.

Note, here, that $m < n$; hence, $k < n$. If the number of equations is less than the number of unknowns, we must have an infinite number of solutions.[5]

One point of major difference between the present case and the counterpart case involving nonhomogeneous equations concerns the solution vector in situations involving $k < n$. In the case of homogeneous equations, we find that

$$\mathbf{x} = \begin{bmatrix} -4\gamma_3 \\ 5\gamma_3 \\ \gamma_3 \end{bmatrix}$$

and observe that *each* entry of $\mathbf{x}$ involves the arbitrary parameter γ_3. Thus, if we set $\gamma_3 = 0$, then $\mathbf{x} = \mathbf{0}$, and the trivial solution is included. Also, in the present illustration, where we have only one parameter γ_3, each solution is a scalar multiple of each other solution. For example, if we let $\gamma_3 = 1$ and then let $\gamma_3 = 2$, we have

$$\mathbf{x}_1 = \begin{bmatrix} -4 \\ 5 \\ 1 \end{bmatrix}; \qquad \mathbf{x}_2 = \begin{bmatrix} -8 \\ 10 \\ 2 \end{bmatrix} = 2\mathbf{x}_1$$

In contrast, if we reproduce the solution

$$\mathbf{x} = \begin{bmatrix} 2 \\ 6-3\gamma_3 \\ \gamma_3 \end{bmatrix}$$

in Section B.2.2 dealing with nonhomogeneous equations, we see that the first entry (2) does *not* involve γ_3.

To sum up, if some $\mathbf{x}^\circ \neq \mathbf{0}$ is a solution, then $\lambda\mathbf{x}^\circ$ (where λ is an arbitrary scalar) is also a solution in the case of homogeneous equations (a fact that was noted in Chapter 5 in the context of matrix eigenstructures), provided that $n - k = 1$. If *more than one* free parameter is found (i.e., $n - k > 1$), then it no longer follows that all solutions are scalar multiples of each other. However, if $\mathbf{x}^\circ \neq \mathbf{0}$ is a solution, it still follows that $\lambda\mathbf{x}^\circ$ is also a solution. This can be easily seen by noting that

$$\lambda\mathbf{A}\mathbf{x}^\circ = \mathbf{A}(\lambda\mathbf{x}^\circ) = \mathbf{0}$$

Again, as recalled from the discussion of eigenstructures in Chapter 5, if $\mathbf{A}$ is nonsingular and $r(\mathbf{A}) = k = n$, then nontrivial solutions cannot exist.

In summary, we can recapitulate the general method for solving either case:

1. nonhomogeneous equations of the form $\mathbf{Ax} = \mathbf{b}$;
2. homogeneous equations of the form $\mathbf{Ax} = \mathbf{0}$.

[5] As observed, in the case of homogeneous equations, either one solution exists (i.e., the trivial solution) or an infinity of solutions exists, depending upon the relationship between matrix rank and number of unknowns.

In the nonhomogeneous equations case the augmented matrix **M** is reduced to echelon form and then, via additional elementary row operations, to identity matrix form for the appropriate submatrix.[6] In the homogeneous equations case the analogous operations are carried out on the coefficients matrix **A**.

Figure B.2 recapitulates the various outcomes in tree diagram form. If we examine the case of nonhomogeneous equations first, we see that the primary outcomes are (a) none, (b) one, and (c) infinitely many solutions. We check $r(\mathbf{A})$ versus $r(\mathbf{M})$ to ascertain which condition prevails.

Assuming that $r(\mathbf{A}) = r(\mathbf{M}) = k$—and, hence, at least one solution exists—we check to see whether $k = n$, the number of unknowns. If so, then a unique solution exists. Next, if $k = m$, all equations are independent, while if $k < m$, some equations ($m - k$ of them) are redundant. Part (a) designates the first case in the tree diagram, while Part (b) designates the second.

If $k < n$, an infinite number of solutions exist. Again, we check to see if $k = m$ or $k < m$ so as to see if the equations are either all independent or not. Similar remarks pertain to the case of homogeneous equations with the exception, of course, that just one (the trivial solution) or infinitely many solutions exist in this instance; that is, if the system is homogeneous, it is consistent.

B.3 INTRODUCTORY ASPECTS OF GENERALIZED INVERSES

In Section B.2 a general method, utilizing reduction to echelon form followed by additional elementary row operations for finding an appropriate identity submatrix, was illustrated for solving sets of simultaneous equations. As was noted, provided that the equations are consistent, a solution—indeed an infinite number of solutions—can be found if $\mathbf{A}^{-1}$, the regular inverse of the coefficients matrix, does not exist.

At this point our interest centers on cases in which $\mathbf{A}^{-1}$, in the usual sense, does not exist, and yet we would still like to solve the set of equations of interest. This is the type of problem that the concept of *generalized inverse* has been developed to solve.

The literature on generalized inverses is of relatively recent origin, and its nomenclature and mathematical notation are not standard across authors. Two basic types of generalized inverse are discussed here:

1. the Moore–Penrose inverse (sometimes referred to as the pseudoinverse), written as $\mathbf{A}^{+}$;
2. the *g* inverse (sometimes referred to as the conditional inverse), written as $\mathbf{A}^{-}$.

However, the reader should be made aware of the fact that many types of generalized inverses exist—each obeying a particular set of properties.[7] Also, different ways have been developed to define these inverses. Our discussion in this appendix merely scratches the surface of an already broad and still expanding topic.

[6] The general procedure for converting the echelon reduced form to an identity submatrix is formalized in Section B.4. As noted there, the complete procedure involves reduction of **M** to what is known as Hermite form (in the case of a square coefficients matrix).

[7] The names and symbols for the Moore–Penrose and the *g* inverse follow those of Good (1969); other authors use different names, symbols, or both.

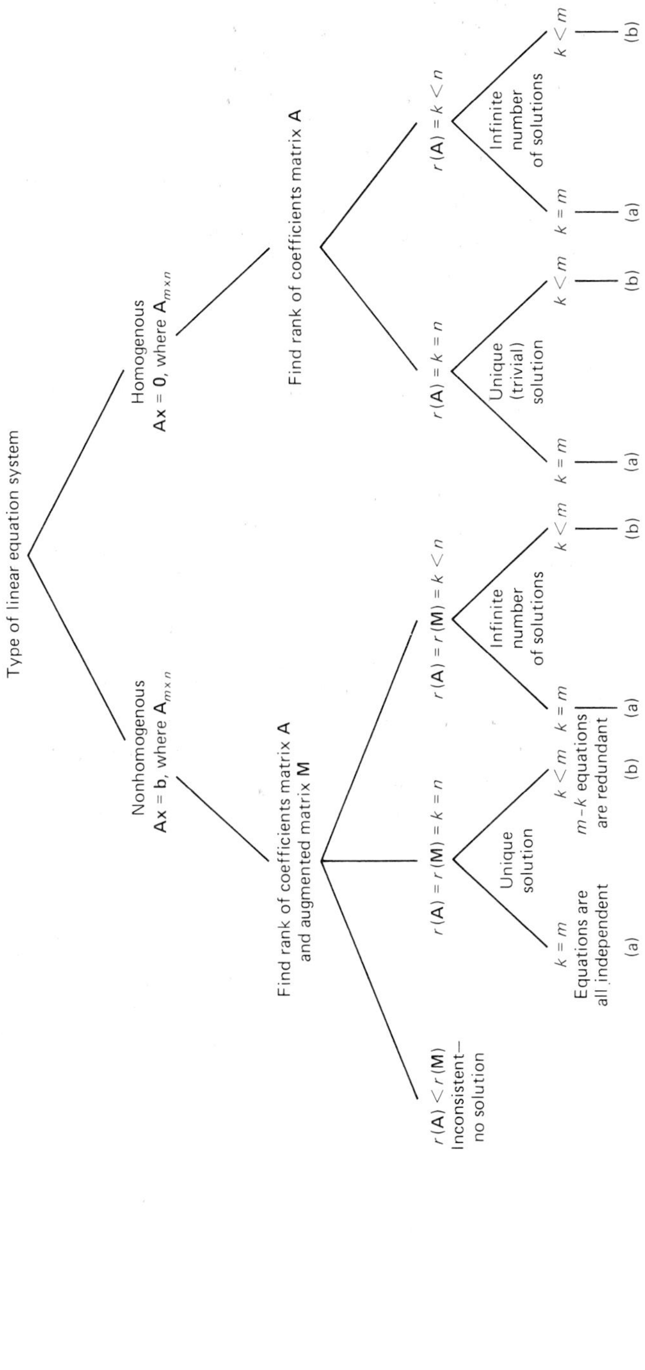

Fig. B.2 A classification of results obtained by solving simultaneous equations.

B.3.1 The Penrose Conditions

Research on generalized inverses goes back at least to 1920 with the work of Moore. Working independently, Penrose (1955) later defined the concept of a *unique* generalized inverse (now often called the Moore–Penrose inverse, denoted by $\mathbf{A}^+$) as a matrix that, in conjunction with the matrix $\mathbf{A}$ from which it is derived, satisfies four conditions:

(i) $\mathbf{A}\mathbf{A}^+\mathbf{A} = \mathbf{A}$ (ii) $\mathbf{A}^+\mathbf{A}\mathbf{A}^+ = \mathbf{A}^+$

(iii) $(\mathbf{A}\mathbf{A}^+)' = \mathbf{A}\mathbf{A}^+$ (symmetry) (iv) $(\mathbf{A}^+\mathbf{A})' = \mathbf{A}^+\mathbf{A}$ (symmetry)

There are alternative contexts in which to discuss $\mathbf{A}^+$. One context concerns the familiar case of solving a set of nonhomogeneous linear equations:

$$\boxed{\mathbf{A}\mathbf{x} = \mathbf{b}}$$

As we know, if $\mathbf{A}$ is nonsingular, a unique solution exists and is given by

$$\boxed{\mathbf{x} = \mathbf{A}^{-1}\mathbf{b}}$$

Moreover, the reader can easily observe that $\mathbf{A}^{-1}$, the *regular* inverse, satisfies the four Penrose conditions. Still, cases might exist where $\mathbf{A}$ is either rectangular or else square and singular so that $\mathbf{A}^{-1}$ does not exist.

Suppose, then, that we define a matrix A of order $m \times n$ where $r(\mathbf{A}) = k \leqslant \min(m, n)$. If a Moore–Penrose inverse $\mathbf{A}^+$ exists, it will be of order $n \times m$; this must be so because $\mathbf{A}\mathbf{A}^+$ is symmetric and, hence, square. It can be proved that for any matrix $\mathbf{A}$, there exists a unique matrix $\mathbf{A}^+$ that satisfies the four Penrose conditions.[8]

However, in solving a set of simultaneous equations, it is not always necessary that the solution be unique, as pointed out in Section B.2 in the context of echelon matrices. Moreover, it might be of interest to consider generalized inverses that obey only one (or more) of the four Penrose conditions. For example, if

$$\boxed{\mathbf{x} = \mathbf{A}^-\mathbf{b}}$$

is a solution to a set of consistent equations, $\mathbf{A}^-$ (also of order $n \times m$ if $\mathbf{A}$ is $m \times n$) need not be unique.

Indeed, in order for $\mathbf{A}^-$, called the *g* inverse, to exist, only the *first* of Penrose's four conditions, namely,

$$\boxed{\mathbf{A}\mathbf{A}^-\mathbf{A} = \mathbf{A}}$$

need be satisfied. Thus, if our interest centers on solving simultaneous equations, and we do not require the generalized inverse to be unique, it may be easier to compute $\mathbf{A}^-$—and usually it is—than to compute $\mathbf{A}^+$, the Moore–Penrose generalized inverse.

Accordingly, we shall wish to examine both the "stronger" (Moore–Penrose inverse) case and the "weaker" (*g* inverse) case. We say "stronger" since all Moore–Penrose generalized inverses are *g* inverses but not the converse.

[8] A proof of this assertion can be found in Graybill (1969, p. 97).

We start with the Moore–Penrose generalized inverse by showing its relevance to basic structure (or singular value decomposition), a concept already discussed in Chapter 5. Not only does the concept of basic structure provide one way to define $\mathbf{A}^+$, but the present discussion should also help illuminate earlier remarks on matrix decomposition into its basic structure.

Then we turn to a discussion of $\mathbf{A}^-$, the g inverse. Our interest in this type of (nonunique) generalized inverse stems from the relative ease with which it can be computed and its close connection with solution methods for simultaneous equations that have already been discussed in Section B.2.

B.3.2 Left and Right General Inverses

In the discussion (Section 5.7) of the basic structure of an arbitrary matrix $\mathbf{A}$, we recall that $\mathbf{A}$, of order $m \times n$, can be decomposed into the triple product

$$\boxed{\mathbf{A} = \mathbf{P}\boldsymbol{\Delta}\mathbf{Q}'}$$

where $\mathbf{P}'\mathbf{P} = \mathbf{Q}'\mathbf{Q} = \mathbf{I}$ and $\boldsymbol{\Delta}$ is diagonal of order $k \times k$, with $k \leqslant \min(m, n)$ positive entries that can be arranged in decreasing order of magnitude.

For the moment, let us place no restrictions on $\mathbf{A}$—it need be neither square nor basic and, hence, the rank of $\mathbf{A}$ may be less than its smaller order. Next, let us consider the following matrix $\mathbf{A}^+$, defined as being of rank k and of order $n \times m$:[9]

$$\boxed{\mathbf{A}^+ = \mathbf{Q}\boldsymbol{\Delta}^{-1}\mathbf{P}'}$$

We obtain $\mathbf{A}^+$ from $\mathbf{A} = \mathbf{P}\boldsymbol{\Delta}\mathbf{Q}'$ by taking the reciprocals of the diagonal entries of $\boldsymbol{\Delta}$ and then transposing the triple product $\mathbf{P}\boldsymbol{\Delta}^{-1}\mathbf{Q}'$ into $\mathbf{Q}\boldsymbol{\Delta}^{-1}\mathbf{P}'$.

Let us see what happens if we then premultiply $\mathbf{A}$ by $\mathbf{A}^+$:

$$\begin{aligned}\mathbf{A}^+\mathbf{A} &= (\mathbf{Q}\boldsymbol{\Delta}^{-1}\mathbf{P}')(\mathbf{P}\boldsymbol{\Delta}\mathbf{Q}') = \mathbf{Q}\boldsymbol{\Delta}^{-1}(\mathbf{P}'\mathbf{P})\boldsymbol{\Delta}\mathbf{Q}' = \mathbf{Q}\boldsymbol{\Delta}^{-1}(\mathbf{I}_{k\times k})\boldsymbol{\Delta}\mathbf{Q}'\\ &= \mathbf{Q}(\boldsymbol{\Delta}^{-1}\boldsymbol{\Delta})\mathbf{Q}' = \mathbf{Q}(\mathbf{I}_{k\times k})\mathbf{Q}' = \mathbf{Q}\mathbf{Q}'\end{aligned}$$

What is found here is the major product moment of the right orthonormal section of $\mathbf{A}$. As we know $\mathbf{Q}\mathbf{Q}'$ is symmetric, since it is a product moment matrix.

Suppose next that $\mathbf{A}$ is postmultiplied by $\mathbf{A}^+$. Without going through the algebra, the result is

$$\mathbf{A}\mathbf{A}^+ = \mathbf{P}\mathbf{P}'$$

where $\mathbf{P}$ is the left orthonormal section of $\mathbf{A}$. Again, $\mathbf{P}\mathbf{P}'$ is symmetric since it is a (major) product-moment matrix.

The reader will observe that $\mathbf{A}^+\mathbf{A}$ and $\mathbf{A}\mathbf{A}^+$ are each symmetric. Moreover, we also have

$$\begin{array}{ll}\text{(i)}\quad \mathbf{A}\mathbf{A}^+\mathbf{A} = \mathbf{A} & \text{(ii)}\quad \mathbf{A}^+\mathbf{A}\mathbf{A}^+ = \mathbf{A}^+\\ \quad\ \mathbf{P}\mathbf{P}'\mathbf{P}\boldsymbol{\Delta}\mathbf{Q}' = \mathbf{P}\boldsymbol{\Delta}\mathbf{Q}' & \quad\ \mathbf{Q}\mathbf{Q}'\mathbf{Q}\boldsymbol{\Delta}^{-1}\mathbf{P}' = \mathbf{Q}\boldsymbol{\Delta}^{-1}\mathbf{P}'\\ \quad\ = \mathbf{A} & \quad\ = \mathbf{A}^+\end{array}$$

[9] As recalled from Chapter 5, decomposition to basic structure is unique; also $\mathbf{A}^+$ is unique, given that $\mathbf{Q}\boldsymbol{\Delta}^{-1}\mathbf{P}'$ is also of rank k.

and, hence, conclude that all four of the Penrose conditions are met. Notice that $r(\mathbf{A}^+) = r(\mathbf{A}) = k$ where $k \leqslant \min(m, n)$.

Next, let us suppose that $\mathbf{A}_{m \times n}$ (with $m > n$) is *basic* (as described in Chapter 5). If so, $k = n$ and $\mathbf{Q}'$ in the triple product $\mathbf{P\Delta Q}'$ will be square, of order $n \times n$, resulting in

$$\boxed{\mathbf{A}^+\mathbf{A} = \mathbf{QQ}' = \mathbf{Q}'\mathbf{Q} = \mathbf{I}_{n \times n}}$$

If this relation is met, we let $\mathbf{A}^+ = \mathbf{L}$; the matrix $\mathbf{L}$ is sometimes referred to as the *left pseudoinverse* of $\mathbf{A}$.

By the same token, if $\mathbf{A}$ is "horizontal," of order $m \times n$ $(m < n)$, and basic, then $k = m$ and $\mathbf{P}$ will be square, of order $m \times m$, and we shall have

$$\boxed{\mathbf{AA}^+ = \mathbf{PP}' = \mathbf{P}'\mathbf{P} = \mathbf{I}_{m \times m}}$$

If this relation is met, we let $\mathbf{A}^+ = \mathbf{R}$; the matrix $\mathbf{R}$ is sometimes referred to as the *right pseudoinverse* of $\mathbf{A}$. Finally, it should be clear that if and only if $\mathbf{A}$ is both square and basic (i.e., nonsingular) will it possess *both* a left and right pseudoinverse and these inverses *will be the same.*[10]

However, suppose we return to the first case in which $\mathbf{A}$ is nonbasic:

$$\boxed{r(\mathbf{A}) = k < \min(m, n)}$$

It is possible, of course, to find a generalized inverse of $\mathbf{A}$ that meets only the first Penrose condition

$$\boxed{\mathbf{AA}^-\mathbf{A} = \mathbf{A}}$$

by defining $\mathbf{A}^-$ in terms of the square roots of *all* of the eigenvalues of either $\mathbf{AA}'$, if $m \leqslant n$, or $\mathbf{A}'\mathbf{A}$, if $m \geqslant n$, *including those eigenvalues that turn out to be zero.*

If so, we refer to this case as a *g* inverse of $\mathbf{A}$ and continue to denote it as $\mathbf{A}^-$. In this version of the generalized inverse the basic diagonal $\mathbf{\Delta}$ of $\mathbf{A}(=\mathbf{P\Delta Q}')$ has $\min(m, n) - k$ zeros, and $\mathbf{P}$ and $\mathbf{Q}'$ are no longer unique. It turns out, however, that

$$\boxed{\mathbf{x} = \mathbf{A}^-\mathbf{b}}$$

is still a solution to the set of consistent equations

$$\boxed{\mathbf{Ax} = \mathbf{b}}$$

and, in this sense, $\mathbf{A}^-$ is still a generalized inverse, specifically a *g* inverse.

[10] Still, it should be pointed out that although $\mathbf{A}^+$ is unique, the matrix $\mathbf{A}$–if basic but singular–will, in general, have an infinity of other matrices that satisfy $\mathbf{LA} = \mathbf{I}$ or $\mathbf{AR} = \mathbf{I}$, as the case may be. However, only one of this infinity of matrices will be the Moore–Penrose inverse. Furthermore, if $\mathbf{A}$ is nonsingular, only one matrix $\mathbf{A}^{-1}$ $(=\mathbf{A}^+)$ exists. Thus, $\mathbf{A}$, if nonsingular, has exactly one inverse, $\mathbf{A}^{-1}$. If $\mathbf{A}$ is singular, it has an infinity of generalized inverses, one of which is the (uniquely specified) Moore-Penrose inverse, $\mathbf{A}^+$.

Why should we ever want to find a version of $\mathbf{Q\Delta}^{-1}\mathbf{P}'$ whose diagonal is of larger order than $k \times k$, where $r(\mathbf{A}) = k$? Again, the motivation may be pragmatic in that it may be easier to compute $\mathbf{A}^-$ (as defined above) even though it is no longer unique.

We now turn to computational methods for finding the Moore–Penrose inverse, after which the g inverse $\mathbf{A}^-$ is discussed.

B.3.3 Some Numerical Procedures for Computing $\mathbf{A}^+$

To illustrate the computation of $\mathbf{A}^+$, let us consider the 3 x 2 matrix

$$\mathbf{A} = \begin{bmatrix} 1 & 3 \\ 1 & 2 \\ 2 & 1 \end{bmatrix}$$

If $r(\mathbf{A}) = n = 2$, we should be able to find a left general inverse such that $\mathbf{LA} = \mathbf{I}_{2\times 2}$.

Using the procedure described in Section 5.7.3, we first find the product-moment matrix with the smaller order, in this case the minor product moment

$$\mathbf{A}'\mathbf{A} = \begin{bmatrix} 6 & 7 \\ 7 & 14 \end{bmatrix}$$

and solve for its eigenstructure

$$\mathbf{D} = \begin{bmatrix} 18.062 & 0 \\ 0 & 1.938 \end{bmatrix}; \qquad \mathbf{Q} = \begin{bmatrix} -0.502 & 0.865 \\ -0.865 & -0.502 \end{bmatrix}$$

We then compute the basic diagonal and its inverse

$$\mathbf{\Delta} = \mathbf{D}^{1/2} = \begin{bmatrix} 4.250 & 0 \\ 0 & 1.392 \end{bmatrix}; \qquad \mathbf{\Delta}^{-1} = \begin{bmatrix} 0.235 & 0 \\ 0 & 0.718 \end{bmatrix}$$

and then solve for $\mathbf{P}$:

$$\mathbf{P} = \mathbf{AQ\Delta}^{-1} = \begin{bmatrix} 1 & 3 \\ 1 & 2 \\ 2 & 1 \end{bmatrix} \begin{bmatrix} -0.502 & 0.865 \\ -0.865 & -0.502 \end{bmatrix} \begin{bmatrix} 0.235 & 0 \\ 0 & 0.718 \end{bmatrix}$$

$$\mathbf{P} = \begin{bmatrix} -0.728 & -0.460 \\ -0.525 & -0.099 \\ -0.439 & -0.882 \end{bmatrix}$$

The matrix $\mathbf{A}$ is now expressed in terms of basic structure as $\mathbf{A} = \mathbf{P\Delta Q}'$. The left general inverse is then

$$\mathbf{A}^+ = \mathbf{L} = \mathbf{Q\Delta}^{-1}\mathbf{P}' = \begin{bmatrix} -0.2 & 0 & 0.6 \\ 0.313 & 0.142 & -0.228 \end{bmatrix}$$

and we have the desired result:

$$\mathbf{LA} = \begin{bmatrix} -0.2 & 0 & 0.6 \\ 0.313 & 0.142 & -0.228 \end{bmatrix} \begin{bmatrix} 1 & 3 \\ 1 & 2 \\ 2 & 1 \end{bmatrix} = \begin{bmatrix} 1 & 0 \\ 0 & 1 \end{bmatrix}$$

However, now let us take the case where $\mathbf{A}$ is nonbasic. To illustrate, consider

$$\mathbf{A} = \begin{bmatrix} 1 & 2 \\ 1 & 2 \\ 2 & 4 \end{bmatrix}$$

In this case we know that $\mathbf{A}$ is not basic since the second column is twice that of the first column and $r(\mathbf{A}) = 1$. We first find

$$\mathbf{A}'\mathbf{A} = \begin{bmatrix} 6 & 12 \\ 12 & 24 \end{bmatrix} = 6 \begin{bmatrix} 1 & 2 \\ 2 & 4 \end{bmatrix}$$

By procedures identical to those just illustrated, the basic structure of $\mathbf{A}$ is then found to be

$$\begin{array}{ccccc} & \mathbf{P} & \Delta & \mathbf{Q}' & \\ \mathbf{A} = & \begin{bmatrix} -0.408 \\ -0.408 \\ -0.816 \end{bmatrix} & [5.477] & [-0.477 \quad -0.894] & = \begin{bmatrix} 1 & 2 \\ 1 & 2 \\ 2 & 4 \end{bmatrix} \end{array}$$

and $\mathbf{A}^+$, of rank $k(\mathbf{A}) = 1$, is

$$\mathbf{A}^+ = \mathbf{Q}\Delta^{-1}\mathbf{P}' = \begin{bmatrix} 0.033 & 0.033 & 0.067 \\ 0.067 & 0.067 & 0.133 \end{bmatrix}$$

The reader can verify that the four Penrose conditions are met, although now it is no longer true that $\mathbf{A}^+\mathbf{A} = \mathbf{I}$.

Computing $\mathbf{A}^+$ after first solving for the basic structure of a matrix is only one of many solution methods. By way of contrast, let us consider another method, due to Penrose himself (1956), that can also be used to find $\mathbf{A}^+$. This method involves implementation of a fairly simple algorithm that entails the following steps:

1. Compute $\mathbf{B} = \mathbf{A}'\mathbf{A}$.
2. Let $\mathbf{C}_1 = \mathbf{I}$, the identity matrix.
3. Compute $\mathbf{C}_{i+1} = \mathbf{I}(1/i)\text{tr}(\mathbf{C}_i\mathbf{B}) - \mathbf{C}_i\mathbf{B}$ for $i = 1, 2, \ldots, k-1$.[11]
4. Compute $k\mathbf{C}_k\mathbf{A}'/\text{tr}(\mathbf{C}_k\mathbf{B})$, to get $\mathbf{A}^+$.
5. Also, it will be found that $\mathbf{C}_{k+1}\mathbf{B} = \phi$; $\text{tr}(\mathbf{C}_k\mathbf{B}) \neq 0$, so that $r(\mathbf{B}) = r(\mathbf{A}) = k$.

[11] The reader should recall that the trace (tr) of a square matrix $\mathbf{A}_{n \times n}$ is equal to the sum of its main diagonal elements:

$$\text{tr}(\mathbf{A}) = \sum_{i=1}^{n} a_{ii}$$

Applying the procedure to the last problem (where $\mathbf{A}$ is nonbasic) gives us

1. $\mathbf{B} = \mathbf{A}'\mathbf{A} = \begin{bmatrix} 6 & 12 \\ 12 & 24 \end{bmatrix}$

2. $\mathbf{C}_1 = \begin{bmatrix} 1 & 0 \\ 0 & 1 \end{bmatrix}; \quad \mathbf{C}_1\mathbf{B} = \begin{bmatrix} 1 & 0 \\ 0 & 1 \end{bmatrix}\begin{bmatrix} 6 & 12 \\ 12 & 24 \end{bmatrix} = \begin{bmatrix} 6 & 12 \\ 12 & 24 \end{bmatrix}; \quad \text{tr } \mathbf{C}_1\mathbf{B} = 30$

3. $\mathbf{C}_2 = \mathbf{I}\,\text{tr}(\mathbf{C}_1\mathbf{B}) - \mathbf{C}_1\mathbf{B} = \begin{bmatrix} 30 & 0 \\ 0 & 30 \end{bmatrix} - \begin{bmatrix} 6 & 12 \\ 12 & 24 \end{bmatrix} = \begin{bmatrix} 24 & -12 \\ -12 & 6 \end{bmatrix}; \quad \mathbf{C}_2\mathbf{B} = \begin{bmatrix} 0 & 0 \\ 0 & 0 \end{bmatrix}$

Since $\mathbf{C}_2\mathbf{B} = \phi$ and $\text{tr}(\mathbf{C}_1\mathbf{B}) \neq 0$, we know that $r(\mathbf{B}) = r(\mathbf{A}) = 1$; we can then go on to find $\mathbf{A}^+$.

4. $\mathbf{A}^+ = \dfrac{1\mathbf{C}_1\mathbf{A}'}{\text{tr}(\mathbf{C}_1\mathbf{B})} = \dfrac{1}{30}\begin{bmatrix} 1 & 1 & 2 \\ 2 & 2 & 4 \end{bmatrix} = \begin{bmatrix} 0.033 & 0.033 & 0.067 \\ 0.067 & 0.067 & 0.133 \end{bmatrix}$

We find, of course, the same solution for $\mathbf{A}^+$ as found earlier. The Penrose procedure, like the basic structure approach, can be used to find $\mathbf{A}^+$, whether or not $\mathbf{A}$ is basic. To complete the discussion, let us apply the Penrose computational procedure to the first case, where $\mathbf{A}$ is basic:

1. $\mathbf{A} = \begin{bmatrix} 1 & 3 \\ 1 & 2 \\ 2 & 1 \end{bmatrix}; \quad \mathbf{B} = \mathbf{A}'\mathbf{A} = \begin{bmatrix} 6 & 7 \\ 7 & 14 \end{bmatrix}$

2. $\mathbf{C}_1 = \begin{bmatrix} 1 & 0 \\ 0 & 1 \end{bmatrix}; \quad \mathbf{C}_1\mathbf{B} = \begin{bmatrix} 6 & 7 \\ 7 & 14 \end{bmatrix}; \quad \text{tr}(\mathbf{C}_1\mathbf{B}) = 20$

3. $\mathbf{C}_2 = \begin{bmatrix} 1 & 0 \\ 0 & 1 \end{bmatrix} 20 - \begin{bmatrix} 6 & 7 \\ 7 & 14 \end{bmatrix} = \begin{bmatrix} 14 & -7 \\ -7 & 6 \end{bmatrix}; \quad \mathbf{C}_2\mathbf{B} = \begin{bmatrix} 35 & 0 \\ 0 & 35 \end{bmatrix}$

4. $\mathbf{A}^+ = \dfrac{2\mathbf{C}_2\mathbf{A}'}{\text{tr}(\mathbf{C}_2\mathbf{B})}, \quad \text{where} \quad \text{tr}(\mathbf{C}_2\mathbf{B}) = 70$

$$= \frac{2}{70}\begin{bmatrix} 14 & -7 \\ -7 & 6 \end{bmatrix}\begin{bmatrix} 1 & 1 & 2 \\ 3 & 2 & 1 \end{bmatrix} = \begin{bmatrix} -0.2 & 0 & 0.6 \\ 0.313 & 0.142 & -0.228 \end{bmatrix}$$

Note that the Penrose computational procedure involves less computation—particularly, no need to find eigenstructures—than the method based on matrix decomposition via basic structure.[12]

[12] Note also that

$$\mathbf{C}_3 = \begin{bmatrix} 1 & 0 \\ 0 & 1 \end{bmatrix}(70/2) - \begin{bmatrix} 35 & 0 \\ 0 & 35 \end{bmatrix} = \phi$$

and, $\text{tr}(\mathbf{C}_2 B) \neq 0$; hence, $r(\mathbf{A}) = 2$.

B.3.4 Some Properties of the Moore–Penrose Inverse

In many respects the Moore-Penrose inverse $\mathbf{A}^+$ acts like a regular inverse $\mathbf{A}^{-1}$. (Indeed, $\mathbf{A}^+$ equals $\mathbf{A}^{-1}$ if $\mathbf{A}$ is nonsingular.) However, even in other cases, $\mathbf{A}^+$ possesses a number of properties, many of which are similar to those displayed by the regular inverse. Some of the more important of these properties are listed below:

1. The Moore–Penrose inverse of the transpose of $\mathbf{A}$ is the transpose of the Moore–Penrose inverse of $\mathbf{A}$: $(\mathbf{A}')^+ = (\mathbf{A}^+)'$.
2. The Moore–Penrose inverse of $\mathbf{A}^+$ is equal to $\mathbf{A}$: $(\mathbf{A}^+)^+ = \mathbf{A}$.
3. The rank of the Moore–Penrose inverse of $\mathbf{A}$ is equal to the rank of $\mathbf{A}$: $r(\mathbf{A}^+) = r(\mathbf{A})$.
4. For any matrix $\mathbf{A}$, $(\mathbf{A}'\mathbf{A})^+ = \mathbf{A}^+(\mathbf{A}')^+$.
5. For any matrix $\mathbf{A}$, $(\mathbf{A}\mathbf{A}^+)^+ = \mathbf{A}\mathbf{A}^+$; $(\mathbf{A}^+\mathbf{A})^+ = \mathbf{A}^+\mathbf{A}$.
6. If $\mathbf{A} = \mathbf{A}'$, then $\mathbf{A}^+ = (\mathbf{A}^+)'$.
7. If $\mathbf{A} = \mathbf{A}'$, then $\mathbf{A}\mathbf{A}^+ = \mathbf{A}^+\mathbf{A}$.
8. If $\mathbf{A}$ is nonsingular, then $\mathbf{A}^{-1} = \mathbf{A}^+$.
9. If $\mathbf{A}$ is an $m \times n$ matrix of rank m, then $\mathbf{A}^+ = \mathbf{A}'(\mathbf{A}\mathbf{A}')^{-1}$ and $\mathbf{A}\mathbf{A}^+ = \mathbf{I}$ (as related to Section B.3.2).
10. If $\mathbf{A}$ is an $m \times n$ matrix of rank n, then $\mathbf{A}^+ = (\mathbf{A}'\mathbf{A})^{-1}\mathbf{A}'$ and $\mathbf{A}^+\mathbf{A} = \mathbf{I}$ (as related to Section B.3.2).

In addition to the properties listed above, the Moore–Penrose inverse figures prominently in the solution of sets of linear equations. More specifically, given the set of nonhomogeneous equations

$$\boxed{\mathbf{A}\mathbf{x} = \mathbf{b}}$$

where $\mathbf{A}$ is of order $m \times n$ and $\mathbf{b}$ is an $m \times 1$ vector of constants, the system of equations is consistent if and only if

$$\boxed{\mathbf{A}\mathbf{A}^+\mathbf{b} = \mathbf{b}}$$

Second, given that the system is consistent (and, hence, has at least one solution), then for each $n \times 1$ vector γ, the $n \times 1$ vector $\mathbf{x}$ is a solution where

$$\boxed{\mathbf{x} = \mathbf{A}^+\mathbf{b} + (\mathbf{I} - \mathbf{A}^+\mathbf{A})\gamma}$$

and every solution to the system can be so written for some $n \times 1$ vector γ.

As just indicated, the Moore–Penrose generalized inverse, assuming it can be found easily, provides a way to solve sets of linear equations. While in principle, we could always compute $\mathbf{A}^+$ in solving sets of simultaneous equations, it is usually the case that we do not need the stronger properties of the Moore–Penrose inverse to get the job done.

However, as recalled, a g inverse $\mathbf{A}^-$ also provides a solution to a set of consistent equations, albeit one that is not unique but, on the other hand, one that is relatively easy to compute. Accordingly, we now turn to a discussion of the g inverse $\mathbf{A}^-$ and its role in solving sets of simultaneous equations.

B.4 THE g INVERSE

If we let $\mathbf{A}$ be an $m \times n$ matrix, a matrix $\mathbf{A}^-$, of order $n \times m$, is defined to be a g inverse of $\mathbf{A}$ if and only if it satisfies the first of the Penrose conditions

$$\boxed{\mathbf{A}\mathbf{A}^-\mathbf{A} = \mathbf{A}}$$

As pointed out earlier, the Moore–Penrose inverse of $\mathbf{A}$ is also a g inverse of $\mathbf{A}$, but the converse does not hold in general. Moreover, in general $\mathbf{A}^-$ is not unique for a given $\mathbf{A}$.

A g inverse is particularly useful in the practical setting of solving sets of simultaneous equations. Fully analogous to the Moore–Penrose inverse, the system of equations $\mathbf{A}\mathbf{x} = \mathbf{b}$ has a solution if and only if

$$\boxed{\mathbf{A}\mathbf{A}^-\mathbf{b} = \mathbf{b}}$$

Second, given that the system is consistent, then for each $n \times 1$ vector γ, the $n \times 1$ vector $\mathbf{x}$ is a solution where $\mathbf{x}$ is

$$\boxed{\mathbf{x} = \mathbf{A}^-\mathbf{b} + (\mathbf{I} - \mathbf{A}^-\mathbf{A})\gamma}$$

Finally, every solution to the system can be so written for some $n \times 1$ vector γ.

The value of a g inverse relates to its relative ease of calculation, particularly by means of the echelon form of a matrix, as considered in Section B.2. However, before discussing a general approach to computing $\mathbf{A}^-$ (in the context of solving sets of equations), we consider the concept of Hermite form. The Hermite form of a matrix provides the key concept for obtaining $\mathbf{A}^-$.

B.4.1 The Hermite Form of a Square Matrix

A square $(n \times n)$ matrix $\mathbf{J}$ is defined to be in (upper) Hermite form if and only if it satisfies the following conditions:

1. $\mathbf{J}$ is upper triangular.
2. Only zeros and ones are on its main diagonal.
3. If a row has a zero on the diagonal, then every entry in the row is zero.
4. If a row has a one on the diagonal, then every other entry is zero in the column in which the one appears.

If $\mathbf{J}$ is of Hermite form, it is also the case that

$$\boxed{\mathbf{J} = \mathbf{J}^2}$$

and $\mathbf{J}$ is said to be *idempotent*. Moreover, for any $n \times n$ matrix $\mathbf{A}$, there exists a nonsingular matrix $\mathbf{G}$ such that

$$\boxed{\mathbf{G}\mathbf{A} = \mathbf{J}_{\mathbf{A}}}$$

and so $\mathbf{A}$ can *always* be reduced to Hermite form via $\mathbf{G}$. (However, $\mathbf{G}$ is nonunique, in general, although $\mathbf{J_A}$ will be.)

Just as was the case in reducing $\mathbf{A}$ to echelon form $\mathbf{H_A}$, we can use elementary row operations to reduce $\mathbf{A}_{n\times n}$ to Hermite form $\mathbf{J_A}$[13] The matrix $\mathbf{G}$ is the nonsingular matrix that brings about the reduction of $\mathbf{A}_{n\times n}$ to Hermite form.

Moreover, once this is done it turns out that the job is finished since $\mathbf{A}^-$ can be defined as

$$\boxed{\mathbf{A}^- = \mathbf{G}}$$

That is, $\mathbf{G}$ is the *g* inverse of $\mathbf{A}$, and all we have to do to find $\mathbf{G}$ is to reduce $\mathbf{A}$ to Hermite form via elementary row operations while performing companion operations on $\mathbf{I}$.

However, since the Hermite form does not exist for rectangular matrices, a slight modification is required to find $\mathbf{A}^- = \mathbf{G}$ when $\mathbf{A}$ is rectangular. If $\mathbf{A}$ is vertical ($m \times n$, with $m > n$), we can append a set of $\mathbf{0}$ column vectors to make $\mathbf{A}$ square. That is,

$$\mathbf{A}_0 = \left[\mathbf{A} \;\vdots\; \phi\right]$$

where $\mathbf{A}_0$ is $m \times m$. Then, if $\mathbf{G}$ is a nonsingular matrix, such that $\mathbf{G}_0\mathbf{A}_0 = \mathbf{J}_{\mathbf{A}_0}$, where $\mathbf{J}_{\mathbf{A}_0}$ is the Hermite form of $\mathbf{A}_0$, we have

$$\mathbf{G}_0 = \begin{bmatrix} \mathbf{G} \\ \cdots \\ \mathbf{G}_1 \end{bmatrix}$$

where $\mathbf{G}$ is the upper $n \times m$ submatrix of $\mathbf{G}_0$. This is the *g* inverse of $\mathbf{A}$. Similarly, if $\mathbf{A}$ is horizontal ($m \times n$, with $n > m$), we can append a set of $\mathbf{0}'$ row vectors to make $\mathbf{A}$ square and proceed to find $\mathbf{G}_{n\times m}$, the left-hand submatrix of $\mathbf{G}_0$.[14]

The strategy should now be clear. In the rectangular case, we make $\mathbf{A}$ square by adding columns or rows of zeros, as the case may be. We then find a nonsingular matrix that reduces $\mathbf{A}_0$ to Hermite form. The matrix $\mathbf{G}$ is the *g* inverse of $\mathbf{A}$.

However, one more facet of the problem has to be introduced before proceeding to find $\mathbf{A}^-$. In Section B.2 a general procedure was introduced for solving sets of simultaneous equations via reduction of either the coefficients matrix $\mathbf{A}$ or the augmented matrix $\mathbf{M}$ to echelon form. As might be surmised, if $\mathbf{A}$ is already square, or made square by appending columns (or rows) of zeros, the Hermite form $\mathbf{J_A}$ of $\mathbf{A}$ can be obtained from its echelon form $\mathbf{H_A}$. This is done by transforming rows of $\mathbf{H}$, via additional elementary row operations, until $\mathbf{J_A}$ is found. The matrix $\mathbf{G}$ that summarizes the *full set of elementary row operations* used in reducing $\mathbf{A}$ to $\mathbf{H_A}$ and then $\mathbf{H_A}$ to $\mathbf{J_A}$ is

[13] As will be shown, if $\mathbf{A}$ is square to begin with (or can be made square by procedures to be described later), we can compute $\mathbf{J_A}$ via additional elementary operations on $\mathbf{H_A}$.

[14] The matrix would appear as

$$\mathbf{G}_0 = \left[\mathbf{G} \;\vdots\; \mathbf{G}_1\right].$$

The next section shows some numerical examples of the general procedure, including a case in which $\mathbf{A}$ is rectangular.

A^-, the desired g inverse. In general, A^- will not be unique.[15] However, the matrix J_A in Hermite form *is* unique for a given matrix **A**.

Before proceeding with the computation of A^-, three additional properties of matrices in Hermite form are of interest to note:

1. The Hermite form J_A of **A** has the same rank as **A**.
2. If **A**, of order $n \times n$, is nonsingular, then J_A is the $n \times n$ identity matrix **I**.
3. The rank of **A** is equal to the number of diagonal elements of J_A that are equal to unity.

With the foregoing comments as background, a general procedure can now be stated for finding A^-:

1. If **A** is rectangular, make it square by appending columns (or rows) of zeros.
2. Via elementary row operations reduce **A** to echelon form and then to Hermite form. At the same time, perform the same operations on **I**, the associated $n \times n$ identity matrix.
3. If **A** is nonsingular, then its Hermite form is **I** and $A^- = A^{-1}$.
4. If **A** is singular, then **I** will be transformed to $G = A^-$ as **A** is being reduced to J_A, its Hermite form. And A^- will be the g inverse of interest.

B.4.2 Some Numerical Examples

Let us now consider some illustrations of finding A^- by means of the method presented above. First, let us take a nonsingular matrix **A** and its companion identity matrix:

$$A = \begin{bmatrix} 1 & 4 & 2 \\ 2 & 5 & 3 \\ 3 & 6 & 5 \end{bmatrix}; \qquad I = \begin{bmatrix} 1 & 0 & 0 \\ 0 & 1 & 0 \\ 0 & 0 & 1 \end{bmatrix}$$

As outlined earlier, the task is to reduce **A** to echelon form H_A and then into Hermite form J_A via a series of elementary row operations. Each elementary row operation that is performed on **A** is also performed concurrently on the associated starting identity matrix **I**. As **A** is reduced to Hermite form J_A, **I** is transformed to A^-, the desired g inverse.

The reader should note the similarity of this procedure to that followed in Section B.2. In the present case J_A, the Hermite form of **A**, takes on the role of the identity submatrix computed from the echelon matrix in Section 3.2.

We can now start the row operations, bearing in mind that these, in general, are not unique. We first subtract twice row 1 from row 2 and subtract 3 times row 1 from row 3:

$$\begin{bmatrix} 1 & 4 & 2 \\ 0 & -3 & -1 \\ 0 & -6 & -1 \end{bmatrix}; \qquad \begin{bmatrix} 1 & 0 & 0 \\ -2 & 1 & 0 \\ -3 & 0 & 1 \end{bmatrix}$$

[15] The reason why A^- (=G) is not unique is simply because, in general, there are different sets of elementary row operations (summarized in G) that can lead to J_A—as a matter of fact, an infinity of such sets.

Subtract twice row 2 from row 3:

$$\begin{bmatrix} 1 & 4 & 2 \\ 0 & -3 & -1 \\ 0 & 0 & 1 \end{bmatrix}; \quad \begin{bmatrix} 1 & 0 & 0 \\ -2 & 1 & 0 \\ 1 & -2 & 1 \end{bmatrix}$$

Multiply row 2 by $-1/3$:

$$\mathbf{H}_A = \begin{bmatrix} 1 & 4 & 2 \\ 0 & 1 & 1/3 \\ 0 & 0 & 1 \end{bmatrix}; \quad \begin{bmatrix} 1 & 0 & 0 \\ 2/3 & -1/3 & 0 \\ 1 & -2 & 1 \end{bmatrix}$$

At this point we note that $\mathbf{A}$ is in echelon form $\mathbf{H}_A$ and that $r(\mathbf{A}) = 3$. The next task is to reduce $\mathbf{H}_A$ to Hermite form $\mathbf{J}_A$. To do this, subtract 4 times row 2 from row 1:

$$\begin{bmatrix} 1 & 0 & 2/3 \\ 0 & 1 & 1/3 \\ 0 & 0 & 1 \end{bmatrix}; \quad \begin{bmatrix} -5/3 & 4/3 & 0 \\ 2/3 & -1/3 & 0 \\ 1 & -2 & 1 \end{bmatrix}$$

Subtract 2/3 of row 3 from row 1 and subtract 1/3 of row 3 from row 2:

$$\mathbf{J}_A = \mathbf{I} = \begin{bmatrix} 1 & 0 & 0 \\ 0 & 1 & 0 \\ 0 & 0 & 1 \end{bmatrix}; \quad \begin{bmatrix} -7/3 & 8/3 & -2/3 \\ 1/3 & 1/3 & -1/3 \\ 1 & -2 & 1 \end{bmatrix} = \mathbf{G} = \mathbf{A}^- = \mathbf{A}^{-1}$$

The reader can then check to see that

$$\mathbf{AA}^- = \mathbf{AA}^{-1} = \mathbf{I}$$

Note also that the Hermite form $\mathbf{J}_A$ of a nonsingular matrix $\mathbf{A}$ is an identity matrix of the same order.

Next, we consider the case where $\mathbf{A}$ is square but not of full rank. In Section B.2.1 we encountered a matrix of this type:

$$\mathbf{A} = \begin{bmatrix} 1 & 1 & 3 \\ 1 & 2 & 6 \\ 0 & 1 & 3 \end{bmatrix}; \quad \mathbf{I} = \begin{bmatrix} 1 & 0 & 0 \\ 0 & 1 & 0 \\ 0 & 0 & 1 \end{bmatrix}$$

If we subtract the first row from the second, we get

$$\begin{bmatrix} 1 & 1 & 3 \\ 0 & 1 & 3 \\ 0 & 1 & 3 \end{bmatrix}; \quad \begin{bmatrix} 1 & 0 & 0 \\ -1 & 1 & 0 \\ 0 & 0 & 1 \end{bmatrix}$$

Next we subtract the second row from the third:

$$\mathbf{H}_A = \begin{bmatrix} 1 & 1 & 3 \\ 0 & 1 & 3 \\ 0 & 0 & 0 \end{bmatrix}; \qquad \begin{bmatrix} 1 & 0 & 0 \\ -1 & 1 & 0 \\ 1 & -1 & 1 \end{bmatrix}$$

and note that we have the echelon form of $\mathbf{A}$, $\mathbf{H}_A$, in which $r(\mathbf{A}) = 2$. Next, we subtract the second row from the first:

$$\mathbf{J}_A = \begin{bmatrix} 1 & 0 & 0 \\ 0 & 1 & 3 \\ 0 & 0 & 0 \end{bmatrix}; \qquad \begin{bmatrix} 2 & -1 & 0 \\ -1 & 1 & 0 \\ 1 & -1 & 1 \end{bmatrix} = \mathbf{A}^-$$

to obtain $\mathbf{J}_A$, the Hermite form of $\mathbf{A}$. We find $\mathbf{A}^-$ as well and note that

$$\mathbf{J}_A = \mathbf{J}_A^2$$

as it should, since $\mathbf{J}_A$ is idempotent.

We then solve for $\mathbf{x}$ as

$$\mathbf{x} = \mathbf{A}^-\mathbf{b}$$

where, from Section B.2.1, $\mathbf{b}' = (8, 14, 6)$, so that

$$\mathbf{x} = \underset{\mathbf{A}^-}{\begin{bmatrix} 2 & -1 & 0 \\ -1 & 1 & 0 \\ 1 & -1 & 1 \end{bmatrix}} \underset{\mathbf{b}}{\begin{bmatrix} 8 \\ 14 \\ 6 \end{bmatrix}} = \begin{bmatrix} 2 \\ 6 \\ 0 \end{bmatrix}$$

However, as noted earlier, when the rank of $\mathbf{A}$ is less than the number of unknowns, there is an infinite number of solutions. Hence, we express $\mathbf{x}$ in the general form:

$$\mathbf{x} = \mathbf{A}^-\mathbf{b} + (\mathbf{I} - \mathbf{A}^-\mathbf{A})\boldsymbol{\gamma}$$

where $\boldsymbol{\gamma}$ is an arbitrary vector. Specifically, we have

$$\mathbf{x} = \underset{\mathbf{A}^-\mathbf{b}}{\begin{bmatrix} 2 \\ 6 \\ 0 \end{bmatrix}} + \left\{ \underset{\mathbf{I}}{\begin{bmatrix} 1 & 0 & 0 \\ 0 & 1 & 0 \\ 0 & 0 & 1 \end{bmatrix}} - \underset{\mathbf{A}^-\mathbf{A}}{\begin{bmatrix} 1 & 0 & 0 \\ 0 & 1 & 3 \\ 0 & 0 & 0 \end{bmatrix}} \right\} \underset{\boldsymbol{\gamma}}{\begin{bmatrix} \gamma_1 \\ \gamma_2 \\ \gamma_3 \end{bmatrix}}$$

$$= \begin{bmatrix} 2 \\ 6 \\ 0 \end{bmatrix} + \begin{bmatrix} 0 \\ -3\gamma_3 \\ \gamma_3 \end{bmatrix} = \begin{bmatrix} 2 \\ 6 - 3\gamma_3 \\ \gamma_3 \end{bmatrix}$$

where γ_3 is considered as an arbitrary parameter. Note that this is the same result as found in Section B.2.1.

Finally, let us consider the case where $\mathbf{A}$ is rectangular. As an illustration, we again return to Section B.2.2 and consider the 4 x 3 vertical matrix

$$\mathbf{A} = \begin{bmatrix} 1 & 1 & 3 \\ 1 & 2 & 6 \\ 0 & 1 & 3 \\ 2 & 3 & 9 \end{bmatrix}$$

After appending a column vector of zeros to make $\mathbf{A}$ square, we have

$$\mathbf{A}_0 = \left[\begin{array}{ccc:c} 1 & 1 & 3 & 0 \\ 1 & 2 & 6 & 0 \\ 0 & 1 & 3 & 0 \\ 2 & 3 & 9 & 0 \end{array}\right]; \quad \mathbf{I}_0 = \left[\begin{array}{cccc} 1 & 0 & 0 & 0 \\ 0 & 1 & 0 & 0 \\ 0 & 0 & 1 & 0 \\ \hdashline 0 & 0 & 0 & 1 \end{array}\right]$$

Note that the associated identity matrix $\mathbf{I}_0$ has an extra row added. Reduction of $\mathbf{A}_0$ to echelon form (with concurrent elementary row operations on $\mathbf{I}_0$) gives us first the echelon form:

$$\mathbf{H}_{\mathbf{A}_0} = \left[\begin{array}{ccc:c} 1 & 1 & 3 & 0 \\ 0 & 1 & 3 & 0 \\ 0 & 0 & 0 & 0 \\ 0 & 0 & 0 & 0 \end{array}\right]; \quad \left[\begin{array}{cccc} 1 & 0 & 0 & 0 \\ -1 & 1 & 0 & 0 \\ 1 & -1 & 1 & 0 \\ \hdashline -1 & -1 & 0 & 1 \end{array}\right]$$

and a further elementary row operation that subtracts row 2 from row 1 leads to

$$\mathbf{J}_{\mathbf{A}_0} = \left[\begin{array}{ccc:c} 1 & 0 & 0 & 0 \\ 0 & 1 & 3 & 0 \\ 0 & 0 & 0 & 0 \\ 0 & 0 & 0 & 0 \end{array}\right]; \quad \left[\begin{array}{cccc} 2 & -1 & 0 & 0 \\ -1 & 1 & 0 & 0 \\ 1 & -1 & 1 & 0 \\ \hdashline -1 & -1 & 0 & 1 \end{array}\right] = \mathbf{G}_0$$

We then write $\mathbf{A}^-$ by dropping the fourth row of the transformed identity matrix to get the 3 x 4 matrix

$$\mathbf{G} = \mathbf{A}^- = \begin{bmatrix} 2 & -1 & 0 & 0 \\ -1 & 1 & 0 & 0 \\ 1 & -1 & 1 & 0 \end{bmatrix}$$

Having found $\mathbf{A}^-$, the *g* inverse, we can go on to solve for **x** in terms of the vector $\mathbf{b}' = (8, 14, 6, 22)$, as from Section B.2.2:

$$\mathbf{x} = \overset{\mathbf{A}^-}{\begin{bmatrix} 2 & -1 & 0 & 0 \\ -1 & 1 & 0 & 0 \\ 1 & -1 & 1 & 0 \end{bmatrix}} \overset{\mathbf{b}}{\begin{bmatrix} 8 \\ 14 \\ 6 \\ 22 \end{bmatrix}} = \begin{bmatrix} 2 \\ 6 \\ 0 \end{bmatrix}$$

Since the rank of **A** is only two, we have an infinite number of solutions, written in general form as

$$\mathbf{x} = \overset{\mathbf{A}^-\mathbf{b}}{\begin{bmatrix} 2 \\ 6 \\ 0 \end{bmatrix}} + \left\{ \overset{\mathbf{I}}{\begin{bmatrix} 1 & 0 & 0 \\ 0 & 1 & 0 \\ 0 & 0 & 1 \end{bmatrix}} - \overset{\mathbf{A}^-\mathbf{A}}{\begin{bmatrix} 1 & 0 & 0 \\ 0 & 1 & 3 \\ 0 & 0 & 0 \end{bmatrix}} \right\} \overset{\boldsymbol{\gamma}}{\begin{bmatrix} \gamma_1 \\ \gamma_2 \\ \gamma_3 \end{bmatrix}}$$

$$= \begin{bmatrix} 2 \\ 6 \\ 0 \end{bmatrix} + \begin{bmatrix} 0 \\ -3\gamma_3 \\ \gamma_3 \end{bmatrix} = \begin{bmatrix} 2 \\ 6-\gamma_3 \\ \gamma_3 \end{bmatrix}$$

as was also obtained in Section B.2.2.

We observe, in passing, that the fourth row in **A** is redundant with the others, since it is the sum of the first two rows.

The procedure just outlined is quite general and can be applied to square or rectangular coefficients matrices, using the modification (to make **A** square) that was just illustrated. Of course, if one or more rows of zeros are appended to make **A** square, then the resulting additional columns of the transformed identity matrix are dropped in finding $\mathbf{A}^-$. Otherwise, the method is the same. In brief, the present method of computing *g* inverses is fully consistent with the echelon procedure of Section B.2. Thus, the echelon procedure (and consequent reduction of **A** to an identity submatrix) of Section B.2 is a general method for solving a specific set of linear equations.

The present method of solving for $\mathbf{A}^-$ is also fully general for finding a *g* inverse and then solving the system

$$\boxed{\mathbf{x} = \mathbf{A}^-\mathbf{b} + (\mathbf{I} - \mathbf{A}^-\mathbf{A})\boldsymbol{\gamma}}$$

for *any* desired **b** of constants. If **A** is nonsingular, then $\mathbf{A}^- = \mathbf{A}^{-1}$, and the second term on the right drops out. If **A** is singular, then $\mathbf{A}^-$ will still exist (as long as the equations are consistent, a test that can be made via the procedure of Section B.2).

Generalized inverses, of various types, are playing increasingly important roles in multivariate analysis. For more extensive discussion of the topic, the reader is referred to books by Pringle and Rayner (1971) and Rao and Mitra (1971).

B.5 SUMMARY

The role of generalized inverses in solving sets of linear equations has been the main subject of this appendix. Generalized inverses provide a counterpart role to regular inverses in cases where the coefficients matrix is singular. We started the discussion by reviewing a general procedure, involving reduction of either a coefficients or an augmented matrix to echelon form, for determining whether a set of simultaneous equations had (a) none, (b) exactly one, or (c) infinitely many solutions.

A general solution procedure, employing elementary row operations, was described and illustrated numerically. After reducing the augmented matrix **M** or the coefficients matrix **A** to echelon form, $\mathbf{H}_M$ or $\mathbf{H}_A$, additional elementary row operations were employed to reduce $\mathbf{H}_M$ (or $\mathbf{H}_A$) to an identity submatrix. Illustrations were provided for both nonhomogeneous and homogeneous sets of equations.

We then turned to a discussion of generalized inverses. The Penrose conditions were introduced, and the Moore–Penrose inverse $\mathbf{A}^+$ was described in the context of basic structure. The related concepts of left and right pseudoinverses were also illustrated, and properties of the (unique) Moore–Penrose were listed.

The appendix was concluded with a companion discussion of the *g* inverse $\mathbf{A}^-$. This (nonunique) generalized inverse need satisfy only the first Penrose condition. In general, $\mathbf{A}^-$ is easier to compute than $\mathbf{A}^+$ and, furthermore, plays a central role in solving sets of simultaneous equations. Several numerical illustrations of one procedure, involving reduction of a square coefficients matrix to Hermite form, were presented and tied in with the general procedure (of Section B.2) that was based on matrix reduction to echelon form.

REVIEW QUESTIONS

1. By means of reduction to echelon form, find the rank of

a. $\mathbf{A} = \begin{bmatrix} 1 & 1 & 2 & 4 \\ 3 & 3 & 6 & 12 \end{bmatrix}$ b. $\mathbf{B} = \begin{bmatrix} 2 & 1 & 3 \\ 1 & 2 & 0 \end{bmatrix}$

c. $\mathbf{C} = \begin{bmatrix} 1 & 4 & 1 \\ 3 & 2 & 7 \end{bmatrix}$ d. $\mathbf{D} = \begin{bmatrix} 1 & 2 & 3 \\ 1 & 2 & 5 \\ 2 & 4 & 8 \end{bmatrix}$

2. Reduce the following matrices to identity submatrices:

a. $\mathbf{A} = \begin{bmatrix} 1 & 2 & 0 & -1 \\ 3 & 4 & 1 & 2 \\ -2 & 3 & 2 & 5 \end{bmatrix}$ b. $\mathbf{B} = \begin{bmatrix} 0 & 2 & 3 & 4 \\ 2 & 3 & 5 & 4 \\ 4 & 8 & 13 & 12 \end{bmatrix}$

3. Using elementary row operations on **A** and **I**, simultaneously, find the inverse of

a. $$\mathbf{A} = \begin{bmatrix} 2 & 3 \\ 1 & 4 \end{bmatrix}$$ b. $$\mathbf{A} = \begin{bmatrix} 1 & 3 & 3 \\ 1 & 4 & 3 \\ 1 & 3 & 4 \end{bmatrix}$$

4. Find a left pseudoinverse of

$$\mathbf{A} = \begin{bmatrix} 1 & 1 & 1 \\ 3 & 4 & 3 \\ 2 & 1 & 5 \\ 3 & 3 & 4 \end{bmatrix}$$

5. Consider the system of equations

$$\begin{aligned} x_1 + 3x_2 - 2x_3 - x_4 + 2x_5 &= 1 \\ 2x_1 + 6x_2 - 4x_3 - 2x_4 + 4x_5 &= 2 \\ x_1 + 3x_2 - 2x_3 + x_4 &= -1 \\ 2x_1 + 6x_2 + x_3 - x_4 &= 4 \end{aligned}$$

Reduce the augmented matrix **M** to echelon form and find **x**.

6. Consider the matrix

$$\mathbf{A} = \begin{bmatrix} 1 & 3 \\ 2 & 1 \\ 3 & 2 \end{bmatrix}$$

(a) Find the Moore–Penrose inverse $\mathbf{A}^+$.

(b) Find a g inverse $\mathbf{A}^-$ via reduction of **A** to Hermite form.

Answers to Numerical Problems

CHAPTER 2

2.1a $\begin{bmatrix} 4 & 1 & -1 \\ 3 & -4 & 2 \\ 5 & -1 & -2 \end{bmatrix} \begin{bmatrix} x \\ y \\ z \end{bmatrix} = \begin{bmatrix} 0 \\ 1 \\ 7 \end{bmatrix}$ 2.1b $\begin{bmatrix} 2 & 3 & 1 \\ 1 & 1 & 7 \\ 3 & 5 & 4 \end{bmatrix} \begin{bmatrix} x \\ y \\ z \end{bmatrix} = \begin{bmatrix} 11 \\ 24 \\ 25 \end{bmatrix}$

2.2a $\begin{bmatrix} 3 & 5 & 1 \\ 3 & 2 & 1 \end{bmatrix}$ 2.2b $\begin{bmatrix} 3 & 6 & 1 \\ 7 & 1 & -1 \end{bmatrix}$ 2.2c $\begin{bmatrix} 3 & 6 & 1 \\ 7 & 1 & -1 \end{bmatrix}$

2.2d $\begin{bmatrix} -1 & -2 & -7 \\ 1 & -1 & 3 \end{bmatrix}$ 2.2e $\begin{bmatrix} -3 & -5 & -1 \\ -3 & -2 & -1 \end{bmatrix}$ 2.2f $\begin{bmatrix} -1 & 0 & -7 \\ 9 & -3 & -1 \end{bmatrix}$

2.3a 26 2.3b $\begin{bmatrix} -2 \\ -4 \\ -8 \end{bmatrix}$ 2.3c (5, 15, 20) 2.3d 23 2.3e 210

2.3f 23/2

2.4a (−7, 8) 2.4b $\begin{bmatrix} 4 & 6 & 8 \\ -2 & 4 & 0 \end{bmatrix}$ 2.4c $\begin{bmatrix} -4 & 12 \\ 3 & -4 \end{bmatrix}$

2.4d $\begin{bmatrix} 0 & 2 & 0 \\ 8 & -2 & -4 \end{bmatrix}$ 2.4e $\begin{bmatrix} 240 & -80 & -120 \\ -80 & 35 & 40 \end{bmatrix}$ 2.4f $\begin{bmatrix} 1 & 3 & 4 \\ 2 & 6 & 8 \\ 4 & 12 & 16 \end{bmatrix}$

2.5a

(i) $(\mathbf{DE})' = \mathbf{E}'\mathbf{D}' = \begin{bmatrix} ea + gb & ec + gd \\ fa + hb & fc + hd \end{bmatrix}$

(ii) $\mathbf{D}'\mathbf{E}' = \begin{bmatrix} ae + cf & ag + ch \\ be + df & bg + dh \end{bmatrix}$ (iii) $\mathbf{E}'\mathbf{D}'$ = See (i) above.

2.5b

(i) $(\mathbf{DE})' = \mathbf{E}'\mathbf{D}' = \begin{bmatrix} 3 & 0 \\ 10 & 4 \end{bmatrix}$ (ii) $\mathbf{D}'\mathbf{E}' = \begin{bmatrix} 3 & 0 \\ 17 & 4 \end{bmatrix}$

(iii) $\mathbf{E}'\mathbf{D}'$ = See (i) above.

2.6a $\mathbf{F} \neq \phi$; $\mathbf{G} \neq \phi$; $\mathbf{FG} = \phi$ **2.6b** $\mathbf{GF} = \begin{bmatrix} -10 & 30 & 50 \\ 0 & 0 & 0 \\ 0 & 0 & 0 \end{bmatrix} \neq \phi$

2.7a 84 **2.7b** −30 **2.7c** 0 **2.7d** 0

2.8a (−2, 6, 10) **2.8b** (−3, 0, 18) **2.8c** 28 **2.8d** (0, 0, 0)

2.9 $\mathbf{X}^2 - \mathbf{X} - 2\mathbf{I} = (\mathbf{X} + \mathbf{I})(\mathbf{X} - 2\mathbf{I}) = \begin{bmatrix} a^2 + bc - a - 2 & ab + bd - b \\ ac + cd - c & bc + d^2 - d - 2 \end{bmatrix}$

2.10a $\begin{bmatrix} 5 & 8 \\ 8 & 13 \end{bmatrix}$ **2.10b** $\begin{bmatrix} 4 & 0 \\ 0 & 9 \end{bmatrix}$ **2.10c** $\begin{bmatrix} 28 & 66 \\ 44 & 105 \end{bmatrix}$ **2.10d** $\begin{bmatrix} 30 & 48 \\ 72 & 114 \end{bmatrix}$

2.11a 0 **2.11b** 0 **2.11c** 0 **2.11d** $a^2 + b^2$

2.12a 148 **2.12b** −32 **2.12c** −128

2.13 $|\mathbf{A}| = 3(10) + 2 - 17 = 15$

2.14 $|\mathbf{M}| = (1)(-8)(-2.75)(-5.8182) = -128$

2.15a $\Sigma Y = 33$ $\bar{X}_1 = 4.33$ $\Sigma YX_2 = 165$ $\Sigma X_3{}^2 - (\Sigma X_3)^2/m = 50.83$

2.15b

$$\mathbf{S} = \begin{bmatrix} 41.5 & 31 & 38 & -37.5 \\ 31 & 39.3 & 33.3 & -43.7 \\ 38.5 & 33.3 & 38.3 & -38.2 \\ -37.5 & -43.7 & -38.2 & 50.8 \end{bmatrix}$$

2.15c

$$\mathbf{C} = \begin{bmatrix} 6.9 & 5.2 & 6.3 & -6.3 \\ 5.2 & 6.6 & 5.6 & -7.3 \\ 6.3 & 5.6 & 6.4 & -6.4 \\ -6.3 & -7.3 & -6.4 & 8.5 \end{bmatrix}$$

2.15d

$$\mathbf{R} = \begin{bmatrix} 1 & 0.76 & 0.96 & -0.82 \\ 0.76 & 1 & 0.85 & -0.98 \\ 0.96 & 0.85 & 1 & -0.86 \\ -0.82 & -0.98 & -0.86 & 1 \end{bmatrix}$$

2.15e

$$\mathbf{A}_d = \begin{bmatrix} -3.5 & -3.3 & -3.8 & 4.2 \\ -1.5 & -2.3 & -0.8 & 3.2 \\ -2.5 & 0.7 & -1.8 & -0.8 \\ 1.5 & -1.3 & 0.2 & 0.2 \\ 2.5 & 2.7 & 3.2 & -2.8 \\ 3.5 & 3.7 & 3.2 & -3.8 \end{bmatrix}$$

2.15f $-3.5 - 1.5 - 2.5 + 1.5 + 2.5 + 3.5 = 0$

CHAPTER 3

3.1a $\|a\| = \sqrt{5}$; $\|b\| = \sqrt{3}$; $\|c\| = \sqrt{\pi^2 + 7}$

3.1b the plane $y = 1 - x$ for all z

3.1c a parabola $z = x^2$ for all y

3.1d a sphere with center (0, 0, 0) and radius 1

3.2a $\begin{bmatrix} 5/2 \\ 3/2 \\ 1/2 \end{bmatrix}$ 3.2c $\|PR\| = 3\sqrt{3/2}$; $\|RQ\| = 3\sqrt{3/2}$; $\|PQ\| = 3\sqrt{3}$

3.3a $k = -1$ 3.3b $k_1 = 1$; $k_2 = 2$; $k_3 = -1$ 3.3c linearly independent vectors; $k_1 = k_2 = 0$

3.4a 1.414 3.4b 0 3.4c $-\frac{1}{2}$ 3.4d $-6 \leqslant a'b \leqslant 6$

3.5 $\dfrac{|a'c|}{\|c\|} + \dfrac{|b'c|}{\|c\|} = \|a_p\| + \|b_p\|$ (if a′c and b′c have same sign)

3.6

			α	β	γ
3.6a	$a' + b'$	$3\sqrt{10}$	$-1/3\sqrt{10}$	$5/3\sqrt{10}$	$8/3\sqrt{10}$
3.6b	$a' - b'$	$\sqrt{110}$	$5/\sqrt{110}$	$-7/\sqrt{110}$	$6/\sqrt{110}$
3.6c	$5a' + 10b'$	$\sqrt{5450}$	$-20/\sqrt{5450}$	$55/\sqrt{5450}$	$45/\sqrt{5450}$
3.6d	$\frac{1}{2}(a' - b')$	$\sqrt{110}/2$	$5/\sqrt{110}$	$-7/\sqrt{110}$	$6/\sqrt{110}$

$\cos\theta_{xy} = \dfrac{-43}{2\sqrt{5995}}$; where $x' = 5a' + 10b'$; $y' = \frac{1}{2}(a' - b')$

3.7a $a^* = \begin{bmatrix} 1/\sqrt{14} \\ 2/\sqrt{14} \\ 3/\sqrt{14} \end{bmatrix}$ $b^* = \begin{bmatrix} 33/\sqrt{1414} \\ -18/\sqrt{1414} \\ 1/\sqrt{1414} \end{bmatrix}$ $c^* = \begin{bmatrix} 4/\sqrt{101} \\ 7/\sqrt{101} \\ -6/\sqrt{101} \end{bmatrix}$

3.7b $a^* = \begin{bmatrix} 2/\sqrt{5} \\ 1/\sqrt{5} \end{bmatrix}$ $b^* = \begin{bmatrix} -1/\sqrt{5} \\ 2/\sqrt{5} \end{bmatrix}$ $c^* = \begin{bmatrix} 0 \\ 0 \end{bmatrix}$

3.7c The third vector vanishes; in a two-dimensional space no more than two vectors can constitute a basis.

3.8 $a^* = \begin{bmatrix} 1/8 \\ 5/8 \end{bmatrix}$

3.9 $a'b = 0$. The orthogonal vector **c** is of the form $\begin{bmatrix} -2k \\ k \\ k \end{bmatrix}$

3.10 The equation for the circle is independent of parameter Ψ, denoting the angle of rotation. New equations for the ellipse with coordinates (u, v) instead of (x, y) are

3.10a $\frac{5}{2}u^2 + \frac{5}{2}v^2 - 3uv = 4$ 3.10b $\frac{7}{4}u^2 + \frac{13}{4}v^2 - \frac{3\sqrt{3}}{2}uv = 4$

3.10c $\frac{7}{4}u^2 + \frac{13}{4}v^2 + \frac{3\sqrt{3}}{2}uv = 4$

3.11 The vectors $(4, -2, 1, 7)$ and $(2, -3, x, y)$ are linearly independent for any x, y since $(4, -2)$ and $(2, -3)$ are linearly independent in a two-dimensional space.

3.12
$$(\mathbf{e}_1, \mathbf{e}_2, \mathbf{e}_3) = \frac{1}{17}(\mathbf{f}_1, \mathbf{f}_2, \mathbf{f}_3)\begin{bmatrix} 7 & -1 & 3 \\ -8 & 6 & -1 \\ 6 & -13 & 5 \end{bmatrix}$$

3.13a $\mathbf{a}'\mathbf{b} = 7; \quad \mathbf{a}^{*\prime}\mathbf{b}^* = \dfrac{3\sqrt{2} - 11\sqrt{6}}{4}$

3.13b $\|\mathbf{a}\| = \sqrt{5}; \quad \|\mathbf{b}\| = \sqrt{13}; \quad \|\mathbf{a}^*\| = \sqrt{5}; \quad \|\mathbf{b}^*\| = \sqrt{13}$

Vector lengths are preserved under rotation.

3.15
$$\begin{bmatrix} 3 & 5 \\ 2 & 4 \end{bmatrix}\begin{bmatrix} 1 & 3 & 3 & 1 \\ 1 & 1 & 3 & 3 \end{bmatrix} = \begin{bmatrix} 8 & 14 & 24 & 18 \\ 6 & 10 & 18 & 14 \end{bmatrix}$$

Area of $(ABCD)$ is 4 and area of $(A'B'C'D')$ is 8.

3.16c $\mathbf{y}'\mathbf{x}_2 = m \operatorname{cov}(Y, X_2); \quad \|\mathbf{y}\| = \sqrt{m}\, s_y; \quad \|\mathbf{x}_2\| = \sqrt{m}\, s_x$

CHAPTER 4

4.1a an identity mapping

4.1b a projection on z, followed by a reversal of direction along z

4.1c a reversal of direction along z

4.1d leaves terminus of vector at same height but moves vector three times as far from z.

4.2a
$$\mathbf{x}_1^* = \begin{bmatrix} 1 \\ 3/2 \end{bmatrix}; \quad \mathbf{x}_2^* = \begin{bmatrix} 4 \\ 4 \end{bmatrix}; \quad \mathbf{x}_3^* = \begin{bmatrix} 5 \\ 6\frac{1}{2} \end{bmatrix}$$

4.2b
$$\mathbf{x}_1^* = \begin{bmatrix} 1 \\ 3/2 \end{bmatrix}; \quad \mathbf{x}_2^* = \begin{bmatrix} 4 \\ 4 \end{bmatrix}; \quad \mathbf{x}_3^* = \begin{bmatrix} 5 \\ 6\frac{1}{2} \end{bmatrix}; \quad \mathbf{x}_4^* = \begin{bmatrix} 2 \\ 4 \end{bmatrix}$$

4.2c
$$\mathbf{x}_5^* = \begin{bmatrix} 1 \\ 5/2 \end{bmatrix}; \quad \mathbf{x}_6^* = \begin{bmatrix} 2 \\ 3 \end{bmatrix}; \quad \mathbf{x}_7^* = \begin{bmatrix} 2 \\ 6 \end{bmatrix}; \quad \mathbf{x}_8^* = \begin{bmatrix} 1 \\ 5\frac{1}{2} \end{bmatrix}$$

Areas are invariant while angles and lengths change.

4.3a
$$\text{(i)} \begin{bmatrix} 2 & 3 \\ 8 & 6 \\ 10 & 12 \end{bmatrix} \qquad \text{(ii)} \begin{bmatrix} 2 & 3 \\ 8 & 6 \\ 10 & 12 \\ 4 & 9 \end{bmatrix} \qquad \text{(iii)} \begin{bmatrix} 2 & 6 \\ 4 & 6 \\ 4 & 15 \\ 2 & 15 \end{bmatrix}$$

4.3b (i) $\begin{bmatrix} 3 & 1 \\ 8 & 2 \\ 13 & 4 \end{bmatrix}$ (ii) $\begin{bmatrix} 3 & 1 \\ 8 & 2 \\ 13 & 4 \\ 8 & 3 \end{bmatrix}$ (iii) $\begin{bmatrix} 5 & 2 \\ 6 & 2 \\ 12 & 5 \\ 11 & 5 \end{bmatrix}$

4.4a $(x_1{}^*, x_2{}^*) = \left\{(x_1, x_2) \begin{bmatrix} 0.707 & 0.707 \\ -0.707 & 0.707 \end{bmatrix}\right\} + (2, -1)$

4.4b $(x_1{}^*, x_2{}^*) = (x_1, x_2) \begin{bmatrix} 0.866 & 0.5 \\ -0.5 & 0.866 \end{bmatrix} \begin{bmatrix} 3 & 0 \\ 0 & 1/2 \end{bmatrix}$

4.4c $(x_1{}^*, x_2{}^*) = (x_1, x_2) \begin{bmatrix} 3 & 0 \\ 0 & 1/2 \end{bmatrix} \begin{bmatrix} 0.866 & 0.5 \\ -0.5 & 0.866 \end{bmatrix}$

4.5a $\begin{bmatrix} 1/2 & 1 \\ 0 & 1/2 \end{bmatrix}$ **4.5b** $\begin{bmatrix} 0.35 & 1.06 \\ -0.35 & -0.35 \end{bmatrix}$

4.5c $\begin{bmatrix} 0.35 & 0.35 \\ -0.35 & 0.35 \end{bmatrix}$ **4.5d** $\begin{bmatrix} 0.35 & 1.06 \\ -0.35 & -0.35 \end{bmatrix}$

4.6a $\mathbf{A}^{-1} = \begin{bmatrix} 5/11 & -2/11 \\ -7/11 & 5/11 \end{bmatrix}$ **4.6b** $\mathbf{A}^{-1}$ is not defined

4.6c $\mathbf{A}^{-1}$ is not defined. **4.6d** $\mathbf{A}^{-1} = \dfrac{-1}{ad + bc} \begin{bmatrix} -d & b \\ c & a \end{bmatrix}$

4.7a 2 **4.7b** 3 **4.7c** 3

4.8a $\begin{bmatrix} 1 & 2 & 3 & 9 & 2 \\ 0 & 1 & 2 & 1 & 4 \\ 0 & 0 & 1 & -17/2 & -7/2 \\ 0 & 0 & 0 & 1 & -8/37 \end{bmatrix}$ **4.8b** $\begin{bmatrix} 1 & 4 & 3 \\ 0 & 1 & -1 \\ 0 & 0 & 1 \end{bmatrix}$ **4.8c** $\begin{bmatrix} 1 & 1 & 1 \\ 0 & 1 & 6 \\ 0 & 0 & 1 \\ 0 & 0 & 0 \\ 0 & 0 & 0 \end{bmatrix}$

4.9a $r(\mathbf{A}) = 2$; $r(\mathbf{B}) = 3$. Regular inverse does not exist.

4.9b $r(\mathbf{A}) = 3 = r(\mathbf{B}) = 3$;

$$\mathbf{A}^{-1} = \begin{bmatrix} -1/3 & -2 & 20/6 \\ 1/6 & 1 & -7/6 \\ 1/6 & 0 & -1/6 \end{bmatrix}$$

4.9c $r(\mathbf{A}) = r(\mathbf{B}) = 2$. Regular (3×3) inverse does not exist.

4.10 $\mathbf{A}^{-1} = 1/4 \begin{bmatrix} -3 & 1 & 7 \\ -1 & -1 & 5 \\ 5 & 1 & -13 \end{bmatrix}$; $|\mathbf{A}| = 4$; $x_1{}^* = -1/2$; $x_2{}^* = -1/2$; $x_3{}^* = 3/2$

4.11 $\mathbf{R}^{-1} = \begin{bmatrix} 10.26 & -9.74 \\ -9.74 & 10.26 \end{bmatrix}$; $|\mathbf{R}| = 0.0975$; $\mathbf{b}^* = \begin{bmatrix} 1.072 \\ -0.128 \end{bmatrix}$

4.12a $\mathbf{x}^{\circ *} = \begin{bmatrix} -377 \\ 1406 \end{bmatrix}$ 4.12b $\mathbf{x} = \begin{bmatrix} -5 \\ 4 \end{bmatrix}$ 4.12c $|\mathbf{T}| = -2$; $|\mathbf{T}^{\circ}| = -2$

4.13

$$\underset{\mathbf{x}^{\circ *}}{\begin{bmatrix} x_1^{\circ *} \\ x_2^{\circ *} \\ x_3^{\circ *} \end{bmatrix}} = \underset{\mathbf{T}^{\circ}}{\begin{bmatrix} 7 & 1 & 10 \\ 0 & 1 & 0 \\ -4 & 0 & -6 \end{bmatrix}} \underset{\mathbf{x}^{\circ}}{\begin{bmatrix} x_1^{\circ} \\ x_2^{\circ} \\ x_3^{\circ} \end{bmatrix}}$$

4.14

$$\underset{\mathbf{x}^{\circ *}}{\begin{bmatrix} x_1^{\circ *} \\ x_2^{\circ *} \\ x_3^{\circ *} \end{bmatrix}} = \underset{\mathbf{T}^{\circ}}{\begin{bmatrix} 3 & 0 & 0 \\ 0 & 6 & 0 \\ 0 & 0 & 9 \end{bmatrix}} \underset{\mathbf{x}^{\circ}}{\begin{bmatrix} x_1^{\circ} \\ x_2^{\circ} \\ x_3^{\circ} \end{bmatrix}}$$

$\mathbf{T}^{\circ}$ is a stretch transformation.

CHAPTER 5

5.1a $\lambda_1 = 5$; $\lambda_2 = 8$; $\mathbf{U} = \begin{bmatrix} 3/\sqrt{58} & 0 \\ -7/\sqrt{58} & 1 \end{bmatrix}$

5.1b $\lambda_1 = -4$; $\lambda_2 = -6$; $\mathbf{U} = \begin{bmatrix} 2/\sqrt{5} & 4/\sqrt{17} \\ 1/\sqrt{5} & 1/\sqrt{17} \end{bmatrix}$

5.1c $\lambda_1 = 2$; $\lambda_2 = -7$; $\mathbf{U} = \begin{bmatrix} \sqrt{2}/2 & -5/\sqrt{41} \\ \sqrt{2}/2 & 4/\sqrt{41} \end{bmatrix}$

5.1d $\lambda_1 = 5$; $\lambda_2 = 4$; $\mathbf{U} = \begin{bmatrix} 1/\sqrt{10} & 1/\sqrt{5} \\ -3/\sqrt{10} & -2/\sqrt{5} \end{bmatrix}$

5.1e $\lambda_1 = 4$; $\lambda_2 = 2$; $\lambda_3 = -2$; $\mathbf{U} = \begin{bmatrix} 1/\sqrt{11} & 3/\sqrt{14} & \sqrt{2}/2 \\ 3/\sqrt{11} & 2/\sqrt{14} & 0 \\ 1/\sqrt{11} & 1/\sqrt{14} & \sqrt{2}/2 \end{bmatrix}$

5.1f $\lambda_1 = 3$; $\lambda_2 = 3$; $\lambda_3 = 3$; $\mathbf{U} = \begin{bmatrix} 1/\sqrt{3} & 1/\sqrt{3} & 1/\sqrt{3} \\ 1/\sqrt{3} & 1/\sqrt{3} & 1/\sqrt{3} \\ 1/\sqrt{3} & 1/\sqrt{3} & 1/\sqrt{3} \end{bmatrix}$

All eigenvectors are of the form $\begin{bmatrix} k \\ k \\ k \end{bmatrix}$

5.2a $\text{tr}(\mathbf{A}) = 13$; $|\mathbf{A}| = 40$ 5.2b $\text{tr}(\mathbf{A}) = -10$; $|\mathbf{A}| = 24$

5.2c $\text{tr}(\mathbf{A}) = -5$; $|\mathbf{A}| = -14$ 5.2d $\text{tr}(\mathbf{A}) = 9$; $|\mathbf{A}| = 20$

5.3a $\begin{bmatrix} 0 \\ k \end{bmatrix}$ 5.3b $\begin{bmatrix} k \\ 0 \end{bmatrix}$ or $\begin{bmatrix} 0 \\ k \end{bmatrix}$ 5.3c Any vector remains invariant under central dilation.

5.3d There is no vector remaining invariant under this transformation since eigenvalues are complex.

5.4 The matrix $\mathbf{A} = \begin{bmatrix} 5 & 1 \\ 1 & 3 \end{bmatrix}$ has eigenvalues given by the diagonal elements of $\mathbf{D} = \begin{bmatrix} 4+\sqrt{2} & 0 \\ 0 & 4-\sqrt{2} \end{bmatrix}$ and associated eigenvectors that are

$$\mathbf{u}_1 = \frac{1}{\sqrt{4-2\sqrt{2}}}\begin{bmatrix} 1 \\ \sqrt{2}-1 \end{bmatrix}; \qquad \mathbf{u}_2 = \frac{1}{\sqrt{4-2\sqrt{2}}}\begin{bmatrix} 1-\sqrt{2} \\ 1 \end{bmatrix}$$

as columns of $\mathbf{U}$. Then, let $\mathbf{A} = \mathbf{UDU}'$.

5.4a $\mathbf{U}\begin{bmatrix} 12+3\sqrt{2} & 0 \\ 0 & 12-3\sqrt{2} \end{bmatrix}\mathbf{U}'$ 5.4b $\mathbf{U}\begin{bmatrix} 6+\sqrt{2} & 0 \\ 0 & 6-\sqrt{2} \end{bmatrix}\mathbf{U}'$

5.4c $\mathbf{U}\begin{bmatrix} 1+\sqrt{2} & 0 \\ 0 & 1-\sqrt{2} \end{bmatrix}\mathbf{U}'$ 5.4d $\mathbf{U}\begin{bmatrix} 88+50\sqrt{2} & 0 \\ 0 & 88-50\sqrt{2} \end{bmatrix}\mathbf{U}'$

5.4e $\mathbf{U}\begin{bmatrix} \frac{1}{4+\sqrt{2}} & 0 \\ 0 & \frac{1}{4-\sqrt{2}} \end{bmatrix}\mathbf{U}'$ 5.4f $\mathbf{U}\begin{bmatrix} \sqrt{4+\sqrt{2}} & 0 \\ 0 & \sqrt{4-\sqrt{2}} \end{bmatrix}\mathbf{U}'$

5.4g $\mathbf{U}\begin{bmatrix} (4+\sqrt{2})^{-1/2} & 0 \\ 0 & (4-\sqrt{2})^{-1/2} \end{bmatrix}\mathbf{U}'$

5.5 $$\mathbf{A}^{-1} = \frac{1}{4}\begin{bmatrix} 1 & 2 & 1 \\ 1 & -2 & 1 \\ 1 & 2 & -3 \end{bmatrix}; \qquad \mathbf{ABA}^{-1} = \begin{bmatrix} 5 & 0 & 0 \\ 0 & 1 & 0 \\ 0 & 0 & 1 \end{bmatrix}$$

5.6 $\lambda_1 = 12$; $\lambda_2 = 6$; $\lambda_3 = 6$. The last two eigenvalues are not unique. One possible solution is

$$\mathbf{U} = \begin{bmatrix} 1/\sqrt{6} & 1/\sqrt{2} & -1/\sqrt{3} \\ -2/\sqrt{6} & 0 & -1/\sqrt{3} \\ 1/\sqrt{6} & -1/\sqrt{2} & -1/\sqrt{3} \end{bmatrix}$$

Then

$$\mathbf{D} = \mathbf{U}'\mathbf{AU} = \begin{bmatrix} 12 & 0 & 0 \\ 0 & 6 & 0 \\ 0 & 0 & 6 \end{bmatrix}$$

5.7a 3 5.7b 3 5.7c 2 5.7d 1

5.8a $$\mathbf{A} = \begin{bmatrix} 1/\sqrt{2} & 1/\sqrt{2} \\ 1/\sqrt{2} & -1/\sqrt{2} \end{bmatrix}\begin{bmatrix} 1 & 0 \\ 0 & 1 \end{bmatrix}\begin{bmatrix} 1/\sqrt{2} & 1/\sqrt{2} \\ -1/\sqrt{2} & 1/\sqrt{2} \end{bmatrix}$$

5.8b $$\begin{matrix} & \mathbf{P} & \boldsymbol{\Delta} & \mathbf{Q}' \\ \mathbf{A} = & \begin{bmatrix} -0.650 & 0.646 \\ -0.616 & -0.173 \\ 0.430 & 0.744 \end{bmatrix} & \begin{bmatrix} 2.022 & 0 \\ 0 & 3.861 \end{bmatrix} & \begin{bmatrix} 0.290 & -0.957 \\ 0.957 & 0.290 \end{bmatrix} \end{matrix}$$

5.8c
$$\mathbf{A} = \begin{bmatrix} 0.707 & 0.707 \\ -0.707 & 0.707 \end{bmatrix} \begin{bmatrix} 1 & 0 \\ 0 & 1 \end{bmatrix} \begin{bmatrix} 1 & 0 \\ 0 & 1 \end{bmatrix}$$

5.8d
$$\mathbf{A} = \begin{bmatrix} 0.114 & 0.634 \\ 0.282 & -0.108 \\ 0.396 & 0.526 \\ 0.553 & -0.551 \\ 0.667 & 0.083 \end{bmatrix} \begin{bmatrix} 10.592 & 0 \\ 0 & 2.969 \end{bmatrix} \begin{bmatrix} 0.994 & -0.107 \\ 0.107 & 0.994 \end{bmatrix}$$

In all four cases, $r(\mathbf{A}) = 2$.

5.9a

(5.8a): $$(\mathbf{A}'\mathbf{A})^2 = \begin{bmatrix} 1/\sqrt{2} & 1/\sqrt{2} \\ 1/\sqrt{2} & -1/\sqrt{2} \end{bmatrix} \begin{bmatrix} 1 & 0 \\ 0 & 1 \end{bmatrix} \begin{bmatrix} 1/\sqrt{2} & 1/\sqrt{2} \\ 1/\sqrt{2} & -1/\sqrt{2} \end{bmatrix}$$

(5.8b): $$(\mathbf{A}'\mathbf{A})^2 = \begin{bmatrix} 0.290 & 0.957 \\ -0.957 & 0.290 \end{bmatrix} \begin{bmatrix} 16.744 & 0 \\ 0 & 227.248 \end{bmatrix} \begin{bmatrix} 0.290 & -0.957 \\ 0.957 & 0.290 \end{bmatrix}$$

5.9b

(5.8a): $$(\mathbf{A}'\mathbf{A})^{1/2} = \begin{bmatrix} 1/\sqrt{2} & 1/\sqrt{2} \\ 1/\sqrt{2} & -1/\sqrt{2} \end{bmatrix} \begin{bmatrix} 1 & 0 \\ 0 & 1 \end{bmatrix} \begin{bmatrix} 1/\sqrt{2} & 1/\sqrt{2} \\ 1/\sqrt{2} & -1/\sqrt{2} \end{bmatrix}$$

(5.8b): $$(\mathbf{A}'\mathbf{A})^{1/2} = \begin{bmatrix} 0.290 & 0.957 \\ -0.957 & 0.290 \end{bmatrix} \begin{bmatrix} 2.022 & 0 \\ 0 & 3.861 \end{bmatrix} \begin{bmatrix} 0.290 & -0.957 \\ 0.957 & 0.290 \end{bmatrix}$$

5.10a
$$\mathbf{A}^{1/2} = \begin{bmatrix} 1/\sqrt{2} & 1/\sqrt{2} \\ -1/\sqrt{2} & 1/\sqrt{2} \end{bmatrix} \begin{bmatrix} 1 & 0 \\ 0 & \sqrt{7} \end{bmatrix} \begin{bmatrix} 1/\sqrt{2} & -1/\sqrt{2} \\ 1/\sqrt{2} & 1/\sqrt{2} \end{bmatrix}$$

$$\mathbf{A}^{-1/2} = \begin{bmatrix} 1/\sqrt{2} & 1/\sqrt{2} \\ -1/\sqrt{2} & 1/\sqrt{2} \end{bmatrix} \begin{bmatrix} 1 & 0 \\ 0 & 1/\sqrt{7} \end{bmatrix} \begin{bmatrix} 1/\sqrt{2} & -1/\sqrt{2} \\ 1/\sqrt{2} & 1/\sqrt{2} \end{bmatrix}$$

5.10b
$$\mathbf{A}^{1/2} = \begin{bmatrix} 1/\sqrt{2} & 1/\sqrt{2} \\ -1/\sqrt{2} & 1/\sqrt{2} \end{bmatrix} \begin{bmatrix} \sqrt{6} & 0 \\ 0 & 2\sqrt{2} \end{bmatrix} \begin{bmatrix} 1/\sqrt{2} & -1/\sqrt{2} \\ 1/\sqrt{2} & 1/\sqrt{2} \end{bmatrix}$$

$$\mathbf{A}^{-1/2} = \begin{bmatrix} 1/\sqrt{2} & 1/\sqrt{2} \\ -1/\sqrt{2} & 1/\sqrt{2} \end{bmatrix} \begin{bmatrix} 1/\sqrt{6} & 0 \\ 0 & 1/2\sqrt{2} \end{bmatrix} \begin{bmatrix} 1/\sqrt{2} & -1/\sqrt{2} \\ 1/\sqrt{2} & 1/\sqrt{2} \end{bmatrix}$$

5.10c
$$\mathbf{A}^{1/2} = \begin{bmatrix} 1/\sqrt{5} & 2/\sqrt{5} \\ -2/\sqrt{5} & 1/\sqrt{5} \end{bmatrix} \begin{bmatrix} 2 & 0 \\ 0 & 3 \end{bmatrix} \begin{bmatrix} 1/\sqrt{5} & -2/\sqrt{5} \\ 2/\sqrt{5} & 1/\sqrt{5} \end{bmatrix}$$

$$\mathbf{A}^{-1/2} = \begin{bmatrix} 1/\sqrt{5} & 2/\sqrt{5} \\ -2/\sqrt{5} & 1/\sqrt{5} \end{bmatrix} \begin{bmatrix} 1/2 & 0 \\ 0 & 1/3 \end{bmatrix} \begin{bmatrix} 1/\sqrt{5} & -2/\sqrt{5} \\ 2/\sqrt{5} & 1/\sqrt{5} \end{bmatrix}$$

5.10d
$\mathbf{A} = \mathbf{U}\begin{bmatrix} 2 & 0 \\ 0 & 2 \end{bmatrix}\mathbf{U}'$, where $\mathbf{U}$ is any set of two orthonormal vectors.

$$\mathbf{A}^{1/2} = \mathbf{U}\begin{bmatrix} \sqrt{2} & 0 \\ 0 & \sqrt{2} \end{bmatrix}\mathbf{U}'; \qquad \mathbf{A}^{-1/2} = \mathbf{U}\begin{bmatrix} 1/\sqrt{2} & 0 \\ 0 & 1/\sqrt{2} \end{bmatrix}\mathbf{U}'$$

5.11a $\mathbf{x}'\begin{bmatrix} 1 & 1 \\ 1 & 1 \end{bmatrix}\mathbf{x}$; rank 1

5.11b $\mathbf{x}'\begin{bmatrix} 1 & -2 \\ -2 & 2 \end{bmatrix}\mathbf{x}$; rank 2

5.11c $\mathbf{x}'\begin{bmatrix} 9 & -3 \\ -3 & 1 \end{bmatrix}\mathbf{x}$; rank 1

5.11d $\mathbf{x}'\begin{bmatrix} 2 & -3/2 \\ -3/2 & 3 \end{bmatrix}\mathbf{x}$; rank 2

5.12a $\mathbf{A} = \begin{bmatrix} 1/\sqrt{2} & 1/\sqrt{2} \\ 1/\sqrt{2} & -1/\sqrt{2} \end{bmatrix}\begin{bmatrix} 2 & 0 \\ 0 & 0 \end{bmatrix}\begin{bmatrix} 1/\sqrt{2} & 1/\sqrt{2} \\ 1/\sqrt{2} & -1/\sqrt{2} \end{bmatrix}$

Rotation–projection on first axis and stretch-rotation

5.12b $\mathbf{A} = \mathbf{U}\begin{bmatrix} \dfrac{3+\sqrt{17}}{2} & 0 \\ 0 & \dfrac{3-\sqrt{17}}{2} \end{bmatrix}\mathbf{U}'$

Rotation–stretch–rotation

5.12c $\mathbf{A} = \begin{bmatrix} -3/\sqrt{10} & 1/\sqrt{10} \\ 1/\sqrt{10} & 3/\sqrt{10} \end{bmatrix}\begin{bmatrix} 10 & 0 \\ 0 & 0 \end{bmatrix}\begin{bmatrix} -3/\sqrt{10} & 1/\sqrt{10} \\ 1/\sqrt{10} & 3/\sqrt{10} \end{bmatrix}$

Rotation–projection on first axis and stretch-rotation

5.12d $\mathbf{A} = \mathbf{U}\begin{bmatrix} \dfrac{5+\sqrt{10}}{2} & 0 \\ 0 & \dfrac{5-\sqrt{10}}{2} \end{bmatrix}\mathbf{U}'$

Rotation–stretch–rotation

5.13a $\mathbf{C} = \begin{bmatrix} 29.52 & 19.44 & 14.44 \\ 19.44 & 14.19 & 10.69 \\ 14.44 & 10.69 & 8.91 \end{bmatrix}$

5.13b $Z_1 = 0.755Y_d + 0.521X_{d1} + 0.396X_{d2}$; $\lambda_1 = 50.50$

$Z_2 = 0.618Y_d - 0.366X_{d1} - 0.696X_{d2}$; $\lambda_2 = 1.72$

$Z_3 = 0.218Y_d - 0.771X_{d1} + 0.598X_{d2}$; $\lambda_3 = 0.39$

5.14a $\mathbf{A} = \mathbf{x}'\begin{bmatrix} -3/\sqrt{10} & 1/\sqrt{10} \\ 1/\sqrt{10} & 3/\sqrt{10} \end{bmatrix}\begin{bmatrix} 10 & 0 \\ 0 & 0 \end{bmatrix}\begin{bmatrix} -3/\sqrt{10} & 1/\sqrt{10} \\ 1/\sqrt{10} & 3/\sqrt{10} \end{bmatrix}\mathbf{x}$

Positive semidefinite

5.14b $\mathbf{x}'\begin{bmatrix} 0.542 & -0.707 & 0.455 \\ -0.643 & 0 & 0.765 \\ 0.542 & 0.707 & 0.455 \end{bmatrix}\begin{bmatrix} -7.372 & 0 & 0 \\ 0 & -3 & 0 \\ 0 & 0 & -1.628 \end{bmatrix}\begin{bmatrix} 0.542 & -0.643 & 0.542 \\ -0.707 & 0 & 0.707 \\ 0.455 & 0.765 & 0.455 \end{bmatrix}\mathbf{x}$

Negative definite

5.14c $\mathbf{x}'\begin{bmatrix} 0.585 & 0.811 \\ -0.811 & 0.585 \end{bmatrix}\begin{bmatrix} 4.081 & 0 \\ 0 & 0.919 \end{bmatrix}\begin{bmatrix} 0.585 & -0.811 \\ 0.811 & 0.585 \end{bmatrix}\mathbf{x}$

Positive definite

5.14d $\mathbf{x}'\begin{bmatrix} 0.707 & -0.408 & -0.577 \\ 0.707 & 0.408 & 0.577 \\ 0 & 0.816 & -0.577 \end{bmatrix}\begin{bmatrix} 6 & 0 & 0 \\ 0 & 6 & 0 \\ 0 & 0 & 0 \end{bmatrix}\begin{bmatrix} 0.707 & 0.707 & 0 \\ -0.408 & 0.408 & 0.816 \\ -0.577 & 0.577 & -0.577 \end{bmatrix}\mathbf{x}$

Positive semidefinite

5.15a

$$\mathbf{W}^{-1/2} = \begin{bmatrix} 0.393 & -0.0424 \\ -0.0424 & 0.345 \end{bmatrix}$$

$$\mathbf{W}^{-1/2}\mathbf{A}\mathbf{W}^{-1/2} = \begin{bmatrix} 21.216 & 13.203 \\ 13.203 & 8.257 \end{bmatrix}$$

$$= \underset{\mathbf{Q}_2}{\begin{bmatrix} 0.849 & -0.529 \\ 0.529 & 0.849 \end{bmatrix}} \underset{\Lambda}{\begin{bmatrix} 29.444 & 0 \\ 0 & 0.029 \end{bmatrix}} \underset{\mathbf{Q}_2'}{\begin{bmatrix} 0.849 & 0.529 \\ -0.529 & 0.849 \end{bmatrix}}$$

$$\mathbf{Y} = \mathbf{X}_d\mathbf{W}^{-1/2}\mathbf{Q}_2 = \mathbf{X}_d \begin{bmatrix} 0.311 & -0.244 \\ 0.146 & 0.315 \end{bmatrix}$$

5.15b Both transformations yield same diagonal matrix.

CHAPTER 6

6.1a $\hat{Y} = -2.263 + 1.550X_1 - 0.239X_2$ $\quad \mathbf{C} = \begin{bmatrix} 14.19 & 10.69 \\ 10.69 & 8.91 \end{bmatrix}$

6.1b $\hat{Y} = -2.263 + 1.550X_1 - 0.239X_2$ $\quad \mathbf{R} = \begin{bmatrix} 1 & 0.951 \\ 0.951 & 1 \end{bmatrix}$

$b_1^* = 1.076;\; b_2^* = -1.309$

6.1c $\hat{Y} = 1.125 - 0.179X_1 + 0.117X_1^2$

$$\mathbf{R} = \begin{bmatrix} 1 & 0.978 \\ 0.978 & 1 \end{bmatrix}; \quad R^2 = 0.954$$

$R^2 = 0.954$ exceeds $R^2 = 0.902$ for Y regressed on X_1 alone.

6.2a

$$\mathbf{R} = \begin{bmatrix} 1 & 0.951 \\ 0.951 & 1 \end{bmatrix}; \quad \mathbf{T} = \begin{bmatrix} 0.707 & 0.707 \\ 0.707 & -0.707 \end{bmatrix}$$

$Z_1 = 0.707X_{s1} + 0.707X_{s2}$; $\quad \lambda_1 = 1.951$

$Z_2 = 0.707X_{s1} - 0.707X_{s2}$; $\quad \lambda_2 = 0.049$

$$\mathbf{F} = \underset{\mathbf{T}}{\begin{bmatrix} 0.707 & 0.707 \\ 0.707 & -0.707 \end{bmatrix}} \underset{\Lambda^{1/2}}{\begin{bmatrix} 1.398 & 0 \\ 0 & 0.208 \end{bmatrix}} = \begin{bmatrix} 0.988 & 0.147 \\ 0.988 & -0.147 \end{bmatrix}$$

A 45° rotation is involved. Eigenvectors (and loadings) differ from those obtained from **C**, the covariance matrix.

6.2b

$$\mathbf{A} = \begin{bmatrix} 53.25 & 41.42 \\ 41.42 & 33.08 \end{bmatrix}; \quad \mathbf{T} = \begin{bmatrix} 0.787 & 0.617 \\ 0.617 & -0.787 \end{bmatrix}$$

$\lambda_1 = 85.79$; $\quad \lambda_2 = 0.54$. $\quad$ Rotation is 38°.

6.2c R^2 is the same if both columns of component scores are used. R^2 is lower ($R^2 = 0.88$) if only the first column of component scores is used.

6.2d

$$\mathbf{C} = \begin{bmatrix} 29.52 & 19.44 & 14.44 \\ 19.44 & 14.19 & 10.69 \\ 14.44 & 10.69 & 8.91 \end{bmatrix} \qquad \mathbf{T} = \begin{bmatrix} 0.756 & -0.618 & -0.218 \\ 0.521 & 0.366 & 0.771 \\ 0.396 & 0.696 & 0.599 \end{bmatrix}$$

$\lambda_1 = 50.51$; $\lambda_2 = 1.72$; $\lambda_3 = 0.398$

6.3a

$$\mathbf{V}_s = \begin{bmatrix} 0.937 & -0.700 \\ 0.348 & 0.715 \end{bmatrix}; \qquad \Lambda = \begin{bmatrix} 29.155 & 0 \\ 0 & 0.028 \end{bmatrix}$$

Discriminant weights change under standardization of data matrix.

6.3b

$$\mathbf{W}^{-1/2}\mathbf{Q}_2 = \begin{bmatrix} 0.905 & -0.612 \\ 0.425 & 0.791 \end{bmatrix}$$

As can be seen from Table 6.5, $\mathbf{W}^{-1/2}\mathbf{Q}_2$ is equal to $\mathbf{V}$, the matrix of eigenvectors obtained from $\mathbf{W}^{-1}\mathbf{A}$.

6.3c

$$\lambda_1 = 3.768; \quad \mathbf{t}_1 = \begin{bmatrix} 0.868 \\ -0.497 \end{bmatrix}$$

In the two-group case only one discriminant is computed.

6.4a

$$\mathbf{C} = \begin{bmatrix} 14.19 & 1.46 \\ 1.46 & 0.25 \end{bmatrix}; \qquad \mathbf{R} = \begin{bmatrix} 1 & 0.774 \\ 0.774 & 1 \end{bmatrix}$$

$\hat{Y} = -2.488 + 1.496X_1 - 1.229X_2$

6.4b $R^2 = 0.907$

APPENDIX A

A.1a $\dfrac{2-2x}{2x-x^2}$ **A.1b** $\dfrac{2(x^2-x-4)}{(x^2+4)^2}$ **A.1c** $\dfrac{1}{2\sqrt{x}}\, e^{x^{1/2}}$ **A.1d** $\dfrac{4\,(x^3-1)^3\,(9x^2-12x^5)}{(2x^3+1)^5}$

A.2 $x = -1$ (minimum); $x = 1$ (maximum)

A.3 $\dfrac{\partial f}{\partial x}(1,3) = 4.2$; $\dfrac{\partial f}{\partial y}(1,3) = 4.6$ **A.4** $\min f(x, y) = -18$

A.5 $f(-2/3, 4/3) = 16/3$ (maximum)

A.6 $\dfrac{\partial g}{\partial \mathbf{x}}(\mathbf{x}) = 2\mathbf{A}x$. If $\mathbf{x}' = (3, 1, 2)$, $\dfrac{\partial g}{\partial \mathbf{x}} = \begin{bmatrix} 34 \\ 28 \\ 40 \end{bmatrix}$

A.7 $\mathbf{x} = \begin{bmatrix} -2 \\ -2 \end{bmatrix}$ (minimum)

APPENDIX B

B.1a $r(\mathbf{A}) = 1$ **B.1b** $r(\mathbf{B}) = 2$ **B.1c** $r(\mathbf{C}) = 2$ **B.1d** $r(\mathbf{D}) = 2$

B.2a
$$\begin{bmatrix} 1 & 0 & 0 & \vdots & 0 \\ 0 & 1 & 0 & \vdots & 0 \\ 0 & 0 & 1 & \vdots & 0 \end{bmatrix}$$

B.2b
$$\begin{bmatrix} 1 & 0 & \vdots & 0 & 0 \\ 0 & 1 & \vdots & 0 & 0 \\ \cdots & \cdots & \cdots & \cdots & \cdots \\ 0 & 0 & \vdots & 0 & 0 \end{bmatrix}$$

B.3a
$$\mathbf{A}^{-1} = \begin{bmatrix} 4/5 & -3/5 \\ -1/5 & 2/5 \end{bmatrix}$$

B.3b
$$\mathbf{A}^{-1} = \begin{bmatrix} 7 & -3 & -3 \\ -1 & 1 & 0 \\ -1 & 0 & 1 \end{bmatrix}$$

B.4
$$\mathbf{A}^{+} = \begin{bmatrix} 2.56 & -2.11 & -1.11 & 2.34 \\ -1.37 & 1.40 & 0.41 & -1.22 \\ -0.78 & 0.55 & 0.55 & -0.67 \end{bmatrix}$$

B.5
$$\mathbf{x} = \begin{bmatrix} -3\gamma_2 + (1/5)\gamma_5 + 6/5 \\ \gamma_2 \\ 3/5(\gamma_5 + 1) \\ \gamma_5 - 1 \\ \gamma_5 \end{bmatrix} = \begin{bmatrix} x_1 \\ x_2 \\ x_3 \\ x_4 \\ x_5 \end{bmatrix}$$

B.6a
$$\mathbf{A}^{+} = \begin{bmatrix} -0.253 & 0.226 & 0.266 \\ 0.412 & -0.107 & -0.067 \end{bmatrix}$$

B.6b
$$\mathbf{A}^{-} = \begin{bmatrix} -0.2 & 0.6 & 0 \\ 0.4 & -0.2 & 0 \end{bmatrix}$$

References

Aaker, D. A. (ed.), *Multivariate Analysis in Marketing: Theory & Application.* Belmont, California: Wadsworth, 1971.

Afifi, A. A., and Azen, S. P., *Statistical Analysis: A Computer-Oriented Approach.* New York: Academic Press, 1972.

Albert, A., *Regression and the Moore–Penrose Pseudoinverse.* New York: Academic Press, 1972.

Anderson, T. W., *Introduction to Multivariate Statistical Analysis.* New York: Wiley, 1958.

Bartlett, M. S., "Multivariate Analysis." *J. Roy. Stat. Soc. Ser. B* **9**(1947), 176–197.

Beveridge, G. S., and Schechter, R. S., *Optimization: Theory and Practice.* New York: McGraw-Hill, 1970.

Bickley, W. G., and Thompson, R. S. H. G., *Matrices: Their Meaning and Manipulation.* London: English University Press, 1964.

Bock, R. D., *Multivariate Statistical Methods in Behavioral Research.* New York: McGraw-Hill, 1975.

Bock, R. D., and Haggard, E. A., "The Use of Multivariate Analysis of Variance in Behavioral Research," in D. K. Whitla (ed.), *Handbook of Measurement and Assessment in Behavioral Sciences.* Reading, Massachusetts: Addison-Wesley, 1968, 100–142.

Bolch, B. W., and Huang, C. J., *Multivariate Statistical Methods for Business and Economics.* Englewood Cliffs, New Jersey: Prentice-Hall, 1974.

Carroll, J. D., "Categorical Conjoint Measurement," in P. E. Green and Y. Wind. *Multiattribute Decisions in Marketing: A Measurement Approach.* Hinsdale, Illinois: Dryden Press, 1973.

Carroll, J. D., "Algorithms for Rotation and Interpretation of Dimensions and for Configuration Matching." Paper presented at the Bell–Penn Multidimensional Scaling Workshop, June 1972.

Carroll, J. D., "A Generalization of Canonical Correlation to Three or More Sets of Variables." *Proc. 76th Annu. Conven. Amer. Psycholog. Assoc.* (1968), 227–228.

Carroll, J. D., "Notes on Factor Analysis." Unpublished paper. Bell Laboratories, Murray Hill, New Jersey, 1965.

Cattell, R. B., "Multivariate Behavioral Research and the Integrative Challenge." *Multivariate Behavioral Res.* **1**(1966), 4–23.

Cliff, N., "Orthogonal Rotation to Congruence." *Psychometrika* **31**(1966), 33–42.

Comrey, A. L., *A First Course in Factor Analysis.* New York: Academic Press, 1973.

Cooley, W. W., and Lohnes, P. R., *Multivariate Data Analysis.* New York: Wiley, 1971.

Corbato, F. J., "On the Coding of Jacobi's Method for Computing the Eigenvalues and Eigenvectors of Real Symmetric Matrices." *J. Assoc. Comput. Mach.* **10**(1963), 123–125.

Coxeter, H. S. M., *Introduction to Geometry.* New York: Wiley, 1961.

Daniel, C., and Wood, F. S., with the assistance of Gorman, J. W., *Fitting Equations to Data.* New York: Wiley, 1971.

Darlington, R. B., "Multiple Regression in Psychological Research and Practice." *Psycholog. Bull.* **69**(1968), 161–182.

Dempster, A. P., *Elements of Continuous Multivariate Analysis.* Reading, Massachusetts: Addison-Wesley, 1969.

Dickman, K., and Kaiser, H. F., "Program for Inverting a Gramian Matrix." *Ed. Psycholog. Meas.* **21**(1961), 721–727.

Draper, N., and Smith, H., *Applied Regression Analysis.* New York: Wiley, 1966.

Dunn, O. J., and Clark, V. A., *Applied Statistics: Analysis of Variance and Regression.* New York: Wiley, 1974.

Dwyer, P. S., *Linear Computations.* New York: Wiley, 1951.

Dwyer, P. S., and MacPhail, M. S., "Symbolic Matrix Derivatives." *Ann. Math. Statist.* **19**(1948), 517–534.

Eckart, C., and Young, G., "The Approximation of One Matrix by Another of Lower Rank." *Psychometrika* **1**(1936), 211–218.

Edwards, A. L., *Experimental Designs in Psychological Research,* 3rd ed. New York: Holt, 1968.

Eisenbeis, R. A., and Avery, R. B., *Discriminant Analysis and Classification Procedures.* Lexington, Massachusetts: Heath, 1972.

Faddeeva, V. N., *Computational Methods of Linear Algebra.* New York: Dover, 1959.

Finn, J. D., *A General Model for Multivariate Analysis.* New York: Holt, 1974.

Francis, J. G. F., "The QR Transformation, Parts I and II." *Comput. J.* **4**(1961), 265–271; (1962), 332–345.

Fuller, L. E., *Basic Matrix Theory.* Englewood Cliffs, New Jersey: Prentice-Hall, 1962.

Goldberger, A. S., *Topics in Regression Analysis.* New York: MacMillan, 1968.

Good, I. J., "Some Applications of the Singular Decomposition of a Matrix." *Technometrics* **11**(1969), 823–831.

Graybill, F. A., *Introduction to Matrices with Applications in Statistics.* Belmont, California: Wadsworth, 1969.

Graybill, F. A., *An Introduction to Linear Statistical Models,* Vol. I. New York: McGraw-Hill, 1961.

Green, B. F., "Best Linear Composites with a Specified Structure." *Psychometrika* **34**(1969), 301–318.

Green, P. E., and Tull, D. S., *Research for Marketing Decisions,* 3rd ed. Englewood Cliffs, New Jersey: Prentice-Hall, 1975.

Green, P. E., and Wind, Y., *Multiattribute Decisions in Marketing: a Measurement Approach.* Hinsdale, Illinois: Dryden Press, 1973.

Guertin, W. H., and Bailey, J. P., Jr., *Introduction to Modern Factor Analysis.* Ann Arbor, Michigan: Edwards, 1970.

Hadley, G., *Linear Algebra.* Reading, Massachusetts: Addison-Wesley, 1961.

Hancock, H., *Theory of Maxima and Minima.* New York: Dover, 1960.

Harman, H. H., *Modern Factor Analysis,* 2nd ed. Chicago: University of Chicago Press, 1967.

Harris, R. J., *A Primer of Multivariate Statistics.* New York: Academic Press, 1975.

Haynes, R. D., Komar, C. A., and Byrd, J., Jr., "The Effectiveness of Three Heuristic Rules for Job Sequencing in a Single Production Facility." *Management Sci.* **19**(1973), 575–580.

Hoerl, A. E., "Fitting Curves to Data," in J. H. Perry (ed.), *Chemical Business Handbook.* New York: McGraw-Hill, 1954, 55–77.

Hohn, F. E., *Elementary Matrix Algebra.* New York: MacMillan, 1964.

Hotelling, H., "Analysis of a Complex of Statistical Variables into Principal Components." *J. Ed. Psychology* **24**(1933), 417–441, 498–520.

Horst, P., "An Overview of the Essentials of Multivariate Analysis Methods," (in R. B. Cattell (ed.), *Handbook of Multivariate Experimental Psychology.* Chicago: Rand McNally, 1966, 129–152.

Horst, P., *Matrix Algebra for Social Scientists.* New York: Holt, 1963.

Horst, P., "Relations Among *m* Sets of Measures." *Psychometrika* **26**(1961), 129–150.

Horst, P., "Least Squares Multiple Classification for Unequal Subgroups." *J. Clinical Psychology* **12**(1956), 309–315.

Huang, D. S., *Regression and Econometric Methods.* New York: Wiley, 1970.

Jeger, M., *Transformation Geometry.* London: Allen & Unwin, 1966.

Johnston, J., *Econometric Methods,* 2nd ed. New York: McGraw-Hill, 1972.

Kendall, M. G., *A Course in Multivariate Analysis.* London: Charles Griffen, 1957.

Kerlinger, F. N., and Pedhazur, E. J., *Multiple Regression in Behavioral Research.* New York, Holt, 1973.

Kettenring, J. R., "Canonical Analysis of Several Sets of Variables." *Biometrika* 58(1972), 433–451.

Klein, F., *Geometry.* New York: Dover, 1948.

Kramer, C. Y., *A First Course in Methods of Multivariate Analysis.* Blacksburg, Virginia: Clyde Kramer, 1972.

Lang, S., *A First Course in Calculus,* Vol. I. Reading, Massachusetts: Addison-Wesley, 1964.

Lawley, D. N., and Maxwell, A. E., *Factor Analysis as a Statistical Method.* London: Butterworths, 1963.

Lingoes, J. C., *The Guttman–Lingoes Nonmetric Program Series.* Ann Arbor, Michigan: Mathesis Press, 1973.

McDonald, R. P., "A Unified Treatment of the Weighting Problem." *Psychometrika* **33**(1968), 351–381.

McNemar, Q., *Psychological Statistics,* 4th ed. New York: Wiley, 1969.

Malinvaud, E., *Statistical Methods of Econometrics.* Chicago: Rand McNally, 1966.

Manning, H. P., *Geometry of Four Dimensions.* New York: Dover, 1956.

Marriott, F. H. C., *The Interpretation of Multiple Observations.* New York: Academic Press, 1974.

Mendenhall, W., *Introduction to Linear Models and the Design and Analysis of Experiments.* Belmont, California: Wadsworth, 1968.

Moore, E. H., "On the Reciprocal of the General Algebraic Matrix." *Bull. Amer. Soc.* **26**(1920), 394–395.

Morgan, J., Sirageldin, I., and Baerwaldt, N., *Productive Americans.* Ann Arbor, Michigan: Survey Research Center, 1965.

Morrison, D. F., *Multivariate Statistical Methods, 2nd ed.* New York: McGraw-Hill, 1976.

Mulaik, S. A., *The Foundations of Factor Analysis.* New York: McGraw-Hill, 1972.

Murdoch, D. C., *Linear Algebra for Undergraduates.* New York: Wiley, 1957.

Neter, J., and Wasserman, W., *Applied Linear Statistical Models.* Homewood, Illinois: Irwin, 1974.

Nunnally, J. C., *Psychometric Theory.* New York: McGraw-Hill, 1967.

Ortega, J. M., and Kaiser, H. F., "The LL^T and QR Methods for Symmetric Tridiagonal Matrices." *Comput. J.* 6(1963), 99–101.

Overall, J. E., and Klett, C. J., *Applied Multivariate Analysis.* New York: McGraw-Hill, 1972.

Paige, L. J., Swift, J. D., and Slobko, T. A., *Elements of Linear Algebra,* 2nd ed. Lexington, Massachusetts: Xerox, 1974.

Pedoe, D. A., *Geometric Introduction to Linear Algebra.* New York: Wiley, 1963.

Penrose, R. A., "On Best Approximate Solutions by Linear Matrix Equations." *Proc. Cambridge Philos. Soc.* 52(1956), 17–19.

Penrose, R. A., "A Generalized Inverse for Matrices." *Proc. Cambridge Philos. Soc.* 51(1955), 406–413.

Perry, M., and Hamm, B. C., "Canonical Analysis of Relations between Socioeconomic Risk and Personal Influence in Purchase Decisions." *J. Marketing Res.* 6(1969), 351–354.

Pettofrezzo, A. J., *Matrices and Transformations.* Englewood Cliffs, New Jersey: Prentice-Hall, 1966.

Press, S. J., *Applied Multivariate Analysis.* New York: Holt, 1972.

Pringle, R. M., and Rayner, A. A., *Generalized Inverse Matrices with Applications to Statistics.* New York: Hafner, 1971.

Quandt, R. E., and Baumol, W. J., "The Demand for Abstract Transport Modes: Theory and Measurement." *J. Regional Sci.* 6(1966), 13–26.

Ralston, A., and Wilf, H. S., *Mathematical Methods for Digital Computers.* New York: Wiley, 1960 (Vol. I); 1967 (Vol. II).

Rao, C. R., *Linear Statistical Inference and Its Applications.* New York: Wiley, 1965.

Rao, C. R., "A Note on a Generalized Inverse of a Matrix with Applications to Problems in Statistics." *J. Roy. Statist. Soc. Ser. B* 24(1962), 152–158.

Rao, C. R., *Advanced Statistical Methods in Biometric Research.* New York: Wiley, 1952.

Rao, C. R., and Mitra, S. K., *Generalized Inverse of Matrices and Its Applications.* New York: Wiley, 1971.

Rorer, L. G. *et al.,* "Configural Judgments Revealed". *Proceedings of the 75th Annual Convention of AMA,* (1967), pp. 195–196.

Rummel, R. J. *Applied Factor Analysis.* Evanston, Illinois: Northwestern University Press, 1970.

Schatzoff, M., "Exact Distributions of Wilks' Likelihood Ratio Criteria." *Biometrika,* **53** (1966), 347–358.

Schönemann, P. H., and Carroll, R. M., "Fitting One Matrix to Another under Choice of a Central Dilation and a Rigid Motion." *Psychometrika,* **35** (1970), 245–256.

Stewart, G. W. *Introduction to Matrix Computations.* New York: Academic Press, 1973.

Tatsuoka, M. M., *Multivariate Analysis.* New York: Wiley, 1971.

Tatsuoka, M. M., and Tiedeman, D. V., "Statistics as an Aspect of Scientific Method in Research on Teaching," in N. L. Gage (ed.), *Handbook of Research on Teaching.* Chicago, Illinois: Rand McNally, 1963, 142–170.

Thurstone, L. L., *Multiple Factor Analysis.* Chicago, Illinois: Univ. of Chicago Press, 1947.

Timm, N. H., *Multivariate Analysis with Applications in Education and Psychology.* Belmont, California: Wadsworth, 1975.

Van de Geer, J. P., *Introduction to Multivariate Analysis for the Social Sciences.* San Francisco: Freeman, 1971.

Ward, J. H., Jr., and Jennings, E., *Introduction to Linear Models.* Englewood Cliffs, New Jersey: Prentice-Hall, 1973.

Widder, D. V., *Advanced Calculus.* Englewood Cliffs, New Jersey: Prentice-Hall, 1947.

Wilde, D. J., and Beightler, C. S., *Foundations of Optimization.* Englewood Cliffs, New Jersey: Prentice-Hall, 1967.

Wilkinson, J. H., *The Algebraic Eigenvalue Problem.* London and New York: Oxford Univ. Press (Clarendon), 1965.

Winer, B. J., *Statistical Principles in Experimental Design.* New York: McGraw-Hill, 1971.

Wish, M., and Carroll, J. D., "Applications of INDSCAL to Studies of Human Perception and Judgment," in E. C. Carterette and M. P. Friedman (eds.), *Handbook of Perception.* New York: Academic Press, 1973.

Wonnacott, R. J., and Wonnacott, T. H., *Econometrics.* New York: Wiley, 1970.

Woods, F. S., *Higher Geometry.* New York: Dover, 1961.

Yefimov, N. V., *Quadratic Forms and Matrices.* New York: Academic Press, 1964.

Index

T

U

V